I0070145

Pulp and Paper Chemistry and Technology   Volume 2
Pulping Chemistry and Technology

Edited by Monica Ek, Göran Gellerstedt, Gunnar Henriksson

Pulp and Paper Chemistry and Technology Volume 2

This project was supported by a generous grant by the Ljungberg Foundation (Stiftelsen Erik Johan Ljungbergs Utbildningsfond) and originally published by the KTH Royal Institute of Technology as the "Ljungberg Textbook".

Pulping Chemistry and Technology

Edited by Monica Ek, Göran Gellerstedt,
Gunnar Henriksson

DE GRUYTER

Editors

Dr. Monica Ek
Professor (em.) Dr. Göran Gellerstedt
Professor Dr. Gunnar Henriksson
Wood Chemistry and Pulp Technology
Fibre and Polymer Technology
School of Chemical Science and Engineering
KTH – Royal Institute of Technology
100 44 Stockholm
Sweden

ISBN 978-3-11-048342-0

Bibliographic information published by the Deutsche Nationalbibliothek

The Deutsche Nationalbibliothek lists this publication in the Deutsche Nationalbibliografie; detailed bibliographic data are available in the Internet at http://dnb.d-nb.de.

© Copyright 2009 by Walter de Gruyter GmbH & Co. KG, 10785 Berlin.
All rights reserved, including those of translation into foreign languages. No part of this book may be reproduced or transmitted in any form or by any means, electronic or mechanic, including photocopy, recording, or any information storage retrieval system, without permission in writing from the publisher. Printed in Germany.
Typesetting: WGV Verlagsdienstleistungen GmbH, Weinheim, Germany.
Printing and binding: Hubert & Co. GmbH & Co. KG, Göttingen, Germany.
Cover design: Martin Zech, Bremen, Germany.

Foreword

The production of pulp and paper is of major importance in Sweden and the forestry industry has a profound influence on the economy of the country. The technical development of the industry and its ability to compete globally is closely connected with the combination of high-class education, research and development that has taken place at universities, institutes and industry over many years. In many cases, Swedish companies have been regarded as the initiator of new technology which has started here and successively found a general world-wide acceptance. This leadership in knowledge and technology must continue and be developed around the globe in order for the pulp and paper industry to compete with high value-added forestry products adopted to a modern sustainable society.

The production of forestry products is based on a complex chain of knowledge in which the biological material wood with all its natural variability is converted into a variety of fibre-based products, each one with its detailed and specific quality requirements. In order to make such products, knowledge about the starting material, as well as the processes and products including the market demands must constitute an integrated base. The possibilities of satisfying the demand of knowledge requirements from the industry are intimately associated with the ability of the universities to attract students and to provide them with a modern and progressive education of high quality.

In 2000, a generous grant was awarded the Department of Fibre and Polymer Technology at KTH Royal Institute of Technology from the Ljungberg Foundation (Stiftelsen Erik Johan Ljungbergs Utbildningsfond), located at StoraEnso in Falun. A major share of the grant was devoted to the development of a series of modern books covering the whole knowledge-chain from tree to paper and converted products. This challenge has been accomplished as a national four-year project involving a total of 30 authors from universities, Innventia and industry and resulting in a four volume set covering wood chemistry and biotechnology, pulping and paper chemistry and paper physics. The target reader is a graduate level university student or researcher in chemistry / renewable resources / biotechnology with no prior knowledge in the fields of pulp and paper. For the benefit of pulp and paper engineers and other people with an interest in this fascinating industry, we hope that the availability of this material as printed books will provide an understanding of all the fundamentals involved in pulp and paper-making.

For continuous and encouraging support during the course of this project, we are much indebted to Yngve Stade, Sr Ex Vice President StoraEnso, and to Börje Steen and Jan Moritz, Stiftelsen Erik Johan Ljungbergs Utbildningsfond.

Stockholm, August 2009   Göran Gellerstedt, Monica Ek, Gunnar Henriksson

List of Contributing Authors

Göran Annergren
Granbacken 14
85634 Sundsvall, Sweden
jag.consulting@sundsvall.mail.telia.com

Birgit Backlund
Innventia AB
Drottning Kristinas väg 61
Stockholm, Sweden
birgit.backlund@innventia.com

Elisabet Brännvall
KTH Royal Institute of Technology
Chemical Science and Engineering
Fibre and Polymer Technology
100 44 Stockholm, Sweden
bettan@kth.se

Per Engstrand
Mid Sweden University
Fibre Science and Communication Network
851 70 Sundsvall, Sweden
per.engstrand@miun.se

Göran Gellerstedt
KTH Royal Institute of Technology
Chemical Science and Engineering
Fibre and Polymer Technology
100 44 Stockholm, Sweden
ggell@polymer.kth.se

Ulf Germgård
Karlstad University
Department of Chemistry
Universitetsg. 2
651 88 Karlstad, Sweden
ulf.germgard@kau.se

Hans Höglund
Mid Sweden University
Fibre Science and Communication Network
851 70 Sundsvall, Sweden
hans.hoglund@miun.se

Björn Johansson
Cleanosol AB
Industriegatan 33 Box 160
29 122 Kristianstad, Sweden
bjorn.johansson@cleanosol.se

Lars Miliander
Trekantsgatan 7
65220 Karlstad, Sweden
lars.miliander@telia.com

Ants Teder
KTH Royal Institute of Technology
Chemical Science and Engineering
Fibre and Polymer Technology
100 44 Stockholm
Sweden
teder@paperdep.kth.se

Hans Theliander
Chalmers University of Technology
Forest Products and Chemical Engineering
412 96 Göteborg, Sweden
hanst@chalmers.se

Contents

1 Overview of Pulp and Paper Processes

Elisabet Brännvall
Royal Institute of Technology, KTH

1.1 Introduction

Pulp technology deals with the liberation of fibres fixed in the wood or plant matrix. Paper technology is the knowledge of how to unify the fibres to form the paper web.

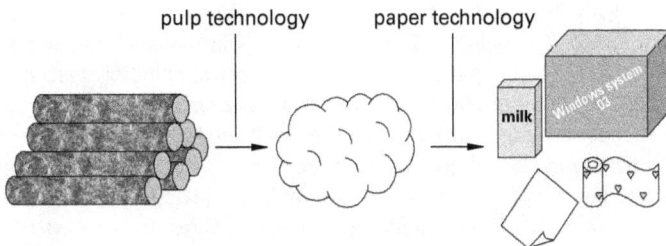

Figure 1.1. The making of paper products from wood includes two main technological fields.

Pulp can be converted to a number of different products with a variety of applications and thereby with a variety of product demands. Paper, as a conveyor of information is perhaps the first thing that comes to mind. Newspapers, magazines, and books require a paper suitable for printing text and pictures on, strong enough to endure browsing and folding and in many cases preferably able to last for generations to come so they can take part of the information. In the production of diapers and tissue paper, the preservation for future generations is not an issue, neither the strength. For these grades, the most important property is the pulp's ability to absorb fluids. Strength, however, is the main feature when it comes to paper products aimed to keep and protect other commodities, dry or liquid. Paper has many advantages compared to other materials competing with paper for package purposes. It is obtained from a renewable raw material (wood or other plants), the pulping and papermaking processes have low effluents to the recipient and the paper packages are easy to recycle.

1.2 Pulping Processes

Pulp consists of fibres, usually acquired from wood. The pulping processes aim first and foremost to liberate the fibres from the wood matrix. In principal, this can be achieved in two ways, either mechanically or chemically. Mechanical methods demand a lot of electric power, but on the other hand they make use of practically the whole wood material, i.e. the yield of the process is high. In chemical pulping, only approximately half of the wood becomes pulp, the other half is dissolved. In a modern chemical pulp mill, however, there is no demand of external energy. For a chemical process to be economically feasible, it has to consist of an efficient recovery system. Spent cooking chemicals and the energy in the dissolved organic material is recovered. The pulp obtained is coloured, the degree of colouring depending on the pulping process. For certain paper grades, the dark pulp has to be bleached. Bleaching leads to brighter (whiter) paper, this gives better contrast between the print and the paper. Other reasons for bleaching is cleanliness, as bleaching removes impurities that otherwise turn up in the paper as dots, and age-preservation, as bleaching can remove chemical structures in the pulp material that otherwise would in time make the paper yellow.

1.2.1 Mechanical Pulping

By grinding wood or wood chips, the fibres in the wood are released and a mechanical pulp is obtained. In the process, some easily dissolved carbohydrates and extractives are lost, but on the whole the pulp yield is little affected. The pulp yield for mechanical pulp is about 90 to almost 100 %, depending on what mechanical pulping method is used. Mechanical pulp fibres are stiff and mostly uncollapsed. Apart from the fibres, the mechanical pulp has a large portion of smaller material called fines. These consist of fragments from the fibre wall and broken fibres and are important for the excellent optical properties of mechanical pulp.

Groundwood pulp is produced by pressing round wood logs against a rotating cylinder made of sandstone. The logs are fed parallel to the cylinder axis and the fibres are scraped off. *Figure 1.2* shows an example of a grinder.

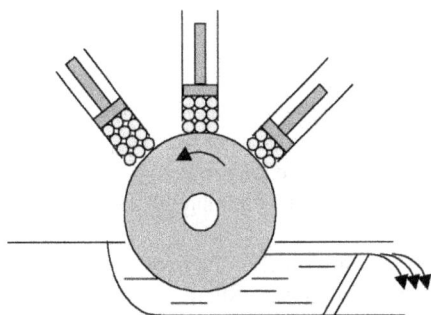

Figure 1.2. A schematic figure of a stone grinder.

Another type of mechanical pulp is refiner pulp. In this method, chips are fed into the centre of two refining discs, *Figure 1.3* One, or both, discs rotate whereby the chips are reduced in chips and fibres abraded off. The disks are grooved, coarser grooves near the centre and finer toward the perimeter. The closer the edge of the disk the wood material comes, the finer the pulp.

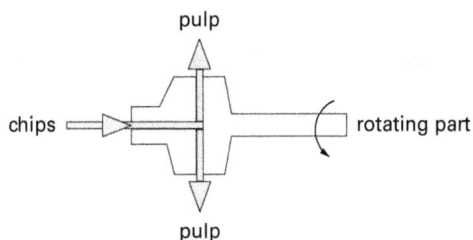

Figure 1.3. A chip refiner with one rotating disc.

A mechanical pulp consists not only of fibres released from the wood matrix. A substantial part of the pulp is made up of so-called *fines*. These are smaller particles, such as broken fibres, material from the fibre surface, and give the mechanical pulp its specific optical characteristics. However, the pulp strength is enhanced if the pulp consists of a higher portion of long fibres. By softening the lignin in the middle lamella, by an increase in temperature, the fracture can take place in the secondary (or primary) wall of the fibre, thus resulting in less fines formation. The chips can be pre-treated with steam, ~120 °C, before feeding to refiner. This type of refiner pulp is called TMP, which stands for thermo-mechanical pulp. By soaking the chips in a sodium sulphite solution the lignin becomes sulphonated and the lignin softening temperature is decreased. In a CTMP process (Chemo-Thermo Mechanical Pulp), the chemical treatment is followed by a steam pre-treatment.

1.2.2 Chemical Pulping

Since the fibres in wood and plants are „glued" together with lignin, the chemical way to produce pulp is to remove most of the lignin and thereby release the fibres. The delignification of wood is achieved by degrading the lignin molecules and introducing of charged groups, keep the lignin fragments in solution and eventually remove them by washing. No pulping chemicals are entirely selective towards lignin, also the carbohydrates of the wood are to varying extent lost. Approximately half of the wood material is dissolved in chemical pulping. No chemical pulping method is able to remove all lignin in the pulping stage, at least not without severe damage to the carbohydrates. The delignification is therefore terminated with some lignin remaining in the pulp. The amount of lignin left in the pulp is estimated by determining the kappa number of the pulp.

Kraft cooking is the dominant chemical pulping method globally. The cooking chemicals used are sodium hydroxide, and sodium sulphide. By leaving out the sodium sulphide and only use sodium hydroxide as the cooking chemical, the process is called *soda cooking*. In the *sulphite cooking process*, sulphurous acid (H_2SO_3) and bisulphite ions (H_2SO_3-) are the active chemicals to degrade and dissolve lignin. Sulphite pulping can be performed at a pH ranging from 1–2 in acid sulphite pulping to 7–9 in neutral sulphite pulping. Organic solvents can be used in delignification, either as the sole degrading chemical, as in suggested organosolv processes, or as reinforcement chemical, as an addition to sulphite, sulphate or soda processes. The solvents used are ethanol, methanol and peracetic acid.

Chemical pulp fibres are more flexible than mechanical pulp fibres. They conform better to each other when forming the paper and offer good strength properties to chemical pulp.

The cooking procedure can be either continuous or batch-wise. *Figure 1.4* shows a continuous cooking system with subsequent bleaching stages. *Figure 1.5* shows a batch-wise cooking configuration. An oxygen delignification stage after cooking is generally applied in the production of bleached pulp. In this stage the lignin content of the pulp is reduced before the pulp enters the final bleaching.

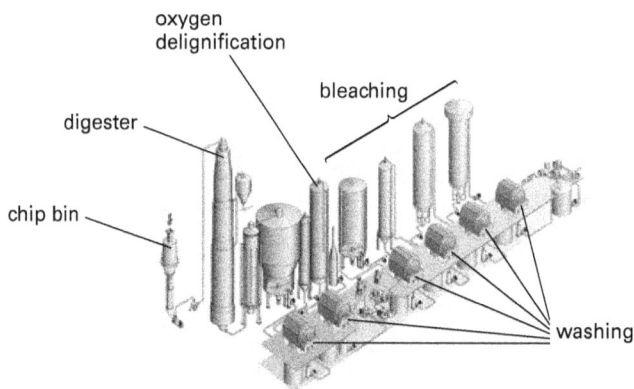

Figure 1.4. A continuous cooking system (Andritz).

Of the chemical pulping methods, kraft pulping dominates. The cooking liquor in kraft pulping, the white liquor, consists of sodium hydroxide, NaOH, and sodium sulphide, Na_2S. The active cooking species are OH^- and HS^-. The hydrogen sulphide is the main delignifying agent and the hydroxide keeps the lignin fragments in solution.

Kraft cooking is also called sulphate cooking. The name „Kraft cooking" derives from the German and Swedish word meaning strength. It was first used in the context of pulp from the sulphate process with a high lignin content. These pulp grades have extremely high strength and are used for packages such as linerboard and for sackpaper. Kraft pulp, however, refers to all pulp processed by the sulphate pulping method, also bleachable grades with very low lignin content.

Figure 1.5. Batch-wise cooking configuration employing two digesters and accompanied with tanks for liquor and pulp (Metso Paper).

1.3 Recovery Cycle in Kraft Pulping

The kraft pulping process includes a recovery cycle. The black liquor displaced from the digester consists of the spent cooking chemicals and the dissolved organic material. The spent chemicals are regenerated to active chemicals and brought back to the cooking stage.

Figure 1.6 shows schematically the stages in the recovery cycle.

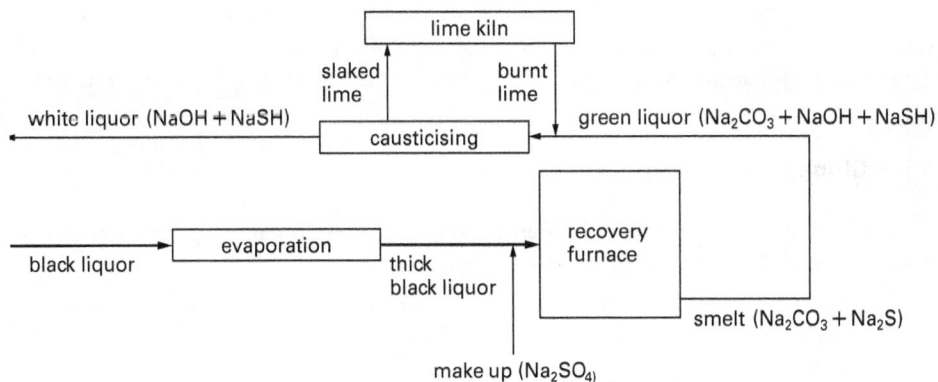

Figure 1.6. The stages in recovery of chemicals and energy in the kraft process.

Evaporating off water increases the dry solid content of black liquor, which is necessary before it is sprayed into a furnace and burnt. In the furnace, the inorganic chemicals in the black liquor are recovered in molten form. The organic contents are combusted and the heat generated is used for steam production. Part of the steam can be used to produce backpressure power.

1.4 Bleaching

Some paper products require a white paper. One reason is the print quality. A whiter paper gives a better contrast between the paper and the print. The cleanliness of the paper is another reason for bleaching. Impurities in the pulp may otherwise turn up on the paper as dots, deteriorating the printing. A third reason for bleaching is the ability of paper to resist ageing. Substances in the pulp can turn the paper yellow and brittle as time goes by, mainly substances connected to lignin. The bleaching agents remove these structures susceptible to discoloration of the paper. However, not all bleaching is aimed at removing lignin. Lignin removing bleaching is principally performed on chemical pulp, whereas the lignin-rich mechanical pulps are subjected to modifications of the lignin structures absorbing light.

The bleaching is usually performed in several stages, using different chemicals in each stage. Chemicals used for bleaching of mechanical pulp are dithionite and hydrogen peroxide. Chemical pulps are bleached, or delignified, with oxygen before entering the actual bleaching plant. There are a number of different chemicals used for the bleaching, for example hydrogen peroxide, chlorine dioxide, ozone and peracetic acid.

1.5 Papermaking

After liberating the fibres from the wood, they have to be consolidated again to form a web of paper. This is achieved on a paper machine. A very dilute slurry of fibres is sprayed on to a moving wire. Apart from fibres, the slurry may contain fillers, retention aid and wet strength additives. The wire is an endless woven wire cloth with a mesh size allowing the water to be drained, but retaining the fibres on the wire. From the wire, the paper web enters the pressing section of the paper machine where water is pressed out by squeezing the paper web between steel rolls. To further increase the dryness of the paper, it is dried in the drying section usually consisting of cylinders with steam within. At the end of the paper machine, the web is reeled.

1.5.1 Stock Preparation

A paper can be made using only fibres. However, to obtain certain properties a number of additives are used as well. Examples are listed below.

- Acids and bases to control pH
- Starch to improve dry strength
- Resins to improve wet strength
- Fillers, such as clay, talc, titanium dioxide, to improve optical properties
- Dyes and pigments to get desired colour

- Optical brighteners to improve the apparent brightness of the paper
- Retention aids to improve the retention of fines and fillers in the paper web
- Slimicides to prevent slime growth

1.5.2 Paper Machine

The paper machine is made up of a number of components with different objective. *Figure 1.7* shows a Fourdrinier paper machine, with a straight outline from beginning to end. The modern paper machines have a more compact structure and the components not as easily recognised.

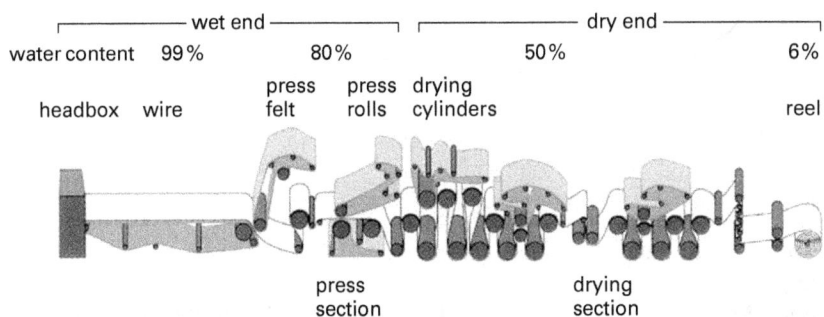

Figure 1.7. A simplified plot of a paper machine.

- *Headbox.* The dilute pulp slurry, ~0.6 % dry solids content, is to be evenly spread out to form the paper web. This is carried out by the headbox, which is a pressurised flow box, distributing the pulp stock onto the moving wire.
- *Wire.* The wire is an endless moving belt, a woven cloth, allowing the water to be drained and retaining the fibres. *Figure 1.8* shows twin-wire forming.

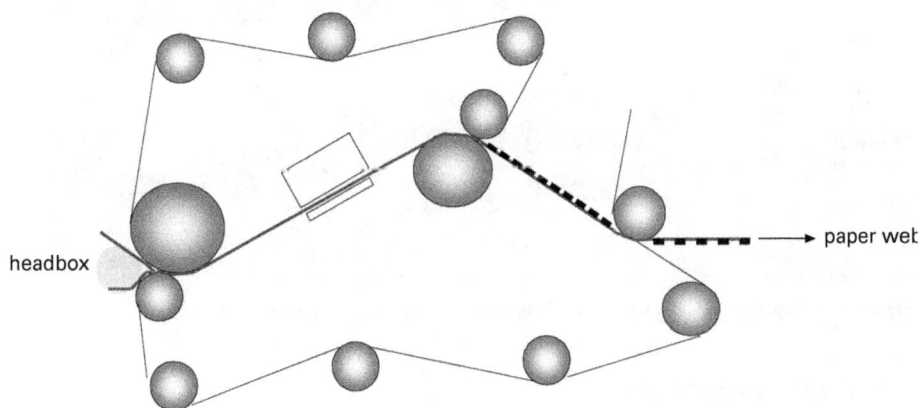

Figure 1.8. The headbox spreads the pulp slurry between the top and bottom wires in twin-wire forming.

- *Press section.* For more efficient water removal, the paper web is couched off the wire and onto a felt and passed between pressing rolls, *Figure 1.9.*

heated roll

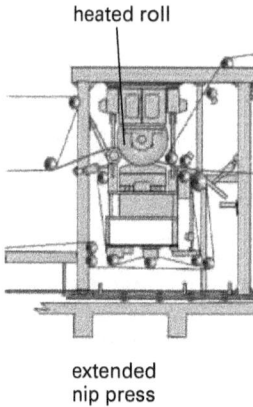

extended
nip press

Figure 1.9. The extended nip makes the press section more efficient.

- *Drying section.* The remaining water is removed by drying the paper on steam heated hot cylinders.
- *Calandering.* Most paper grades are subjected to calandering whereby the paper web passes through nips between iron rolls. This decreases the thickness of the paper and evens out variations in grammage along and across the paper web. The second objective of calandering is to improve the surface properties of the paper, mainly to make it smoother.
- *Coating.* Many printing papers demand a very even surface. By applying a coating mixture, the hollows on the surface of the paper sheet are filled.
- *Reel.* Finally, the paper web is wound onto a reel.

Figure 1.10. Paper wound onto reels and subsequently cut to suitable width (Metso Paper).

1.6 Paper Properties

In a simplified manner, the important paper properties can be divided into *strength* and *printability*.

The strength of the fibre raw material has the main influence on paper strength. The longer softwood fibres give stronger paper than the short hardwood fibres. The refining of pulp, also called the beating operation, modifies the fibres to conform better in the sheet and thereby increase the bonding strength of the sheet. Addition of fillers reduces the paper strength.

The printability is to a great extent related to paper surface properties. Short hardwood fibres give a much smoother paper surface than long softwood fibres and therefore better printability. The purpose of calandering and coating is mainly to improve the surface smoothness and the printability of the paper. The paper strength is also a printability parameter since sufficient paper web strength is required to avoid web break when the paper is running through the printing press.

1.7 Distribution of Different Pulping Methods

Of all the pulp produced worldwide, almost three quarters are chemical pulp, *Figure 1.11*. Of the chemical pulp, the major part is produced by the kraft process.

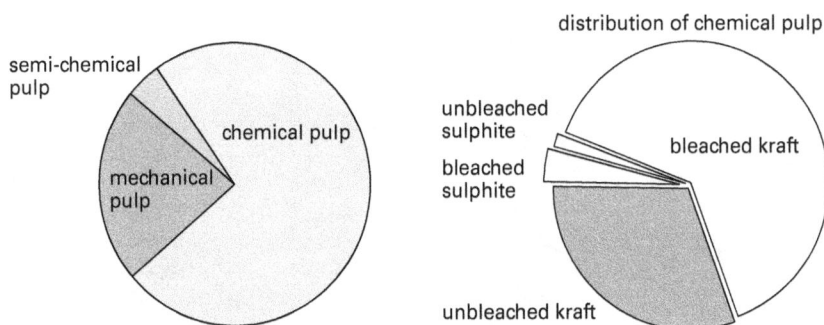

Figure 1.11. To the left the distribution between, chemical, mechanical and semi-chemical pulp production globally. The kraft pulping method is the domination pulping process worldwide, right diagram. (Figure FAO) In a semi-chemical pulping process, a smaller portion of the lignin is removed chemically and the fibres are separated mechanically in refiners.

1.8 Consumption of Different Paper Grades

As can be seen from *Figure 1.12,* paper is primarily used as a communicating media, conveying information through text and pictures.

Other media have appeared in the last decades but paper still holds its stand and even continues to increase in importance. As computers made their entry, predictions were made of the paperless office. These fears, however, have not been realised. On the contrary, paper consumption continues to increase and to some extent, the development of electronic devices has contributed to the increased consumption, *Figure 1.13*.

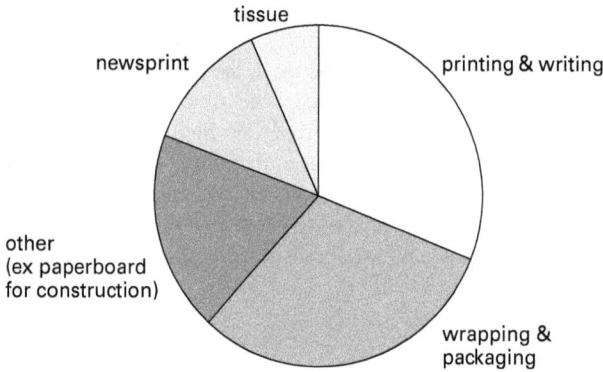

Figure 1.12. The global distribution of consumption of different paper grades (Figures FAO).

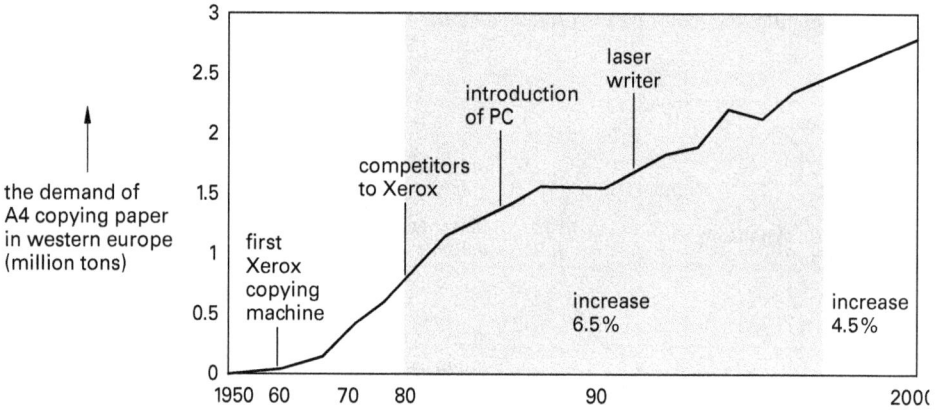

Figure 1.13. The demand of copying paper in western Europe (MoDo Paper).

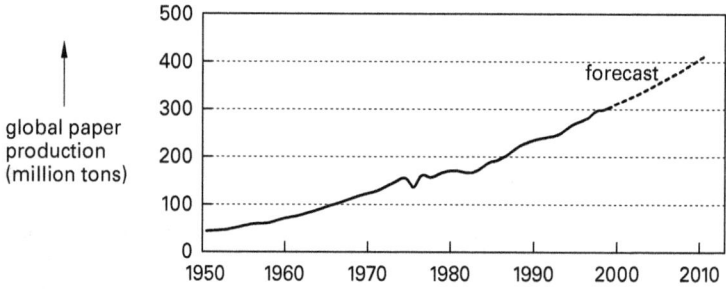

Figure 1.14. The worldwide paper production (Skogsindustrierna).

The consumption of paper is strongly dependent on the economical level of the society. As shown in *Figure 1.15*, the higher the gross domestic product a country has, the higher the consumption of paper

- 1999 Paper demand, kg per capita -

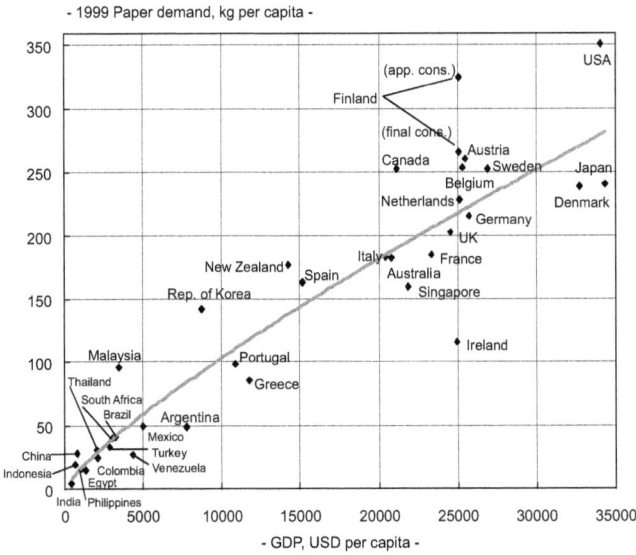

Figure 1.15. The more money a person makes, the more paper he or she uses.

The paper industry is concentrated and dominated by a few large producers. The global market is divided into continental markets where different producers have their main market share, *Fig 1.16*.

Leading Global Paper & Paperboard Producers

Regional capacity distribution, 2004/1 Q

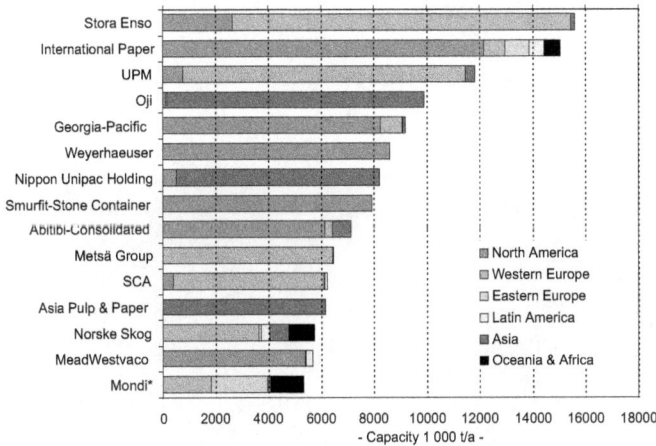

Figure 1.16. Leading global paper and paperboard producers (Skogsindustrierna).

2 Wood Handling

Elisabet Brännvall
Department of Fiber and Polymer Technology, Royal Institute of Technology, KTH

2.1 Introduction

Large quantities of wood are handled at a pulp mill, the wood consumption ranges between 2000 and 12000 tons per day. The management of the wood yard has a significant impact on the final pulp quality. Chip damages end up as reduced pulp quality and since the wood cost is the highest single cost component in pulp production, fibre losses have a severe effect on the economy of the mill.

In *Figure 2.1,* the wood handling steps at the pulp mill are shown. When the logs arrive to the mill, the weight or volume is determined. After de-loading the timber trucks or railway carriages, the logs are slashed (cross-cut) to standard lengths. If the logs are frozen, de-icing is required in order to facilitate barking. Sand and stones are removed in the de-icing step as well as after barking. This is necessary in order to minimise the wear on the chipper knives. For the same purpose a metal detector is usually placed before the chipper infeed to detect any metal scrap. In the chipper, the logs are reduced to smaller pieces, chips, and conveyed to the storage. Chip storage can be either open-air storage in a chip pile or in silos. Before the pulping, the chips are screened to ensure an even size distribution. The smallest fraction, pin chips and fines, is discarded and used as fuel. The overthick chips are reduced in size and fed with the accepted chip size fraction to the digester or refiner.

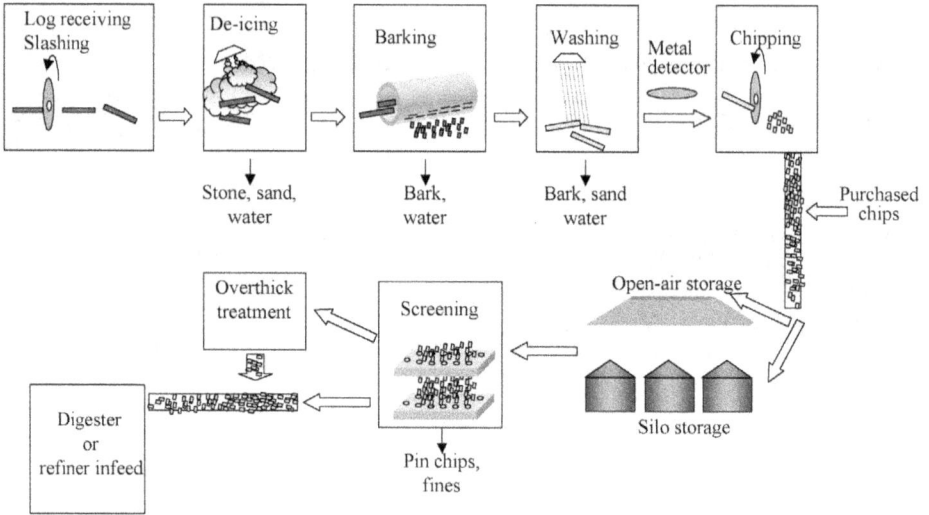

Figure 2.1. Handling of wood at the pulp mill.

2.2 Barking

2.2.1 Bark Morphology

The bark of the tree approximately amounts to 15 % of the tree's dry weight. It can roughly be divided into outer and inner bark. The outer bark is dead tissue and acts primarily as a protection for the tree. The living inner bark transports water and nutrients to the cambium layer of the wood, *Figure 2.2.*

Bark consists of many different kinds of cells. The cork cells of the outer bark die at an early stage. They are cemented together into a tight tissue that resists water and gas. The sieve elements in the inner bark transport water and nutrients whereas the parenchyma cells store nutrients. Two different types of cells for the mechanical support are found -bast fibres, measuring up to 3 mm in length, and small stone cells.

Chemically, the bast fibres resemble the wood fibres as they are made up of cellulose, hemicellulose, and lignin. The other cell types have a high content of extractives. In addition, the mineral content of the bark is higher compared to wood tissue.

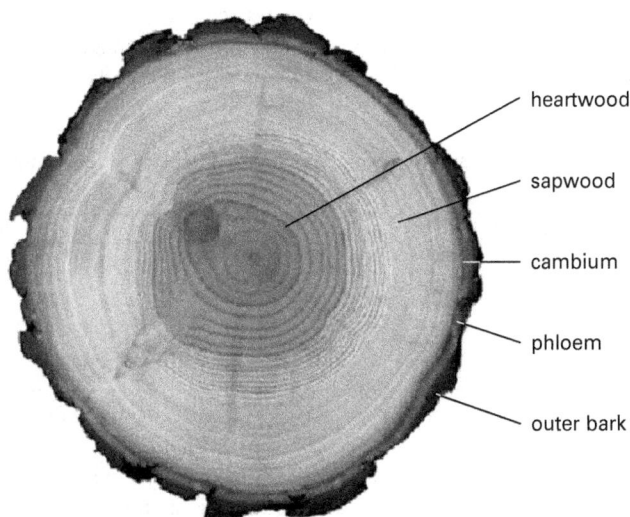

Figure 2.2. The structure of the wood stem.

2.2.2 Motives for Barking Wood

Both of the terms *bark* or *debark* wood applies to the same procedure, to remove the bark from the tree. Whether to use the term barking or debarking is a matter of preference, the term barking is chosen here. The motives for barking wood before pulping are many and are listed below:

1. *Low yield.* The fibre proportion of bark is very low. There would be little sense in trying to get hold of these fibres in the pulping process, only about 20 % of the bark can be retrieved by chemical pulping. Generally, any other types of cells present in the tree are of no interest or a nuisance in the pulp and paper mills.
2. *Damage to equipment.* Hard granules of sand can be embedded in the bark and cause wear in chipping equipment.
3. *High content of extractives.* The high content of extractives in bark gives higher consumption of cooking and bleaching chemicals. They give rise to pitch problems in both pulp and paper mill.
4. *Drainage obstacles.* The small bark cells may also cause problems in de-watering the paper web formed on the paper machine.
5. *Dirt on paper.* Some of the bark cells are not degraded in neither cooking nor bleaching and end up as dark spots on the paper web.

2.2.3 Barking Theory

The aim of barking is to remove outer as well as inner bark. Preferably, the separation of bark from wood occurs in the cambium layer between sapwood and bark. In favourable circumstances, the cambium is rather soft and with low shearing strength making the log easy to bark. The force needed to remove the bark depends for one thing on the wood specie. *Table 2.1* gives examples of trees and the effort needed for barking. As an example, birch is approximately twice as hard to bark than pine or spruce. It is therefore not suitable to bark birch and softwoods together.

Table 2.1. Force needed to bark wood is species dependent. (Papermaking Science and Technology, Gullichsen and Fogelholm).

Barking force needed	Example of species
Easy to bark	Southern pines Maple Oak Most tropical hardwoods
Normal to bark	Spruce Beech Hemlock
Difficult to bark	Birch Elm
Very difficult to bark	Black poplar Ironwood
Almost impossible to bark	Basswood Hickory

Some wood species need certain pre-treatment before the actual barking. Certain eucalyptus species and acacia have very tight fitting smooth bark that is rubbed off with difficulty. Special bark shredders are used in order to make cuts and tear in the bark creating positions where barking can start. The bark of some eucalyptus types is torn off as long chunks that roll and are entangled into each other. These coils of bark roll around and are discarded together with the logs and may eventually end up in the chipper and continue through the whole process.

The strength with which the bark adheres to the wood is related to the cutting season. In the summertime, when the tree has its growing season, the bark is much less tightly attached to the wood as compared to the season when no growth occurs, *Figure 2.3*.

Adhesion bark-to-wood, N/cm^2

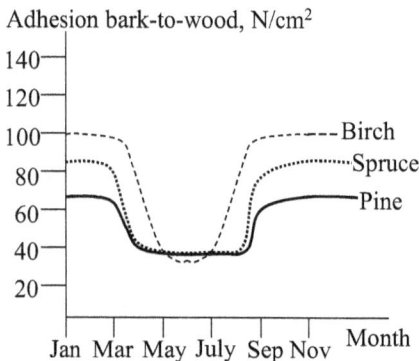

Figure 2.3. The adhesion of bark to wood, measured as the shearing strength, during the year. In the short summer period in the Nordic countries, the resistance against barking is much lower compared to the winter months.

In the wintertime, the bark may become even more tightly bonded to the wood as temperature falls and the log freezes, illustrated in Fig 2.4. In the figure another factor greatly influencing the resistance against barking is exemplified, the moisture content of the log. If the log dries during storage, it becomes more difficult to bark.

The degree of barking can be expressed either as percentage of bark-free surface or the weight-% of bark in chips.

Bark-free surface, %

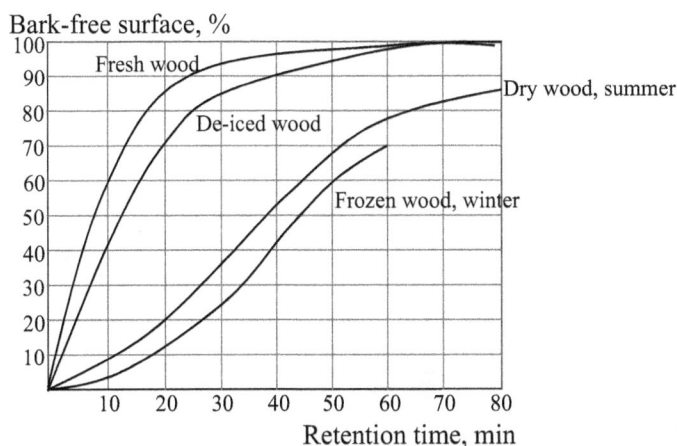

Figure 2.4. The retention time in a drum barker needed to reach a certain barking degree, expressed as bark-free surface of the log. (Papermaking Science and Technology, Gullichsen and Fogelholm).

The requirements for barking depend on the pulp to be manufactured, Table 2.2. Neutral sulphite semi-chemical pulp (NSSC) is quite insensitive to bark. This pulp is used in the corrugated media of paperboard and the requirement for cleanliness is not very high. Bleached pulp may have good tolerance towards bark, as the chemical treatments in the bleaching stages can, to a certain degree, be able to degrade and remove specks caused by bark elements. Unbleached qualities of chemical pulp, without the bark eliminating treatments in a bleach plant, are therefore more sensitive to the occurrence of bark in chips. Mechanical pulps have also higher requirements for barking as the bleaching chemicals used are not efficient in removing specks caused by bark.

Table 2.2. The barking requirements for different pulp types. (Papermaking Science and Technology, Gullichsen and Fogelholm)

Pulping process	Barking degree Bark-free surface [%]	Bark content of chips [%(wt)]
NSSC (semi-chemical pulp)	70	3
Bleached softwood kraft pulp	85–92	1
Mechanical pulp	98–99	0.5–0.2
Unbleached sulphite pulp	98–99	0.5–0.2

Although bark causes problems in pulping it is not possible to bark to a barking degree of 100 % without unacceptable wood losses. The loss of wood increases with the degree of barking. Barking of easily barked wood leads to less wood losses compared to wood difficult to bark. Also the process conditions in the barking equipment influence the wood loss.

Bark very difficult to remove is that found in bark pockets. Bark pockets occur in connection to branch wood, which often becomes ingrown and forms knots. The branch wood bark is thereby embedded in the wood, beneath the surface.

2.2.4 Barking Equipment

In pulp mills, barking is practically always performed in rotary drum barkers, *Figure 2.5*. The drum barkers consist of a long cylinder. It can be either horizontal or with a slight inclination, the end where unbarked logs are fed in lying on a higher level than the end discharging the barked logs. The drum walls are equipped with log lifters in the longitudinal direction. The log lifters can be made of steel or rubber. The separated bark is discharged from the drum through longitudinal or inclined bark slots in the drum wall.

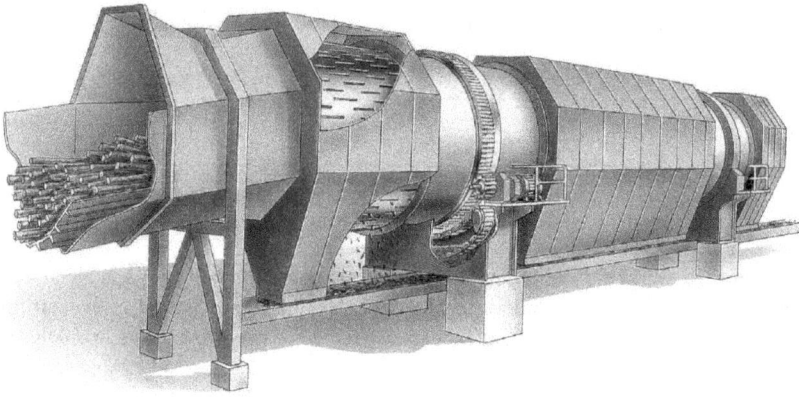

Figure 2.5. Example of a rotary parallel drum barker (Metso Paper).

The drum barking can be performed either as tumble barking or as parallel barking. In Scandinavia, the tumble barker is the most common type of barker. *Table 2.3* summarises the characteristics of these barking equipment.

Table 2.3. The different appearances of parallel and tumble barkers.

	Parallel barking	Tumble barking
Diameter (Ø) of barker cylinder	Ø< length of log	Ø > length of log
Length of barker cylinder	10–15 × Ø	4–7 × Ø
Degree of filling	~25 %	~50 %
Rotation speed	8–15 rpm	4–7 rpm

The tumble barker drum is short and wide, having a diameter between 4 and 6 m and the length 20–40 m. The logs for tumble barking are cut to shorter lengths than the diameter of the drum. The direction of the logs within the drum is random, as the term tumble barking denotes. The friction achieves the barking as logs rub against each other. The tumble drum barker rotates slowly and the volume of the drum filled with logs can vary from 25 to 60 %. The degree of filling influences the retention time of the logs in the drum. The retention time, and thereby the barking efficiency, increases with increased filling up to a certain point. Too high filling degree prevents the free movement of the logs and barking is reduced. The optimum filling degree is about 50 %.

The parallel drum barkers are narrower and longer, having a diameter of 3 to 4.5 m and the length usually in the range of 40 to 60 m. The logs to be barked are full-length trees, the log length exceeding the diameter of the drum barker. As the name implies, the logs are fed through the drum barker parallel to each other, aligned with the drum axis. They roll on one another as the drum rotates as well as rub against the inner surface of the drum. The parallel drum barker rotates relatively fast. The volume of the drum filled with logs is low, around 25 %.

Parallel barking handles the logs more gently thus resulting in lower wood losses in the barking operation *Figure2.6.*

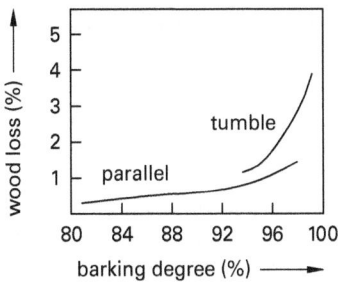

Figure 2.6. Increased barking degree results in higher wood losses. Parallel barking is a milder process giving less wood losses than tumble barking.

Parallel barking has lower wood losses for other reasons as well. There is no need to reduce the tree length, so-called slashing, which tumble barking demands. The slashing of logs also leads to some wood losses. Additionally the tumbling of the logs, wear and splinter the end of the logs and causing loss of wood.

When operating certain barking equipment, the wood losses during barking are dependent much on the rotation speed of the drum. The more revolutions per minute the drum rotates, the faster a given degree of barking is obtained however with a higher wood loss as a consequence, *Figure 2.7a.*

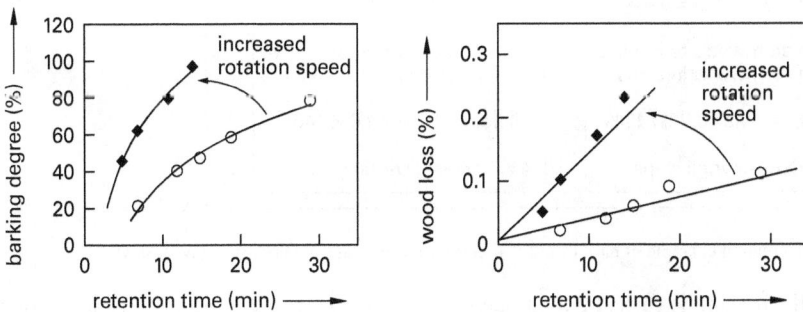

Figure 2.7a. The degree of barking can empirically be described by the equation $1-e^{-t/k}$ where t is the retention time and k is determined by rotation speed, degree of filling and wood conditions. The wood loss is linearly dependent on retention time and influenced by rotation speed.

In tumble barking, the wood losses are to a great deal dependent on the log length. Longer logs tend to break and the wood at the uneven surfaces created at the break is easily rubbed off.

Barking can be performed in either dry drums, semi-wet drums or wet drums. The dry drum barker is the dominant type used in Scandinavia. In the climatic conditions prevailing in this area, the drums are furnished with possibility of de-icing the logs. As mentioned earlier, frozen logs become much more difficult to bark and therefore steam is used to melt the ice. Furthermore it was shown that dried logs are more difficult to bark than fresh logs. By wetting the logs, barking becomes easier. In semi-wet drums, water is sprayed on the logs whereas wet drums are partially soaked in water. However, although wetting makes barking easier the increased water content of the removed bark gives rise to a problem -the removal of water from the bark, see below *2.2.6 Use of bark.*

2.2.5 Barking of Sawn Logs

Much of the harvested timber ends up in sawmills for lumber production. Sawmills however, do not make use of the entire log. The outer parts of the log, the slabs, are a residue coming to good use in pulp mills. The logs are usually barked and the slabs chipped at the sawmill. The slabs constitute about 40 % of the wood of the entire log and the fibres are of good quality as they are longer in the outer parts than in the juvenile core of the tree. However, the slabs will contain all the bark left after barking the log. The ratio of bark to wood is higher for sawmill chips and therefore the barking operation at the sawmill is of importance for the pulp mill. Should for example the barking degree at the sawmill decrease from 90 % to 80 % the consequence would be twice as much barks in the sawmill chips, *Figure 2.7b.*

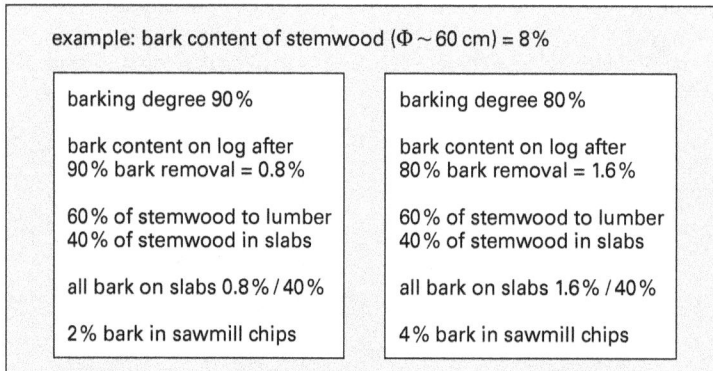

---

example: bark content of stemwood (Φ ~ 60 cm) = 8 %

barking degree 90 %	barking degree 80 %
bark content on log after 90 % bark removal = 0.8 %	bark content on log after 80 % bark removal = 1.6 %
60 % of stemwood to lumber 40 % of stemwood in slabs	60 % of stemwood to lumber 40 % of stemwood in slabs
all bark on slabs 0.8 % / 40 %	all bark on slabs 1.6 % / 40 %
2 % bark in sawmill chips	4 % bark in sawmill chips

Figure 2.7b. Calculation example of bark in sawmill chips after decrease in barking degree at sawmill.

In sawmills, barking is carried out on one log at a time. *Figure 2.8* shows an example of a one-log barker. The layout figure on top, shows the infeed, which centres the log and feeds it towards the actual barker where it is received by the spiked rolls. The purpose of the rolls is to transport the log into the barker and to hold the log steady and prevent it from rotating. The photo below shows the barker with the spiked rolls and the barking tools inside.

infeed   barker   outfeed

2300

2500

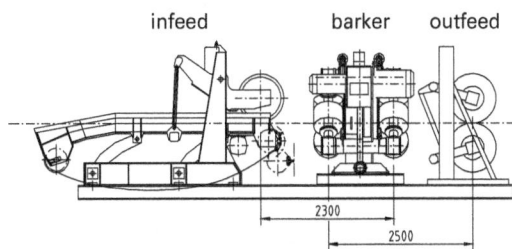

Figure 2.8. One-log barker used primarily in sawmills (Copyright Söderhamn Eriksson Cambio).

Figure 2.9, shows the barking technique performed by the barker. The upper set of figures demonstrates a log moving toward the barking tools. When the log hits the tools, they climb up to its surface. Ideally, the bark is to be sheared off and split from the wood at the cambium layer, as depicted in the lower set of drawings.

Figure 2.9. The barking technique performed by the one-log barker. The log is held by the spiked rolls and transported forward, while the barking tools rotate and shave off the bark (Copyright Söderhamn, Eriksson and Cambio).

2.2.6 Use of Bark

The bark removed is used in the mill for heat and power production. It constitutes an important part of the fuel need of the pulp and paper industry, *Fig 2.10.*

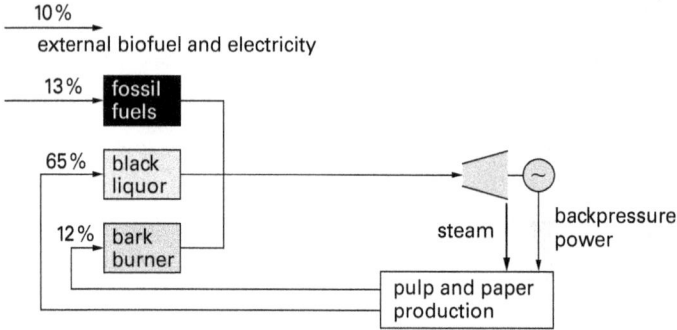

Figure 2.10. The distribution of the total consumption of energy by the Swedish pulp and paper industry 2000. The external biofuel includes purchased bark. (ÅF-IPK, Skogsindustrierna).

The heat value of bark depends greatly on the moisture content, *Figure 2.11*. At moisture content above 60 % it is difficult to burn the bark without fossil fuel support. From this point of view, wet barking is disadvantageous and the wet bark has to be processed in a bark press before combustion. Additionally, the wastewater from the bark press needs to be recovered and cleaned.

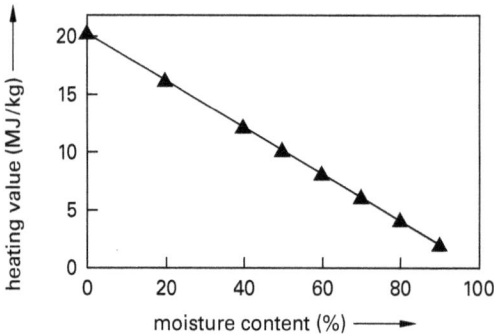

Figure 2.11. The heating value of bark decreases considerably with the moisture content (Papermaking Science and Technology Gullichsen, Fogelholm).

The heat value varies also for bark from different tree species, *Table 2.3*. To get the heating value for moist bark, the equation below is used.

$$HV_{wet} = HV_{dry} \cdot (1 - m_w) - r_{25} \cdot m_w \tag{2.1}$$

HV_{wet} = heating value moist fuel MJ/kg fuel

HV_{dry} = heating value dry fuel MJ/kg fuel
m_w = mass of water kg water/kg fuel
r_{25} = evaporation heat for water at 25 °C = 2.44 MJ/kg water

Table 2.3. Bark from different tree species have different heating values (Papermaking Science and Technology Gullichsen, Fogelholm).

Tree specie	Heating value dry bark, [MJ/kg bark]	Typical moisture content [%]	Heat value moist bark, [MJ/kg bark]	Heat content, [GJ/ADMT pulp]
Birch	22.7	55	8.9	4.0–5.7
Pine	20.0	60	6.5	1.8–3.7
Spruce	28.6	60	6.0	

The bark collected before and from the barking drum also contains sand and stones. These are removed in a stone and sand trap, where the heavier stone and sand fragments sediment. Wet bark is pressed to increase the dry solid content. The supply and consumption of bark varies with time depending on the production level at the mill. Bark is stored at the mill site either in a silo or under a roof to ensure the heating value of the bark.

2.3 Chipping of Wood

With contribution by Tord Bergman, Iggesund Tools

2.3.1 The Theory behind Chipping

With the exception of groundwood mechanical pulping, where logs are used, all pulping processes need smaller pieces of wood, so-called chips. In chemical processes, this is a prerequisite in order for chemicals and heat to be as evenly and as fast as possibly distributed throughout the wood. For refiner mechanical pulping, the construction of the disc refiners does not allow whole logs to be fed into them and furthermore mechanical pulping methods frequently include pretreatment of chips with heat and in some cases chemicals.

The dimensions of the chip are defined according to *Figure 2.12*. The commercial chip size varies, depending on chipper settings and wood specie, but to give an idea of the dimension averages, the length, (L) normally is 20 30 mm, the thickness, (t) 3–8 mm, and the width, (W) somewhere around 15–30 mm.

The chipping is accomplished by forcing a sharp-edged knife through the log, *Fig 2.13*. The chip length is determined by the length of the cut of the log, i.e. the distance between chipper disc and knife, *T-dimension*. The cutting edge of the knife is pressed against the wood, expelling both cutting and shearing forces. The further the knife penetrates into the log the higher the shearing force built up.

When the shearing force is sufficient, a slice of wood will be shaved off. How deep down the shearing takes place determines the thickness of the chip. However, the shearing takes place at a randomly selected plane whereby the thickness varies. The chipping results in long thin segments of wood. These eventually break up into smaller pieces giving the final width of the chips. There is a relationship between chip length and chip thickness, longer chips tend to be thicker as well.

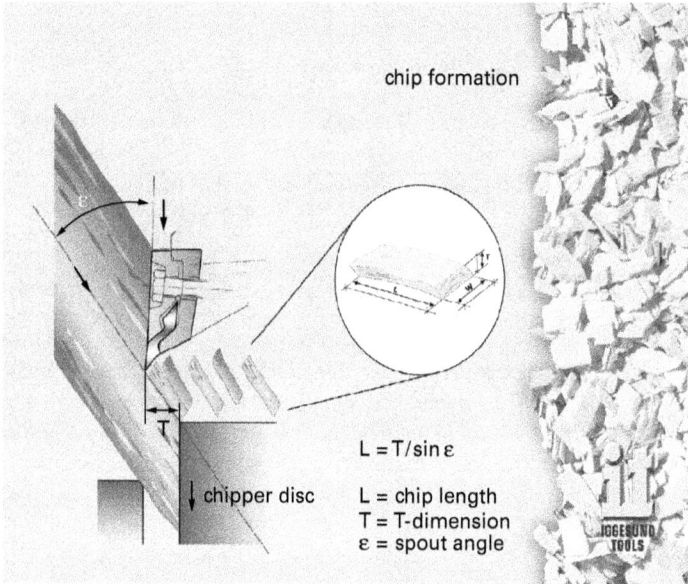

Figure 2.12. The dimensions of wood chip (Iggesund Tools©).

$$L = T/\sin \varepsilon$$

L = chip length
T = T-dimension
ε = spout angle

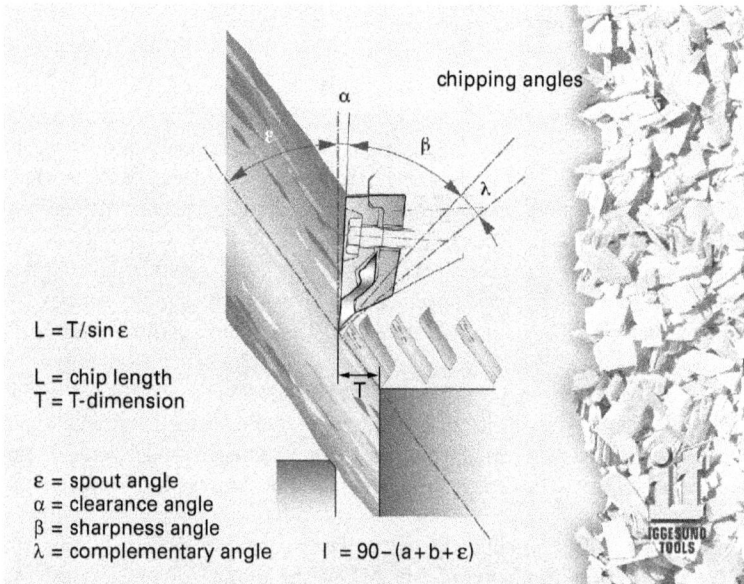

$$L = T/\sin \varepsilon$$

L = chip length
T = T-dimension

ε = spout angle
α = clearance angle
β = sharpness angle
λ = complementary angle

$$l = 90 - (a + b + \varepsilon)$$

Figure 2.13. Chipper geometry. The angles vary with different chipper manufacturers, the β-angle usually is between 32° and 38°, the α-angle between 2° and 4°, and the ε–angle is typically 32–38° (Iggesund Tools©).

Since the wood is cut right across the fibres, chipping results in a shortening of the average fibre length. *Figure 2.14* shows chips from softwood, with an average fibre length of 3 mm, cut to different chip length. The shorter the chips, the more pronounced the shortening of the fibre length. Therefore the nominal chip length in a chipper is not set under 18–20 mm, and a normally value is 20–30 mm.

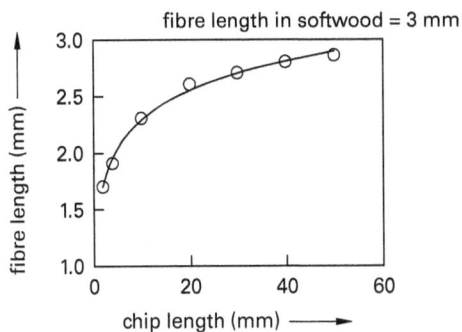

Figure 2.14. The relationship between the average fibre length in softwood chips and the length of the chip. The shorter the chip length, the shorter the average fibre length.

Shortening of the fibre is not the only damage induced when chipping. When wood is compressed above its elastic limit, a plastic deformation will occur. It is visible as cracks and strain damage, *Figure 2.15*. On the fibre level, the damage consists of misaligned zones. The fibrils in these zones have changed their original direction, in the course breaking hydrogen bonds and increasing the tension in bonds. The individual fibres are not necessarily damaged, why a reduction in fibre strength cannot be detected. However, these damaged zones of the fibres are more accessible to chemicals. In the subsequent cooking and bleaching, the chemical reactions will occur much faster and the resulting pulp strength will be reduced.

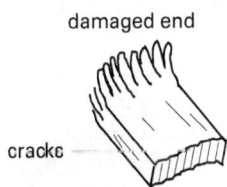

Figure 2.15. The end of the chip subjected to shearing forces is damaged and cracks are introduced to the chip.

Hand-cut laboratory chips have neither damaged ends nor cracks. Liquor penetrates much easier into technical chips through the cracks introduced and thereby the actual thickness of the chips is decreased. As shown in *Figure 2.16,* the apparent thickness of technical chips is much higher compared to solid undamaged laboratory made chips.

The width of the chip is determined after the actual chipping. The slice of wood produced by the chipper falls free and breaks into smaller pieces at random positions along the slice. Great width has no major importance for the processes ahead, but breaking the chips gives small pieces (pin chips). Pin chips, however, do cause problems. A high amount of pin chips usually indicates that the amount of even smaller particles, down to fines, is large. Pin chips and fines tend to clog up the circulation in digesters. Additionally, yield is lost and the quality of the pulp produced is decreased.

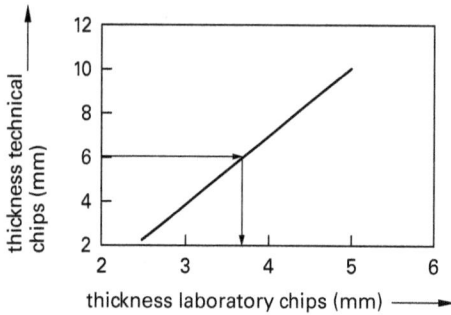

Figure 2.16. The relationship between the thickness of laboratory hand-made chips and technical chips. As an example, kraft cooking technical chips of 6 mm in thickness gives the same reject amount at given cooking conditions as laboratory chips of less than 4 mm thickness.

2.3.2 Factors Affecting Chip Quality

A sharp knife is a prerequisite for chips of good quality. Worn knives give an uneven chip surface. This gives rise to higher amounts of pin chips and fines. The plastic deformation of chip ends increases as well. Another reason behind the occurrence of chips with uneven surface is too large distance between anvil and knife-edge. *Figure 2.12* and *Figure 2.13* shows the TurnKnife® system with turnable edges which allows that each knife can be used twice. A typical life span of a knife-edge is about 1–2 days, depending on wood species and climate factors. Using high quality material in the knife can increase the life span of the knife-edge.

The λ angle in *Figure 2.13* has an impact on chip quality. The λ angle is the complementary angle to the angle of the log and the cutting surface of the knife. A higher λ, at a constant chip length, results in thinner chips and lower fines content, or vice versa, i.e. constant chip thickness at a higher chip length. The complementary angle (λ) can be adjusted by changing either the spout angle (ε) or the knife sharpness angle (β). In practice, the easier method is to change the sharpness angle of the knife, since the spout angle is determined by the equipment settings, i.e. the chipper model. However, the smaller the sharpness angle, the more easily the knife-edge is damaged. In modern chippers the spout angle is now decreased to 32–33˚, instead of the older model's 36.5-38˚.

Another prerequisite for good chip quality is the feed of logs into the chipper. In order not to get a varying size distribution of chips the logs are not to move about at the infeed. The logs have to be fixed in their position length-wise so all cuts become as similar as possible. The speed with which the logs are fed into the chipper should equal the suction speed of the logs, i.e. the speed with which the knives pull in the logs.

The disc speed (m/s) affects the amount of pin chips and fines created, *Figure 2.17*.

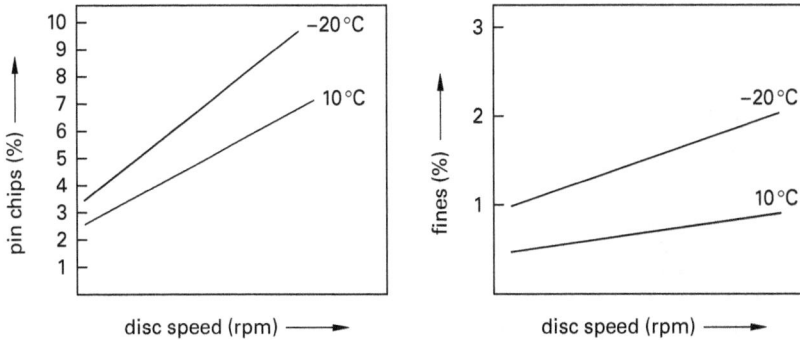

Figure 2.17. A higher disc speed and frozen logs increase the amount of pin chips and fines created in the chipping process.

As seen from the same figure, the temperature of the log has a great influence on the amount of pin chips and fines. Frozen wood results in thinner chips and higher amount of pin chips and fines.

2.3.3 Chippers

The most common chippers in use are the disc chippers. They consist of a rotating disc with 10-16 radial arranged knives. The log is fed endwise toward the disc through a spout. The feeding system can be either gravity feed as in *Figure 2.18,* or horizontal feed, *Figure 2.19.*

Figure 2.18. Disc chipper with gravity feed (Metso Paper©).

The gravity feed is best suited for short logs. Horizontal feed chippers use a conveyor to feed the log toward the chipper, and they are best suited for longer logs.

About 30 % of the wood chips used in Nordic countries for pulp production comes from the sawmills. In sawmills, the chipping is achieved at the same time as the timber is reduced to blocks. The equipment is therefore called reducers and the aim is to remove surface wood and get an even surface of the blocks. And at the same time chips are produced from the surplus wood, *Figure 2.20.*

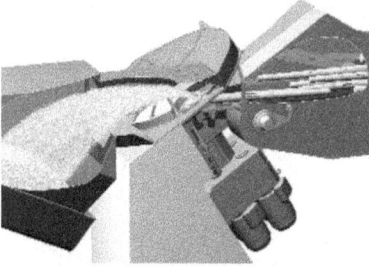

Figure 2.19. The horizontally fed HHQ-Chipper™ (Andritz©).

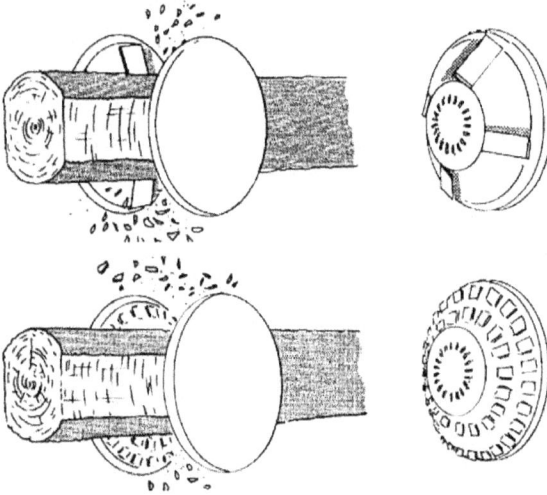

Figure 2.20. Two types of reducers used in sawmills. Above a long-knife disc, cutting chips across the fibre direction. Below a spiral disc, cutting chips parallel to fibre direction.

After the first reducer stage the log is rotated 90° and goes through a second stage of the same type of reducer. After that the square-shaped block is saw into sawn timber.

In chip manufacturing with reducers the chip length is controlled by the log feed speed and the rotation speed of the reducer.

2.4 Storage of Chips

To ensure continuous production at the pulp mill, a certain amount of raw material has to be stocked at the mill site. The amount of roundwood delivered to the mill can vary over time. Roundwood is stored as a buffer before the wood room (barking, chipping, screening). The intake to the wood room is practically constant over time. Additionally, there is storage of chips between wood room and digester or refiner. Pulp mills usually have enough chips stored for 5–10 days production.

Chips can either be stored in outdoor chip piles or chip silos. *Figure 2.21* shows an outdoor pile with an automated belt conveyor loading chips to the pile. The chips are taken from the pile's bottom by screw feeders to the chip conveying belt, which feeds the digester or refiner.

Figure 2.21. Chip storage in outdoor pile. Chips are fed to the pile on belt conveyors. Screw feeders supply the chip conveying belt at the bottom of the pile (Metso Paper©).

Storage of chips in a silo is advantageous for several reasons. There is no wind separation, which occurs in piles and can result in finer material blowing to and concentrating at a certain part of the pile. The chips are shielded from snow and rain as well as sand and other foreign particles. The turnover of chips runs with the principal first in-first out. The only disadvantage is the limited capacity. Normal silo volumes are 15–20 000 m³. *Figure 2.22* shows a chip silo.

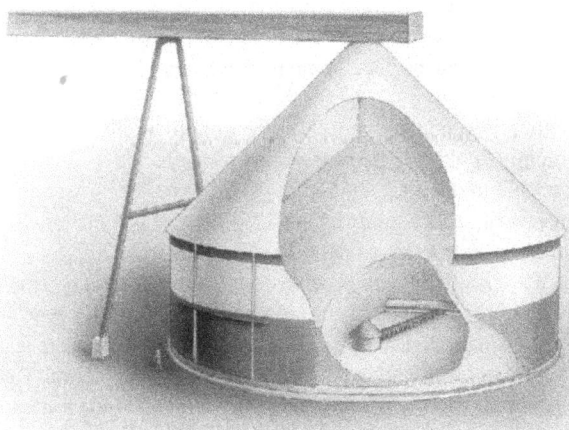

Figure 2.22. A chip silo (Metso Paper©).

When chips are placed in silos or piles, certain reactions will occur in the living cells. These reactions consume oxygen and release heat. The temperature within a pile is approximately 50°C. For bacterial attack or moulds to grow three criteria is needed -carbon, oxygen and water. The carbohydrates and lignin in wood provide the first requirement. White rot bacteria degrade cellulose, hemicellulose as well as lignin. The brown rot bacteria attack only the carbohydrates. Although the bacteria are aerobic and need oxygen they are not very sensitive to the amount, they can survive in an atmosphere with less oxygen than in the air. At least a humidity of 25 % is necessary for the well being of the bacteria. Air-dried wood is not attacked. The loss of wood as a function of storage time is shown in *Figure 2.23*.

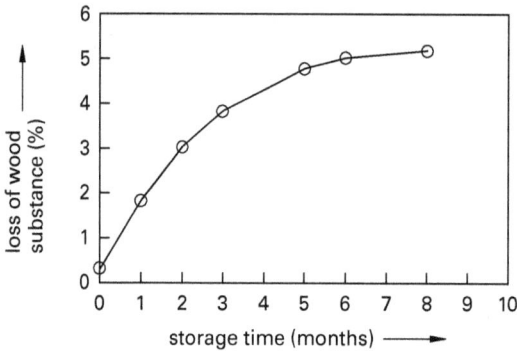

Figure 2.23. Wood substance is lost as a consequence of bacterial attack. Storage times at mills are normally less than two weeks and in any case never exceed two months.

Also the extractives content decreases with storage. Especially the terpenes are volatile and within a month more than two-thirds of the terpene amount is lost. The production of tall oil, an important by-product at softwood chemical pulp mills, is reduced as a consequence of long storage time. Chips stored for a month will result in only half the quantity of tall oil produced compared to pulping fresh chips. A large and important fraction of the raw materials are the sapwood from sawmill chips. Sawmill chips contain lower proportions of terpenes since they lack the terpene rich heartwood, and consequently the admixture of sawmill chips influence the tall oil production.

However, a decrease in the extractives content can be favourable to avoid pitch problems further on in the process. For sulphite cooking a decrease is a pre-requisite since this type of chemical cooking is much less capable to dissolve remaining extractives. For the extractive reactions to have time to take place in the short storage time in silos, warm moist air at about 60 °C is circulated. Within 2–4 days the extractive content is sufficiently low.

During storage, the esters of fatty acids (triglycerides and resin) decrease rapidly as they are enzymatically hydrolysed. This leads to an initial increase of free acids that however are decreased as they are subjected to autooxidation. The resin acids on the second hand are stable in the conditions prevailing during chip storage. The neutral and unsaponifiable components are also quite stable, although this is depending on wood specie and some degradation can take place.

2.5 Screening

Uniform chip size is a prerequisite for uniform pulping. The chips produced, however, are not always as uniform in size as required. For example worn knives in the chipper or the difference in wood quality have an impact on the chip quality.

1. Frozen wood results in more pin chips and fines
2. Dry wood (>70 %) gives more pin chips and fines
3. Wood with branches produces more overthick chips
4. Small timber results in more pin chips as well as more overthick chips, since small timber usually is more branched

There will always be some quantity produced of overthick chips, pin chips, and fines (sawdust). To ensure pulping uniformity and avoid circulation problems, the chips will therefore be screened before the pulping operation. The chips are separated according to the categories:

1. Oversize chips (Ø > 45 mm)
2. Overthick chips (thickness > 8 mm)
3. Accept
4. Pin chips (Ø < 7 mm)
5. Fines (sawdust)

Oversize chips normally goes to a rechipper, and the chips are recirculated back to the screen. Overthick chips are reduced in thickness and retrieved to the process, and goes with the accept to the pile. The pin chips are fed to the process in controlled amounts but the fines are of no value in the pulp process and for that reason used as fuel in the bark boiler.

2.5.1 Different Screening Equipment

Gyrating screens. The gyrating screens consist of perforated plates placed one above the other. The chips are spread out on the top plate and a rotating or vibrating motion shakes the chips. Oversize chips will be retained on the top plate, unable to pass through the holes. Accept chips, pin chips and fines fall down to the next plate. The size of the holes in this plate allows fines and pin chips to be separated and fall further down. On the lowest screen, the pin chip fraction is retained. The gyrating motion and the slight inclination of the screen plates moves the chips retained on each plate toward the output. *Figure 2.24* shows one type of gyrating screen.

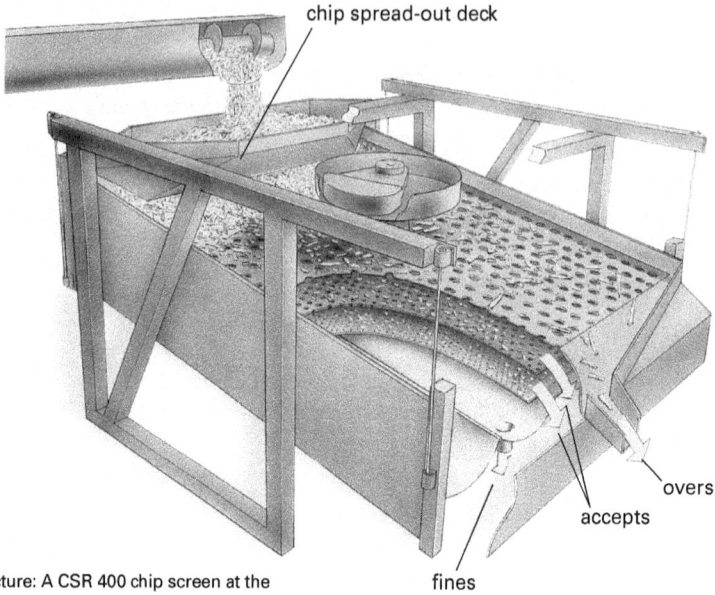

chip spread-out deck

overs

accepts

fines

Cover picture: A CSR 400 chip screen at the
TMP plant of Industrias Forestales S.A., Chile

Figure 2.24. Chip screen with perforated plates shaking the chips to achieve separation of oversize chips, pin chips and fines from accept chips (Metso Paper©).

Disc screens. Disc screens are rotating shafts with discs mounted on them. The discs on one shaft interweave between the discs of the next shaft, *Figure 2.25*. The spacing between discs of adjacent shafts are the slots through which chips thin enough can escape. This type of screen is commonly used for removing the overthick chips. The rotating motion of the discs moves the chips forward.

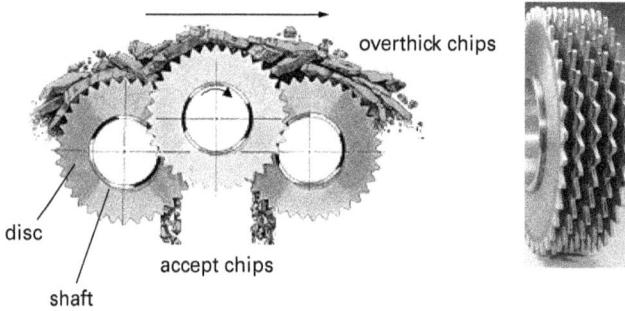

overthick chips

disc

accept chips

shaft

Figure 2.25. Disc screen used separation of the overthich chips. (Acrowood©)

Roll screens. The roll screens are solid rods with grooves cut into their surface. The rotating motion moves the chips forward and smaller particles (pin chip and fines) can pass between the rods, *Figure 2.26.*

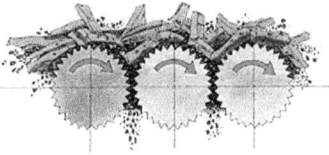

Figure 2.26. Roller screen for separation of pin chips and fines (Rader©).

Bar screens. Figure 2.27 shows a bar screen. The bars are alternately oscillated. The motion is gentle and as the chips are fed to the bars they tumble and turn and if any dimension fits between the bars they fall between the bars. Fines and pin chips are separated at the very beginning, whereas bigger chips travel along for a while before fitting through the bars. Overthick chips are transported by a slight inclination to the end of the screen.

Figure 2.27. Bar screen for separation of overthick chip (Rader©).

Jet screen. Chips can be separated using air impact. As the chips slide across an air gap, heavier material falls into a scrap trap, oversize chips fly a distance short enough to guide them to the gate to the reject screw. Accept chips are blown a bit further and the lighter material far away.

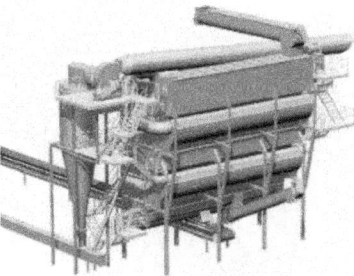

Figure 2.28. JetScreen™ (Andritz©).

2.5.2 Overthick Treatment

The overthick chips are reduced in thickness and retrieved to the accept chips. When cutting the thick chips the size reduction naturally is to occur in the grain direction in order to obtain thinner chips rather than shorter thick chips. One way to accomplish this is by a chip cleaver.

The principal of the chip cleaver is that the knife housing revolves with a certain speed whereby the chips are oriented towards the wall of the housing by the centrifugal forces. The inner counter housing revolves with a greater speed than the knife housing but in the same direction thus pushing the overthick chips towards the knives, *Figure 2.29*.

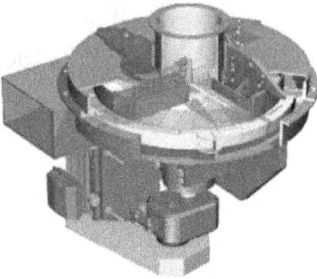

Figure 2.29. HQ-Sizer (Andritz$^©$).

Another way to handle overthick chips is to treat them between two rolls rotating toward one another, *Figure 2.30*. The rolls are profiled, either as smooth curves or pyramid peaks.

Figure 2.30. Chip conditioner rolls (Rader$^©$).

The chips are drawn through the nip between the rolls and cracks and fissures are introduced to the chip by the force applied, *Figure 2.31*.

Figure 2.31. Conditioned chips (Rader$^©$).

3 Mechanical Pulping Chemistry

Göran Gellerstedt
Department of Fibre and Polymer Technology, Royal Institute of Technology, KTH

3.1 Introduction

The disintegration of wood by grinding of logs or refining of chips results in the formation of mechanical pulps termed stone groundwood (SGW) or thermomechanical pulp (TMP) respectively. In the Nordic countries, the predominant wood species for making mechanical pulp is spruce with minor quantities of aspen also being used. Mechanical pulps are used predominatly as major or sole component in graphic papers such as newsprint and LWC. For certain end-uses, such as in liquid board and in absorbing papers, small amounts of chemicals like sodium sulfite, are added to the wood prior to refining and the resulting pulp is termed chemithermomechanical pulp (CTMP). In all cases, the yield of pulp is in the order of 95-98 %, i.e. all the wood components are retained in the fiber with only minor losses of various water soluble substances pre-dominantly acetylated galactoglucomannan.

Brightness and opacity are important quality parameters for mechanical pulps. When fresh spruce wood is used, brightness values in the order of 60–63 % ISO are usually encountered for the unbleached pulp. Inferior wood quality or inefficient debarking may, however, result in low-er brightness values. Therefore, bleaching is often necessary either to restore an even brightness level or to raise the brightness for making higher quality paper products. For spruce mechanical pulps, an upper brightness limit around 80 % ISO can be achieved. For such brightness levels, alkaline hydrogen peroxide is the preferred bleaching agent while for lower brightness increas-es, bleaching with dithionite (hydrosulfite) can also be used. Both of these bleaching agents are used in such a way that a minimum of wood substance is released; so called lignin-retaining bleaching. All types of mechanical pulps suffer from a rapid brightness reversion caused by ei-ther heat or day-light irradiation. For this reason, bleached mechanical pulp cannot replace bleached chemical pulp in paper products where brightness permanence is required.

3.2 Brightness Changes

In different types of mechanical pulps, the light scattering coefficient can be widely different. Consequently, since the brightness of the pulp, according to the Kubelka-Munk equation [1], is dependent on both the light scattering and the light absorption, the amount of chromophoric substances (the light absorption) in the pulp does not alone determine the brightness level. A theoretical example is shown in *Figure 3.1*. Here, reflectance values measured at 457 nm (= brightness) in the range of from 55 to 80 have been plotted against two different values for the light scattering coefficient (*s*) with the respective light absorption coefficient (*k*) being cal-culated using the Kubelka-Munk equation. Obviously, a given amount of chromophoric materi-al, absorbing at 457 nm, can give rise to brightness values which may be several units apart or, alternatively, pulps with a given brightness may contain different amounts of chromophoric ma-terial.

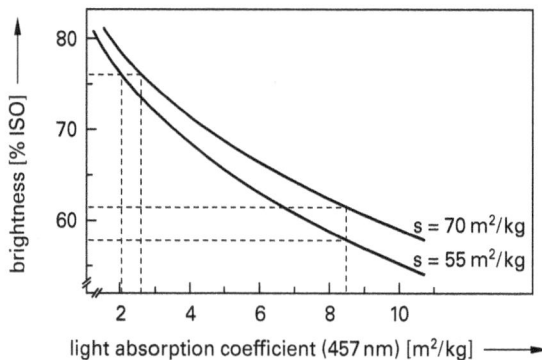

$$\frac{k}{s} = \frac{(1 - R_0)^2}{2R_0} \quad [1]$$

k = light absorption coefficient
s = light scattering coefficient
R_0 = reflectance (R_{457} = brightness)

brightness [% ISO]

light absorption coefficient (457 nm) [m²/kg]

$s = 70 \, m^2/kg$
$s = 55 \, m^2/kg$

Figure 3.1. A theoretical example showing how different light scattering coefficients (*s*) can affect the brightness according to the Kubelka-Munk equation.

Fresh spruce wood is a bright wood species having a light absorption coefficient (k measured at 457 nm) of around 5 m²/kg. During the defibration process, new chromophores are formed resulting in an increased light absorption. Also in the further processing of the pulp through the mill, chromophores are formed and the final unbleached pulp can have a k_{457}-value of around 7–8 m²/kg. In a peroxide bleaching stage, most of the chromophoric material can be eliminated but on subsequent exposure of the pulp or paper to heat or light, a rapid brightness reversion takes place. The brightness changes taking place when wood is converted into mechanical pulp are schematically shown in *Figure 3.2*, expressed as changes in the k_{457}-value.

Light absorption coefficient
at 457 nm, m²/kg

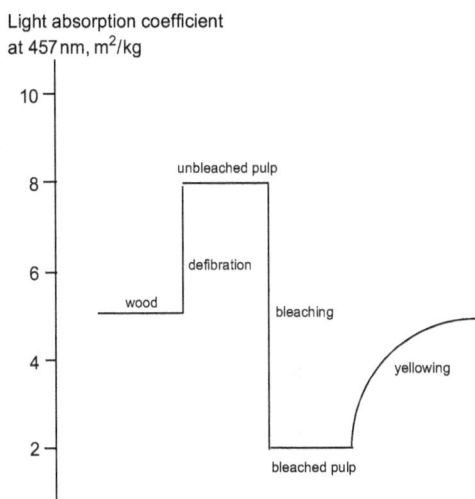

Figure 3.2. Schematic picture showing the brightness changes (expressed as k_{457}-values) taking place when going from wood to unbleached and bleached mechanical pulp.

3.3 The Color of Wood

Most wood species have a yellowish to brownish color with the major contribution coming from specific lignin structures. In addition, the bark contains a variety of reactive components which may directly or indirectly, after oxidation, contribute to color. In mechanical pulping, the brightness of the wood is of great importance since it has been demonstrated that a bright wood will give a bright pulp and vice versa. Furthermore, an efficient debarking is essential since, otherwise, the colored bark components will result in a inferior pulp brightness. The presence of phenolic components in both bark and wood makes the pulp susceptible to the presence of transition metal ions and phenol-metal ion complexes as well as catalysis of air (oxygen) oxidation of phenolic groups into quinones may occur.

3.3.1 Chromophores in Lignin

The major chromophore present in softwood lignin is coniferaldehyde which has an abundance of around 5 units per 100 phenylpropane units. Its UV-maximum in solution is around 350 nm but in the solid state a strong red-shift to around 400 nm can be observed. Thus, this structure is yellow and can be assumed to be the predominant contributor to the wood color. Other types of conjugated carbonyl structures present in smaller or trace amounts include α-keto structures, or-tho- and para-quinones and dienone structures (precursor to β-1 structures). The quinones, in particular, contribute to color since they have light absorption in the visible range of the spec-trum whereas the other types of structures are assumed to have only marginal effects on the brightness. The latter are, however, reactive and indirectly they may contribute to e.g. the light sensitiveness of mechanical pulps (Section 3.8). The natural abundance of iron in wood is around 10–50 ppm. Strong evidence indicates that at least part of this iron is present as iron-cat-echol complex thus contributing to the wood color since such complexes have a light absorption around 550 nm. In *Figure 3.3*, some major conjugated lignin structures are shown.

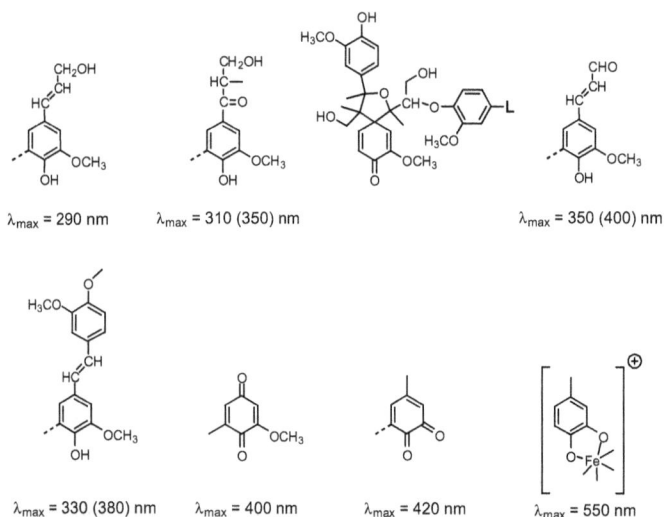

Figure 3.3. Conjugated structures present in softwood lignins. The pertinent absorption maximum in solution and in the solid state (within brackets) is given.

3.4 Reactive Groups in Lignin

In addition to the chromophoric groups discussed in Section 3.3, native lignin contains a variety of free phenolic hydroxyl groups. In spruce wood, the total number of such groups is around 13 per 100 phenylpropane units as shown in *Figure 3.4*. Except for the „normal" type of phenolic group, lignin also contains trace amounts of hydroquinone and catechol structures. These are of particular interest since both types can easily be converted into the corresponding quinone structure on exposure to oxygen under mild conditions. This type of reaction, autoxidation, is catalysed by transition metal ions and may readily take place e.g. during the refining of wood or storage of pulp (Section 3.8.2).

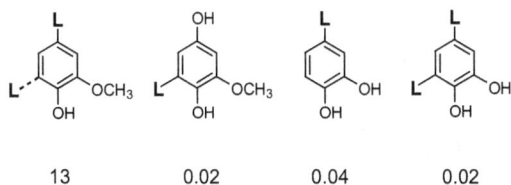

Number per 100
phenylpropane units: 13   0.02   0.04   0.02

Figure 3.4. Types and amounts of phenolic end-groups in softwood (spruce) lignin.

In the presence of water under neutral or alkaline conditions, the primary quinone structure will react further with oxygen and more intensely colored quinones are formed. If other phenolic constituents, such as water soluble bark phenols, are also present, a combination of autoxidation and phenol condensation may occur resulting in the formation of new strongly colored structures attached to the lignin backbone. The reactions are outlined in *Figure 3.5.*

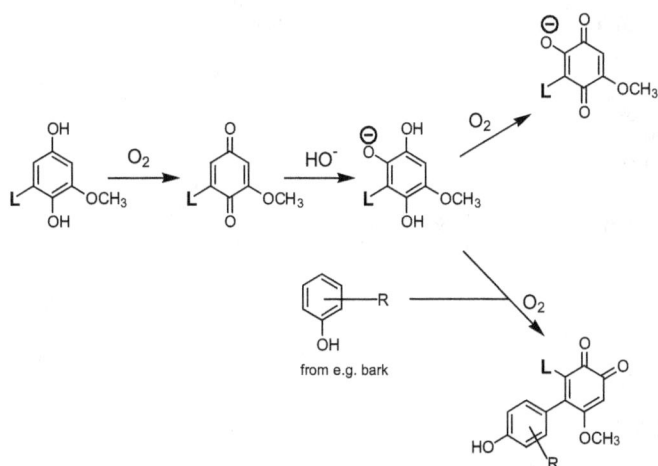

Figure 3.5. Autoxidation of a lignin hydroquinone structure and further conversion of the product into either a hydroxylated quinone or a condensed quinone in the presence of oxygen, water and, in the latter case, a reactive phenol. The reactions are catalysed by the presence of transition metal ions and proceed faster at an alkaline pH.

The conversion of wood to TMP is carried out in the presence of steam and at a high temperature, around 140–170 °C, usually in two refining stages. Under these conditions, new chromophores are formed in the lignin, predominantly in the region below 400 nm as shown in *Figure 3.6.* Thus, the number of conjugated carbonyl and double bond structures in lignin increases as a result of refining since such groups usually absorbs light in the region of ~300–400 nm.

In addition, the number of free phenolic hydroxyl groups increases to some extent (from 13 to ~14 units per 100 phenylpropane units) due to homolytic cleavage of some of the β-O-4 structures in lignin as shown in *Figure 3.7*. In this reaction, additional α-carbonyl groups are also formed.

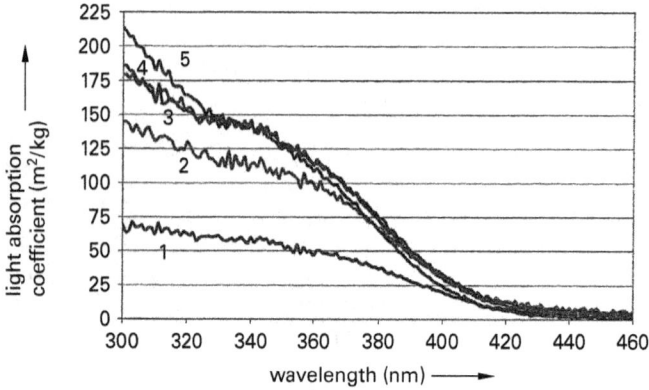

Figure 3.6. Light absorption coefficient spectra for spruce wood meal (1), TMP after the first stage refiner (2), TMP after the second stage refiner (3), TMP after latency treatment (4), paper furnish (5).

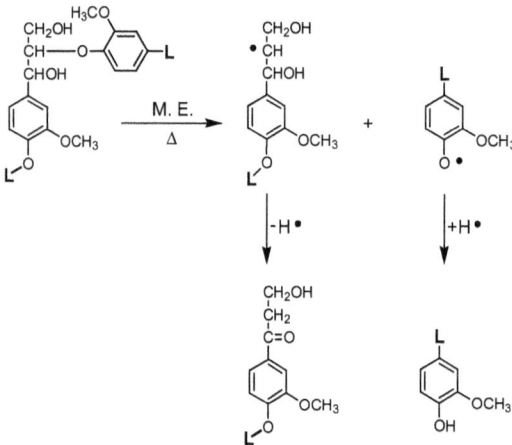

Figure 3.7. Homolytical cleavage of a β-O-4 structure in lignin during wood refining (M.E. = mechanical energy) and formation of a new phenolic hydroxyl group together with an α-carbonyl structure.

Under more intensive milling conditions such as those encountered in the preparation of Milled Wood Lignin, MWL, the cleavage of β-O-4 structures (and possibly other ether linkages in lignin) can be substantial and the number of phenolic groups in spruce MWL is usually in the order of 20 per 100 phenylpropane units.

3.5 Sulfonation of Wood

In the manufacturing of chemithermomechanical pulp, CTMP, the wood is impregnated at an elevated temperature with aqueous sodium sulfite (1–4 % sodium sulfite on wood) prior to refining. Thereby, a more selective liberation of whole fibers can be achieved together with an efficient elimination of resin components due to the mild alkaline conditions. In addition, a certain brightnening effect is obtained due to the reducing properties of the sulfite/bisulfite system.

The treatment of wood with sulfite around 130 °C at a neutral or slightly alkaline pH results in a very rapid initial reaction and within a few minutes, a degree of sulfonation of approximately 3 sulfonic acid groups (~50 mmoles/kg wood) per 100 phenylpropane units can be achieved without any noticeable dissolution of substance, *Figure 3.8.*

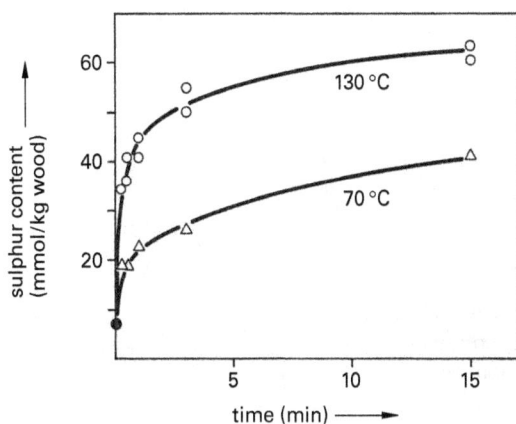

Figure 3.8. Sulfur uptake in spruce wood on treatment with 3 % sodium sulfite at two different temperatures.

With higher charges of sulfite (5–10 % sodium sulfite) and a reaction time of around 5–30 min, further sulfonation of the wood can take place with an upper limit of around 15–20 sulfonic acid groups per 100 phenylpropane units. Subsequent refining of highly sulfonated wood results in the production of chemimechanical pulp (CMP) with a pulp yield of around 90 %.

The mild reaction conditions encountered in the production of CTMP results in a sulfonation of only the most reactive structures present in the wood material. From studies of different morphological regions of CTMP fibers it has been shown that the sulfonation reaction is highly heterogeneous with a much higher degree of sulfonation of the primary wall material as compared to the true fiber wall or the middle lamella fraction. This is illustrated in *Figure 3.9.* for a series of CTMP prepared with different degrees of sulfonation.

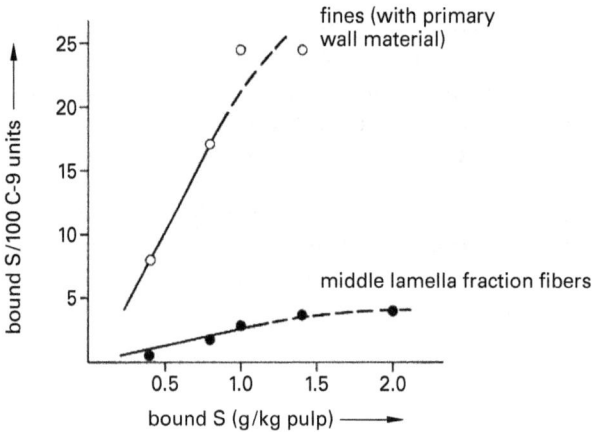

Figure 3.9. Number of sulfonic acid groups per 100 phenylpropane (C-9) units in fines and in middle lamella and fiber fractions respectively as a function of the total amount of sulfonic acid groups in the pulp. Sulfonic acid groups calculated as sulfur.

3.5.1 Sulfonation Chemistry

Among the various lignin structures, only a few are able to react with sulfite under the mild conditions of CTMP manufacturing. The most prominent of these is the coniferaldehyde structure which contains an electron deficient α-carbon atom and, consequently, will react with a strong nucleophile such as the sulfite anion. In a competing or consecutive reaction, the coniferaldehyde structure may also react at its γ–aldehyde group with formation of a hydroxy-sulfonic acid structure (*Figure 3.10*). In practice, only a portion of the coniferaldehyde structures seems to be eliminated, however, and in both unbleached and bleached CTMP, such structures can still be detected. The high reactivity of the primary wall material (*Figure 3.9*) indicates that coniferaldehyde structures are particularly abundant in this part of the wood. Thus, the sulfonation reaction will result in an increased hydrophilisation and swelling of this part of the fiber wall which, in turn, will promote a more selective fiber-fiber separation.

Among other reactive lignin structures, ortho- and para-quinones, being α, β-unsaturated carbonyl structures, may also react with sulfite with destruction of the chromophoric system. Furthermore, the presence of small amounts of sulfite in the wood during defibration will result in a reduction of transition metal ions into their lowest valency states (*Figure 3.11*). This is an advantage and in the pretreatment of wood with sulfite, a simultaneous addition of a chelating agent like DTPA results in an increased pulp brightness due to chelation of metal ions by DTPA rather than by the lignin.

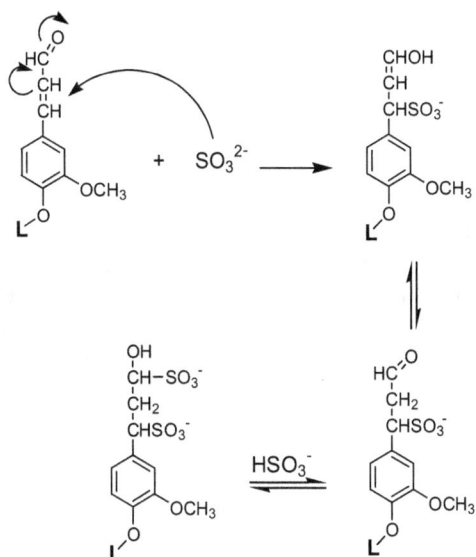

Figure 3.10. Reactions of a coniferaldehyde structure with sodium sulfite.

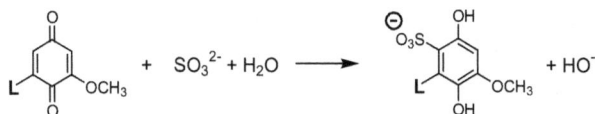

$$2Mn^{3+} (Fe^{3+}) + SO_3^{2-} + 2HO^- \longrightarrow 2Mn^{2+} (Fe^{2+}) + SO_4^{2-} + H_2O$$

Figure 3.11. Sulfonation of a quinone structure and reduction of transition metal ions by sodium sulfite.

3.6 Chemistry of Bleaching

In the bleaching of mechanical pulps, a brightening effect without any loss of substance is desired. This can be achieved either reductively with sodium dithionite at a rather neutral pH or oxidatively with alkaline hydrogen peroxide. Other bleaching agents such as sodium borohydride have been tried but are not used commercially due to the high cost. Bleaching with dithionite can be used to increase the brightness up to approximately 10 brightness units whereas with hydrogen peroxide much higher brightness gains can be achieved with an upper limit for spruce mechanical pulps of around 80 % ISO. In many cases, dithionite is used only to adjust the pulp brightness to an even level regardless of the quality of the wood raw material. Then, the chemical reaction is done simply by adding dithionite to the pulp suspension. Bleaching with hydrogen peroxide, on the other hand, is usually carried out as a true bleaching stage with pretreatment of the pulp with a chelating agent to eliminate the transition metal ions as efficiently as possible followed by addition of the bleaching chemicals. These are except hydrogen peroxide and alkali also sodium silicate, used as a further protector of the peroxide but also as a buffering agent. The reaction is carried out in a bleach tower at a temperature of around 60–70 °C and a retention time of 1–2 hours.

3.6.1 Dithionite Bleaching Chemistry

The reactions between pulp and dithionite is restricted to the reduction of certain types of carbonyl structures such as ortho- and para-quinones. To some extent, a reduction of double bonds in e.g. coniferaldehyde structures may also take place. The result will be a partial bleaching since much of the color attributed to coniferaldehyde structures will still be present in the pulp which consequently will retain a yellowish color. The stability of aqueous dithionite is limited and, therefore, such solutions usually contain bisulfite ions in addition to the dithionite ion. On prolonged storage, this will result in a reduced capacity of the bleaching solution since the bisulfite ion is a weaker reducing agent. In addition, the decomposition reactions will also give rise to thiosulfate and sulfate ions as outlined in *Figure 3.12*.

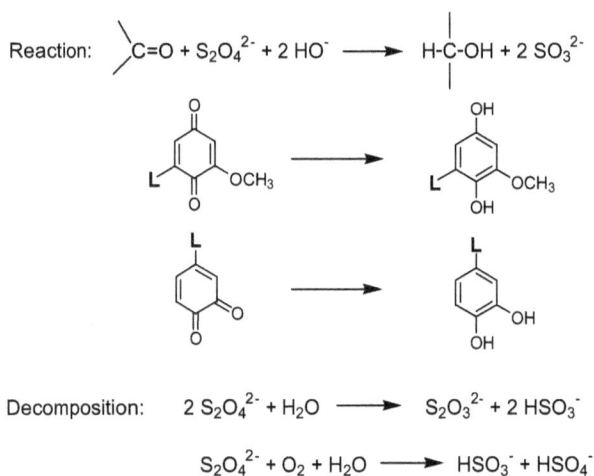

Figure 3.12. Predominant pulp reactions and decomposition reactions of aqueous dithionite solution.

3.6.2 Peroxide Bleaching Chemistry

In the presence of alkaline hydrogen peroxide, mechanical (spruce) pulps can be bleached to brightness levels around 80 % ISO with a charge of ~4–5 % peroxide. The bleaching conditions must be carefully chosen since the addition of alkali alone to a mechanical pulp will result in a darkening reaction. Furthermore, hydrogen peroxide is susceptible to decomposition reactions and in particular, the presence of manganese ions will accelerate the decomposition into oxygen and water. For this reason, the pulp must be pretreated with a chelating agent, usually diethylenetriaminepentaacetic acid (DTPA). On subsequent washing, the majority of the transition metal ions originally present in the pulp can be removed as shown in *Figure 3.13*. Despite these measures, a certain decomposition of hydrogen peroxide can take place with formation of oxygen and water via an intermediate formation of the radicals, superoxide and hydroxyl. Once formed, the highly reactive hydroxyl radical may either react with superoxide to form oxygen, with hydrogen peroxide to form superoxide or with an organic substrate. In the latter case, e.g. aromatic rings in lignin can be oxidized and degraded. The superoxide, on the other hand, is far less reactive but can undergo dismutation to form hydrogen peroxide and oxygen (*Figure 3.14*).

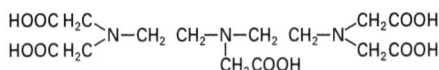

$pK_1 = 1.79 \quad pK_2 = 2.56 \quad pK_3 = 4.42 \quad pK_4 = 8.76 \quad pK_5 = 10.42$

DiethyleneTriaminePentaAcetic acid (DTPA)

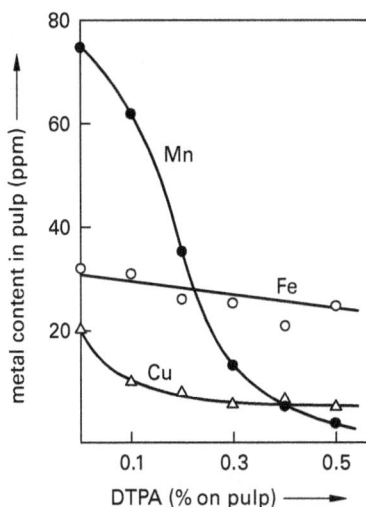

Figure 3.13. Reduction of transition metal ions in a mechanical pulp with different charges of chelating agent, DTPA. The chemical formula for DTPA and its pK_a-values are shown to the left.

$$H_2O_2 + HO_2^- \longrightarrow O_2^{\cdot-} + HO^\cdot + H_2O \quad \text{(cat. by Mn-, Fe-ions)}$$

$$HO^\cdot + O_2^{\cdot-} \longrightarrow O_2 + HO^-$$

$$HO^\cdot + HO_2^- \longrightarrow O_2^{\cdot-} + H_2O$$

$$HO^\cdot + R \longrightarrow \text{oxidative degradation}$$

$$2\,O_2^{\cdot-} + H_2O \longrightarrow O_2 + HO_2^- + HO^-$$

Figure 3.14. Decomposition of hydrogen peroxide and alternative possibilities for the further reactions of hydroxyl and superoxide radicals. R denotes an organic substrate.

The alkaline conditions encountered in peroxide bleaching establishes an equilibrium between hydrogen peroxide and its anion at the beginning of the bleaching stage since the starting pH in the bleaching liquor usually is around pH = 12 (*Figure 3.15*). Due to a rapid release of acetic acid from the pulp (galacto)-glucomannan and the formation of acidic groups in lignin on reaction with hydrogen peroxide, the pH will, however, drop to around pH = 10.5 within a few minutes. The further pH-drop to the final value of around pH = 8.5–9.0 takes place at a much lower rate since the presence of silicate in the bleaching liquor gives a buffering effect. The unavoidable pH-drop occurring during mechanical pulp bleaching is a drawback since the actual bleaching agent is the hydrogen peroxide anion. Obviously, this species is only present as a small fraction of the total concentration of hydrogen peroxide when the pH drops below approximately pH = 10.

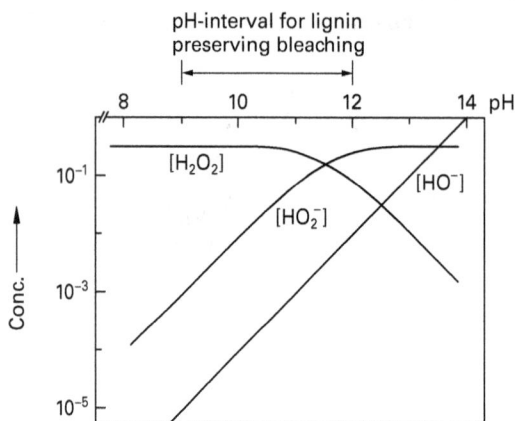

Figure 3.15. The aqueous hydrogen peroxide system with the approximate interval for lignin preserving bleaching of mechanical pulp denoted in the diagram.

The desirable brightening effect encountered in mechanical pulp bleaching is due to the elimination of chromophoric groups in the lignin portion but without much yield loss since the conditions are chosen such that no lignin depolymerisation takes place (Section 3.7). Measurements of reflectance spectra of unbleached and peroxide bleached mechanical pulp reveals that the elimination of light absorbing structures takes place both in the UV and the visible light regions of the spectrum (*Figure 3.16*). The strong maximum shown in the figure at around 375 nm can be attributed to the elimination of extended chromophores e.g. of the coniferaldehyde type whereas the broad and flat absorption with a maximum around 475 nm in the visible region should at least in part be due to the elimination of quinone structures.

Mechanical pulp bleaching employing low charges of hydrogen peroxide (~1 %) results in an efficient utilization of the applied charge and a rapid brightness increase takes place. The exact reaction conditions do not seem to be critical and, obviously, the pulp contains a certain amount of chromophores that are highly reactive and accessible to the alkaline peroxide solution. Under such conditions, it is not possible, however, to react brightness values above ~70–75 % ISO. When higher charges of peroxide are used accompanied with higher charges of alkali, the stoichiometry changes and under such conditions much more peroxide is consumed in order to reach a given brightness but higher brightness values can also be obtained (*Figure 3.17*). In addition to a more efficient elimination of chromophores, factors such as an increased decomposition of hydrogen peroxide and reactions between peroxide and various sub-structures within the lignin macromolecule may play a role.

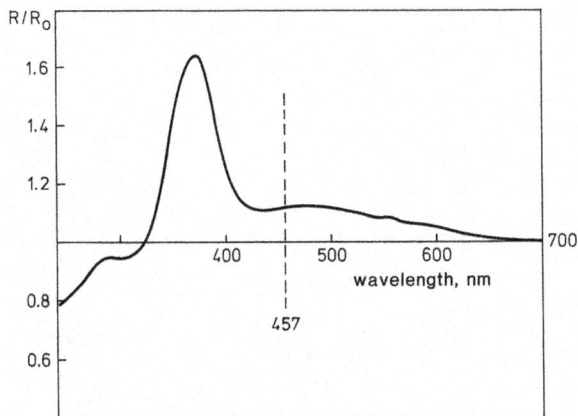

Figure 3.16. Reflectance spectra of unbleached (R_0) and peroxide bleached (R) mechanical pulp measured from 240 to 700 nm. The change in brightness is indicated by the dotted line.

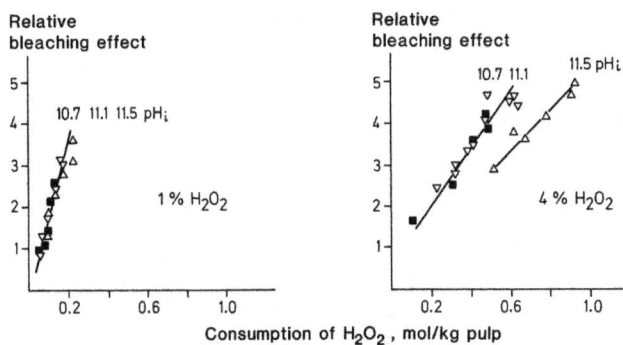

Figure 3.17. Bleaching of mechanical pulp with alkaline hydrogen peroxide. The effects of the charge of peroxide and the initial bleaching liquor pH (pH_i) on the bleaching response and stoichiometry are shown.

The hydrogen peroxide anion is a strong nuclephile. Thus, in analogy to the reactivity of sulfite, structures in the pulp bearing an electron deficient carbon atom will be the most prone to react. Among prominent structures of this type, coniferaldehyde structures and other types of conjugated carbonyl structures can be found, preferentially located as part of the lignin macromolecule. Coniferaldehyde itself, is readily attacked by the hydrogen peroxide anion with formation of an α-hydroperoxide (*Figure 3.18*). This will react further via an epoxide to form a new hydroperoxide intermediate which finally is decomposed into an aromatic aldehyde structure and 2 moles of formic acid. If the original coniferaldehyde structure contains a free phenolic hydroxyl group, the aromatic aldehyde can be further transformed into a hydroquinone structure and 1 mole of formic acid. The latter reaction is usually referred to as the „Dakin reaction" and the reaction mechanism is detailed in *Figure 3.19*.

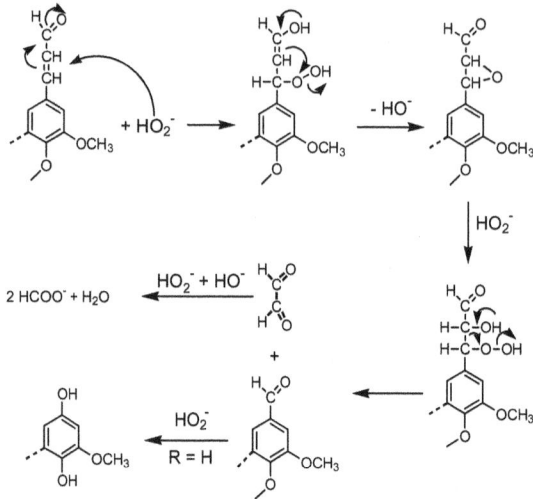

Figure 3.18. The oxidation of coniferaldehyde structures with alkaline hydrogen peroxide.

The Dakin reaction:

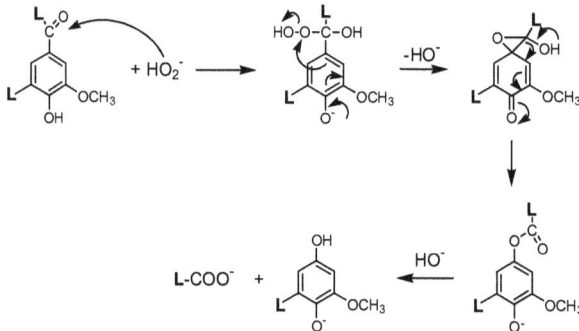

Figure 3.19. The oxidation of aryl-α-carbonyl structures with alkaline hydrogen peroxide, the „Dakin reaction".

A third important type of carbonyl structures in mechanical pulps are the quinones. Although only present in trace amounts, their strong color in the visible light region make them strong contributors to the overall color of the pulp. Most types of „simple" quinones are, however, very reactive towards alkaline hydrogen peroxide and are easily oxidized into colorless end-groups usually containing carboxyl groups as shown in *Figure 3.20*. In addition, methanol and low molecular weight acids are formed. When the concentration of the hydrogen peroxide anion is low, however, as in the final part of bleaching, quinones can also react in a competing reaction resulting in the formation of a hydroxylated quinone structure. Thereby, a small portion of the original quinone can be converted into a product with low reactivity towards further attack by peroxide and with an intense color in the visible light range.

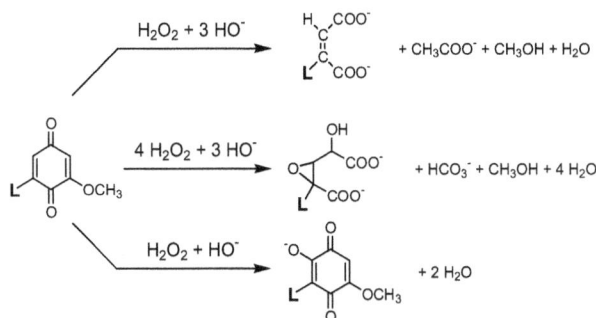

Figure 3.20. The oxidation of quinone structures with alkaline hydrogen peroxide.

The strong influence of transition metal ion like manganese on the reactions between lignin and alkaline hydrogen peroxide has been studied both with model compounds and in pulp bleaching experiments. Thus, the oxidation of a simple aryl-α-carbonyl structure, acetoguaiacone, with alkaline peroxide in the absence of any metal ions proceeds stoichiometrically with formation of methoxyhydroquinone and acetic acid. If, however, a catalytic amount of manganese(III) is added to the reaction system, the oxidation rate as well as the consumption of hydrogen peroxide increases substantially as shown in *Figure 3.21*. At the same time, the yield of methoxyhydroquinone decreases and several other products form. Obviously, a catalytic decomposition of hydrogen peroxide with formation of hydroxyl and superoxide radicals takes place as was shown in *Figure 3.14*. As a result of the increased concentration of hydroxyl radicals in the system, a direct oxidation of the aromatic ring in the starting material as well as in the product will compete with the Dakin reaction resulting in an overall reaction rate increase.

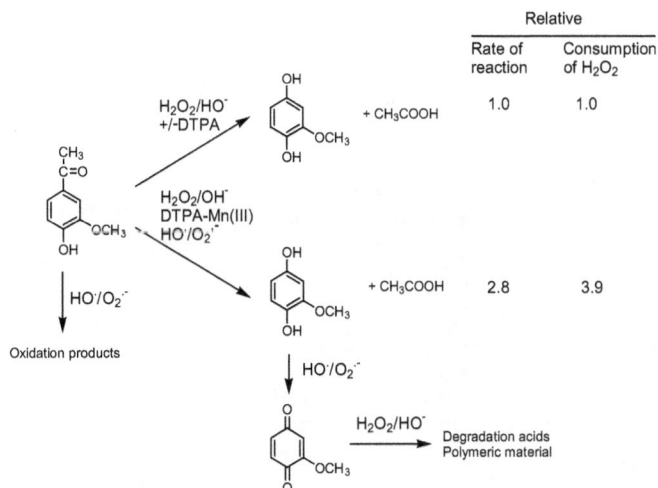

Figure 3.21. Oxidation of acetoguaiacone, a lignin model compound, with alkaline hydrogen peroxide in the absence and presence of manganese(III).

In corresponding peroxide bleaching experiments with mechanical pulp in the absence and in the presence of manganese(III), after careful elimination of transition metal ions with DTPA, the importance of a metal free system for an optimal bleaching performance have been confirmed. This is illustrated in *Figure 3.22* and it can be seen that, in analogy to the model experiment above, an increased consumption of hydrogen peroxide as well as an inferior brightness after bleaching will be the result of insufficient elimination of manganese from the system.

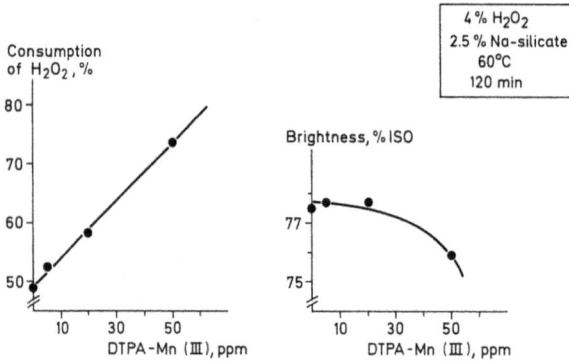

Figure 3.22. The influence of manganese(III) on the bleaching of mechanical pulp with alkaline hydrogen peroxide.

3.7 Structural Changes in Mechanical Pulping

The predominant portion of the wood material is retained in the mechanical pulp fibers both after refining and subsequent bleaching. In the refining process, a yield loss of 3–5 % is usually encountered with hemicelluloses being the predominant contributor (*Table 3.1*). Among these, acetylated galactoglucomannan dominates but small amounts of xylan and pectin are also dissolved.

Table 3.1. Dissolution of different wood constituents in the production of TMP.

Wood component	Amount, kg/t TMP
Acetic acid	1–2
Lignans	2–3
Extractives	4–6
Hemicelluloses, pectin	18–21
Lignin	3–5
Others	6–8

Under the alkaline conditions prevailing in a hydrogen peroxide bleaching stage, a rapid hydrolysis of the ester-linked acetate groups in galactoglucomannan takes place. Approximately 20 kg/t of TMP can be released as acetic acid. A substantial portion of the deacetylated and less water soluble hemicellulose is, however, redeposited on the fibers thus partly compensating for the weight loss from acetyl groups. Furthermore, alkaline hydrolysis of pectin will result in the release of methanol together with the dissolution of considerable quantities of polygalacturonic acid (pectin without methyl ester groups). As a result of these changes, a further yield loss of around 3 % (on wood) can be found after a peroxide stage.

The presence of acidic groups in pulp fibers is of importance since these will affect the swelling capacity of the pulp and they will also act as binding sites for cations. Mechanical pulp fibers, both unbleached and bleached, contain considerable amounts af acidic groups. In unbleached fibers, these are predominantly located in the hemicelluloses and to a smaller extent in the pectin although CTMP also will contain sulfonic acid groups in the lignin. After peroxide bleaching, a substantial increase of the amount of carboxyl groups can be found. The remaining pectin in its polygalacturonate form is now the major contributor together with the hemicelluloses. In addition, a substantial amount of carboxyl groups are formed in the lignin as a result of the oxidative bleaching (*Table 3.2*). For bleached CTMP, an additional amount of lignin-bound sulfonic acid groups are also present.

Table 3.2. Approximate amounts (mmol/kg pulp) of carboxyl groups and their origin in mechanical pulp fibers.

Pulp	Wood component			Sum
	Hemicell.	Pectin	Lignin	
TMP	75	20	~0	95
P-bleached TMP	70	80	70	220

3.8 Yellowing of Mechanical Pulps

The yellowing of mechanical pulps caused by either light or heat is a severe technical problem that to a large extent has prevented the wider use of such pulps in a variety of paper grades. In particular the light-induced discoloration is rapid and caused by the fact that lignin is able to absorb light in the UV-region of day-light as shown in *Figure 3.23*.

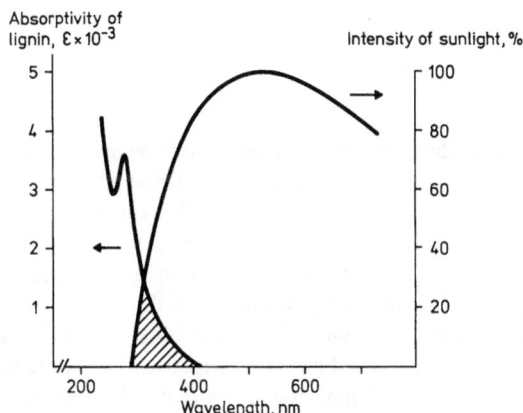

Figure 3.23. Light absorption spectrum of lignin and emission spectrum of sunlight. Within the marked area, light absorption takes place.

The native lignin present in wood or mechanical pulp contains several types of functional groups with ability to absorb light in the wavelength region of approximately 300–400 nm. Some of these are shown in *Figure 3.24*, but others such as coniferyl alcohol end-groups are also present. In addition, certain aromatic rings may directly absorb light in the lower end of the sunlight spectrum.

The general mechanism of light absorption involves a primary absorption of a quantum of light resulting in the excitation of one electron which is lifted from the singlet ground state to an excited singlet state. The energy absorbed can be released in different ways as shown in *Figure 3.24*. If the spin of the excited electron is reversed, an excited triplet state located on a lower energy level is created by intersystem crossing. Since the life time of triplet states usually are longer than those of excited singlet states, the probability of a chemical reaction occurring from a triplet state is also higher. An excited carbonyl group usually reacts in this way.

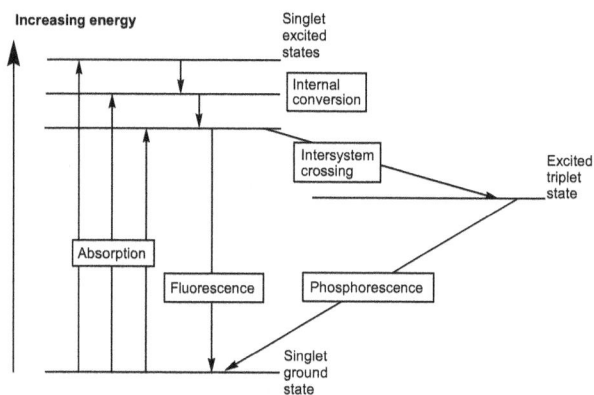

Figure 3.24. Molecular energy levels in photochemical processes and various ways of releasing the absorbed energy.

3.8.1 Chemistry of Photo-induced Yellowing

On light irradiation of a mechanical pulp, new chromophores in both the UV- and the visible light region are rapidly formed as shown in *Figure 3.25* . The strongest light absorption increase is centered around 330 nm and can be ascribed to the formation of e.g. aryl-α-carbonyl structures in lignin. At longer wavelengths, a broad light absorption extending far into the visible region of the spectrum can be seen with a flat maximum around 430 nm. Obviously, quinone structures together with more undefined colored structures are gradually formed.

The general mechanism for this type of lignin oxidation involves the absorption of light by a carbonyl group in lignin. By intersystem crossing, an excited triplet is formed and this, in turn, will abstract a hydrogen from e.g. a benzyl alcohol or a phenol structure in lignin. Thereby, two radical species are formed which in further reaction steps will be converted into carbonyl groups by the action of oxygen (*Figure 3.26*). The latter is converted into hydrogen peroxide.

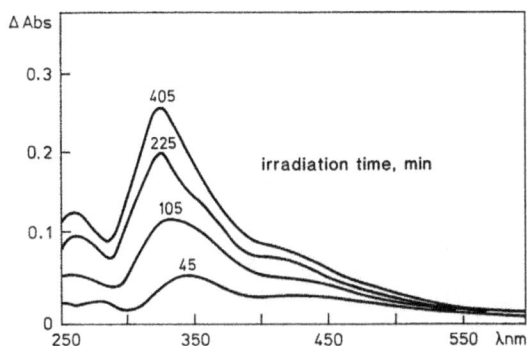

Figure 3.25. Changes in the light absorption coefficient (Δ Abs) as a function of wavelength on irradiation of a peroxide bleached mechanical pulp with imitated daylight.

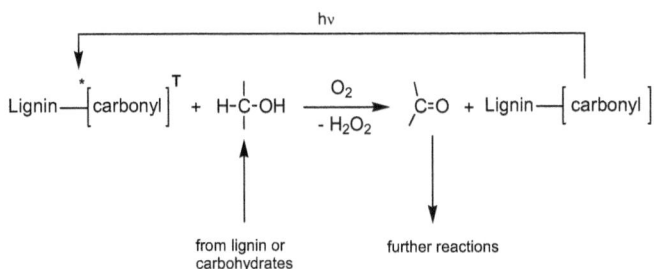

Figure 3.26. General mechanism for the photochemical oxidation of an alcohol or phenol to a carbonyl structure by the action of a photo-sensitizer; a carbonyl structure.

In principle, the reaction shown in the figure can take place on any available alcohol structure in the pulp with the only prerequisite being that the originally present carbonyl group is able to absorb light. This has been demonstrated in a model experiment in which filter paper made from bleached chemical pulp was impregnated with either veratrumaldehyde or vanillin and irradiated with imitated sunlight. In both experiments, the aldehyde can absorb light but only vanillin, being a phenol, can be converted into colored products. Nevertheless, a strong brightness decrease can be observed in both experiments and, in addition, the viscosity of the filter paper also decreases (*Figure 3.27*).

Obviously, veratrumaldehyde or vanillin absorbs light and by hydrogen abstraction from the cellulose, a cellulose-derived radical is formed which subsequently is converted into a carbonyl group. Thereby, further degradation reactions are facilitated resulting in both depolymerisation and color formation. When vanillin is present, a competition between the cellulose and the vanillin for the abstraction of the proton takes place resulting in a somewhat lower degree of viscosity loss but a higher discoloration intensity.

The kinetics for the light-induced yellowing has been studied by irradiation experiments on bleached mechanical pulp. As shown in *Figure 3.28*, the formation of new chromophores in the pulp takes place in two steps with the initial discoloration being much faster than the second.

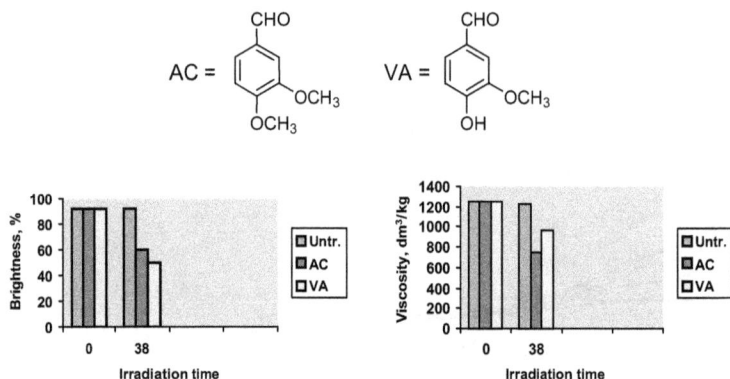

Figure 3.27. Photo-induced degradation of bleached chemical pulp (filter paper) by the action of a photo-sensitizer; an aromatic aldehyde. AC = veratrumaldehyde, VA = vanillin.

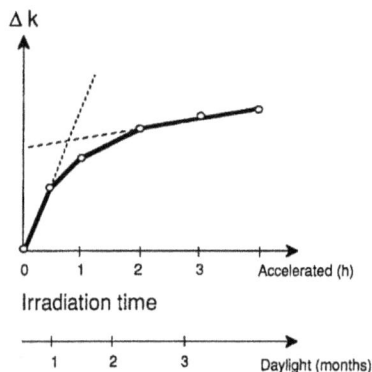

Figure 3.28. Change of the light absorption coefficient (at 457 nm) in bleached mechanical pulp as a function of irradiation time with imitated sunlight. Approximate irradiation time on exposure to normal daylight is included for comparison.

Based on a large number of model compound studies, it has been suggested that the initial reactions in the yellowing process predominantly involves a photo-oxidation of simple phenolic end-groups of the catechol, hydroquinone and diguaiacylstilbene types (*Figure 3.29*). Such structures are easily converted into the corresponding quinone and, despite a very low abundance in the lignin, the strong coloration in the visible light range can easily explain the rapid part of the pulp yellowing.

"Rapid" phase

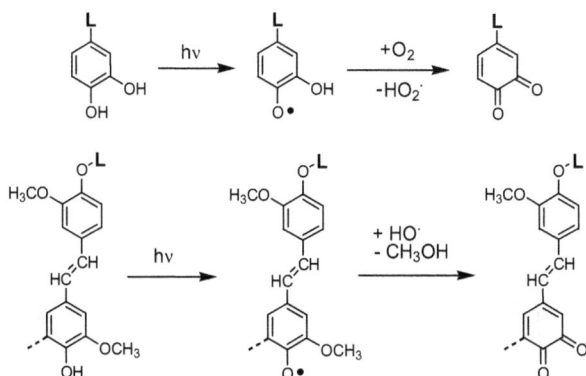

Figure 3.29. Photo-induced reactions in lignin thought to be responsible for the „rapid" phase of yellowing of mechanical pulps.

On prolonged daylight irradiation, the brightness of mechanical pulp continously decreases albeit at a slower rate as compared to the initial. Here, several contributing reactions may occur including unspecific oxidation of aromatic end-groups with a free phenolic hydroxyl group and oxidation of benzyl alcohol groups according to the general mechanism shown in *Figure 3.26*. In the latter case, the formation of an α-carbonyl group present in a β-O-4 structure gives a structural unit which may directly react further in the presence of light to give a β-aryl ether cleavage and formation of colored products (*Figure 3.30*).

"Slow" phase

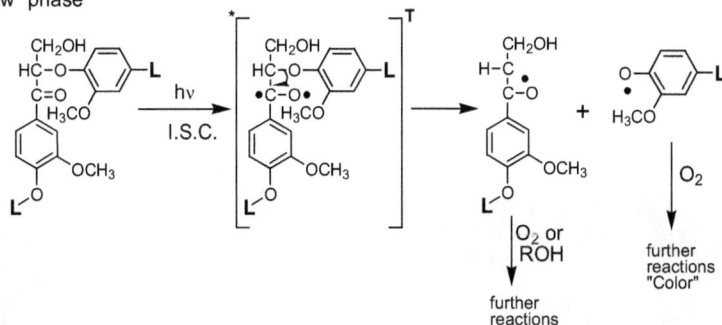

Figure 3.30. Photo-induced cleavage of a β-O-4 structure in lignin containing an α-carbonyl group (I.S.C. = intersystem crossing). The reaction may contribute to the „slow" phase of yellowing of mechanical pulp.

3.8.2 Chemistry of Heat-Induced Yellowing

In the absence of light, mechanical pulps can undergo a noticable yellowing which, again, is a drawback for a further use in higher quality papers. At elevated temperatures, the reaction occurs rapidly and can be measured as a brightness change already in the manufacturing process as shown in *Figure 3.6*. Under storage conditions, either as pulp or paper, a further brightness decrease may take place, particularly for bleached grades. The chemistry involved in this yellowing is an autoxidation of reactive phenolic end-groups in the lignin such as catechol and hydroquinone groups, viz. the same type of reactions as those described in Section 3.4.

3.9 Further Reading

General Literature

Forsskåhl, I. (2000) Brightness Reversion. In *Forest Products Chemistry* (Ed. P. Stenius), Book 3. In *Papermaking Science and Technology* (Series Eds. J. Gullichsen and Hannu Paulapuro). Fapet Oy.

4 Mechanical Pulping

Hans Höglund
Mid Sweden University

4.1 Introduction

Mechanical pulps with yields of 96–98 % are produced by two different commercial processes, in which wood is processed either in the form of logs treated in grinders, *Figure 4.1* or in the form of wood chips that are converted to pulp by mechanical work in a refiner *Figure 4.2*. Wood source for mechanical pulp in Europe and Canada, where approximately 75 % of all mechanical pulps are produced, is mainly spruce. Almost 40 % of the pulp production from wood fibres in Scandinavia is mechanical pulp (year 2001). The mechanical pulp accounts for approximately 20 % of the total world production of pulp (year 2001).

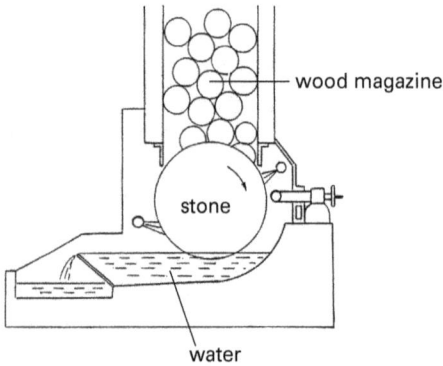

Figure 4.1. Grinder.

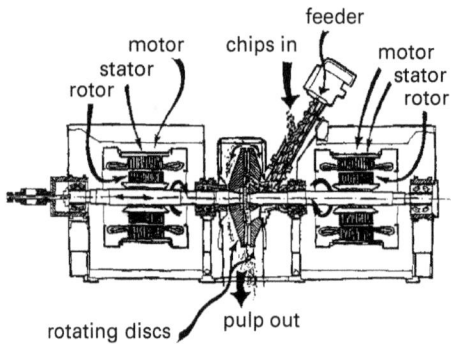

Figure 4.2. Double disc refiner.

4.1.1 The Processes – Short Description

In the groundwood process, debarked logs are pressed against a rotating grindstone, which is simultaneously cleaned and cooled by using shower water. Fibres and fibre fragments are torn from the wood under intensive treatment from the grits of the stone. Refiner pulps are produced in different types of disc refiners. In the refiner, wood chips are first disintegrated to fibres and fibre fragments, these are subsequently worked in a narrow gap between two parallel patterned refiner discs, either one or both of them are rotating.

A number of process variants have been established for the two methods. It is mainly the temperatures at preheating and processing of the wood material that distinguish the different methods. Groundwood pulps (GWD) have for a long time been produced in non-pressurised systems. In more recent systems the logs are preheated and processed under pressurised steam. The temperature in systems for production of pressurised groundwood (PGW) can vary from 100 °C to 140 °C. Refiner pulp is usually produced in a so-called thermomechanical pulping processes (TMP). After a short preheating of the chips in steam in the interval 115–155 °C the chips are refined in a pressurised system. Refiner pulp, which is produced without preheating, is called refiner mechanical pulp (RMP). Chemi-thermomechanical pulp (CTMP) is a special type of TMP. In CTMP processes the chips are initially impregnated with small amounts of lignin softening chemicals, and then heated with steam to approximately the same temperature interval as in the TMP process.

The initial softening of the wood material has a large impact on the fracture mechanics in the wood and on the fibre structure in the subsequent processing. Processes therefore give pulps with different quality profiles that are related to the different heating temperatures before the mechanical processing. The quality profile can in addition be influenced by impregnating the wood with lignin softening chemicals, as in the CTMP process, or with small amounts of chemicals added in the refiner. With pre-treatment and processing at increasingly higher temperatures the fractures in the wood structure is successively more concentrated to the lignin rich areas, i.e. to the middle lamella area, (see *Figure 4.13*) which have been most effectively softened. This is especially obvious with preheating and processing at temperatures well above the softening temperature of the lignin, which for spruce chips is at 125–145 °C at the processing conditions in question. The final pulp quality is controlled by the energy input to the mechanical processing of the fibre material. The intensity and the speed of the processing have a large influence on the final properties of the fibres as well.

4.1.2 Main Quality Characteristics

Mechanical pulps consist of a mix of fibres, fibre fragments and fines which have been created during the processing, *Figure 4.3*.

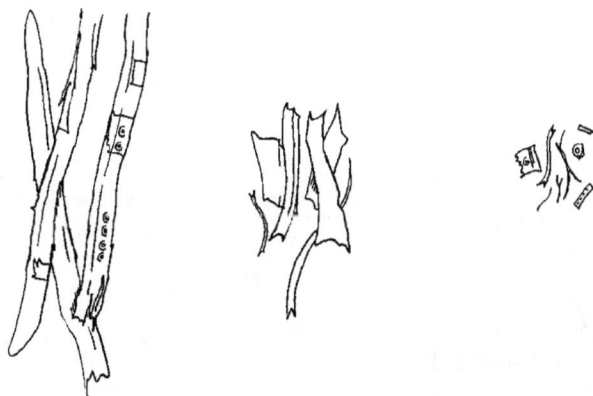

Figure 4.3. Mechanical pulps consist of a mixture of fibres, fibre fragments (middle fraction) and fines.

The wide distribution of size fractions and the facts that the material is lignin rich and stiff give mechanical pulps their special character. Since unbroken and unfractured fibres have a high bending stiffness, mechanical pulps must contain a relatively large amounts of fines with high binding potential in order to form sheets with sufficient strength. Fines and more flexible

60

fibre fragments act as binders between the long and stiff fibres. Mechanical pulps are usually used in furnishes together with chemical pulp. By adding chemical pulp fibres, which are more flexible, the bonding strength can be increased in the sheet.

Mechanical and chemi-mechanical pulps have particularly two quality advantages which are used in the production of paper products. Firstly, the sheets from mechanical pulp have considerably higher opacity than sheets from chemical pulps, because the light scattering surface at a given strength is larger, *Figure 4.4*.

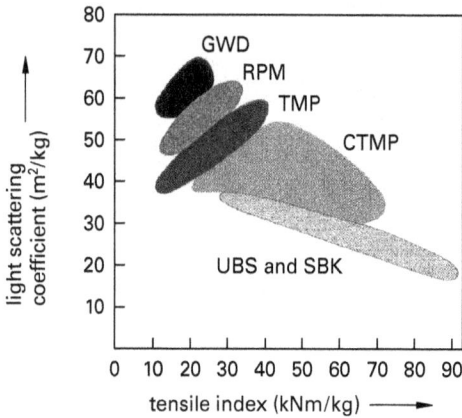

Figure 4.4. Softwood pulps of interest in printing papers. SBK = Semibleached kraft, UBS = Unbleached sulphite (ISO sheets).

Secondly, the higher fibre stiffness in mechanical and chemi-mechanical pulps contributes to sheet structures giving lower densities (higher bulk), *Figure 4.5*.

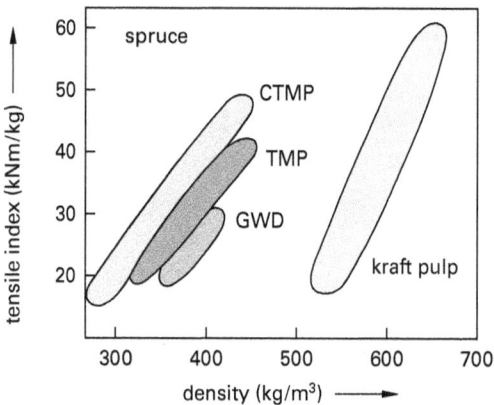

Figure 4.5. Sheets from spruce CTMP have a high bulk (low density) at a certain strength.

A high bulk, i.e. high calliper at a given grammage, contributes to improve the bending stiffness of the sheet.

The disintegration of single fibres to fibre fragments and fines in mechanical pulping provides a high specific surface area. With increasing treatment (energy input) in the processes the amount of fibre fragments and fines is increasing, at the same time the fibres become more slender and more flexible. Increasing amount of fibre fragments and fines increases both the bounded and the unbounded surface area in a sheet with mechanical pulp. The high fraction of unbounded surface area in the relatively loose sheet structure, that can be built with the relatively long stiff fibres, gives the sheet its high light scattering ability, *Figure 4.6*.

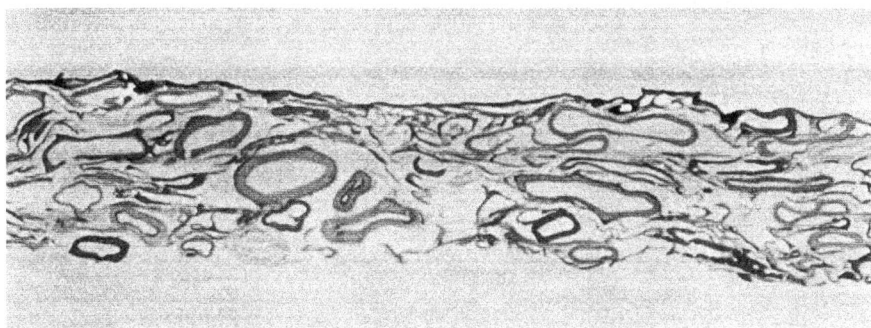

Figure 4.6. Cross section of a printed newspaper.

The high light scattering ability (high opacity) is used in the production of thin so-called wood-containing, uncoated printing papers, such as news print and magazine papers and as base paper in coated printing papers. These are by far the largest fields of applications for mechanical pulps.

Another major area for mechanical and chemi-mechanical pulps is in products, for which high bulk is of importance, for example in the middle layer of box board and in certain tissue qualities and in fluff based absorption products. The high stiffness of the mechanical fibre is used to build bulky structures, *Figure 4.7*.

In order to achieve a high bulk it is especially important with a large fraction of long fibres. Chemi-mechanical long fibre pulps (like CTMP) at high yields (>90 %) are consequently especially useful for those qualities. During production of chemi-mechanical pulps, the wood lignin is softened both chemically and thermally, which gives a pulp characterised by a large and more flexible long fibre fraction and a considerably lower shives content (see page 33) than a pulp from a pure mechanical pulping process. To reach a desired strength level, a chemi-mechanical pulp does not need to have as large fraction of binding material as a pure mechanical pulp.

The physical properties of a mechanical or a chemi-mechanical pulp can to a large degree be related to its dewatering properties and to the fraction of fibres, fibre fragments and fines. The drainage resistance for mechanical and chemi-mechanical pulps is traditionally measured with a freeness tester. Typical freeness levels (Canadian Standard Freeness, abbreviated CSF) for pulps for different paper qualities can be seen in, *Figure 4.8*.

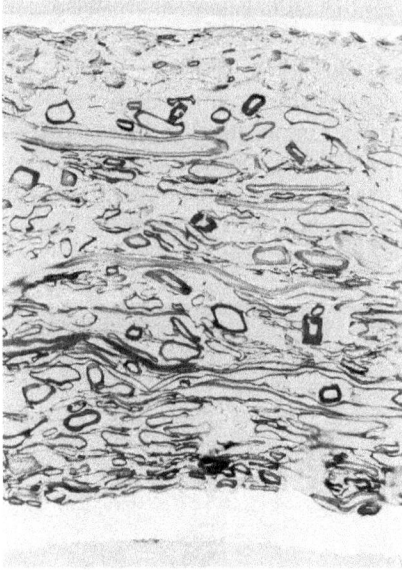

Figure 4.7. Cross section of a box board with CTMP in the middle and chemical pulps in the outer layers.

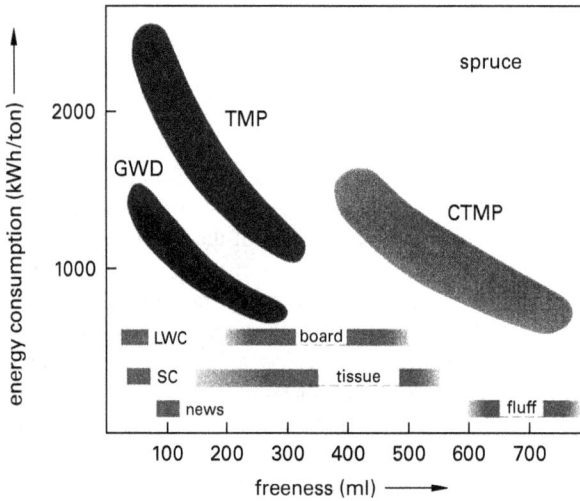

Figure 4.8. Energy consumption in grinding and refining at manufacturing of mechanical pulps of different grades (LWC=light weight coated magazine paper, SC=super calendered magazine paper).

The lower the freeness, the more difficult the pulp is to dewater. The amount of different particle types can be measured with either a Bauer McNett fractionator, in which the particles are separated by a sieving procedure or by an image analysis system in which a camera registers the particles in a diluted pulp suspension. Mechanical and chemi-mechanical pulps intended for different end products are produced in a wide range of distribution of the different particle types, *Figure 4.9.*

Bauer McNett classifications of GWD, TMP and
CTMP from spruce

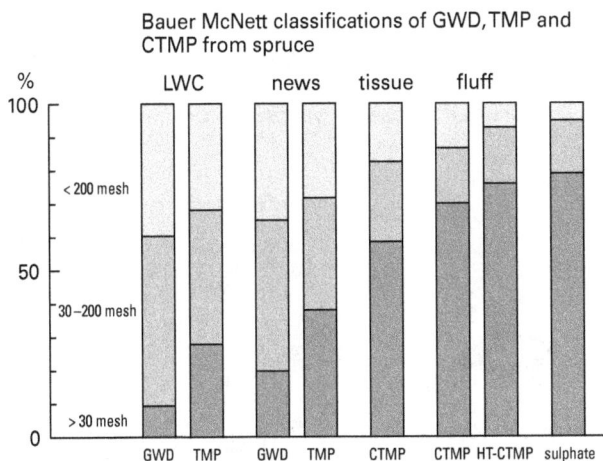

Figure 4.9. Representative fibre distributions in mechanical and chemimechanical pulps for different paper grades. (HT CTMP is a special pulp, that is manufactured in refining of chips that are preheated to a temperature > 170 °C).

Pulps used for manufacturing of high quality magazine papers with a high surface smoothness consist to a large degree of fine material and to a small degree of long fibres. Pulps intended for bulky boxboard, tissue or fluff products have a considerably higher amount of long fibres. With increasing energy input in the pulping process the amount of fine material is increased, *Figure 4.8.* With an increasing amount of fines the pulp becomes more difficult to dewater.

4.2 Prerequisites of the Wood Physics

4.2.1 The Fibre Material

The wood fibre has by nature developed as a component in a bending stiff construction (the tree), in which it needs to endure high dynamic and mechanical hardship in changing climate. In order to be useful for papermaking it needs to be made flexible and to get a defibrillated binding prone surface. During chemical pulping in which roughly half the amount of wood substance is dissolved the fibre can easily be defibrillated and made flexible during the beating process, and it will easily collapse during pressing and drying in the paper machine. The degree if collapse can be defined by the ratio of the cross section of a collapsed and uncollapsed („intact") fibre (see *Figure 4.43*). The mechanical pulp fibre, in which only a few percent of substance has been dissolved, remains stiff if not exposed to intensive treatment. During such treatment a partial collapse of the fibre is desirable as well as a rupturing and reduction of the fibre wall thickness. The chance of partially collapsing the fibre is increased with increasing degree of thermal and chemical softening under the pre-treatment and under the mechanical treatment. During the treatment, the fibre dimensions are reduced, when the outer parts of the fibre are peeled off and are being converted to fines. The flexibility increases at the same time as the amount of fine material is increasing. Since the demand on this process step is that fines are created at the same time as the fibre length is preserved the degree of softening is of great importance and only certain wood qualities can be used.

Softwoods and especially spruce have showed themselves very suitable as raw materials for mechanical pulping.

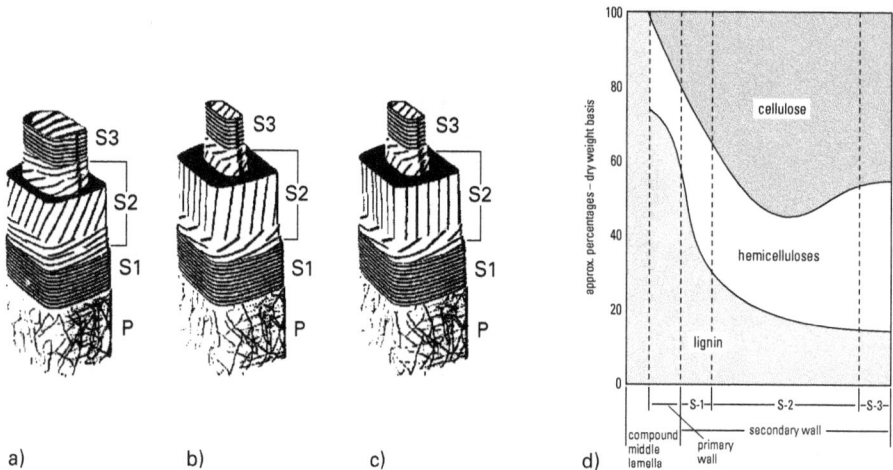

Figure 4.10 a–d. Cell wall models of Norway spruce tracheids
a) earlywood tracheid; b) latewood tracheid from the juvenile wood; c) latewood tracheid from the mature wood; d) distribution of chemical constituents in the cell wall.

Spruce fibres from earlywood *Figure 4.10*, *Table 4.1* are relatively long and have a favourable ratio between fibre wall thickness and fibre thickness, which make them relatively easy to fully or partially collapse during the mechanical treatment. Latewood has considerably less favourable dimensions.

Table 4.1. Typical dimensions of some softwood and hardwood fibres.

	Length, mm	Width, µm	Wall thichness, µm
Spruce, earlywood		35	2,3
Spruce, latewood	3,0	20	4,5
Aspen	1,0	22	2,7
Birch	1,1	20	3,0
Beech	1,2	24	4,5

The fibre dimensions given in *Table 4.1* are mean values. The scatter around these values is large in this biological material, *Figure 4.11*.

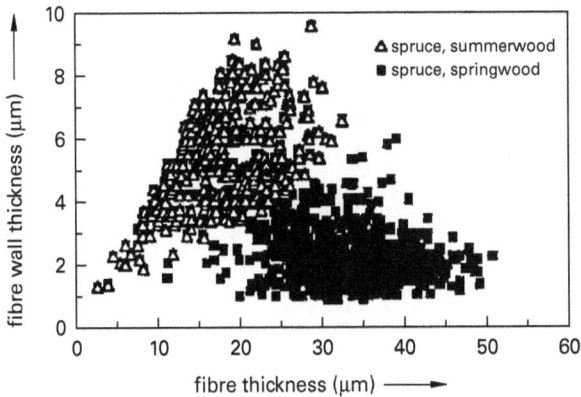

Figure 4.11. Distribution of fibre thickness (measured in radial direction) and fibre wall thickness in spruce wood.

Thin walled earlywood fibres can of course collapse by far easier than thick walled latewood fibres. The latter can therefore be a severe disturbance in the sheet structure of lightweight magazine papers, since the thickness of the sheet is of the same order as the thickness of a fibre, *Figure 4.12*.

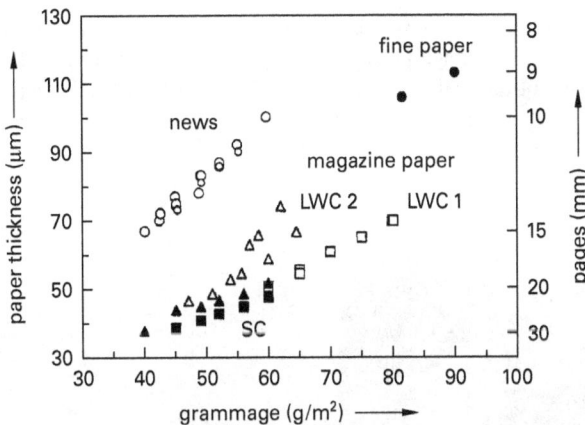

Figur 4.12. Low grammage wood containing papers approach the thickness of one native spruce wood fibre.

In order to use hardwood for mechanical pulping the fibre material must be thin walled. Examples of such hardwoods are aspen and poplar. Thick walled hardwoods, such as birch or beech fibres are on the other hand clearly unsuited for mechanical pulping only. The treatment necessary to increase the flexibility results in fibre length reduction without attaining the desirable changing in the fibre structure. However, most types of hardwood can be used for chemimechanical pulping with yields of 85–92 % after the fibre walls have been softened and a certain amount of substances have been dissolved.

4.2.2 Softening of Lignin

The degree of softening of the wood lignin during the pretreatment controls where in the wood structure the rupture takes place during grinding and refining. If the middle lamella lignin, which binds the fibre together to wood structure, have retained much of its original stiffness most of the rupture will be in the fibre walls themselves *Figure 4.13*.

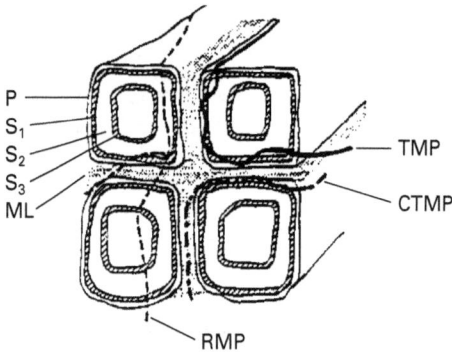

Figure 4.13. Schematic diagram showing typical fracture zones in softwood as affected by different mechanical pulping processes. The cell wall layers are indicated as P (primary wall), S_1–S_3 (secondary walls) and ML (middle lamella).

The lignin content in the exposed fibre surfaces will be relatively low (c.f. *Figure 4.10 d*), giving a surface which is easy to defibrillate and with a high bonding potential. The surfaces are characterised by high roughness with protruding thread-like fibrils of cellulosic material. On the other hand the average fibre length will be short, since many fractures will run across the fibre walls.

fibre produced by pressurized refining at
115 °C   170 °C

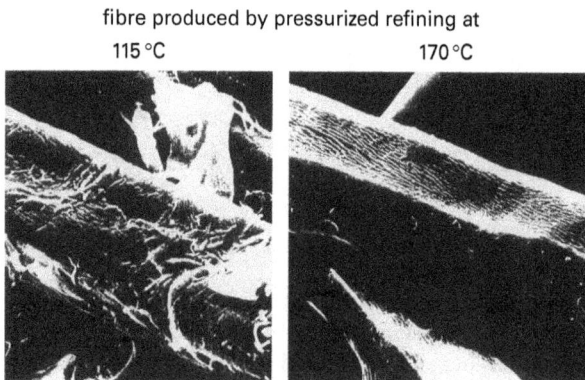

Figure 4.14. Softwood fibres refined after preheating to different temperatures.

The wood fibres can be liberated with low energy input and little fibre damages in mechanical and chemimechanical processes if the middle lamella lignin is well softened. Preheating and treatments are then made at temperature considerably above the softening temperature of the lignin. The fractured areas become smooth and have a high fraction of lignin *Figure 4.14*.

The softening temperature can be determined by mechanical spectroscopy in different types of dynamic-mechanical measuring equipment. It is defined as the temperature at which the so-called internal friction reaches its maximum, *Figure 4.15*.

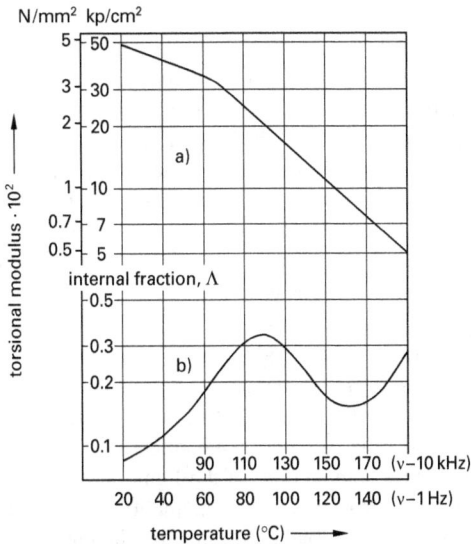

Figure 4.15. Torsional modulus – a measure of stiffness – and internal friction on wet spruce wood at different temperatures and strain frequencies. The softening temperature is defined as the temperature where the internal friction has its peak value.

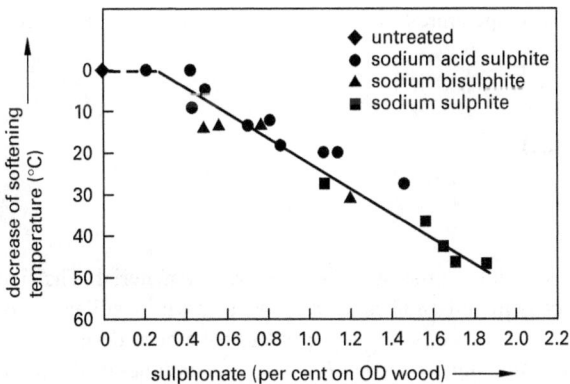

Figure 4.16. The softening temperture decreases with sulphonation of wood lignin.

Since wood is a visco-elastic material the softening temperature will be a function of the strain rate. As the strain rate increases the temperature must increase if the stiffness of the material is not to increase at higher dynamic loads. An increase of the strain rate of 10 times corresponds to a temperature increase of approximately 7 °C. The measured softening temperature of wood coincides with the softening temperature of the lignin. At the pertinent process temperatures for production of mechanical pulps the softening temperatures for softwood is in the range of 125–145 °C. The softening temperature for hardwood lignin is approximately 20 °C lower. The softening temperature can be lowered by sulphonation of the lignin and this fact can be used for the production of chemi-mechanical pulps like CTMP, *Figure 4.16*.

With increasing softening of the wood material the energy to create new surfaces in the wood structure is reduced. Even though the fibres are easily released from the wood structure when preheating and treatments are made at temperatures well above the softening temperatures of the lignin it is not guaranteed that the total energy consumption will be low for the production of the mechanical pulp of a certain quality. This is particurlarly true for production of pulp for printing papers. It has been proven difficult to process the fibres with high energy efficiency to high flexibility and to create a defibrillated surface if the fibres first have been released under mild condition at temperatures well above the lignin softening temperature. On the contrary, in the production of printing paper pulps, the initial refining stage is usually made at temperatures directly under the softening temperature of the lignin. Better conditions are then achieved for the subsequent treatments of the fibre to make it more flexible and at the same time more fibrillated with a less lignin rich surface. The latter is being regarded to make the surface more willing to make bonds. In addition the creation of light scattering fibre material is more effective if the initial release has been made at temperatures below the softening temperature of the lignin. On the other hand, treatments well below the softening temperature of the lignin in non-pressurised systems easily reduces of the fibre length, which leads to lower pulp quality. This means that pulps for printing paper production are with advantage made after heating the chips in a pressurised system at a temperature slightly lower than the softening temperature of the lignin.

On the other hand production of pulps for which fibre length is prioritised the process temperatures at preheating and treatment are chosen above the softening temperature of the lignin. A pulp with a large amount of long fibres, which gives a sheet structure with high bulk, can then be produced. Chemimechanical pulp in which the lignin has been impregnated by a chemical, for example sulphite, and the softening temperature has been lowered, show interesting quality properties in this respect.

4.3 Groundwood Production

4.3.1 Systems

The groundwood process is our oldest pulping process with wood as raw material. The first groundwood pulps were produced already in mid-1800 using the principles we are still using today. Since then a number of different constructions for pulp grinders have been designed and tested. Two types of designs are most commonly used today *Figure 4.17*, continues chain grinders and batch-wise hydraulically pressed grinders

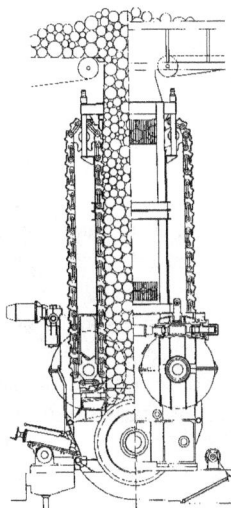

Figure 4.17a. The principle of the Voit chain ginder.

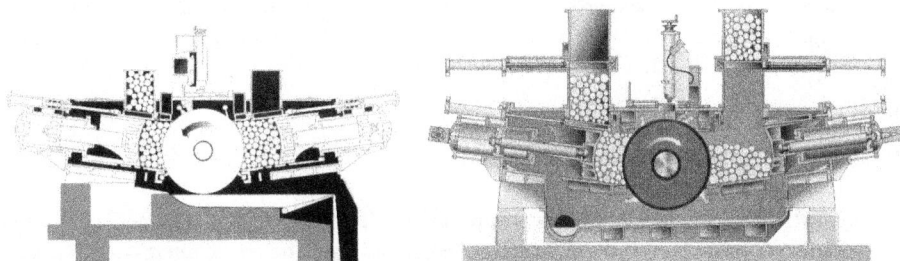

Figur 4.17b. Modern atmosperic Valmet grinder.   **Figure 4.17c.** The Valmet Pressurizedpocket (PGW) grinder.

Of the older types of grinders the so-called chain grinder is the most common (*Figure 17a*). In this type, the logs are continuously fed to an open magazine where heavy chains mounted on the opposing gable push the wood (logs) against the rotating stone. However, the type of grinders in which logs from two magazines are hydraulically pressed against the stone (*Figure 17 b, c*) dominates the market today. In this grinder design the wood magazine are filled batchwise. The diameters of the stones in more recent grinders are approximate 1.8 m. The grinding zone varies between 1.0–1.6 m. The stones are rotated at approximately 300 rpm, giving a peripheral speed of 25–35 m/s. The largest grinders are equipped with 7 MW motors and have a production capacity of 80–100 ton/day for pulp for magazine paper production.

The stones are mounted on a cylindrical core and are made of ceramic segments with grits of silicon carbide or aluminum oxide. The size of the grits varies in the range between 0.15–0.45 mm. The grit particle size is an important variable to control the pulp quality. For production of low freeness pulp for magazine papers, stones with small particle size grits are used, whereas for production of higher freeness, stones with large particle sizes are used. The quality of the groundwood pulp is determined by the pressure between the grits and the wood and the peripheral speed of the stone, beside the size and shape of the grits. The surface of the stone is provided with a pattern of land areas and grooves. With the pattern the contact area between stone and wood is regulated. The grooves also function as channels for supply of water and removal of released fibre material. The surface of the stone is sharpened and cleaned when the applied pattern has been worn too much and when the grits have been heavily rounded or clogged by debris. After sharpening, new sharp grits are at the surface giving a stepwise change in pulp characteristics. The pulp quality therefore varies with the so-called sharpening cycle. This variation can be minimised if the surface is kept clean by high pressure nozzles with water, which can be used during normal production.

The ground wood process is rather sensitive for variation in the wood quality and wood freshness. With fresh straight spruce logs the best quality is achieved.

An important development of the groundwood technology was introduced in end of the 1970's, when the pressurised grinders were launched *(Figure 17c)*. In the pressurised groundwood process the grinding is done in a system that is pressurised by air to 3–5 bars. In that way a higher boiling point is achieved and the temperature of the shower water can be raised. Depending on process design the temperature of the shower water varies between 70–120 °C. Since treatment and the release of fibres in pressurised grinding is done at higher temperatures and thereby with a more softened wood material it is mostly the long fibre related pulp properties that have been improved.

Groundwood and pressurised groundwood pulps are today mainly used in different types of printing paper (magazine papers) where a high amount of light scattering material is desirable, but they are also used in different boxboard qualities.

4.3.2 Fibre Release and Grinding Mechanism

Even though the groundwood process may in principle seem to be quite simple there are a number of rather complicated and difficult to control processes going on on the micro- and fibre level when the wood is turned into pulp. The process can be divided into three phases. First the wood fibre is softened mechanically and thermally, when the wood is compressed by the passing grits *Figure 4.18*.

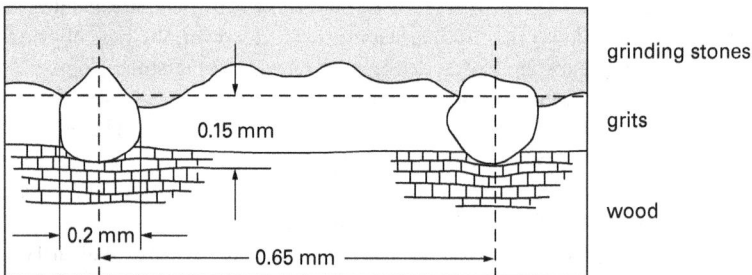

Figure 4.18. Typical dimensions of the surface of a grinding stone. The grits treat the fibres perpendicular to their longitudiual axis.

Secondly, the fibres are released when the softened fibres in a layer closest to the rotation stone are torn away. Finally there is a certain after-treatment of the released fibres when they are removed from the grinding zone.

During the grinding high frequency pressure pulses are transferred from the passing grits via the thin water film to the wood. The strain rate is approximately 50 kHz, i.e. $50 \cdot 10^3$ grits are passing over a certain point in the wood surface per second. At a typical wood feeding speed to the rotation stone (appr. 1,5 mm/sec), approximately 10^3 grits pass a fibre in the woods outer layer before it is torn out. The pressure pulses protrude a few tenth of a millimetre into the wood structure. The protrusion depth depends on the size and shape of the grits and the pressure between the wood and the stone. The transferred pressure pulses break some bonds in the fibre wall even before the fibre is released. At the same time the visco-elastic wood material is heated by the internal friction and the generated heat softens the lignin. That is a prerequisite to later release of the fibre from the wood without too much fibre length reduction.

The amplitude of the transferred pressure pulses is obviously increasing when the fibres are approaching the actual grinding zone. On the other hand, the temperature of the wood is reaching its maximum some distance into the wood since the fibres closest to the grinding zone is cooled by water from the applied water film. In the heated zone the temperature can easily reach 100 °C or above causing the water of the wood to evaporate quickly. If the wood is too dry or if water is not fed continuously to the grinding zone the fibres will dry. In pressurised grinders this is prevented, which is the most likely explanation for the higher quality of the pressurised groundwood pulps.

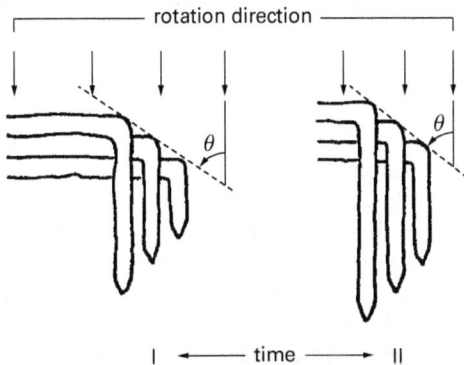

Figure 4.19. Two stages in the peeling of loosened fibres in grinding.

During the fibre release fibres are combed from the wood surface as described in *Figure 4.19*. The fibre orientation in the wood is seldom fully parallel with the grinding surface. The fibres are therefore gradually pulled out after a large number of grits have acted on them. During this process the fibres are bent 90 degree before they are released from wood structure. It can be easily understood that most of the fibres easily can be broken even though they are well softened. The fibres are most likely to be further shortened by the treatment in the transport out from the grinding zone

4.4 Manufacturing of Refiner Pulps

4.4.1 The TMP-System

The first systems for production of mechanical pulps from chips in a refiner were developed during the 1930's by Arne Asplund. In the Asplund process, wood chips are refined in pressurised systems at temperatures well above the softening temperature of the lignin. The pulp is used for fibreboard manufacturing. Not until the end of 1950 successful trials were made with refiner pulps for papermaking. The pulp was made in atmospheric refiners in a system without pre-treatment of the chips with steam and was called RMP (Refiner Mechanical Pulp). The development of the technology for refiner pulps for printing paper production picked up speed during the 1960's when the principles for the present processes for production of TMP (Thermo-Mechanical Pulp) started to take shape. In the TMP-process the chips are preheated and refined in a steam pressurised system.

4.4.2 Preheating and Refining

The different operations in the TMP-process are shown in, *Fig.4.20.*

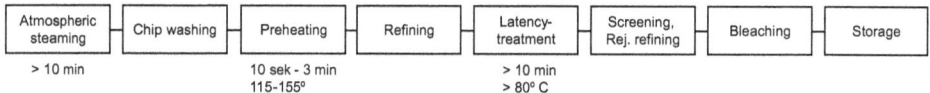

Atmospheric steaming	Chip washing	Preheating	Refining	Latency-treatment	Screening, Rej. refining	Bleaching	Storage
> 10 min		10 sek - 3 min 115-155°		> 10 min > 80° C			

Figure 4.20. Unit operation in a typical TMP process.

The process starts by atmospheric steam treatment in which snow and ice are also thawed by the recovered steam from the refiners. Next, the chips pass through a chips washer to remove sand and other foreign objects and particles heavier than the chips. This is necessary to protect the process equipment, such as the discs of the refiner, from extensive wear. Before the refining, the chips are heated in different kinds of preheaters to 115–155 °C. In older systems it is common with preheating times up to 3 min. Newer systems are often run with considerably shorter preheating times (down to 10 seconds). The initial defibration in a refiner is in the coarse patterned feeding zone, the so called breaker bar zone, where the chips are converted to single fibres or fibre bundles, *Figure 4.21.*

These are then slung by the centrifugal force into a narrow disc gap between finely grooved refiner discs where the fibres are intensively „refined" at high pulp concentrations (30–50 %). Water is continuously added to compensate for the liquid that evaporates when the fibres are refined.

Most mechanical energy is converted in the refiner to heat via visco-elastic and plastic losses. The energy needed to create new fracture surfaces in the wood structure, i.e. new surface area on the fibre material in the pulp, is relatively small compared to the total energy input in the refining process. However, a large total energy input seems to be necessary in order to increase the flexibility of the stiff wood fibres in refining.

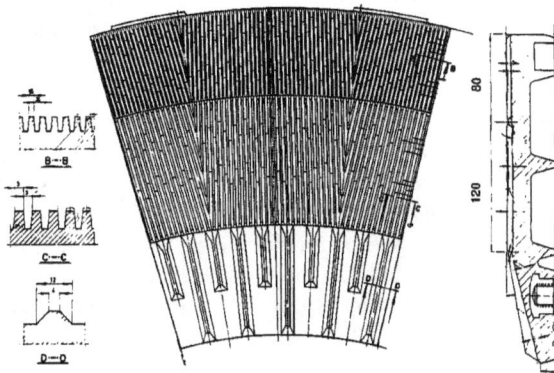

Figure 4.21. Refiner segment. Measure in mm.

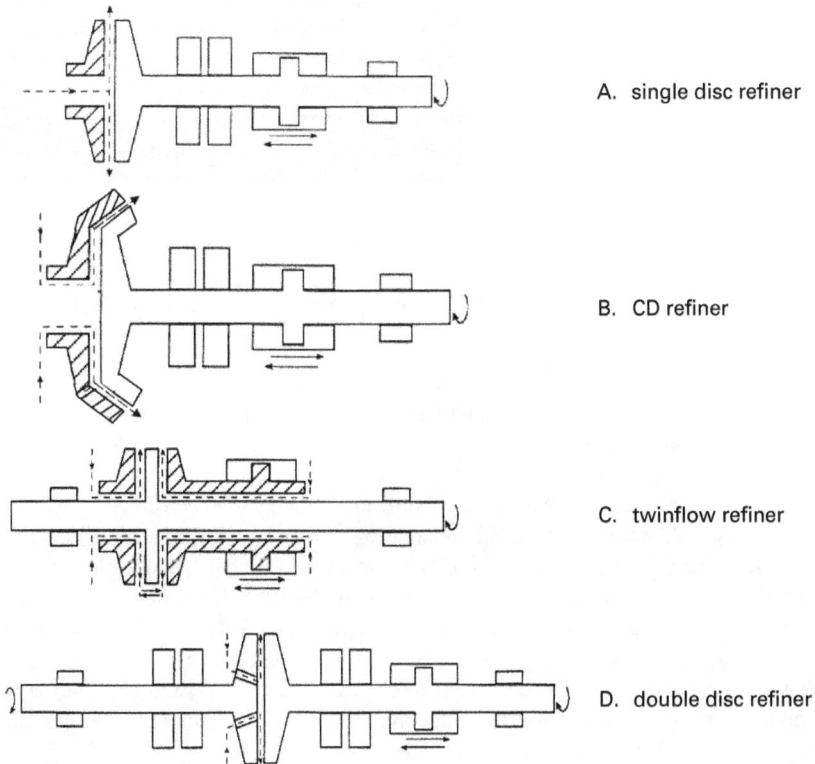

A. single disc refiner

B. CD refiner

C. twinflow refiner

D. double disc refiner

Figure 4.22. Different types of chip refiners.

The conditions during the initial defibration have a large impact on the final pulp quality and on the total energy consumption. Several variations of preheating systems have therefore been developed to adjust the conditions to best suit the subsequent refining. If a pulp with high light scattering ability is desired the preheating system should be designed so that the temperature at the feed to the breaker bars zone is in the low range. A special type of pressurised TMP-system (PRMP) is as an example run completely without preheating. On the other hand should the chips be well softened before refining, if one desires higher long fibre content, a high preheating temperature and a long preheating time is then used.

Systems with refining in one or two steps are most common. A number of different types of refiners have been developed the last decades, *Figure 4.22*.

Since the 60's large improvements in the quality and performance have been made, at the same time as the size of the refiners have grown. Today the largest refiners have a production capacity of 350 ton/day in one stage refining for newsprint quality pulp with motor powers of approximately 30 MW, *Figure 4.23*.

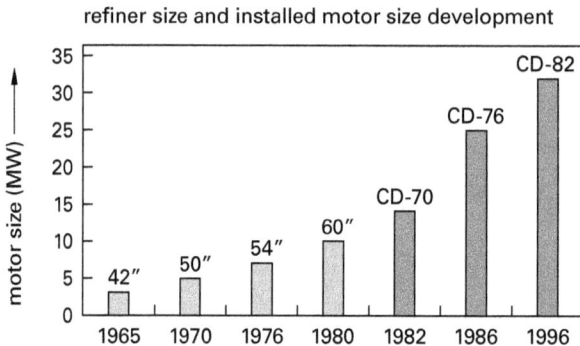

refiner size and installed motor size development

Figure 4.23. The size and capacity of commercial refiners has increased dramatically over the last 20 years. CD = Conical disc.

Two stage refining systems are built in one line for refining of 700 ton/day. In Europe with 50 Hz power supply most refiners are driven with sycron motors at 1500 rpm. In order to work a higher number of revolutions the refiner can be equipped with a gearbox. In US and Canada with 60 Hz power supply most refiners are driven at 1800 rpm.

There is a distinction between single disc and double disc refiners in which either one or both discs rotate (*Figures, 4.22 a and d*). To handle increasingly higher production volumes special designs for single refiners have been developed. In the so-called Conical Disc Refiners (CD refiners) the refining zone consists of a flat and a conical part (*Figure 4.22 b*). In this way a more stable machine construction can be designed, which is a prerequisite in order to use larger refiner disc surfaces. In addition, the fibre flow rate is reduced in the outer zone. In flat disc refiners when the diameters are increased the centrifugal forces can otherwise get too large at the periphery. In another design to handle larger production volumes the refiner has been provided with refining zones on both sides of the rotating disc (*Figure 4.22 c*).

The major part of the energy consumed in refining is converted to heat (steam). The steam can be recovered and used in different unit operations in a mill. In an integrated mill the steam can, in addition to the TMP production, be used in the dryer section of the paper machine.

4.4.3 Latency Treatment

After refining the fibres must be treated to remove so-called latency. It is created after refining at high consistency and temperature when bent fibres and fibres entangled in fibre bundles are cooled after the discharge from the refiner. The bent shape becomes permanent when the lignin becomes stiffer. The strength properties of a sheet become low if the fibres are stiffened in the bent and entangled form. The fibres can however be straightened if they are stirred at lower consistency at a certain temperature. Typical condition for latency removal in a TMP process is intensive agitation in a chest at concentrations lower than 6 % under at least 10 minutes and at temperature above 80 °C. The straitening of the fibres can be seen for example in a reduction of the dewatering value (freeness), *Figure 4.24*.

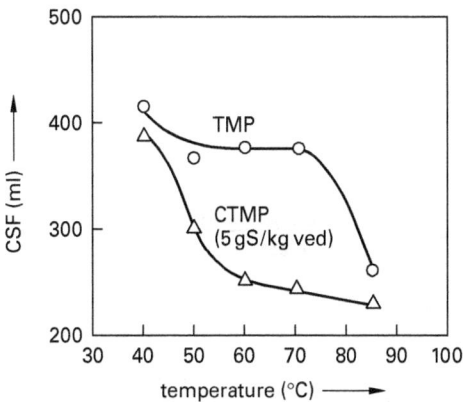

Figure 4.24. Freeness values at latency treatment at different disintegration temperatures. Table freeness drop indicate the limit temperature for latency removal.

The figure reveals that to remove latency the fibres need to be adequately softened.

4.4.4 Screening

After latency removal, the pulp is cleaned by screens. Some typical screen room designs are depicted in, *Figure 4.25*.

The screen room has several important tasks in mechanical pulp production. Shives removal, fibre fractionation, removal of heavy debris and refining of the reject should be performed as separate unit operations.

Shives removal is of course one of the main functions of the screen room. Shives are removed in screens in which also the coarsest fibre fractions and the longest fibre material is removed for separate reject refining. For screening closed pressurised screens are used, *Figure 4.26*, in which the pulp fibres are separated by stationary vertical screening baskets with holes (commonly 1.0–1.5 mm in diameter) or slots (usual slot widths 0.15–0.25 mm).

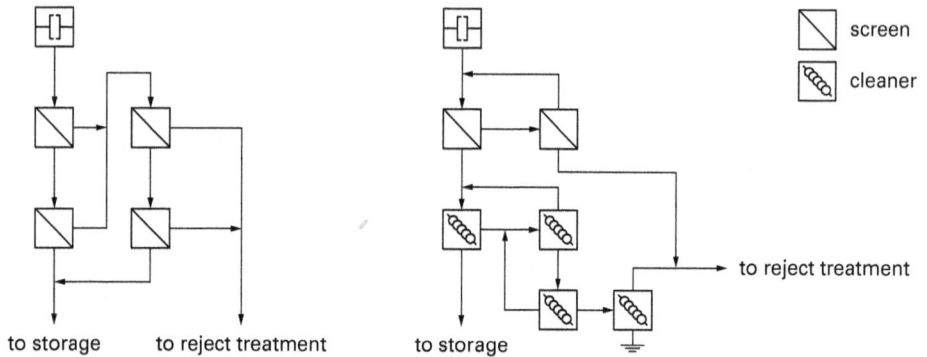

Figure 4.25. Principal layouts of reject plants.

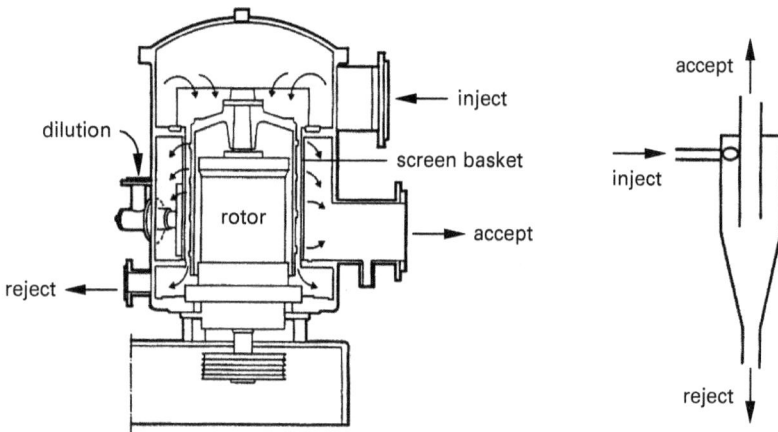

Figure 4.26. Principal sketch of a screen and a Vortex cleaner.

The screening is usually done at consistencies around 1.5 %. More recent developed high consistency screens can work at consistencies up to 4 %. The basket's holes or slits are kept clean by a rotor that also facilitates the pulp feed along the basket.

The working principle of screens is either as barrier screens or as fractionators. Shives are effectively removed by small rejects removal in barrier screening on slotted screen baskets, *Figure 4.27*. Long coarse fibres are removed by fractionating in hole baskets. However with the present technology higher reject removal is needed to achieve low shives content on baskets with holes.

When screening according to the barrier principle, the fibre material is separated by slotted baskets (slot width 0.15–0.25 mm). The turbulence at the basket surface is high and controlled by the rotor design and speed and different coarsely grooved patterns in the basket's surface, *Figure 4.28*.

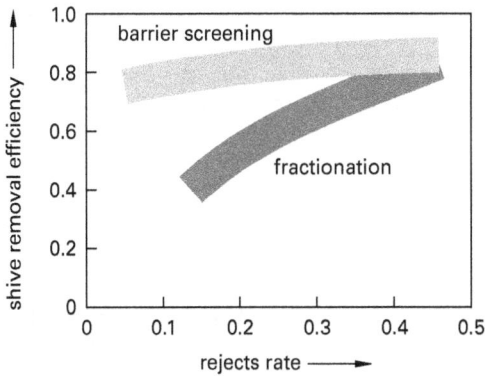

Figure 4.27. Shive removal at barrier screening (spotted screen basket) and fractionation (screen basket with holes).

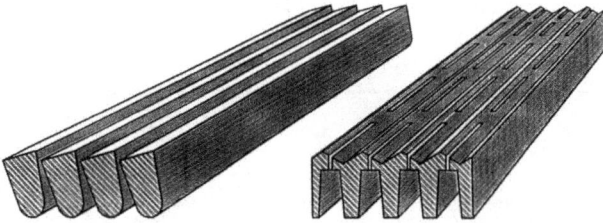

Figure 4.28. Different configuration of slotted screen baskets.

The fine slits are a physical barrier preventing particles larger than a certain dimension, mainly shives, to pass. With present equipment coarse stiff fibres can not be separated from more flexible fibres, if not the slit widths are made very small giving unrealistic low production rates. In order to separate long coarse fibres the turbulence on the screen basket surface needs to be reduced so that they can be trapped in the network that then is formed on the surface. For this type of fractionation, screen baskets with holes are used. The surface of the basket in fractionation is smoother compared to barrier screening. The probability for a certain fibre to be trapped on the surface of the basket is reduced with fibre dimension and with increasing flexibility.

Vortex cleaners are used to separate contaminations with high specific density, for example sand particles, *Figure 4.26*. They sink to the bottom in the vortex that is formed when the pulp suspension enters in the periphery at the top of the cleaner. In vortex cleaners, a certain degree of separation of bark particle and coarse fibre material with small specific surface area and binding potential can be achieved. The last step in the vortex cleaner system is normally the position from where contamination can leave the screening system (*Figure 4.25*). One disadvantage with vortex cleaners is the low concentration used (below 0.8 %), which generates costs for pumping, dewatering etc.

The reject refiner is of course a key component in the screening system. Its function is to reduce the shives content to a minimum and to process the rejected coarse fibre material to high flexibility. The screening system can never be efficient without an efficient reject refiner. In a well operating screening room higher strength levels are reached for the separated long fibre fraction after the reject refining compared to pulp right after the chip refiner (see *Figure 4.44*). Reject pulp can be seen as a type of reinforcement pulp in a TMP furnish. An efficient reject refining is also necessary to reach good surface characteristics for high quality printing papers in which poorly treated thick walled fibre material with low flexibility easily causes disturbances in the sheet structure.

4.4.5 The CTMP-System

For production of CTMP the „TMP-plant" is equipped with a unit for impregnation of the chips with chemicals, *Figure 4.29*.

Softwood chips are usually impregnated with 2–3 % of a weak alkali sodium sulphite solution (pH = 9,5), while for hardwood a stronger alkaline impregnation solution is used in combination with sodium sulfite (Na_2SO_3) or hydrogen peroxide (H_2O_2). Other conditions are more or less the same in present standard TMP and CTMP-systems. In one variation of the CTMP-system (HT CTMP) considerably higher preheating temperatures ($T > 170 °C$) are used to produce pulp with very long fibres with as low energy input as possible.

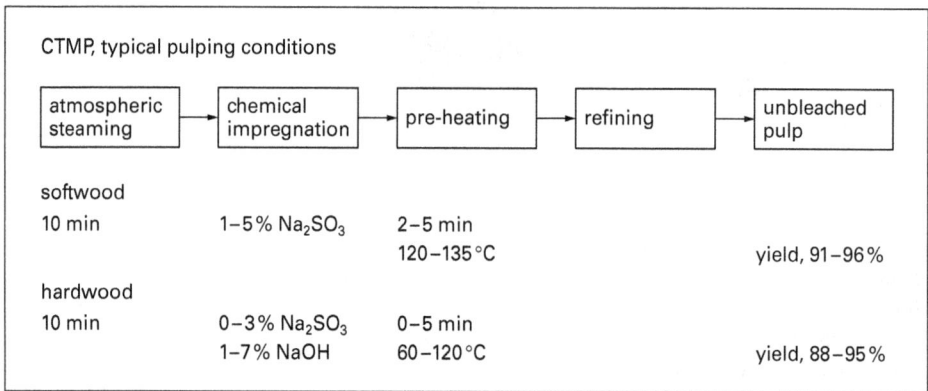

CTMP, typical pulping conditions

atmospheric steaming → chemical impregnation → pre-heating → refining → unbleached pulp

softwood
10 min   1–5% Na_2SO_3 2–5 min
                                          120–135 °C   yield, 91–96 %

hardwood
10 min   0–3% Na_2SO_3 0–5 min
                1–7% NaOH   60–120 °C   yield, 88–95 %

Figure 4.29. Unit operations in typical CTMP processes.

Before the impregnation stage the chips are steamed to temperatures close to 100 °C with the intention to create a steam atmosphere inside in the voids of the chips. In the compression screws of the impregnation unit, the chips are pressed at the same time as they are fed into the impregnation fluid, *Figure 4.30*.

The liquid is sucked into the chips when they are expanding and the impregnation liquid, which is kept colder than the chips, condenses the steam in the voids. The residence time in the impregnation liquid is kept short (<0.5 minutes). Most CTMP-plants are equipped with a retention bin after the impregnation (residence time approximately 30 minutes), to facilitate for the impregnation chemicals to penetrate the fibre walls before pre-heating and refining.

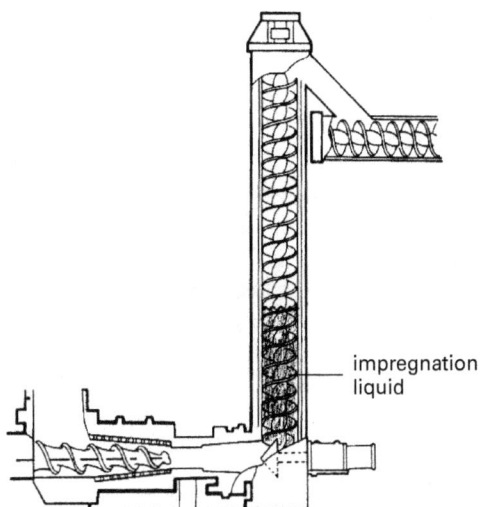

Figure 4.30. Chip impregnator.

CTMP as well as TMP from spruce can be peroxide bleached to about 80 % ISO brightness. When producing CTMP from bright hardwoods, for example aspen, in which the lignin content is lower compared to softwoods, brightness up to 85 % ISO can be accomplished with peroxide bleaching. CTMP is usually washed to low resin contents in systems with counter current flows. This is necessary if the pulp is to be used for liquid or food packaging or in absorption products, which are the most common product areas for spruce CTMP.

4.4.6 Fibre Release and Refining Mechanisms

The refining process begins by feeding randomly orientated chips into the centre between the refiner plates with a high-speed screw. In the refiner the initial defibrating is taken place in the breaker bar zone (*Figure 4.21*), where the chips are converted to fibres or fibre bundles. These are then thrown into the narrow disc gap between the refiner plates.

Experience has shown that the pulp characteristics to a large extent are determined by the process conditions in the early stage of the process. The way fibre release is performed under the initial processing seems to be totally determining for the final pulp quality and even for the total energy consumption for different types of processes. This is in spite that only a minor part, approximately 25 %, of the total energy consumption in the refining process is used for the fibre release. One can imagine that the initial fractures in the wood structure to a large extent also govern the continuing fracturing of the wood structure in the subsequent processing. Too high temperatures compared to the pertinent lignin softening temperature during the initial defibration give large fractions of unbroken, lignin covered and smooth fibres, which are hard to fibrillate and make flexible to desirable extent during the following refining process. During the production of pulp for printing papers a temperature at, or under, the softening temperature of the lignin is desirable during the initial processing. Several investigations show that processing at high strain rates, for example in double disc refiners or single disc refiners run at high revolution, is particularly favourable for production of pulp for printing papers. The light scattering abilities of the pulps are then good at relatively low energy input, *Figure 4.31*.

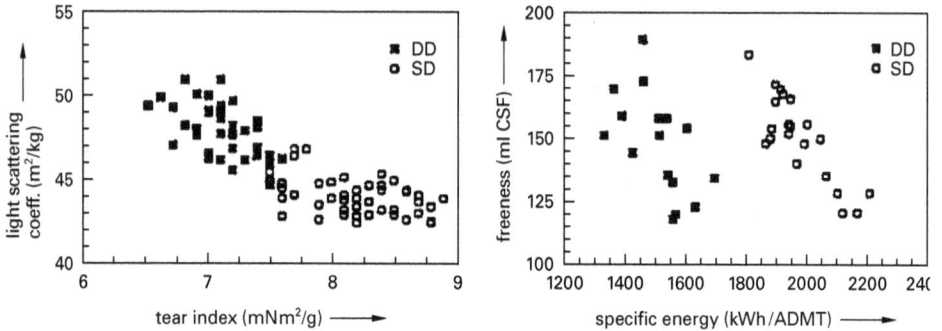

Figur 4.31. Mechanical pulps from double disc refiners (DD) are characterized of a higher light scattering and a lower tear index (lower long fibre content) than pulp from single disc refiners (SD).

The largest energy input in the refining process is achieved in the disc gap between the parallel refiner discs on the already in the breaker bar zone released fibres. The energy input is large (*Figure 4.8*) and has large effect on the process economy. Process oriented research has therefore for a long time been concentrated on learning how the processing can be performed more energy efficiently. From pictures taken of the events in the refining zone it has been concluded that the fibres are mainly tangentially orientated. Hanging over one of the refiner discs bars the fibres are processed in the axial direction by the opposing disc, which naturally also is fully or partially covered by fibres. Fibre and fibre bundles, which are exposed to cyclic loads from the refiner disc bars, are kneaded against each other in a complicated way difficult to control under their passage through the disc gap. Fibre-fibre friction has an important roll for the efficiency in the energy transfer. The largest strain amplitude is reach at leading edges of the refiner bars. As the disc is worn, the leading edges of the bars are rounded and the refining efficiency is reduced and the energy input is increased slightly. So is the shives content.

The fibre flow through the disc gap is not a strict plug flow, since the pulp fibres can be pressed backwards by the pressure pulses. The degree of re-circulation affects the residence time in the refining zone, which makes it difficult to calculate with any precision. However, different calculations and measurements are indicating a few tenths of a second. The resident time in the breaker bar zone is considerably longer, most likely several seconds. During the refining the fibre dimensions is reduced, both because it partially or fully collapses and because material from the outer surface is detached producing fine material, *Figure 4.32*.

The later have been confirmed by characterisation of TMP refined at higher energy input. Since the pulp fine material is dominated by material from the outer surface of the fibre the lignin content becomes higher than in other fibre fractions (cf. *Figure 4.10 d* and *Figure 4.13*)

During refining, large amount of steam is produced in the refining zone, as the cyclic pressure pulses are transformed into heat by viscoelastic losses in the wood material. Under TMP conditions approximately 1 ton steam per MWh energy input is produced. At some place in the refining zone, a maximum steam pressure is built, the so-called steam peak *Figure 4.33*.

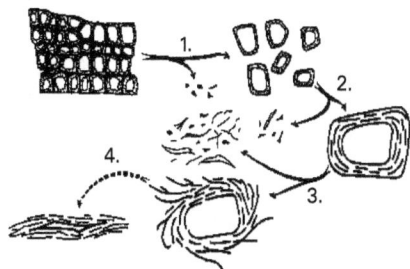

Figure 4.32. A hypothetical division of the chip refining process into different process stages.

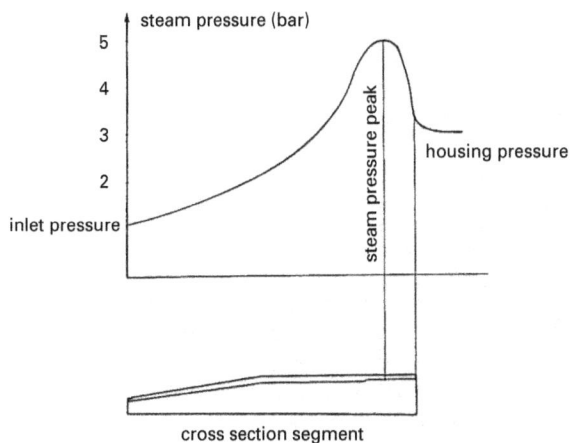

Figure 4.33. Schematic distribution of steam pressure in the refiner gap.

In large refiners, temperatures around 170–180 °C have been measured between the refiners discs even though the temperature at the inlet and discharge is considerably lower (115–155 °C). The consequence is that steam can flow both to the centre of the discs and to their periphery. The position of the steam peak can be changed by the design of the disc patterns. It is foremost the amount of open groove volume in the different sectors that are determining.

The developments of the pattern on refiner discs have largely been built on empirical knowledge, since the process in the refining zone is rather complex and not fully understood. Refiner discs must be designed so that fibres and steam easily flow at the same time as a specified pulp quality can be produced with demands on certain distributions between long fibres, fibre fragments and fines etc. Usually discs with a coarser pattern in the centre and a finer pattern in the periphery are used. The grooves, which mostly are shallower closer to the periphery, are equipped with a number of dams. Typical values for bar- and grove dimensions are given in *Figure 4.21*. The intention with the dams is to force fibres in the grooves to pass the edges of the disc bars, where the most essential of an energy efficient refining is done. A certain amount of dams is necessary to ensure a low shives content. The positions of the dams govern also the steam distribution and the flow to the in and out feed respectively. Samples taken from different positions along the refining zone show that the most intensive refining, i.e. the largest change in the fibre material, seems to occur in the periphery at the finest patterned refining zone, *Figure 4.34*.

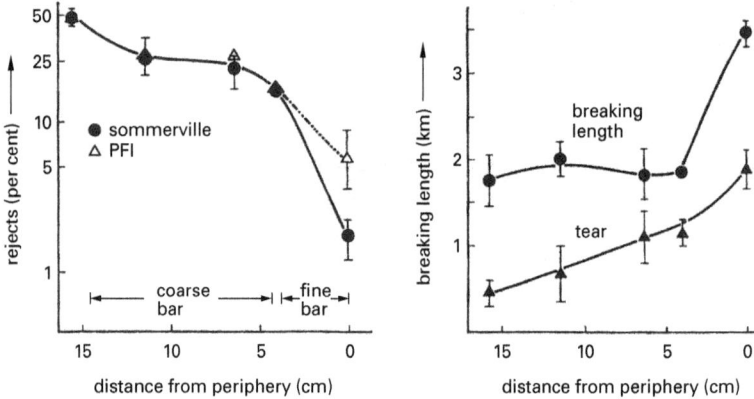

Figure 4.34. Shive content and strength of pulp taken from different positions in the refiner gap. The most efficient treatment seems to be effected at the periphery in the refiner gap.

For a certain type of refiners with a given disc design the refining results are given by parameters which control the fibre flow and the residence time in the refining zone i.e. production, pulp consistency, revolutions and pressure difference between the in and the out feed. For all refiners, the pulp consistency is controlled by injecting water in the refining zone so that it in the discharge is above 30 %. Lower pulp consistencies easily give too high fibre flow in a too narrow disc gap at desirable energy input. This reduces the fibre length considerably with quality degradation as a consequence. Single disc refiners are usually run at higher pulp consistency than double disc refiners. The residence time in the refining zone can be controlled by the pressure difference between the in and out feed. The pressure difference can be both positive and negative. A longer residence time gives a less severe treatment and a pulp with more long fibres and higher strength. Milder refiner conditions are on the other hand not always desirable if a reduction of the specific energy consumption is wanted. On the contrary, the number of pressure pulses with low amplitude must be reduced. Under such conditions the energy input is more or less totally converted to heat without any irreversible structural changes in the fibre structure. Conditions giving fibres with good quality is usually not benefiting from a low energy input in the processes. Because of that it has been difficult to develop processes in which the pulp quality is maintained with considerable lower energy consumption than what is shown in *Figure 4.8*. The span in electrical energy consumption between different process alternatives can however be relatively large, That is especially clear for production of pulp for the lowest freeness levels, i.e. pulp for SC and LWC qualities.

When refining in a two stage process, the quality profile can be much less affected in the second stage compared to the first stage, *Figure 4.35*.

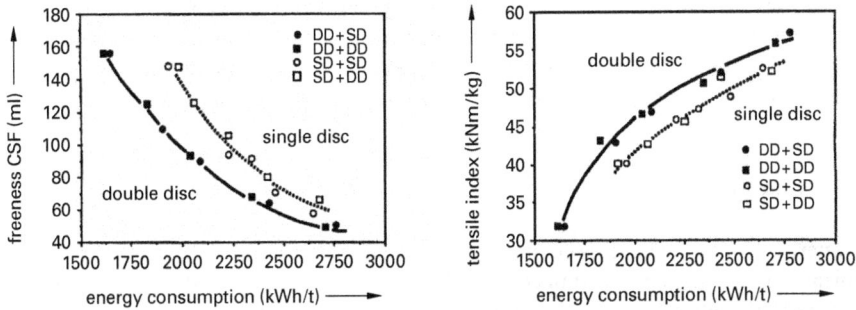

Figure 4.35. Freeness and tensile index for mechanical pulps refined in two stages with single (SD) and double disc (DD) refiners. The character of the pulps is determined by the conditions in first refining stage.

4.5 The Quality Profile for Different Products

It has already been mentioned that there are specifically two quality aspects that can be utilised when mechanical pulp is used in different paper products. Foremost are the pulps ability to give high opacity in thin printing papers. The cause of that is the high light scattering coefficients at a given strength level (*Figure 4.4*). Another interesting property is the ability to give high bulk (low density) at a given tensile index (*Figure 4.5*). That gives advantages in paper properties for example in a middle layer in boxboard or in certain tissue qualities.

However mechanical pulp can not be used in paper products with very high strength demands, for example in liner or in sack paper. Mechanical pulps are also very sparingly used in qualities with high demands on brightness stability, for example fine paper, since lignin rich pulp readily yellows by ultraviolet radiation.

Mechanical pulps are also unique in the respect that their light scattering ability and tensile strength are improved by more refining (lower freeness) because both the bonded and unbound area is increase by refining, *Figure 4.36*.

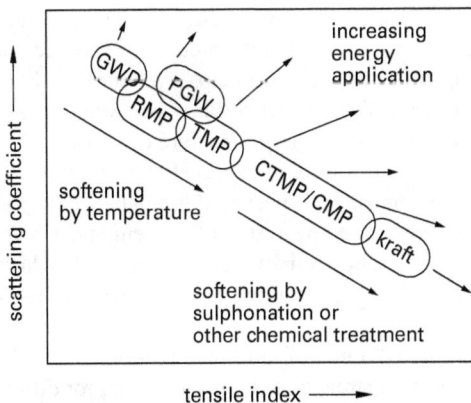

Figure 4.36. Mechanical pulps are unique in the respect that both their light scattering ability and tensile strength are improved by more refining (lower freeness).

With increased softening in the mechanical pulp processes or a lowering of the yield the amount of bonded area increases at the expense of the amount of unbounded area. This results in an improved strength but a reduced light scattering ability. When refining chemical pulps the bonded area and the strength is increase, but the light scattering ability is reduced.

4.5.1 Printing Papers

The dewatering ability, freeness, is commonly used as characterisation of the degree of refining. Mechanical pulp for newsprint is refined to slightly below a freeness of 100 ml. Typical distribution of the fibre material measured as weight percent long fibre, middle fraction (broken fibre parts) and fines for mechanical, chemi-mechanical and chemical pulps intended for newsprint production is seen in, *Figure 4.37*.

BAUER McNETT CLASSIFICATION OF NEWSGRADE

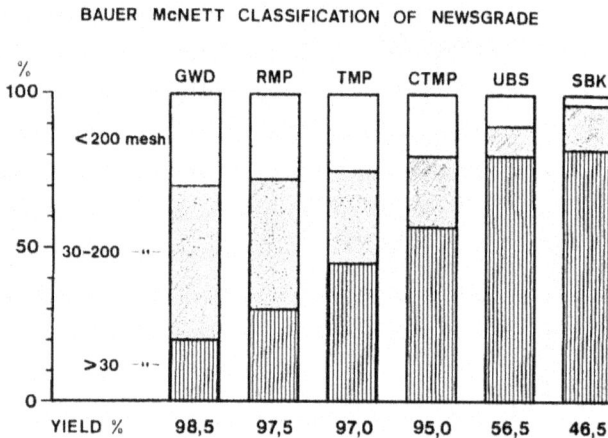

Figure 4.37. Representative fibre distributions in pulps of news grades.

For the mechanical pulps, the long fibre fraction is increasing while the fines fraction is decreasing with increasing degree of softening of the fibre material in the processes. The fraction of long fibre is low for all mechanical pulps compared to the chemical pulps. This is particularly true if consideration is taken to the difference in yield. The number of fibres per weight is approximately double for chemical pulp compared to mechanical pulp. The fibre compositions as well as the relation between strength and optical properties for TMP are such that the pulp can be used as the only pulp for newsprint at ordinary grammages. However, if groundwood pulp is used as mechanical pulp the ratio of long fibre becomes too low to reach the strength demands for a high speed newsprint paper machine. It is mainly reflected in wet and tear strength, which are dependent on very long fibre, *Figure 4.38*. Groundwood pulp must therefore always be reinforced with a certain amount of chemical pulp.

The interplay between fines (BMN < 200 mesh) and a mix of long fibre (BMN > 30 mesh) and middle fraction (BMN 30–200 mesh) in developing strength and light scattering for different types of mechanical pulp is revealed in, *Figure 4.39 a–c*.

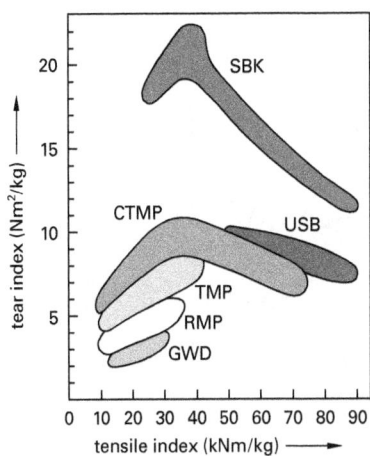

Figure 4.38. Relationship between tear and tensile index. SBK = Semibleached kraft, UBS = Unbleached sulphite. (ISO sheets).

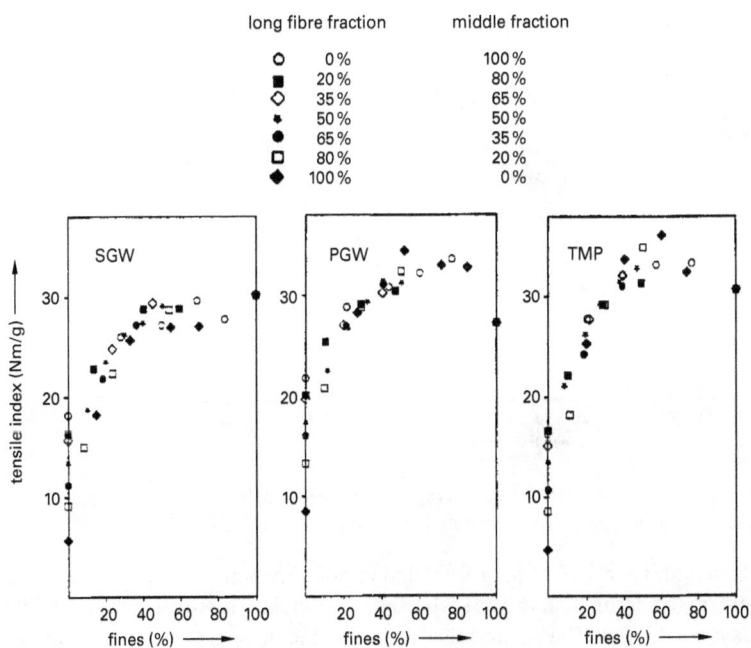

Figure 4.39a. The interplay between fines (BMN < 200 mesh) and long fibre (BMN > 30 mesh) and middle fraction (BMN 30–200 mesh) in development of tensile index.

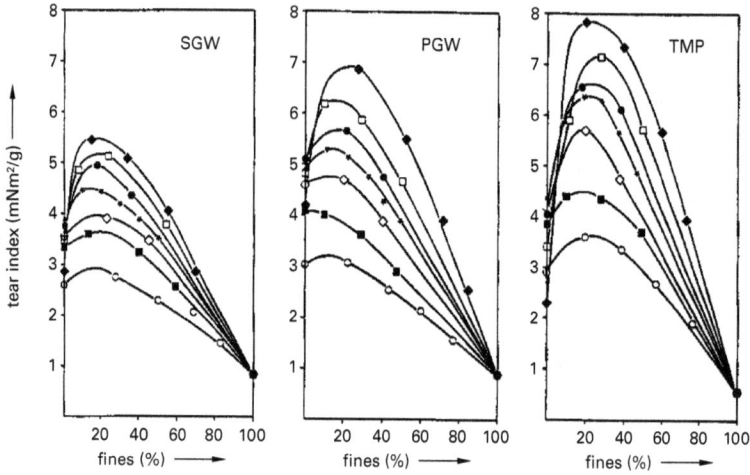

Figure 4.39b. The interplay between fines (BMN < 200 mesh) and long fibre (BMN > 30 mesh) and middle fraction (BMN 30–200 mesh) in development of tear index.

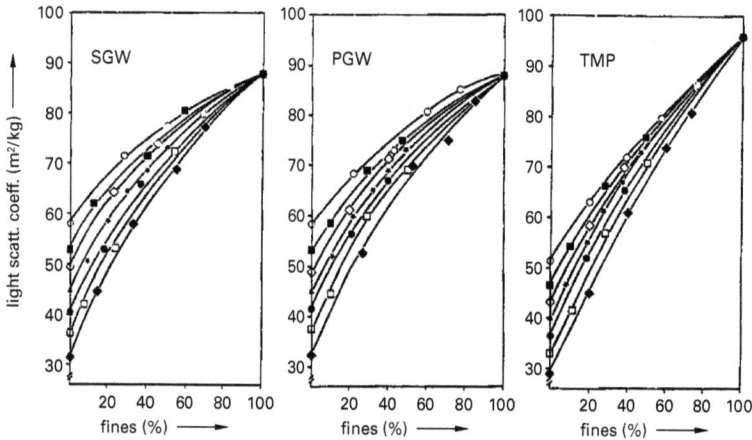

Figure 4.39c. The interplay between fines (BMN < 200 mesh) and long fibre (BMN > 30 mesh) and middle fraction (BMN 30–200 mesh) in development of light scattering coefficient.

Tensile and tear strength for a mix of long fibre and middle fraction only increase with increasing fines content up to approximately 30 % whereas a continuing improvement in the light scattering properties are achieved. The figures also show that the quality of the long fibre and middle fraction are different for the different process types. Their qualities have a large impact on properties that are possible to reach with a reasonable amount of fines. The strength properties at the relevant ratio between long fibre and middle fraction in a groundwood pulp for printing papers is slightly higher compared to TMP, *Figure 4.38 a–b*. This reflects the fact that the long fibre fraction of groundwood is normally processed to a higher fibre flexibility. On the other hand the strength properties increase faster when fines are mixed with long fibre/middle fraction from a TMP furnish. This is because TMP fines contain a larger part of thread-like material, such as fibrils, whereas groundwood pulp has a larger fraction of short torn fibres. On the other hand, the light scattering coefficient reacts in the same way when fines from the two processes are added to the same long fibre fraction, *Figure 4.40*.

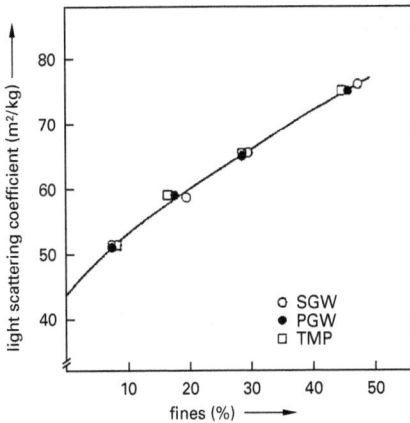

Figure 4.40. The Light scattering coefficient increases in the same way when fines from different processes are added to the same long fibre fraction.

With the actual ratios between long fibre and middle fraction a larger contribution to light scattering in a fines free pulp will be given for groundwood pulp than for TMP. That is why a higher light scattering coefficient is always reached for groundwood pulp at a given fines content. A prerequisite to get a high light scattering from a TMP process is to have access to a thin walled type of wood, which gets a long fibre fraction with a high light scattering ability, *Figure 4.41*.

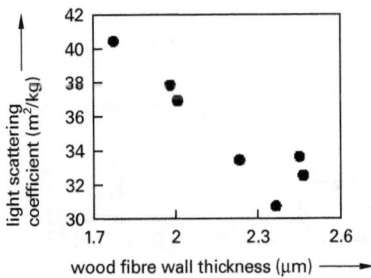

Figure 4.41. The light scattering coefficient of handsheets of long fibres from a number of industrially produced mechanical pulps.

With thick walled hardwood fibres the level of the light scattering coefficient will always be quite low.

For higher demands on surface smoothness and light scattering ability for high quality wood-containing printing papers, in particular in magazine papers, the demand for refining of the mechanical pulp is increasing. The norms for freeness levels will become lower (*Figure 4.8*). The need for more refining and more flexibility of the mechanical long fibres is easy to comprehend when one compares their dimensions with the thickness of a thin printing paper, *Figure 4.12*. This effect is even more pronounced in low basis weight magazine papers and LWC papers. In these the mechanical pulp content is varying in the range 15–50 %. When the amount of chemical pulp is lowered the fraction of mechanical long fibre in the furnish increases in order to replace the well-collapsed chemical fibres. The need for a well refined long fibre fraction is especially important for TMP because these long fibres soon become a part of the load carrying sheet structure in printing papers with low amounts of chemical pulp, *Figure 4.42*.

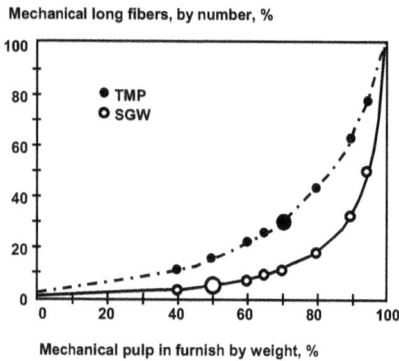

Figure 4.42. The amount of long fibres, by number, from the mechanical pulp can be quite high in TMP based LWC.

The degree of collapse of TMP long fibre is strongly related to the fibre wall thickness, *Figure 4.43*.

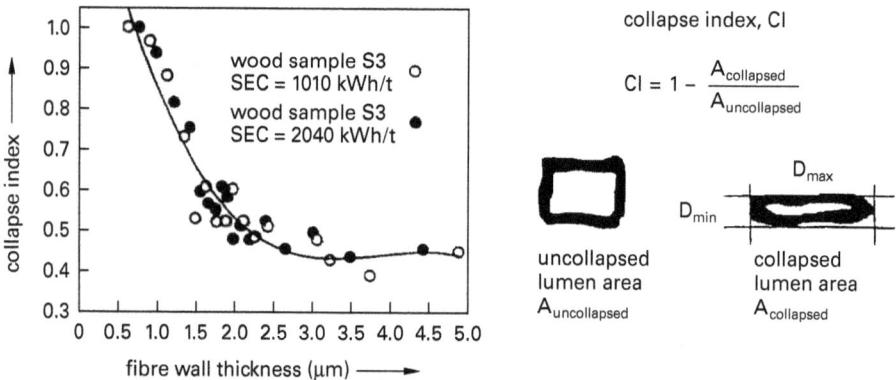

Figure 4.43. The degree of collapse of TMP long fibres is strongly related to the fibre wall thickness.

To achieve a proper treatment of the long fibre fraction it is also very important to have an efficient separation and refining of shives and coarse fibres in the screen room. TMP reject can be refined to a high strength level and get a considerably higher light scattering coefficient than a chemical pulp at a certain strength, *Figure 4.44*.

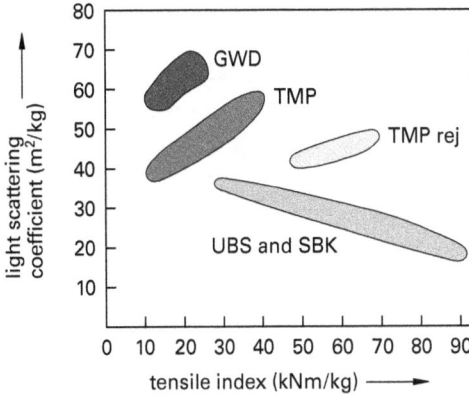

Figure 4.44. Properly treated reject fibres add high strength to the furnish at a high light scattering level. (ISO sheets)

4.5.2 Box Board, Tissue

When using mechanical or chemimechanical pulp in boxboard or in tissue production it is mainly the ability of the high yield fibre to form bulky structures that is utilised. In that respect it is desirable to have as high long fibre fraction as possible at a given strength level. Chemi-mechanical pulps from spruce of the CTMP type have proven to have the most interesting character profile, *Figure 4.5*. Boxboard and tissue pulps from spruce can be produced at freeness levels up to 600 ml with very low shives content, *Figure 4.45*.

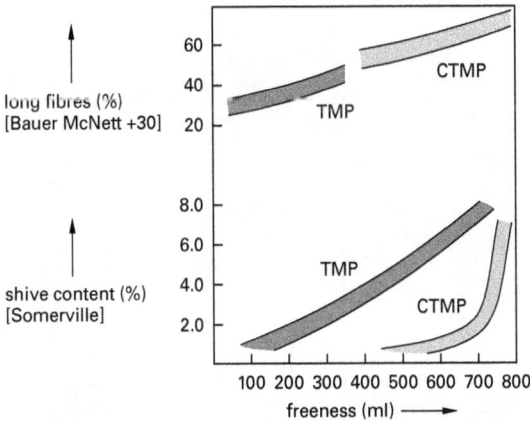

Figure 4.45. CTMP is characterized of a high content of long fibre at a low shive content.

The demand on fibre distributions in CTMP for boxboard and tissue is quite the same. These qualities which are usually bleached with peroxide to brightness levels of 70–80 % ISO are produced with low pitch contents (< 0,10 % DCM).

In the HT CTMP process it is possible to manufacture pulps with an extremely low shive content at a low energy input, *Figure 4.46*.

As the energy consumption is low the long fibre content is very high (*Figure 4.9*) as well as the fibre stiffness and bulk.

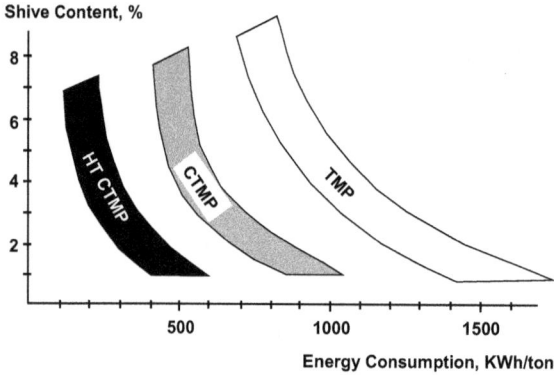

Figure 4.46. Fibres are separated very efficiently at low energy input in the HT CTMP process.

5 Chemistry of Chemical Pulping

Göran Gellerstedt
Department of Fibre and Polymer Technology, Royal Institute of Technology, KTH

5.1 Introduction

In chemical pulping, the wood lignin is made water soluble by the action of bisulphite ions (sulphite pulping) or hydroxide ions in combination with hydrosulphide ions (kraft pulping). Although other processes such as soda pulping with aqueous sodium hydroxide with or without the presence of anthraquinone are commercially used, they play a minor role for the pulping of wood. For the pulping of annual plants, soda pulping is the preferred process, however. The sulphite process was invented by Tilghman (Great Britain) in 1866 and the kraft process by Dahl (Germany) in 1879. The latter type of pulp was, however, only used in unbleached and semi-bleached paper products where fibre strength was an important parameter and sulphite pulping remained the predominant process for making chemical pulp until around the second World war. At that time, an efficient way of bleaching the dark brown kraft pulp to full brightness by employing chlorine dioxide was found. Since the kraft process gave stronger pulps, was less dependent on the wood raw material and had an efficient chemical recovery process, a successive replacement of small inefficient sulphite mills with large and energy efficient kraft mills started and, today, the kraft process is totally dominant for the production of chemical pulps for a large variety of end uses.

5.2 Effects of Hydrosulphide Ions in Alkaline Pulping

The delignification of wood in aqueous alkali proceeds rapidly provided that the cooking liquor also contains hydrosulphide ions. It was early recognized that the delignification rate increased with the charge of sodium sulphide as shown in *Figure 5.1*.

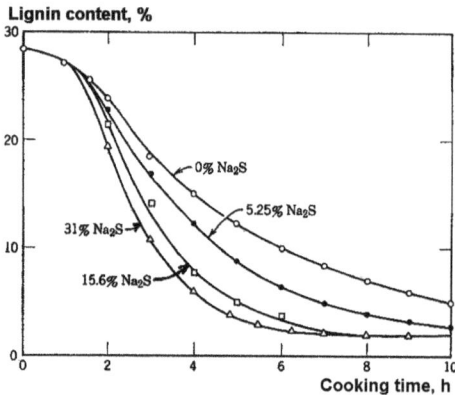

Figure 5.1. Dissolution of lignin from spruce wood at a cooking temperature of 160 °C showing the influence of sulphide charge. (Hägglund 1951).

Furthermore, the kraft process gives a much more comprehensive delignification whereas sodium hydroxide alone, even after very long cooking times, leaves a substantial amount of residual lignin in the fibres. The dissolution of carbohydrates, on the other hand, is not influenced by the presence of hydrosulphide ions and substantial amounts of polysaccharide components are dissolved from the fibres in both soda and kraft pulping (*Figure 5.2*).

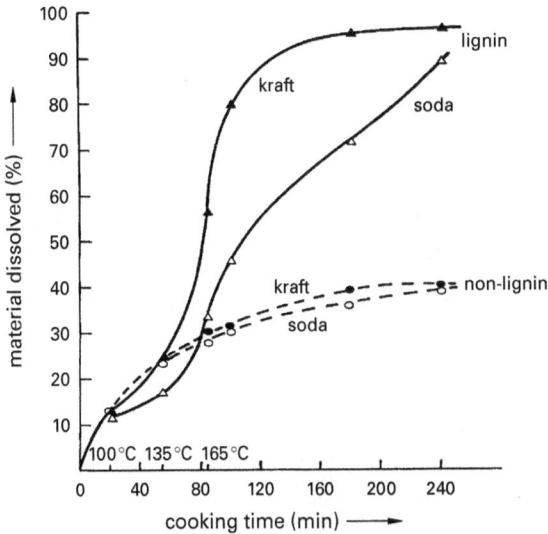

Figure 5.2. Removal of lignin and carbohydrates („non-lignin") in kraft and soda pulping of spruce wood. (Stone and Clayton 1960).

5.3 Delignification Selectivity

The selectivity of dissolution of carbohydrates and lignin during a kraft cook proceeds in three distinct phases with the first one merely being an extraction of both types of components (initial phase). When around 20 % of both carbohydrates and lignin have gone into solution, the kinetics changes dramatically and a rather selective lignin dissolution takes place until approximately 90 % of all lignin has been dissolved (bulk phase). The final portion of the lignin can, however, only be removed with great difficulty and at the expense of a large carbohydrate loss (final phase). In practice, the cook is interrupted at the transition point to the final phase in order not to loose in pulp quality or yield (*Figure 5.3*).

The predominant loss of carbohydrates in kraft pulping is due to the fact that most of the hemicellulose components are degraded and dissolved in the alkaline liquor. Glucomannan and xylans behave differently, however. As shown in *Figure 5.4a*, the glucomannan is rapidly dissolved irrespective of the charge of alkali whereas the dissolution of xylan from both softwood and hardwood (*Figure 5.4b*) becomes more extensive at a higher charge of alkali. Typical values for the total changes in component yields as the result of a kraft cook are shown in *Table 5.1* where the large loss of carbohydrates is further illustrated.

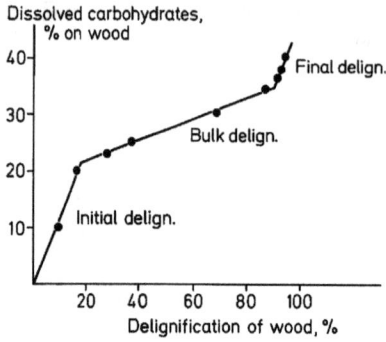

Figure 5.3. Selectivity in the dissolution of carbohydrates and lignin on kraft pulping of softwood.

Table 5.1. Typical yield values (% on wood) for the individual wood components after kraft cooks of pine and birch respectively. Values for wood within brackets.

Wood component	Pine	Birch
Cellulose	35 (39)	34 (40)
Glucomannan	4 (17)	1 (3)
Xylan	5 (8)	16 (30)
Other carbohydrates	~0 (5)	~0 (4)
Lignin	3 (27)	2 (20)
Extractives	<0.2 (4)	0.5 (3)

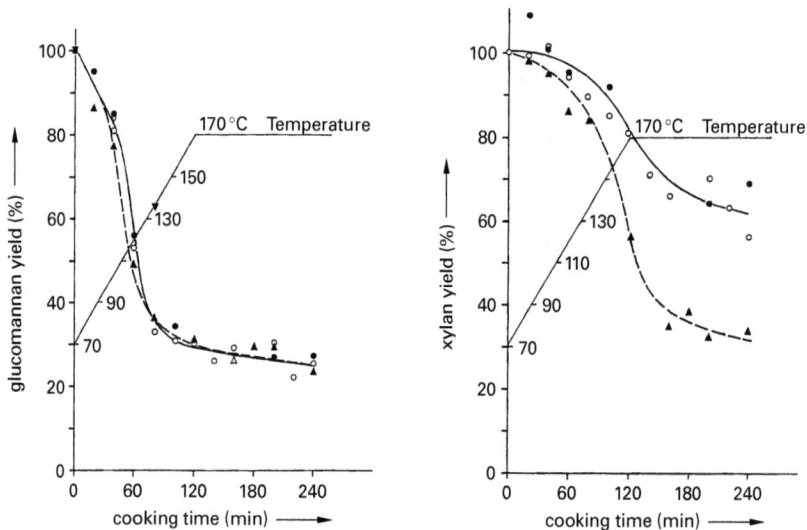

Figure 5.4a. Dissolution of glucomannan and xylan respectively on kraft cooks of pine. Influence of alkalinity (solid and dotted line = low and high charge respectively; Aurell and Hartler 1965).

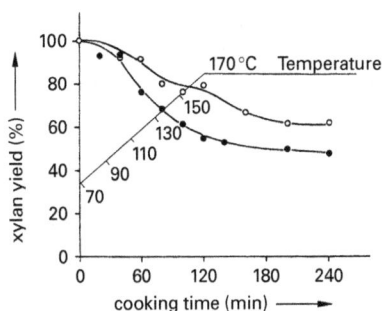

Figure 5.4b. Dissolution of xylan from kraft cook of birch. Influence of alkalinity (unfilled and filled circles denote low and high charge of alkali respectively; Aurell 1964).

Due to the differences in wood morphology and lignin structure, the rate of delignification is higher in hardwood species as compared to softwood (*Figure 5.5*). In addition, a more comprehensive degree of delignification can be achieved for hardwood species under technical pulping conditions. The presence of syringyl units in lignin, obviously, plays an important role and it has been shown that the fiber portion of birch wood, constituting approximately 85 % of the wood weight, contains an almost pure syringyl lignin in the secondary wall. In the middle lamella parts, on the other hand, there is a dominance of guaiacyl lignin. This heterogeneity results in a fast delignification in the bulk phase although the fiber-fiber separation, nevertheless, takes place at a comparatively low residual lignin content.

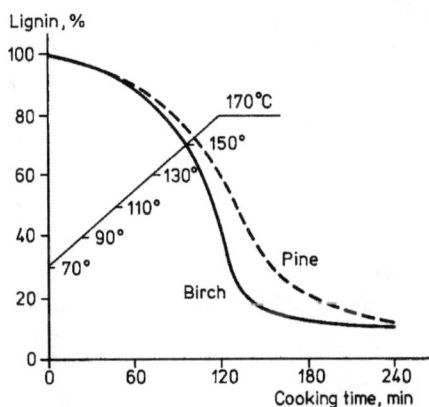

Figure 5.5. Lignin dissolution from pine and birch respectively during kraft cooking. (Aurell 1963).

5.4 The Topochemical Effect

The dissolution of lignin during kraft cooking does not proceed uniformly. Due to structural differences and/or to a lower accessibility of lignin, the middle lamella lignin requires a higher temperature in order to become dissolved. This so called topochemical effect is illustrated in

Figure 5.6 in which a combination of lignin bromination and SEM-EDXA has been used to follow the delignification in various parts of the fiber. A visualisation of the effect is given in *Figure 5.7* at various degrees of delignification. The topochemical effect results in a kraft pulp fiber in which the outer part of the fiber has a higher residual lignin concentration as compared to the average value in the pulp.

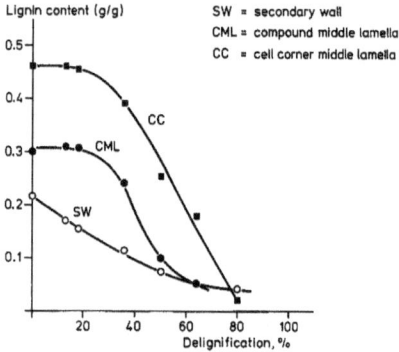

Figure 5.6. Topochemical effects during kraft cooking of Douglas-fir. (Saka et al 1982)

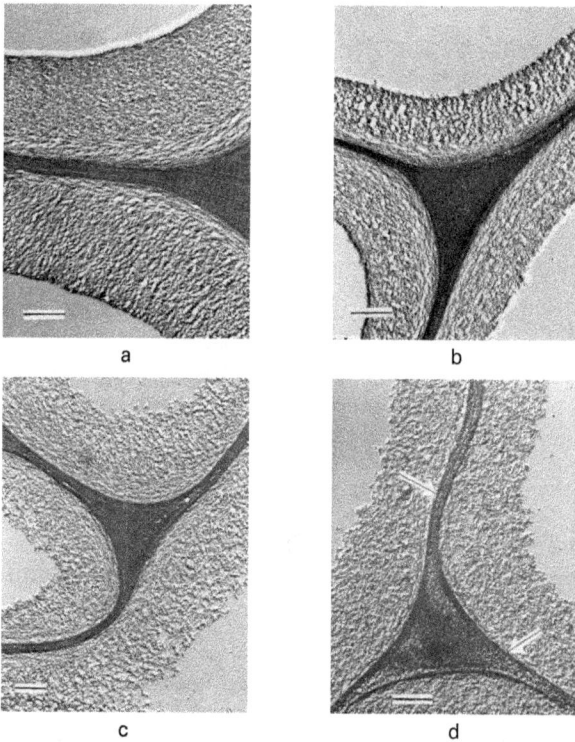

Figure 5.7. SEM pictures showing changes in the lignin skeleton during the successive kraft delignification of spruce wood. Degree of delignification: $a = 0\ \%$, $b = 33\ \%$, $c = 54\ \%$, $d = 71\ \%$. (Parham 1974).

5.5 Lignin Chemistry

The delignification chemistry encountered in kraft and soda pulping has been thoroughly investigated both with lignin model compounds, simulating the various sub- structures present in
lignin, and by analyses of the polymeric lignin. Thereby, it has been clearly shown that a fragmentation of the polymeric lignin and the introduction of hydrophilic groups are necessary pre-
requisites for the dissolution. The cleavage reactions take place in the predominant chemical
linkage connecting the phenylpropane units together, the β-O-4 linkage. By the action of hydrosulphide ions, the phenolic β-O-4 structures are to a large extent fragmented. If no hydrosulphide ions are present, as in soda pulping, the delignification efficiency is poor and the rate of
delignification becomes much lower as shown in *Figure 5.1* and *Figure 5.2*.

5.5.1 Cleavage of β-O-4 Structures

The chemistry of cleavage of phenolic β-O-4 structures is shown in *Figure 5.8*. Thanks to the
alkaline conditions, the phenolic benzyl alcohol structure forms an equilibrium with the corresponding quinone methide. In the presence of hydrosulphide ions, a further equilibrium is present resulting in a benzyl thioalcohol structure. Once formed, the latter can attack the β-carbon
atom in a nucleophilic reaction with formation of a thiirane (episulphide) and a new phenolic
end-group. The episulphide structure, in turn, is not stable and elemental sulphur is expelled to
form polysulphide in the cooking liquor. Under the alkaline conditions, the latter will successively disproportionate and hydrosulphide and thiosulphate ions are formed (*Figure 5.9*).

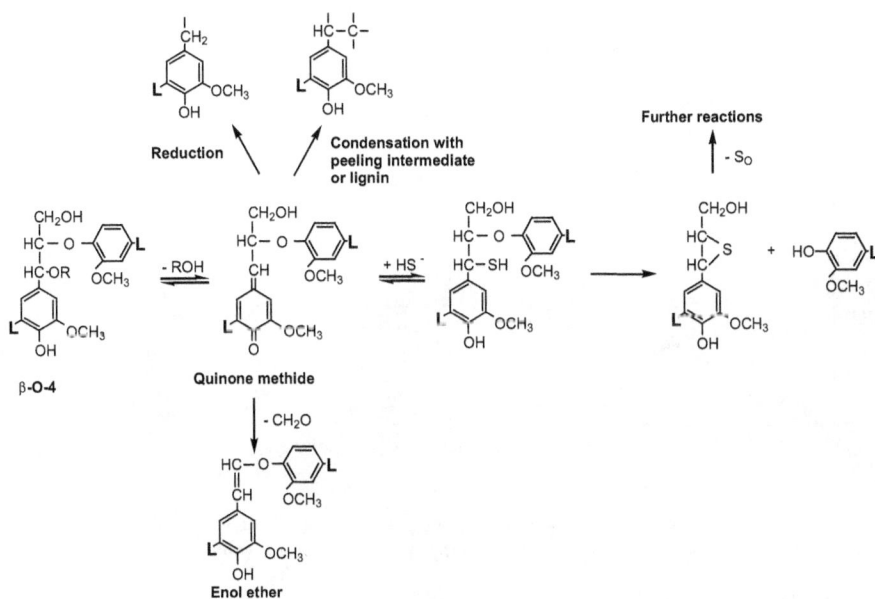

Figure 5.8. Reaction scheme for the cleavage of phenolic β-O-4 structures in lignin during kraft pulping conditions. Competing reactions are also indicated in the figure. L denotes a lignin residue.

In competing reactions, the quinone methide intermediate, shown in *Figure 5.8*, can either loose the γ-hydroxymethyl group and form an enol ether structure without cleavage of the β-ether linkage or react with other nucleophiles present in the pulping system such as lignin or carbohydrate structures thereby creating new stable carbon-carbon linkages. As a further alternative, a reduction of the quinone methide with formation of an α-methylene group may take place.

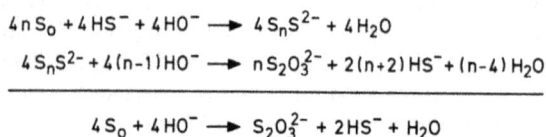

$$4 n S_0 + 4 HS^- + 4 HO^- \longrightarrow 4 S_n S^{2-} + 4 H_2O$$

$$4 S_n S^{2-} + 4(n-1) HO^- \longrightarrow n S_2O_3^{2-} + 2(n+2) HS^- + (n-4) H_2O$$

$$4 S_0 + 4 HO^- \longrightarrow S_2O_3^{2-} + 2 HS^- + H_2O$$

Figure 5.9. The disproportionation of sulphur into hydrosulphide and thiosulphate ions in kraft pulping.

In the cleavage reaction of phenolic β-O-4 structures, the primary product, the episulphide, looses sulphur and is converted, at least in part, to a coniferyl alcohol structure. The reaction can be monitored by analysis of the cooking liquor and, as shown in *Figure 5.10*, both kraft and kraft-anthraquinone (AQ) pulping (see Section 5.7) give rise to the formation of coniferyl alcohol. When the cooking temperature rises above approximately 100 °C, a rapid increase in the formation of coniferyl alcohol takes place and an apparent maximum is reached around 150 °C. At higher temperatures, the rate of decomposition of coniferyl alcohol into vinylguaiacol and other degradation products favourably competes with the formation and a decline is observed. The formation of coniferyl alcohol should be accompanied by a similar formation of sulphur (as polysulphide sulphur) and, as shown in *Figure 5.11*, this is also the case.

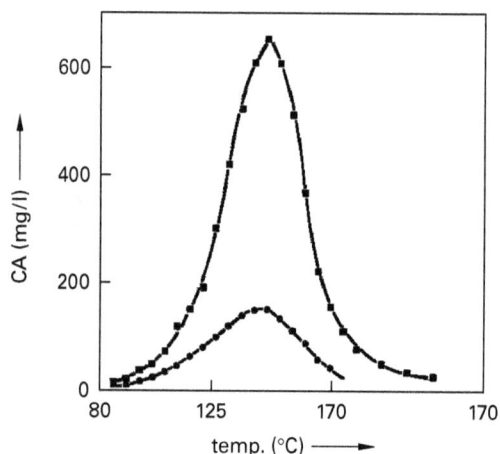

Figure 5.10. The presence of coniferyl alcohol (CA) in the pulping liquor as a function of pulping temperature (and time) in laboratory pulping of softwood (circles = kraft; squares = kraft-AQ). (Mortimer 1982).

Although the kinetics for the rate of delignification in the initial phase of kraft pulping does not indicate any dependancy of either hydroxide or hydrosulphide concentration, the pulping selectivity is nevertheless much influenced by a high concentration of hydrosulphide ions in this particular part of the cook as shown in *Figure 5.12*. In other parts of the cook, the influence is much less.

Sulfur, % of sulfide in cooking liquor

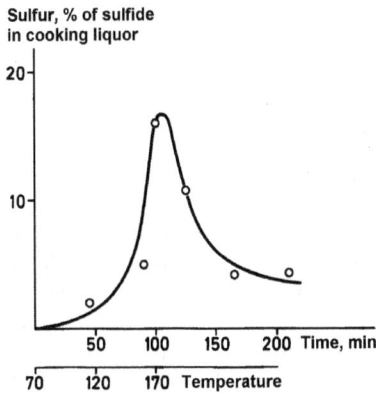

Figure 5.11. The presence of elemental sulphur in the pulping liquor as a function of pulping time in laboratory pulping of softwood.

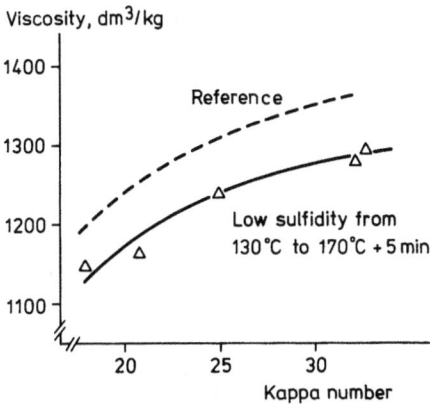

Figure 5.12. A laboratory flow-through kraft cook of softwood with a lowered concentration of hydrosulphide ions in the cooking liquor during the time interval of 60–105 min after the start at 70 °C corresponding to a temperature interval of 130–170 °C + 5 min at 170 °C. Pulping selectivity measured as viscosity as a function of kappa number (Sjöblom et al 1983).

Non-phenolic β-O-4 structures in lignin can also be cleaved during the course of a kraft cook. In contrast to the phenolic type, this cleavage does not, however, involve hydrosulphide ions and the mechanism only depends on the presence of an α-hydroxyl (or a γ-hydroxyl) group which, in the alkaline cooking liquor, can become ionized. In analogy to the mechanism of the phenolic structure, a nucleophilic attack will result in a cleavage of the β-ether linkage and an epoxide (oxirane) structure is formed together with a new phenolic lignin end-group (*Figure 5.13*).

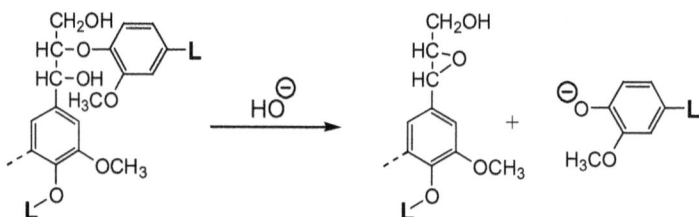

Figure 5.13. The cleavage of non-phenolic β-O-4 structures in kraft (and soda) pulping.

The epoxide, in turn, is not stable but can react further with nucleophiles present in the cook. In model compound studies it has been shown that e.g. accessible sugars may compete with the hydrosulphide ions and become linked to the phenylpropane unit. Other nucleophiles such as lignin units may react in a similar way.

In experiments with lignin model compounds, the kinetics for the cleavage of β-O-4 (and α-O-4) structures has been thoroughly evaluated. Thus, it has been shown that phenolic α- and β-aryl ether structures are cleaved without any influence of either the hydroxide or hydrosulphide ion concentration. The non-phenolic β-O-4 structures, on the other hand, are dependent on the hydroxide ion concentration. The rates of cleavage are much different from each other and, as shown in *Figure 5.14*, the non-phenolic β-O-4 structures will to a large extent decide the efficiency of lignin fragmentation since these are cleaved more slowly when the concentration of alkali decreases in the later part of the cook. In addition, the absolute rate of cleavage is much lower for these units.

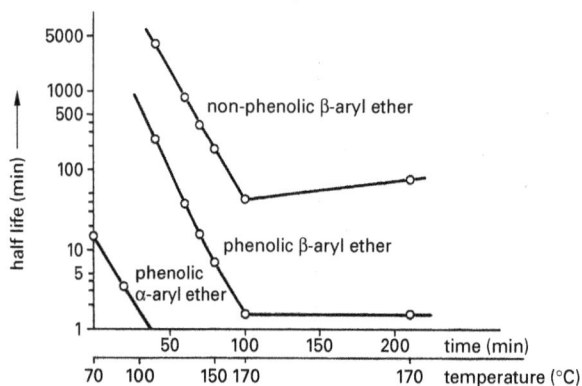

Figure 5.14. Rate constants (calculated as half life times for pseudo-first order reactions) for the cleavage of α- and β-O-4 structures in lignin as a function of pulping time in laboratory experiments with appropriate model compounds. (Ljunggren 1980).

On pulping of wood, the efficiency of cleavage of the β-O-4 structures in lignin can be analysed by several different methods. Acidolysis or thioacidolysis can be applied directly on the fiber material as well as on the dissolved lignin whereas e.g. NMR techniques require an isolation of the residual lignin present in the fibers prior to analysis. Also the concomitant formation of phenolic hydroxyl groups can be directly analysed on the fiber material (Volume 1). Such analyses confirm that a comprehensive lignin fragmentation takes place during the course of a kraft cook as shown in *Figure 5.15*. The lignin that successively becomes dissolved in the cook contains, however, also an appreciable amount of remaining β-O-4 structures indicating that the cleavage reaction becomes much less efficient once a lignin fragment has been solubilized. Thus, lignins isolated from either a normal or a flow-through cook, show similar amounts of remaining β-O-4 structures.

Each β-O-4 structure that is cleaved during the course of the kraft cook gives rise to a new aromatic end-group with a free phenolic hydroxyl group. The latter is ionized in the alkaline pulping liquor and thus facilitates the dissolution of lignin fragments. Analysis of the phenolic groups reveal that in order to become solubilized, a large number of such groups must be created in a certain lignin fragment. This is shown in *Figure 5.16* where it can be seen that irrespective of the size of the fragment (see Section 5.8), an average content of phenolic hydroxyl groups around 60 per 100 phenylpropane units must be present for solubilization. In the remaining fiber lignin, a slow increase in the content can be seen as the cook proceeds. Also here, it must be assumed that the value found is an average and that different portions of the residual lignin have largely different values.

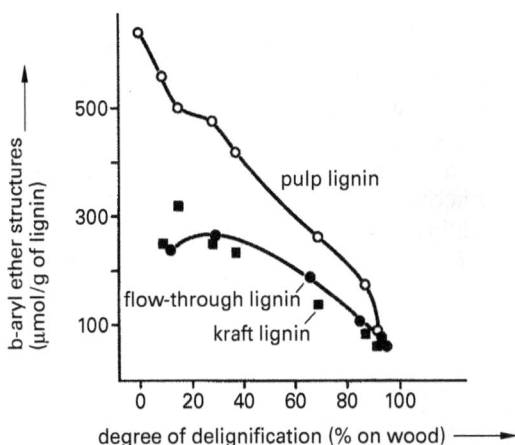

Figure 5.15. Changes in the content of β-O-4 structures as a function of delignification. Analysis by acidolysis. Dissolved lignins are from a normal laboratory cook (kraft lignin) and from a flow-through cook.

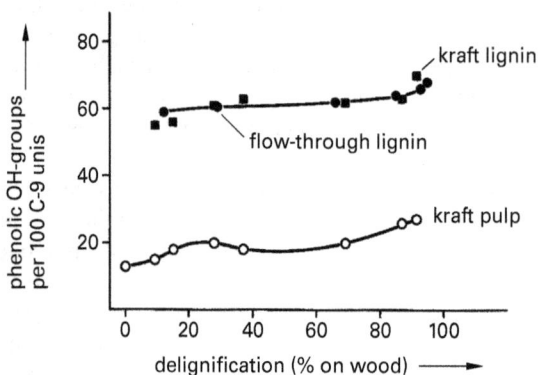

Figure 5.16. The content of phenolic hydroxyl groups present in kraft pulp lignin and in dissolved lignin from normal and flow-through kraft cooks. Data from laboratory experiments.

5.5.2 Competing Reactions in β-O-4 Structures

Despite the efficiency of the β-O-4 cleavage reaction in kraft pulping, competing reactions may occur. These are outlined in *Figure 5.8* and involve elimination as well as reduction and condensation reactions. In all cases, the intermediate quinone methide is the starting point. From this structure, an elimination of the γ-hydroxymethyl group as formaldehyde can take place giving rise to an enol ether structure which in turn is rather stable towards further reactions. In lignin structures lacking an aromatic ring substituent, acidolysis can be used to identify the enol ether. As shown in *Figure 5.17*, even in a well impregnated laboratory kraft cook, enol ether structures are formed, especially in the dissolved lignin. It is unknown, however, whether this has any technical significance.

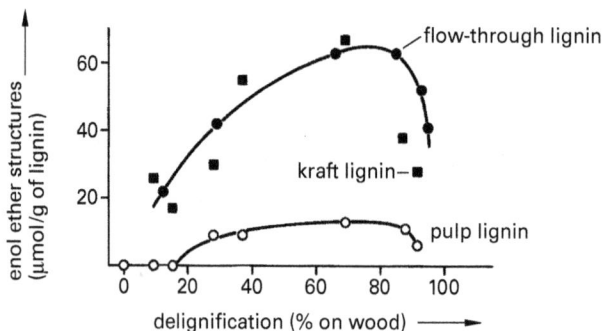

Figure 5.17. Formation of enol ether structures in kraft pulp lignin and in the dissolved lignin from normal and flow-through kraft cooks. Data from laboratory experiments.

The participation of condensation reactions during kraft (and soda) pulping has been the subject of numerous investigations. In experiments with model compounds, it can easily be demonstrated that various types of condensation reactions either between lignin structures or between lignin and carbohydrate structures can occur. In the complex pulping system, on the other hand, the presence of structural units in the lignin, resulting from condensation reactions, has not been unequivocally demonstrated. Obviously, the heterogeneity of the pulping system and the fact that polymeric substances will have difficulties in reacting with each other can explain these observations.

The presence of reduced carbon atoms of the methylene type both in the dissolved and the residual kraft pulp lignin has led to the conclusion that reduction may be an important reaction pathway in kraft pulping. Again, model experiments have been used to support the observations. Also for this type of reaction, it is not known with certainty if, indeed, reduction takes place in the polymeric lignin during the kraft cook since, recently, the native wood lignin has been shown to contain small amounts of sub-structures containing methylene groups.

5.5.3 Reactions with other Sub-Structures

In other types of lignin sub-structures such as the β-5 and the β-1 structures, the phenylpropane units are held together with carbon-carbon linkages. This is also the case with 5.5 structures whereas in 4-O-5 structures, an aromatic ether linkage is present. In all these sub-structures, the linkage between the phenylpropane units is completely stable during pulping conditions. The phenolic variants of the former two structures may form quinone methide intermediates in equilibrium with the corresponding benzyl (thio)alcohol structures. The further reactions of the quinone methide are, however, restricted to an elimination of either a proton or formaldehyde and formation of stilbene structures (*Figure 5.18*).

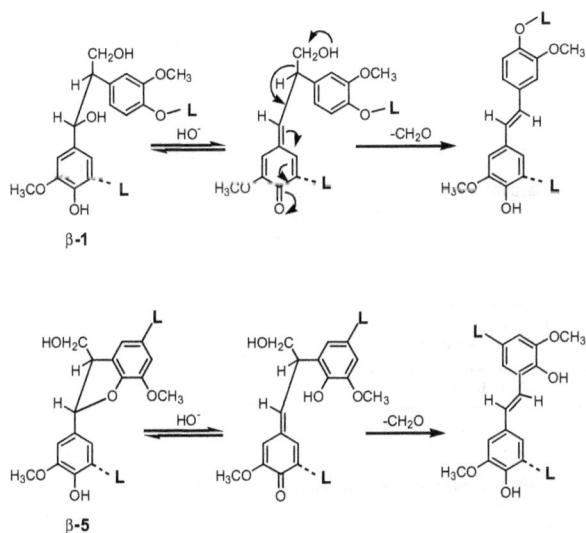

Figure 5.18. Formation of stilbenes from β-5 and β-1 lignin sub-structures during kraft (and soda) pulping.

Dibenzodioxocin (5-5-O-4) structures are found as sub-structures in the native (softwood) lignin with a frequency of around 5 per 100 phenylpropane units. Being an α,β-diaryl ether structure, the α-aryl ether linkage in the phenolic variant is easily cleaved to form a quinone methide in alkaline conditions followed by the same reactions as those elucidated in *Figure 5.8.*

5.5.4 Formation of Malodorous Compounds

The presence of strong nucleophiles and the high temperature employed in kraft pulping induces a cleavage of the methyl-aryl ether linkages in lignin. Although the cleavage reaction takes place to a minor extent only, the products formed are highly detrimental from an environmental point of view since they constitute the origin of the smell from kraft mills. The primary reaction involves an attack of a hydrosulphide ion on the aromatic methyl ether group and formation of methylmercaptan (MM) together with a phenolic hydroxyl group. Under the alkaline conditions prevailing in the cook, a methylmercaptide anion is formed and this in turn can attack a new methyl group with formation of dimethylsulfide (DMS) (*Figure 5.19*). Both methylmercaptan and dimethylsulfide are volatile and, in particular, methylmercaptan is also highly toxic. When exposed to air (oxygen), a portion of the methylmercaptan may become oxidized with formation of dimethyldisulfide (DMDS), a further malodorous and volatile product. By these reactions, a total of approximately 1–2 % of the methoxyl groups are converted into hydroxyl groups during a kraft cook.

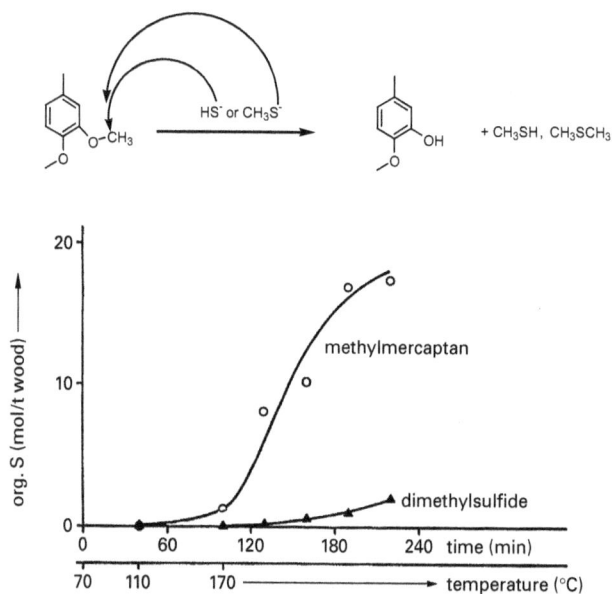

Figure 5.19. Formation of malodorous compounds in kraft pulping. Data at 40 % sulfidity.

5.6 Carbohydrate Reactions

The reactions of carbohydrates in a kraft cook are comprehensive and will result in a large yield loss as shown in *Figure 5.2* and *Figure 5.3*. Most of this yield loss takes place very early in the cook as shown in *Figure 5.3*. The predominant carbohydrate chemistry encountered in kraft pulping is well known and includes a successive depolymerisation of the polysaccharides starting from the reducing end group. This so called peeling reaction starts immediately when the wood comes in contact with the alkaline pulping liquor and proceeds rapidly even at temperatures around 100 °C. At considerably higher temperatures, ~170 °C or more, a random alkaline hydrolysis of glucosidic bonds may also take place to some extent.

5.6.1 The Peeling Reaction

The peeling reaction has a particular importance for polysaccharides containing a substituent in the 4-position like in glucomannans, xylans and cellulose (*Figure 5.20*). In the aqueous environment during pulping, the reducing end group in the polysaccharide will form an equilibrium between the hemiacetal and the open aldehyde form. In the presence of alkali, a further equilibrium will be established between the aldehyde and the keto form. From both these forms, a subsequent β-elimination reaction may occur. In the former case, the reaction will result in a stabilisation of the polysaccharide chain (*Figure 5.21*) whereas the other type of reaction will give rise to a cleavage of the glucosidic bond in the 4-position. Thereby, the first sugar unit in the chain is eliminated and a new reducing end unit is liberated and able to go through the same reaction sequence again. Once, the β-elimination has taken place, the liberated product can react further via a benzilic acid rearrangement to form an isosaccharinic acid. This product is stable and (softwood) kraft pulping liquors usually contain in the order of 100 kg of glucoisosaccharinic acid per ton of pulp.

The result of the peeling reaction is a depolymerisation of the polysaccharide and at a certain point, the remaining chain may become completely solubilized before being further degraded. For glucomannans with a degree of polymerisation (DP) of around 100 in the native state, this reaction is very efficient and irrespective of the charge of alkali in the cook, this type of polysaccharide is efficiently degraded and dissolved as shown in *Figure 5.4a* and *Table 5.1*. For cellulose, on the other hand, the reaction does not result in any significant degree of dissolution since the starting DP of cellulose is high, around 10 000.

In xylans, the peeling reaction is much less efficient due to the substituents, arabinose and 4-O-methylglucuronic acid in softwoods and 4-O-methylglucuronic acid in hardwoods. This is also visualized in *Figure 5.4a* and *Figure 5.4b* for xylan in softwood and hardwood respectively. Here, a pronounced influence of the charge of alkali can be seen illustrating that the entire polymer has a sufficient solubility for a direct dissolution in the pulping liquor. The presence of a substituent in the 3-position favours a direct β-elimination from the open aldehyde form as illustrated in *Figure 5.21*. Thereby, a stabilisation of the polysaccharide chain and formation of a metasaccharinic acid takes place preventing the peeling reaction from occurring. To some extent, this stabilization reaction will also take place in polysaccharides like in glucomannans and cellulose as indicated in *Figure 5.20*. For these polysaccharides, the kinetics is such that, on the average, the peeling reaction will result in the loss of around 50–100 sugar units before a „stopping" reaction occurs.

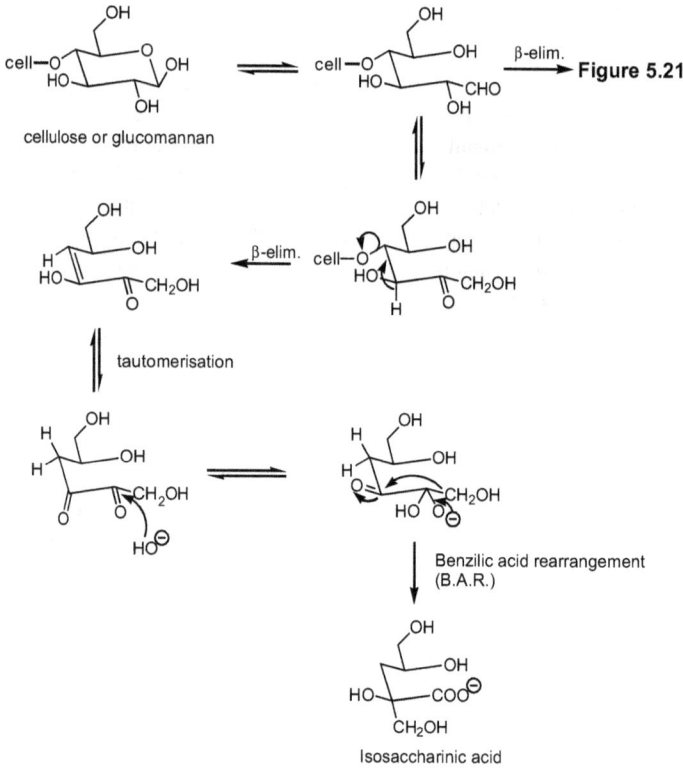

Figure 5.20. Mechanism for the peeling reaction in kraft (and soda) pulping.

5.6.2 Alkaline Hydrolysis

A certain further loss of cellulose may take place when the kraft (or soda) cook approaches the maximum temperature due to alkaline hydrolysis of glucosidic linkages. The reaction is initiated by the C-2 hydroxyl group which, if ionized, can attack the C-1 carbon atom and expell the attached anhydroglucose unit (*Figure 5.22*). The result will be a cleavage of the cellulose chain and thus a large reduction of DP, formation of a new reducing end group and a rapid „secondary" peeling reaction. It must be emphasized, however, that the alkaline hydrolysis is rather heterogeneous and preferentially occurs in the amorphous parts of the cellulose microfibrils. As a result of both peeling and alkaline hydrolysis, a total loss of cellulose of around 10 % (on wood) is generally encountered in kraft pulping.

Figure 5.20

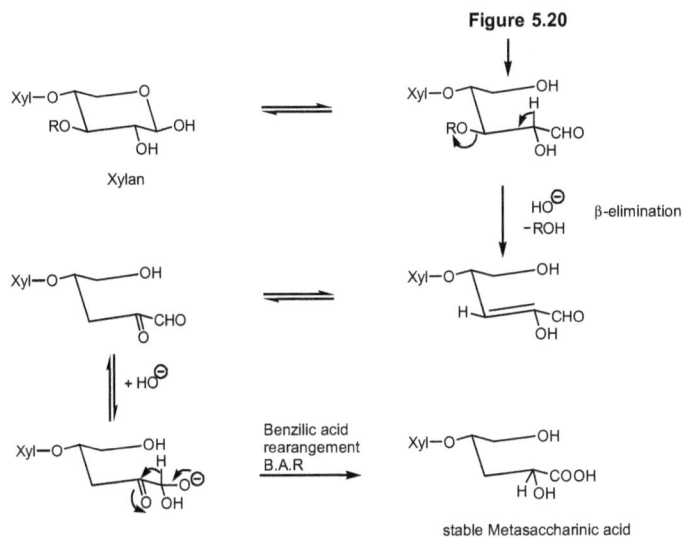

Figure 5.21. The β-elimination in polysaccharide end groups resulting in a stopping reaction and formation of a metasaccharinic acid. RO = arabinose unit.

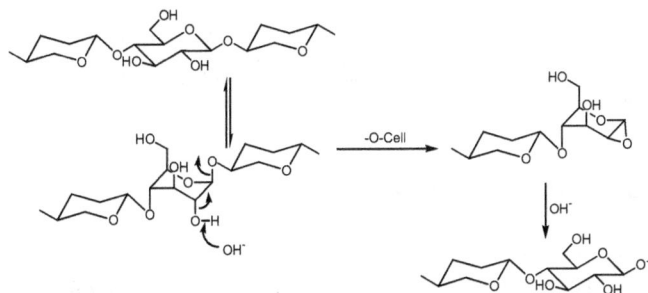

Figure 5.22. Alkaline hydrolysis of a glucosidic linkage via epimerisation and epoxide formation.

5.6.3 Hexenuronic Acid

In both softwood and hardwood, the xylan chain is substituted with 4-O-methylglucuronic acid groups located in the C-2 position and linked through an α-glucosidic linkage. Under alkaline conditions, the methoxyl group can be eliminated as methanol resulting in the formation of a hexenuronic acid (HexA) group (*Figure 5.23*).

The elimination of methanol and the formation of hexenuronic acid groups proceeds rapidly in the kraft pulping process as shown in *Figure 5.24*. Since the hexenuronic acid group is rather stable in alkali, the xylan that remains in the pulp after the cook will contain an appreciable amount of such groups. These will influence the pulp properties and contribute to the bleachability of the pulp.

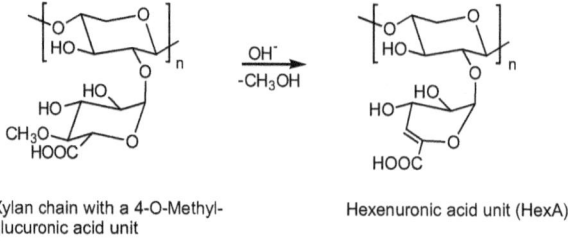

Xylan chain with a 4-O-Methyl-
glucuronic acid unit

Hexenuronic acid unit (HexA)

Figure 5.23. The elimination of methanol from 4-O-methylglucuronic acid groups in xylan during kraft (and soda) pulping.

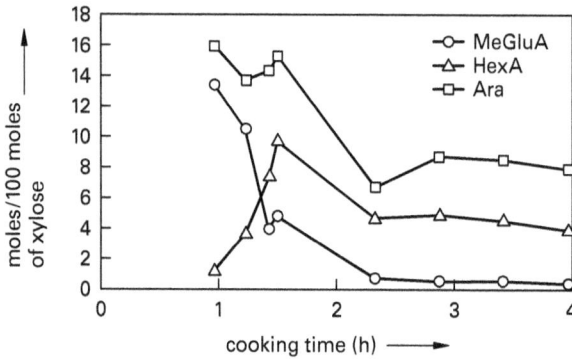

Figure 5.24. Presence of 4-O-methylglucuronic acid (MeGluA), hexenuronic acid (HexA) and arabinose (Ara) during the course of a kraft cook of pine. (Buchert et al 1995).

5.6.4 Stabilization Reactions with Polysulphide

The comprehensive yield loss encountered in kraft pulping is a serious drawback and many attempts have been done to find ways of increasing the pulp yield by minimizing the peeling reaction. In principle, both reductive and oxidative reactions can be used to stabilize the reducing end-groups in polysaccharides. Sodium borohydride is an efficient reducing agent for the stabilisation reaction and gives rise to the formation of an alditol end-group thus preventing any peeling. Consequently, a large yield increase can be obtained. For cost reasons this reaction has found no technical application, however.

The oxidation of a polysaccharide reducing end-group will result in the formation of the corresponding aldonic acid group. Again, a stabilization against peeling will be obtained. By a catalytic partial oxidation of the hydrosulphide ions present in kraft pulping liquor into polysulphide e.g. by air (oxygen), such an oxidation can be achieved and this technology is used by some pulp mills in the world. In the beneficial carbohydrate reaction, it is believed that the enediol form of the reducing end-group, in equilibrium with the corresponding open aldehyde form, is oxidized to a diketo structure and transferred via a benzilic acid rearrangement to the aldonic acid (*Figure 5.25*).

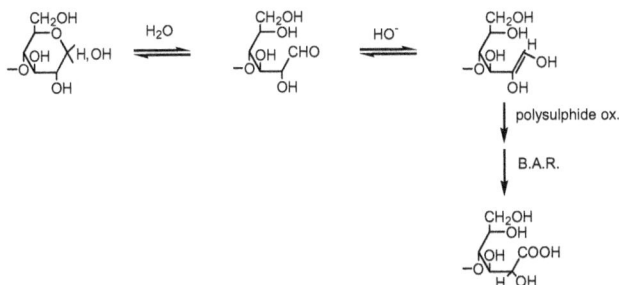

Figure 5.25. Mechanism for the oxidative stabilization of a reducing end-group by polysulphide in alkaline pulping. B.A.R. = benzilic acid rearrangement.

5.7 Pulping with Anthraquinone (AQ)

The finding that a small amount of anthraquinone (AQ) can act as a pulping catalyst giving rise to both an enhanced delignification rate and a yield increase in kraft as well as in soda pulping has revitalized the search for pulping alternatives to kraft. Although it is possible to use the soda-AQ system to produce a „kraft-like" pulp, the main use of AQ today is as an additive in kraft pulping. Thereby, a stabilisation of the polysaccharides against peeling can be achieved and in certain countries, notably Japan, where the wood is expensive the resulting yield increase has a great impact on the overall pulp mill economy. From a chemistry point of view, the function of AQ is that of an almost ideal catalyst since the charge can be kept at around 1 kg per ton of wood. After the cook, some 20 % of the AQ can still be found in the pulping liquor while the remainder becomes irreversibly linked to various lignin structures (*Figure 5.26*).

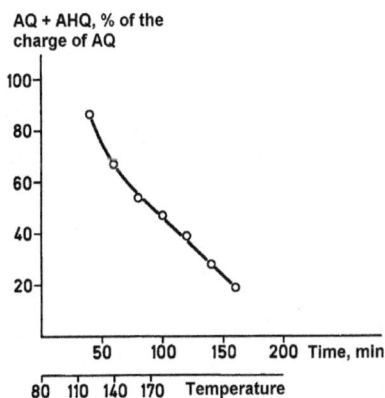

Figure 5.26. Consumption of anthraquinone (AQ) and anthrahydroquinone (AHQ) in soda-AQ pulping. (data from Kubes et al 1980).

The chemistry of AQ in soda or kraft pulping starts with a rapid oxidation of the reducing end-groups in the polysaccharides according to the mechanism outlined in *Figure 5.25*. In this

110

reaction, anthrahydroquinone (AHQ) is formed. The latter will react with quinone methide intermediates formed in the lignin. When part of a β-O-4 structure, the adduct between AHQ and
the quinone methide will decompose with formation of AQ. At the same time, the β-O-4 linkage is cleaved to form coniferyl alcohol (cf *Figure 5.10*) and new phenolic end-groups as shown
in *Figure 5.27*. The rapid consumption of AQ early in the cook and the subsequent lignin reactions with AHQ to reform AQ can be followed analytically and shown to proceed as two consecutive steps (*Figure 5.28*).

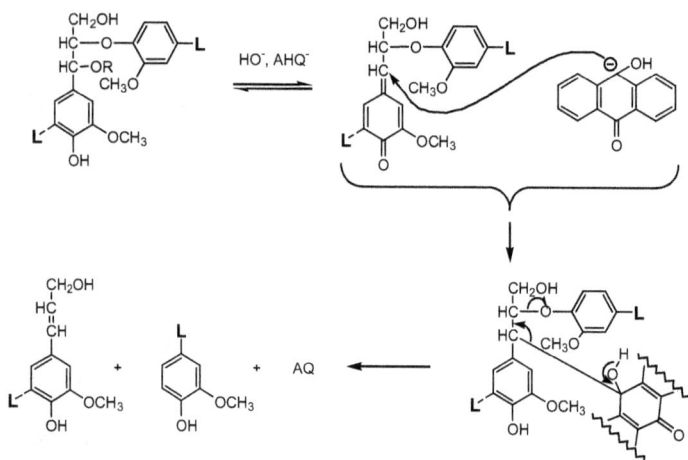

Figure 5.27. Mechanism for the alkaline cleavage of phenolic β-O-4 structures in lignin by anthrahydroquinone
(AHQ).

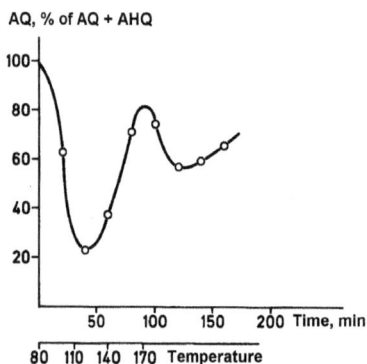

Figure 5.28. The relative amount of AQ present in soda-AQ pulping liquor as function of pulping time in laboratory experiment with softwood. (data from Kubes et al 1980).

5.8 The Final Delignification Phase in Kraft Pulping

When the kraft cook has reached a degree of delignification of approximately 90 %, the selectivity changes (*Figure 5.3*) and prolonged cooking will result in a deterioration of the pulp quality and a rapidly decreasing yield. At the same time, the rate of delignification slows down. Major reasons for this change include the fact that the number of β-O-4 structures successively decreases as the cook proceeds as shown in *Figure 5.15*. Consequently, less possibilities exist for the necessary lignin fragmentation. In addition, lignin is attached to the polysaccharides predominantly through ether linkages, some of which are stable during the conditions of the cook. Thus, towards the end of the cook, the lignin that goes into solution has been shown to contain small but increasing amounts of xylose and glucose units whereas mannose units seem to accumulate in the pulp. The presence of lignin-carbohydrate co-polymers (LCC) in the fiber wall will retard the lignin dissolution since these will have high molecular weight averages and presumably less solubility in the cooking liquor. Furthermore, the molecular size of the lignin that successively is dissolved in the cook increases as shown in *Figure 5.29*. An additional factor that may affect the solubility of lignin is the possibility of formation of larger lignin aggregates from individual molecules, possibly linked together vith calcium ions as bridges.

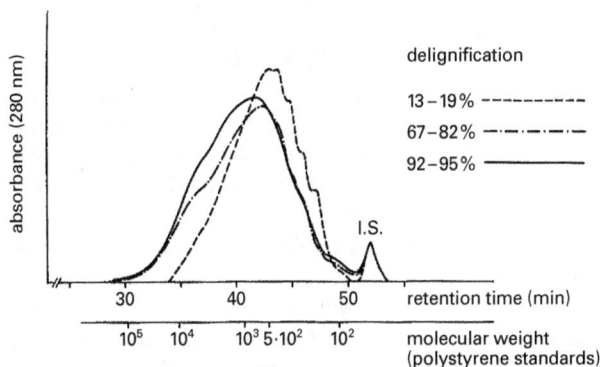

Figure 5.29. Gel permeation chromatograms of dissolved kraft lignin taken out at various intervals of delignification from a flow-through kraft cook of softwood. I.S. = internal standard.

The successively larger lignin molecules that are formed during the kraft process will have increasing difficulties in diffusing out of the fibre wall unless the average pore size increases. Although certain changes in the pore size intervals indeed take place during a kraft cook, as shown in *Figure 5.30*, these do not at all promote a complete elimination of lignin from the fibre wall and, consequently, a „trapping" effect is obtained. As a result, kraft pulp fibers show a „leaching effect" of lignin that can be substantial and correspond to several kappa number units.

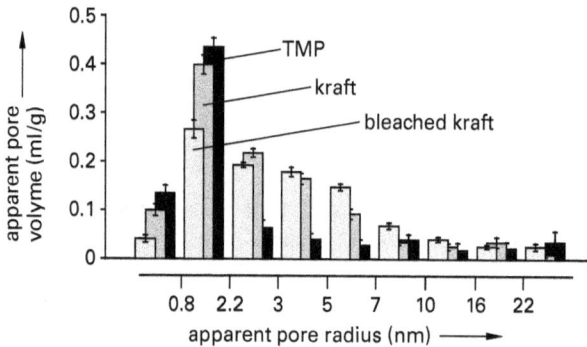

Figure 5.30. Apparent pore volume at different pore size intervals for mechanical pulp (TMP), unbleached kraft pulp (kappa 78) and bleached kraft pulp (kappa 0.5) determined by inverse size exclusion chromatography (ISEC). (Berthold 1996).

5.9 Oxidizable Structures in the Pulp

Traditionally, the remaining lignin in the unbleached kraft pulp has been measured as the kappa number although that number not only reflects the lignin content but all oxidizable structures present in the pulp. Measurements of the consumption of acidic permanganate by the pulp, i.e. the kappa number, can be done not only on pulp, however, but also on various other structural units and model compounds representing structures thought to be present in a pulp sample. Thereby, the individual consumption of permanganate by lignin, hexenuronic acid and other structures such as hydroxyl groups, double bonds, carbonyl groups and combinations of these can be measured separately. As shown in *Figure 5.31*, lignin has a high reactivity towards permanganate but also other structures and functional groups are able to react under the conditions of the kappa number measurement. Notably, hexenuronic acid groups present in the pulp xylan also show a high reactivity and, consequently, such groups will contribute to the kappa number as well. By comparing different analytical techniques, it can be shown that various chemical pulps also contain other components able to consume permanganate, presently denoted „non-lignin" structures (*Figure 5.32*). These may well contain double bonds and/or carbonyl groups located in both lignin and polysaccharides.

5.10 Structure of the Residual Pulp Lignin

The structure of the residual lignin determines to a great extent the further reactions and reactivity of the pulp in the bleaching plant although hexenuronic acid and „non-lignin" structures also may have important influence. The structure of the unbleached pulp lignin can differ depending on the exact pulping conditions. Thus, the extent of cleavage of β-O-4 structures in lignin during the cook can vary widely depending on the charge of hydrosulphide and hydroxide ions as shown in *Figure 5.33* even if the cook has been carried out to the same kappa number.

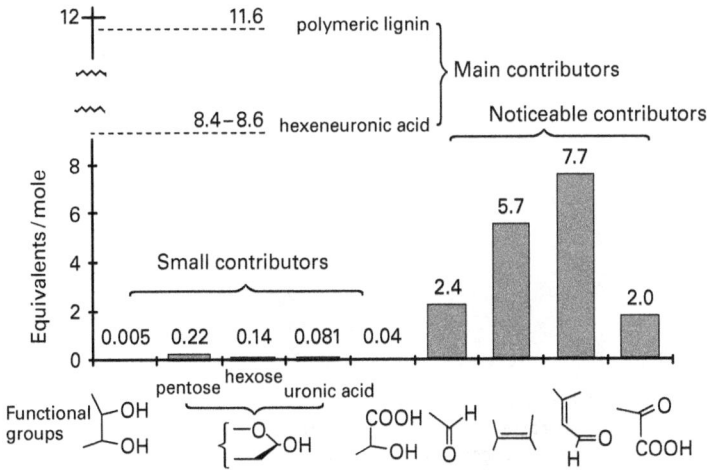

Figure 5.31. Consumption of acidic potassium permanganate (equivalents/mole) in the oxidation of various structures/functional groups. Reaction conditions identical to those used in the standard kappa number measurement.

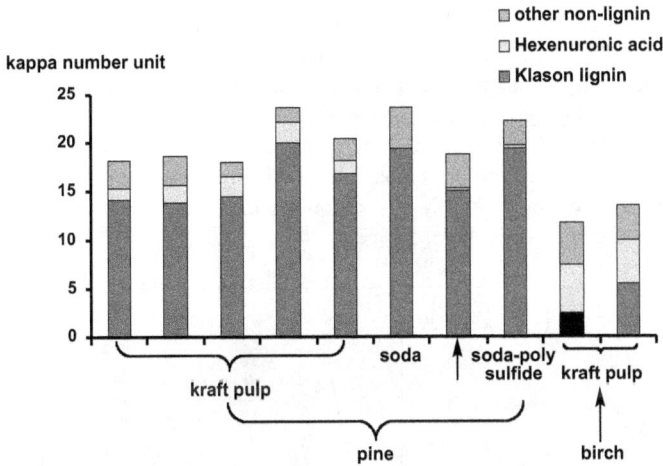

Figure 5.32. Contribution to the kappa number in various unbleached pulps from lignin (Klason lignin), hexenuronic acid and „non-lignin" structures.

Obviously, the chemistry of lignin dissolution in a kraft cook is not only dependent on the cleavage of β-O-4 structures. A low H-factor, i.e. a high charge of hydroxide and hydrosulphide ions, is beneficial thus indicating that a prolonged pulping time, in addition to an increased degree of β-O-4 cleavage, also results in an increase of non-desirable reactions able to prevent or to slow down the overall lignin dissolution.

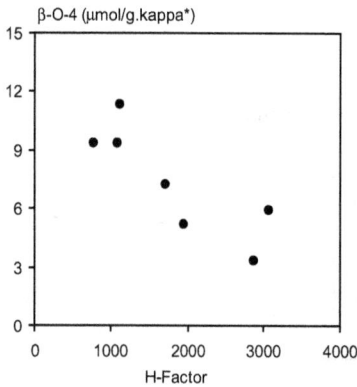

Figure 5.33. Remaining amount of β-O-4 structures in the residual lignin of softwood kraft pulp as a function of the H-factor. Pulping to a kappa number of 20 after correction for the contribution from hexenuronic acid using various charges of hydroxide and hydrosulphide ions. Analysis by thioacidolysis.

In addition to reduced possibilities for lignin fragmentation towards the end of a kraft cook, the fact that lignin is chemically linked to polysaccharides is a further factor that affects the dissolution negatively. Such linkages are present in the native wood but may also be formed during the kraft cook, e.g via a reaction between a sugar unit and either an epoxide or a quinone methide in a β-O-4 structure. At the end of the cook, the lignin that goes into solution has been shown to contain an increased amount of sugars, predominantly glucose, xylose and arabinose thereby indicating linkages between lignin and xylan and cellulose respectively. A further confirmation of the presence of such linkages has been obtained through analysis by GPC of unbleached kraft pulp from pine and birch (*Figure 5.34*). Here, it can be seen that in a softwood pulp, a high UV-absorption (i.e. lignin content) accompanies both the low- and the high molecular weight material whereas in a birch pulp, the UV-absorption to a large extent is linked to the low molecular weight material, i.e. the hemicelluloses.

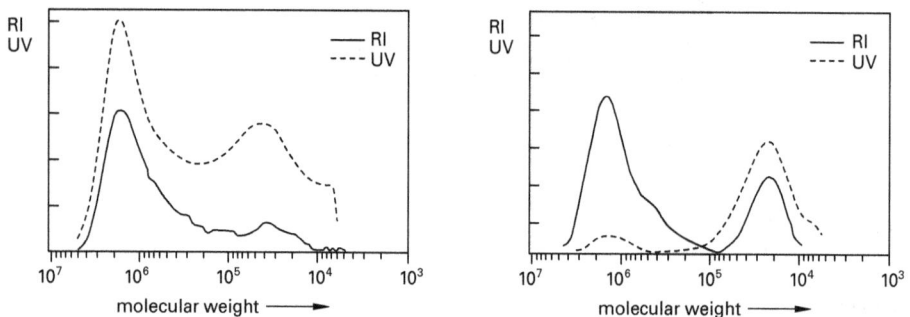

Figure 5.34. GPC analysis of unbleached kraft pulp from pine (left) and birch (right) after dissolution in dimethylacetamide/lithium chloride. RI = refractive index, UV = ultraviolet detection. Molecular weight calibration with pullulan standards. (Karlsson 1997).

5.11 Sulphite Pulping

The dissolution of lignin and the liberation of cellulosic fibers by the action of bisulphite ions can be done with either calcium, sodium, magnesium or ammonium as the counter ion. In the early technology, calciumoxide was suspended in water together with an excess of sulphur dioxide thus creating the pulping liquor. A major disadvantage was the fact that no chemical recovery system could be developed and, consequently, the spent liquor had to be discharged resulting in great environmental problems. In many cases this was solved by evaporation of the liquor and production of lignosulphonates for use e.g. as dispersing agent. In the further development of sulphite pulping, magnesium oxide or sodium hydroxide were introduced as bases and for these, efficient chemical recovery systems have been developed. The fact that pine is less suitable as compared to spruce as wood raw material has also resulted in the development of various multi-stage pulping technologies usually employing sodium hydroxide as the base.

In the production of chemical sulphite pulps, acidic conditions are employed although alkaline sulphite pulping is a potential alternative. The pH of the pulping liquor is in the range of 1.5–4.0 depending on process and product requirements. For the production of corrugated board (the middle wave layer), however, a neutral sulphite pulping liquor can be used to produce high yield NSSC fibres. The sulphite system contains two equilibrium reactions (reactions 1 and 2) and it has been found that in acidic sulphite pulping, a certain amount of bisulphite ions must be present in the liquor at all times in order to obtain a lignin sulphonation reaction. Otherwise, the wood residue will turn dark without any dissolution of lignin due to a predominance for lignin condensation reactions (*Figure 5.35*).

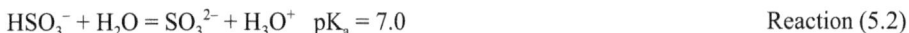

$$SO_2 \cdot H_2O + H_2O = HSO_3^- + H_3O^+ \quad pK_a = 1.9 \hspace{4cm} \text{Reaction (5.1)}$$

$$HSO_3^- + H_2O = SO_3^{2-} + H_3O^+ \quad pK_a = 7.0 \hspace{4cm} \text{Reaction (5.2)}$$

5.11.1 Lignin Chemistry

The necessary lignin dissolution in sulphite pulping is caused by a sulphonation of the lignin thereby successively increasing the water solubility. In addition, acid hydrolysis of ether linkages in both lignin and carbohydrates takes place to various extents depending on the pH. The degree of sulfonation is also dependent on the pH of the cooking liquor as shown in *Figure 5.36*. Thus, under strongly acidic conditions, high degrees of sulphonation can be achieved whereas under neutral or slightly alkaline conditions, the sulphonation seems to reach a plateau value at around 30 sulphonate groups per 100 phenylpropane units (measured as S/OCH_3 in the figure). It can also be seen that, under neutral conditions, a very rapid initial sulphonation takes place resulting in the introduction of around 15 sulphonate groups per 100 phenylpropane units within only a few minutes (cf. chemistry of CTMP).

The reaction between a phenylpropane unit and acidic bisulphite proceeds via the protonation of the benzylic hydroxyl (or ether) group followed by elimination of water and addition of the bisulphite ion (*Figure 5.37*). Both phenolic and non-phenolic phenylpropane units can react thus explaining the high overall degree of sulphonation that can be achieved.

$$SO_2 \cdot H_2O + H_2O = HSO_3^- + H_3O^+ \quad pK_a = 1.9 \quad (1)$$
$$HSO_3^- + H_2O = SO_3^{2-} + H_3O^+ \quad pK_a = 7.0 \quad (2)$$

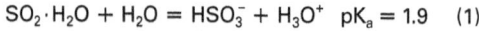

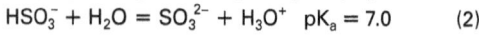

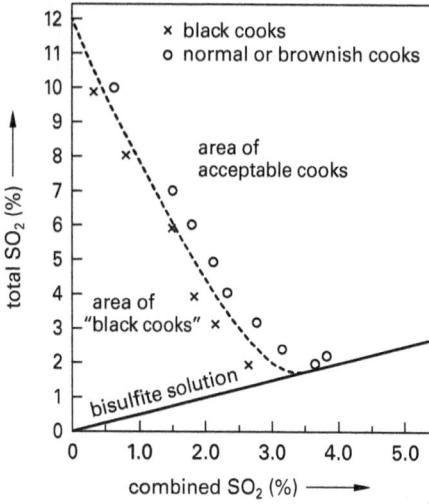

Figure 5.35. Influence of the sulphite pulping liquor composition on the formation of lignin condensation reactions („black cooks"). Charge of sulphur dioxide/bisulphite in per cent on wood. (Kaufmann 1951).

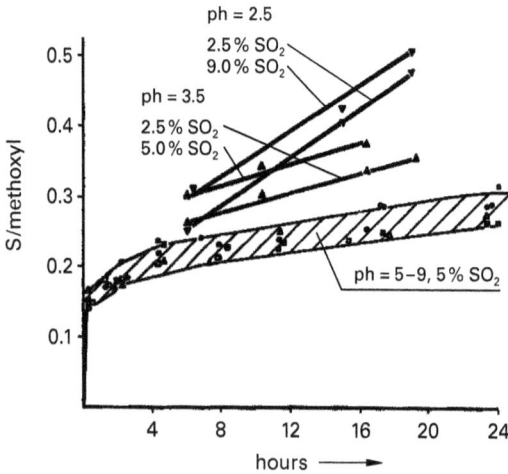

Figure 5.36. Degree of lignin sulphonation after sulphite cooks of spruce wood at various pH-values and charges of sulphur dioxide as a function of time at 135 °C. (Lindgren 1952)

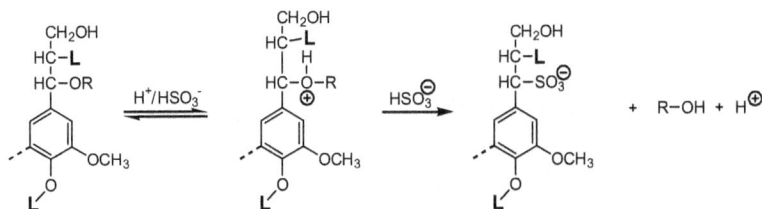

Figure 5.37. Mechanism for the sulphonation of lignin in acidic sulphite pulping.

Under neutral or slightly alkaline conditions, on the other hand, the reaction becomes more selective and only the phenolic phenylpropane units react. In this case, the mechanism is analogous to the reaction in kraft pulping, viz a formation of a quinone methide intermediate followed by addition of a sulphite ion. As shown in *Figure 5.36*, the initial portion of this reaction proceeds very rapidly thus indicating that the content of phenylpropane units in lignin having a free phenolic hydroxyl group should be in the order of 15 per 100 phenylpropane units. In fact, values around 12 have been found. At longer cooking times at a neutral pH, a further slow sulphonation can take place. In experiments with lignin model compounds, it has been shown that sulphonated β-O-4 structures may undergo a β-ether cleavage by attack of sulphite on the β-carbon atom. Thereby, two sulphonate groups can be introduced in the same side chain as shown in *Figure 5.38*. At the same time, a new phenolic lignin end group is created which in turn may become sulphonated.

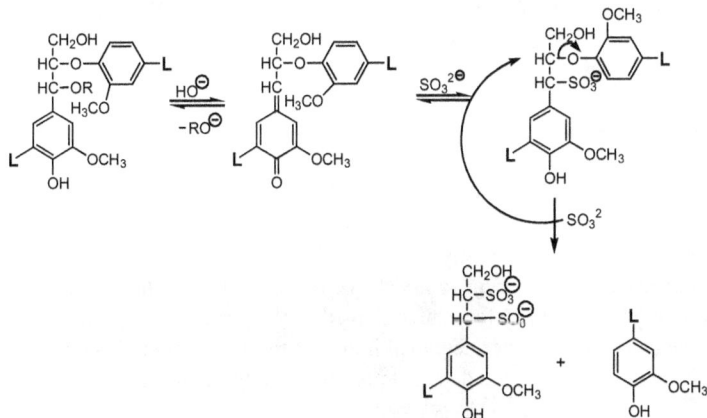

Figure 5.38. Mechanism for the sulphonation of lignin in neutral sulphite pulping.

In acidic sulphite pulping, the major competing reaction to sulphonation is condensation. Due to the presence of free aromatic carbon atoms in the para position to a methoxyl group (C-6 position) a facile formation of carbon-carbon linkages to an adjacent benzylic carbon atom is possible provided that the latter can form a carbonium ion via protonation-elimination of the oxygen function (*Figure 5.39*).

118

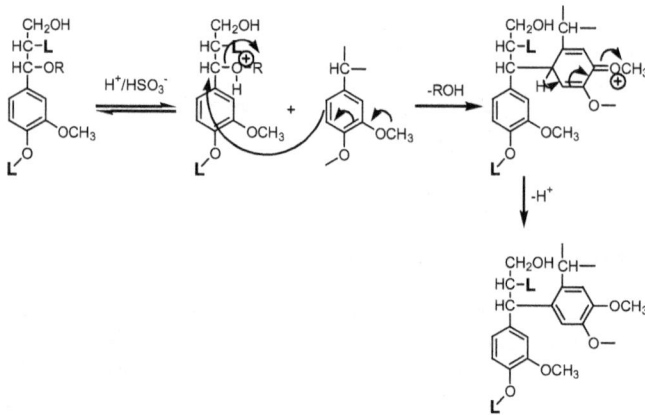

Figure 5.39. Mechanism for the acid catalysed condensation in lignin.

In certain lignin and lignin-like structures, such as lignans, the steric arrangement is such that condensation is favoured over sulfonation and both pinoresinol and hydroxymatairesinol give rise to intramolecular condensation reactions on treatment with acidic sulphite. Furthermore, the presence of pinosylvin (3,5-dihydroxystilbene) and pinosylvin methylether (3-hydroxy-5-methoxy-stilbene) in the heartwood of pine results in a competition between sulphonation of lignin and condensation between lignin and the C-6 position in these extractives. As a result, acid sulphite pulping of pine is less suitable than spruce since the degree of sulphonation will be lower and thus the lignin dissolution less favourable.

5.11.2 Carbohydrate Chemistry

The major reaction of the polysaccharides in acidic sulfite pulping is the acid catalysed hydrolysis of glucosidic linkages resulting in the formation of large amounts of monomeric sugars in the pulping liquor. In particular, the more easily accessible hemicelluloses are degraded and some 70 % of the glucomannan and 50 % of the xylan can be lost in a spruce sulphite pulp. The hexoses present in the pulping liquor can be fermented into ethanol. During the sulphite cook, a portion of the dissolved sugars can be further degraded into furfural (pentoses) and hydroxymethylfurfural (hexoses) in reactions which are dependant on the acidity, temperature and cooking time.

5.12 Further Reading

5.12.1 General Literature

Grace, T.M., and Malcolm, E.W. (Eds.) (1989) *Pulp and paper manufacture, volume 5. Alkaline pulping*. The Joint Textbook Committee of the Paper Industry, TAPPI, CPPA.

Rydholm, S. (1965) *Pulping processes*. Interscience Publishers, John Wiley & Sons, Inc., London.

Sjöström, E. (1993) *Wood chemistry*. Fundamentals and applications. Academic Press, Inc., San Diego, CA.

5.12.2 Ph.D. Theses

Berthold, F. (1997) *Chemical aspects on the degradation of lignin during modified kraft pulping processes*. KTH, Stockholm.

Berthold, J. (1996) *Water absorption and uptake in the fibre wall as affected by polar groups and structure*. KTH, Stockholm.

Karlsson, O. (1997) *Studies on the extent of bonding between lignin and polysaccharides in pulps*. KTH, Stockholm.

Li, J. (1999) *Towards an accurate determination of lignin in chemical pulps. The meaning of kappa number as a tool for analysis of oxidizable groups*. KTH, Stockholm.

Lindström, M.E. (1997) *Some factors affecting the amount of residual phase lignin during kraft pulping*. KTH, Stockholm.

Wafa Al-Dajani, W. (2001) *On the bleachability of alkaline pulps. The influence of residual lignin structure*. KTH, Stockholm.

Wännström, S. (1987) *Studies on structural features of residual kraft pulp lignin*. KTH, Stockholm.

5.12.3 References

Aurell, R. (1963) Några jämförande synpunkter på sulfatkokning av tall- och björkved. *Svensk Papperstidn.* 66: 978–989.

Aurell, R. (1964) Kraft pulping of birch. Part 1. The changes in the composition of the wood residue during the cooking process. *Svensk Papperstidn.* 67: 43–49.

Aurell, R. and Hartler, N. (1965) Kraft pulping of pine. Part 1. The changes in the composition of the wood residue during the cooking process. *Svensk Papperstidn.* 68: 59–68.

Buchert, J., Teleman, A., Harjunpää, V., Tenkanen, M., Viikari, L. and Vuorinen, T. (1995) Effect of cooking and bleaching on the structure of xylan in conventional pine kraft pulp. *Tappi J.* 78(11): 125–130.

Hägglund, E. (1951) *Chemistry of wood*. Academic Press, Inc. Publishers, New York, p. 478.

Kaufmann, Z. (1951) Über die chemischen Vorgänge beim Aufschluss von Holz nach dem Sulfitprozess. *Ph.D. Thesis*, ETH, Zurich. (See Rydholm, S. (1965) *Pulping Processes*. Interscience Publishers, John Wiley & Sons, London, p. 467.)

Kubes, G.J., Fleming, B.I., MacLeod, J.M., and Bolker, H.I. (1980) Alkaline pulping with additives. A review. *Wood Sci. Technol.* 14: 207–228.

Lindgren, B.O. (1952) The sulphonatable groups of lignin. *Svensk Papperstidn.* 55: 78–89.

Ljunggren, S. (1980) The significance of aryl ether cleavage in kraft delignification of softwood. *Svensk Papperstidn.* 83: 363–369.

Mortimer, R.D. (1982) The formation of coniferyl alcohol during alkaline delignification with anthraquinone. *J. Wood Chem. Technol.* 2: 383–415.

Parham, R.A. (1974) Distribution of lignin in kraft pulp as determined by electron microscopy. *Wood Science* 6: 305–315.

Saka, S., Thomas, R.J., Gratzl, J.S., and Abson, D. (1982) Topochemistry of delignification in Douglas-fir wood with soda, soda-anthraquinone and kraft pulping as determined by SEM-EDXA. *Wood Sci. Technol.* 16: 139–153.

Sjöblom, K., Mjöberg, J., and Hartler, N. (1983) Extended delignification in kraft cooking through improved selectivity. Part 1. The effects of the inorganic composition of the cooking liquor. *Pap. Puu* 65: 227–240.

Stone, J.E., and Clayton, D.W. (1960) The role of sulphide in the kraft process. *Pulp Pap. Mag. Can.* T307–T313.

6 Pulping Technology

Elisabet Brännvall
Department of Fibre and Polymer Technology, Royal Institute of Technology, KTH

6.1 Introduction

The aim in chemical pulping is to liberate the fibres from the wood matrix by delignifying the wood. This is achieved by treating wood chips at a somewhat elevated temperature in a solution containing the pulping chemicals until a certain degree of delignification is reached. The process is referred to as cooking but it is undesirable for the cooking liquor to boil, and thereby generate vapour. For this reason the cooking is performed in a pressurized system that also is a condition to achieve the temperature needed. After cooking, the pulping chemicals are recovered.

Figure 6.1 shows a simplified plot of the liquor cycle in kraft cooking, also called sulphate cooking. White liquor is the cooking liquor, composed of sodium hydroxide, NaOH, and sodium sulphide, Na_2S. Black liquor is spent cooking liquor containing the dissolved organic substances and the used inorganic cooking chemicals. In order to transform the inorganic substances back to the active cooking chemicals the black liquor is evaporated and burnt, resulting in a smelt. After dissolution in water, the smelt is turned into green liquor. Causticising converts the green liquor into white liquor.

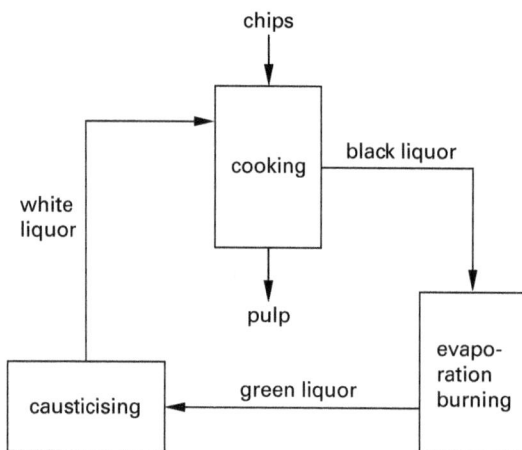

Figure 6.1. A simplified representation of the liquor cycle in kraft cooking.

The degree of delignification is measured by determining the kappa number of the pulp. The kappa number value gives an estimate of the lignin content in pulp. Lower kappa number equals lower amount of lignin in pulp. The total yield of the cook is determined as the amount of pulp produced compared to amount of wood charged. The pulp viscosity can be related to the degree of polymerisation of the carbohydrates and can be used to monitor the degree of carbohydrate degradation. The lower the viscosity value, the more the carbohydrates have been degraded resulting in shorter carbohydrate chain length.

Kraft cooking is the dominant chemical pulping method globally. Other chemical cooking methods exists however. By leaving out the sodium sulphide in the white liquor and only use sodium hydroxide as the cooking chemical the process is called soda cooking. In the sulphite cooking process, sulphurous acid (H_2SO_3) and bisulphite ions ($H_2SO_3^-$) are the active chemicals to degrade and dissolve lignin. Sulphite pulping can be performed at a pH ranging from 1–2 in acid sulphite pulping to 7–9 in neutral sulphite pulping. Organic solvents can be used in delignification, either as the sole degrading chemical, as in suggested organosolv processes, or as reinforcement chemical, as an addition to sulphite, sulphate or soda processes. The solvents used are ethanol, methanol and peracetic acid.

There are two main cooking procedure, batchwise or continuous, using different digesters and equipment, consequently the vessels are either batch or continuous digesters.

6.2 Cooking Equipment

6.2.1 Batch Digesters

Many different kinds of pressure vessels have been used for batch cooking, cylinder shaped or spherical, stationary or rotating. The most common batch digester in use is the stationary vertical cylinder with a conical or spherical bottom. *Figure 6.2* illustrates the conventional batch cooking system with indirect heating. The digester is filled with chips and cooking liquor. The digester is warmed up by a liquor circulation system connected to a heat exchanger. Liquor is displaced through the sieve girdle in the middle of the digester and is circulated to the top and bottom. When desired degree of delignification is reached the digester is discharged to the blow tank.

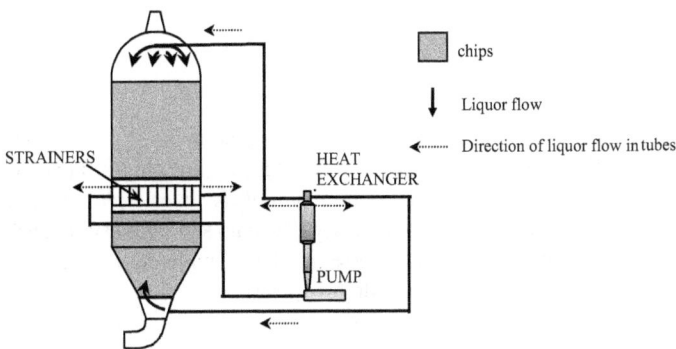

Figure 6.2. Batch digester with indirect heating. The liquor is drawn out through the strainer plates, heated in the heat exchanger, and returned to the digester through the top and bottom inlets.

An ordinary pulp mill may be equipped with a set of four digesters or more, with a size of 150–400 m³ each. Batchwise cooking contribute to some advantages like,

1. Reliable production
2. More flexible production -easy to start / stop allowing stand-by time -possibility to quick changes between, different pulp qualities or softwood / hardwood
3. More efficient turpentine recovery

The conventional batch cooking system is now quite out of date, but has subsequently been under a great development. Modern batch cooking involves a number of operations with a view to obtain high pulp quality and energy efficiency. The idea is to reuse spent cooking liquor, so-called black liquor, and retrieve useful active chemicals and inherent energy to the next batch. The sub-operations are described below.

1. *Chip filling.* The air from within the empty digester is removed by evacuating with a fan. Screened chips are fed to the chip chute connected to the top of the digester *Figure 6.3.* In

order to achieve a better packing degree, usually some form of chip packing is used. Common methods are steam chip spreaders, direct steam injection at the bottom of the digester to shake the chip mass or liquor filling and re-circulation during chip charge. In the figure, steam packing is shown, giving the chip flow a circular motion so the chips fall with an even radial distribution. This improves packing by 20 %, the other two methods mentioned above enables 5–10 % more chips to be packed into the digester compared to natural filling. Normal packing degrees are 0.5 m³ of chips/m³ of volume resulting in 180–200 kg/m³ for softwoods and 220–240 kg/m³ for hardwoods.

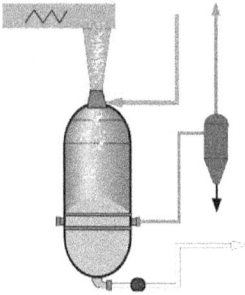

Figure 6.3. Chip filling of batch digester. (Metso Paper©).

2. *Impregnation liquor fill.* The filling of warm black liquor, impregnation liquor, starts when a pre-determined amount of chips has been charged into the digester. The chip filling continues while warm liquor is pumped in through the bottom part of the digester, *Figure 6.4.* When the chip filling is completed, the control system stops the chip screws and chip packing and the digester is capped automatically. The digester is filled completely with liquor. In order to remove any remaining air, the system is slightly pressurized.

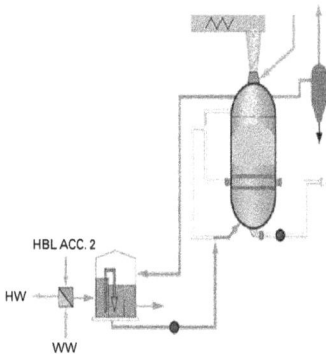

HBL ACC. 2

HW

WW

Figure 6.4. Filling of impregnation liquor. (Metso Paper©).

3. *Hot liquor fill.* Immediately after impregnation, hot black liquor is pumped into the digester through the bottom, followed by filling of either a mix of hot white and black liquor or only white liquor, instantaneously causing a displacement of the warm impregnation liquor through the top of the digester, *Figure 6.5.* This brings the contents of the digester to almost

cooking temperature. The displaced warm impregnation liquor is at that moment taken to the chemical recovery system via evaporation plant

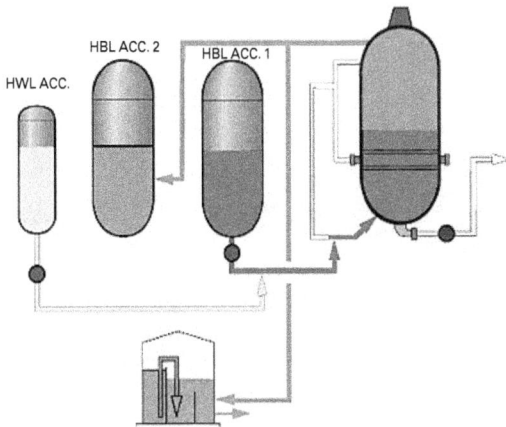

Figure 6.5. Hot liquor fill. HWL = hot white liquor, HBL = hot black liquor. (Metso Paper©).

4. *Heating and cooking.* If the temperature in the digester after the hot liquor fill almost reaches cooking temperature, only a small amount of steam is required to heat up to desired temperature. A compact direct steam heater as shown in *Figure 6.6* can in such case do the heating. An alternative is to have a heat exchanger using indirect steam. In each case, the circulation is achieved by pumping liquor from the middle of the digester through the sieve girdle to the top and bottom of the digester.

5. *Displacement.* When desired degree of delignification is reached, the circulation is stopped. Cool displacement liquor is pumped into the bottom and displaces the hot spent cooking liquor out of the digester top, *Figure 6.7.* The displacement will bring the temperature in the digester to below 100 °C, ensuing that the delignification process is halted, and to prevent the liquid phase to vaporize and damage the liquid-filled fibres.

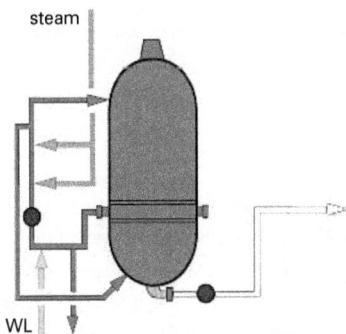

Figure 6.6. Circulation of cooking liquor for heating. After the hot liquor fill, the content of the digester is very near of the required cooking temperature. Only a small amount of steam for heating is therefore required and there is no need for expensive heat exchangers, instead a compact direct steam heater can be used. (Metso Paper©).

126

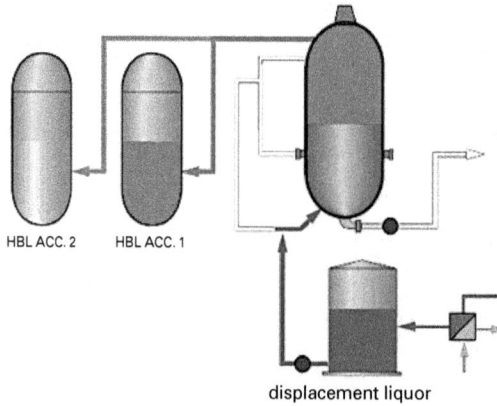

displacement liquor

Figure 6.7. Displacement of cooking liquor by cool liquor. The hottest cooking liquor is taken to accumulator no. 1 and the warm second part, with lower temperature to accumulator no. 2. (Metso Paper©).

6. *Blow.* The digester is emptied by diluting the pulp to suitable consistency with displacement liquor and discharged with a centrifugal pump, *Figure 6.8.*

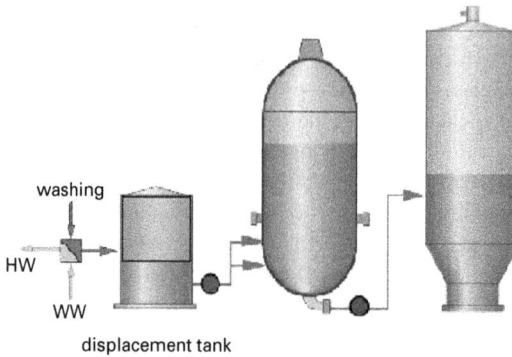

Figure 6.8. Discharge of the pulp, also denoted blow, as the pulp is admitted from the pressure within digester to atmospheric pressure in the flash tank. (Metso Paper©).

6.2.2 Continuous Digesters

The most widespread continuous digesters in use are tall and slim vertical flow digesters with an average production capacity of 1000 to almost 3000 ADMT (air dry metric ton) pulp/day.

The process is continuous, from the chip bin to pulp blow. All the operations in the continuous pulp line are performed, one after the other, in their own specific device and equipment. The continuous digester arrangement comes in single or two vessel configuration, depending on whether the impregnation of chips by cooking chemicals is performed in a separate vessel or in the first part of the digesting vessel. In either case, the chip feeding system is the same, *Figure 6.9.*

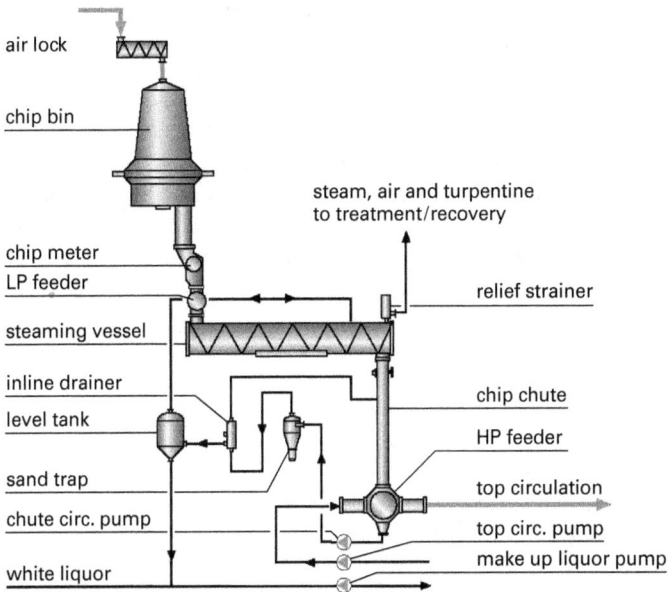

Figure 6.9. An example of a chip feeding system in continuous cooking (Kvaerner Pulping).

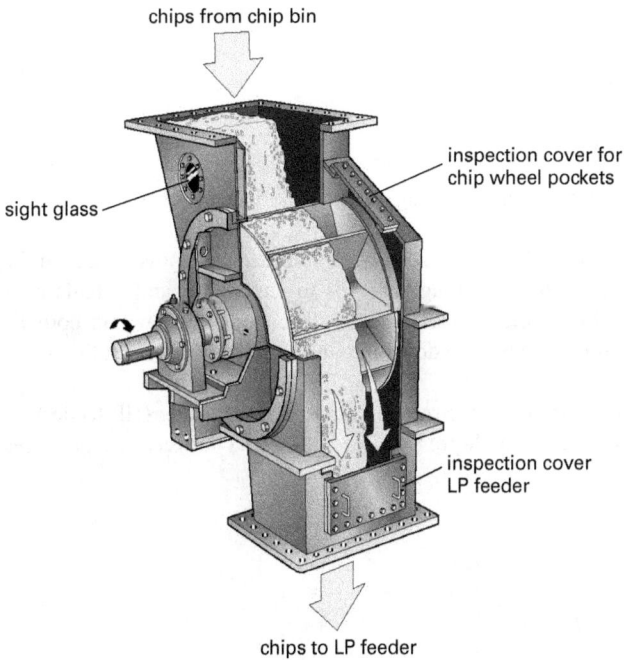

Figure 6.10. Chip meter (Kvaerner Pulping).

The chips are transported on a belt conveyor to the chip bin. Steam, either fresh or flash steam from the flash tank, is lead into the bin in order to pre-steam the chips. After the bin, the chips enter the chip meter, a rotating metering device, furnished with compartments into which the chips fall, *Figure 6.10*.

The chip meter controls the production level in the continuous pulp line. The chip meter, acting as the throttle, regulates all other flows in the pulp line, which therefore correlate with respect to the speed of the meter. As the chip meter rotates, it discharges its content of chips into the low-pressure feeder (LP feeder). This is also a rotating device divided into sections and it acts as a seal between the atmospheric pressure in the chip bin and the pressure further on in the process, *Figure 6.11*.

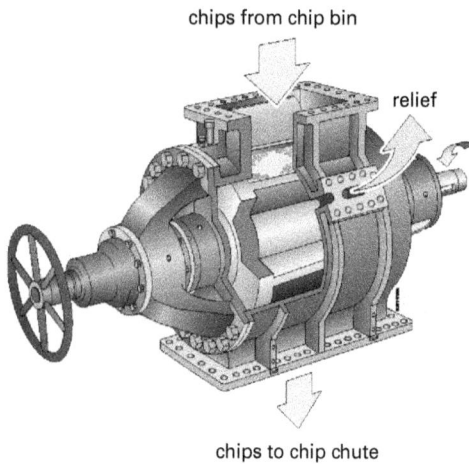

chips from chip bin

relief

chips to chip chute

Figure 6.11. LP feeder (Kvaerner Pulping).

The piece of equipment further on is the steaming vessel, a horizontal tube with a feeding screw, *Figure 6.12*.

Low-pressure steam, approximately 1–1.5 bar, is injected in order to remove all remaining air within the chips. The aim of the two steaming stages is to remove air and volatile compounds from within the chips and raise the chip temperature. The feeding screw transports the chips through the steaming vessel until they drop down into the chip chute and meet the first liquor circulation, *Figure 6.13*.

At the bottom of the chute the chips enter the rotating high-pressure feeder (HP feeder). As in the case of the low-pressure feeder, the high-pressure feeder is a seal between two pressures, *Figure 6.14*.

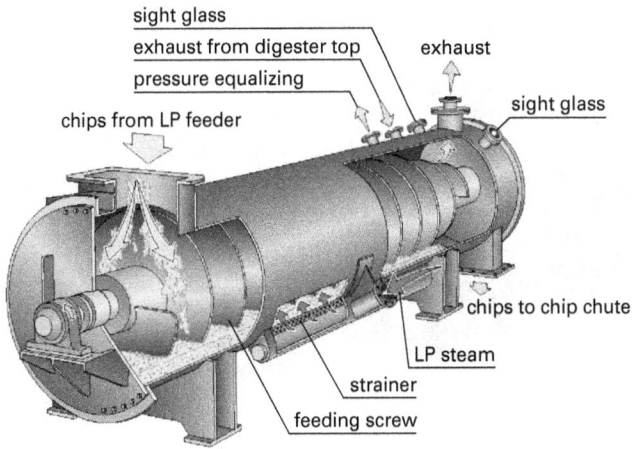

Figure 6.12. Steaming vessel (Kvaerner Pulping).

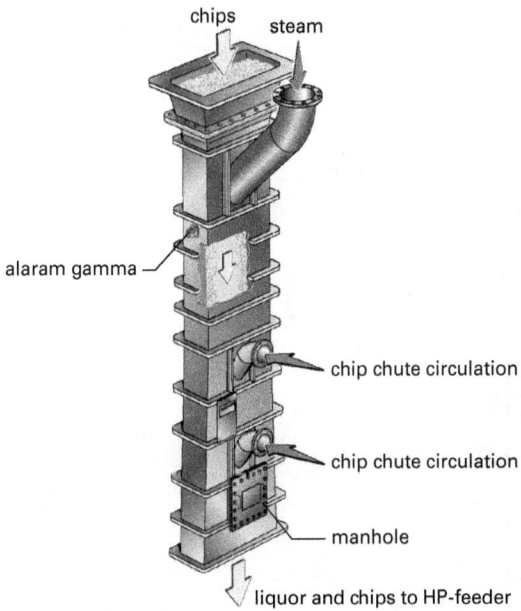

Figure 6.13. Chip chute (Kvaerner Pulping).

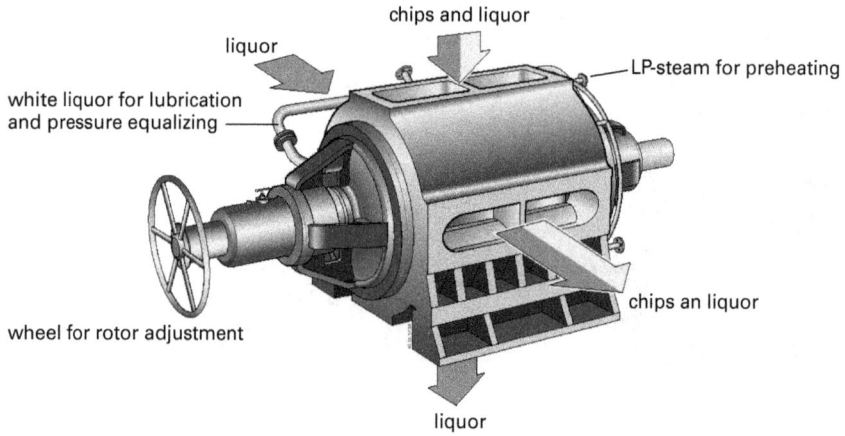

Figure 6.14. HP feeder (Kvaerner Pulping).

The high-pressure feeder is equipped with several through-going channels. As a channel gets into upright position it connects to the chip chute and the channel fills with chips. It then rotates to horizontal position and connects to the circulation pipe. The pressure of the circulation pipe

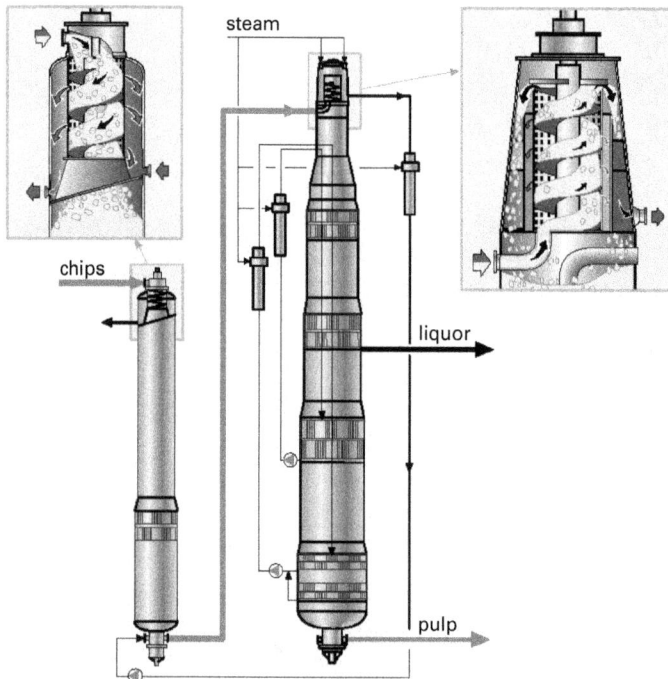

Figure 6.15. Continuous cooking configurations with pre-impregnation and digester vessels. (Kvaerner Pulping).

is that of the digester, approximately 10 bar. Liquor, either from impregnation or digester, enters the channel and transports the chips up to the top of the impregnation or digester vessel.

In *Figure 6.15,* a two-vessel cooking configuration is seen. The chips first enter the impregnation vessel, where liquor penetrates and diffuses into them. The pre-impregnation vessel offers a longer time for penetration and diffusion at a temperature lower than the cooking temperature, whereby chemical reactions are limited. Single-vessel cooking configurations are used for chips easily impregnated in the transfer circulation to the digester.

The top separator can have different appearances. Common designs include a moving screw, as shown in *Figure 6.15*, other more simple top separators are only comprised of a screen. The objective of the top separator is to separate chips and excess transfer liquor. The transfer liquor returns in its loop back to the high-pressure feeder to pick up another batch of chips. The chips already in the digester land on the chip column and start their continual movement down the digester.

The temperature in the cooking zone is 160–170 °C in the case of softwood chips, 150–160 °C for hardwood. The main delignification is obviously achieved in this zone. Strainers are placed around the circumference of the digester at the end of the cooking zone, *Figure 6.16.*

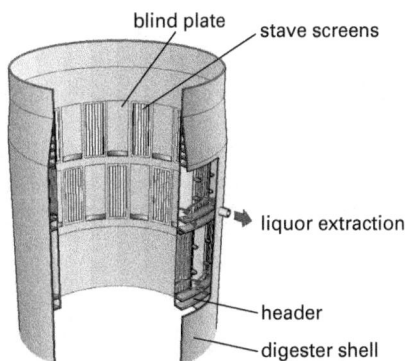

Figure 6.16. The screen plates within a continuous digester. (Kvaerner Pulping).

After the cooking zone, the chips enter the washing zone. Here they meet washing liquor moving counter-currently to the chip column. The washing liquor is introduced at the bottom of the digester to cool the pulp. Then the washing liquor is extracted from the digester through the sieve plates located at the lower part of the digester. It passes through a heat exchanger before re-entering the digester through the central pipe with its outlet just above the bottom strainers. It then moves upward through the vessel in the opposite direction of the chip column, removing dissolved organic substances and spent cooking chemicals. By doing this it turns into black liquor and is extracted through the strainers and transferred to the recovery plant.

At the bottom of the digester, scrapers continually discharge pulp *Figure 6.17.*

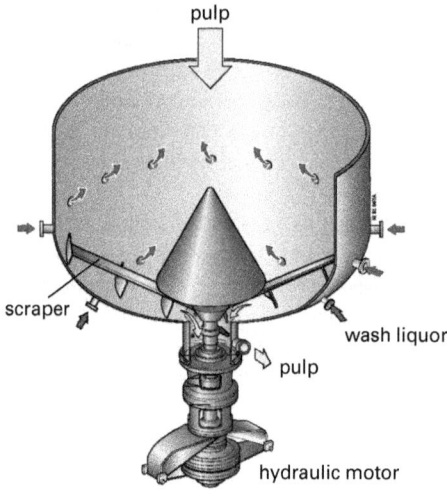

Figure 6.17. The outlet device for digester. (Kvaerner Pulping).

6.2.3 Equipment for High Kappa Number Pulps

A large portion of the pulp production in Sweden consists of unbleached pulp used for different kinds of packages mainly sack paper and kraftliner. Contrary to bleachable grade pulp, where a low kappa number in general is an advantage, linerboard and sack paper manufacturers aspire for high yield and consequently high kappa number, *Figure 6.18*.

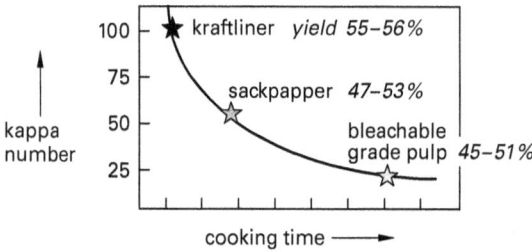

Figure 6.18. Pulp for linerboard and sack paper has higher kappa numbers than pulp for bleaching.

One consequence of terminating the cook at a high kappa number is that the mechanical action in the blow, when the pulp is discharged from the digester, is not sufficient to separate the fibres from one another. The cook has not reached the defibration point so the amount of lignin remaining between the fibres is sufficient to keep them "glued" together. The approximate kappa number at which the defibration point is reached is shown in *Figure 6.19*. The defibration point however, depends greatly on the cooking process.

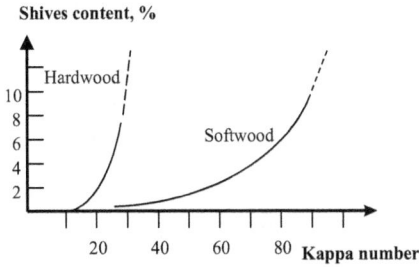

Figure 6.19. The defibration point is defined as the kappa number at which the shives content approaches zero. For hardwood, the defibration point is approximately at kappa number 20 and for softwood around 45.

In order to separate the fibres from each other more extensive mechanical action has to be supplied than what is achieved in the blowing. For this purpose, defibrators are placed after the digesters. In continuous cooking, in-line defibrators are placed directly after the digester. In batch cooking, a hot-stock refiner, also called blow line refiner, is used on the cooked chips after the blow tank, *Figure 6.20*.

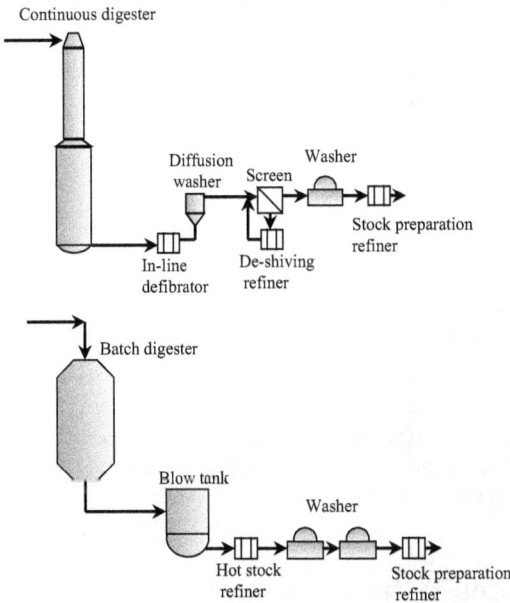

Figure 6.20. Schematic illustration of linerboard pulping lines. In the production of pulp with high kappa number, defibrators or refiners are used to turn the cooked chips into pulp.

The in-line defibrator, receiving the pulp directly from the continuous cooker, works under high pressure. The hot-stock refiner used in the batch cooking line works under atmospheric conditions. The refining in these defibrators differs from that which pulp is subjected to in refiners preceding the paper machine. The refining in the paper mill is to develop the pulp properties, whereas the main goal when refining after cooking is to separate fibres still bonded in the wood matrix and receive pulp from the cooked chips. If the defibration action in the refiners is carried beyond the energy needed merely to separate the fibres, parts of the fibre wall will be torn off. This creation of fines can cause problems in washing of the pulp as they make the pulp more difficult to dewater. For this reason, the defibration can be made in two steps. The defibration after the cook is terminated before there is any risk of fines formation followed by washing of the pulp. After washing, a post-refining step can be applied in order to complete the defibration. A de-shiving refiner can be used to reduce the shives size and content.

6.2.4 Kraftliner Pulps

The pulps of highest yield and highest kappa number are used to produce paperboard facings (kraftliner or linerboard) of corrugated board. The kraftliner is composed of two layers, top and base layer, having different properties. The thin top layer is to provide a smoother surface suitable for print, whereas the base layer is for strength and stiffness. Kraftliner can be produced by having two production lines, one for the top layer, the other for the base layer. The base layer pulp is targeted to a kappa number of 90–110, whereas the top layer pulp can be delignified further down to kappa number around 75. Another arrangement is to produce pulp with kappa number 90 or above and then treat pulp aimed at top and base layers differently in the refiners. The top layer pulp receives more refining energy, thus resulting in fibres that are more flexible and a higher amount of fines, both effects contributing to better printability.

6.2.5 Sack Paper Pulps

The kappa number for sack paper grade pulp is somewhere between 30–60, consequently also with lower yield compared to kraftliner. There may be a defibrator right after the digester although at lower kappa numbers it is not necessary, as the defibration point has been reached in the cook and the pulp defibrates in the blow. After the washing stage, however, the pulp is subjected to high consistency refining. Refining at a consistency of 30–40 % causes the fibres to collapse, to be bent and twisted and compressed axially. Hardly any fines are produced and any bundles of fibres are defibrated and shives reduced. The high consistency refining increases the tear strength and toughness of the paper. In connection to the paper machine, the pulp is refined conventionally in order to develop the tensile strength.

6.3 Nomenclature and Definitions in Kraft Pulping

6.3.1 Active Cooking Chemicals

The cooking liquor in kraft pulping, the so-called white liquor, contains the compounds sodium hydroxide, NaOH, and sodium sulphide, Na_2S. The active cooking species are hydroxide and hydrogen sulphide ions, OH^- respectively HS^-. In solution, Na_2S is dissolved into hydroxide and hydrogen sulphide ions according to Equation (6.1) with the equilibrium constant K_{b1}, Equation (6.2).

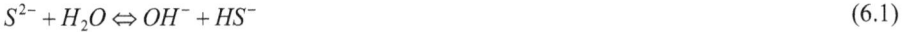

$$S^{2-} + H_2O \Leftrightarrow OH^- + HS^- \tag{6.1}$$

$$K_{b1} = \frac{\left[OH^-\right]\left[HS^-\right]}{\left[S^{2-}\right]} \tag{6.2}$$

The hydrogen sulphide can be hydrolysed further according to Reactions (3–4):

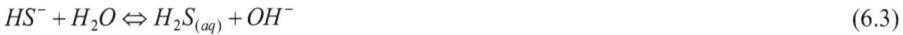

$$HS^- + H_2O \Leftrightarrow H_2S_{(aq)} + OH^- \tag{6.3}$$

$$K_{b2} = \frac{\left[OH^-\right]\left[H_2S\right]}{\left[HS^-\right]} \tag{6.4}$$

Theoretically, the sulphide could exist as S^{2-}, HS^-, and dissolved H_2S in aqueous solutions. However, under the conditions prevailing in alkaline cooking, reaction Equation (6.1) is displaced to the right and thereby sulphide is present as hydrogen sulphide ions. The relative amounts of either compound depend on the concentration of hydroxide ions and the numerical value of the equilibrium constants, as the three sulphide compounds are in equilibrium with each other. The equilibrium constants depend on the temperature and the ionic strength.

Apart from the active chemicals, white liquor also contains other compounds in smaller amounts. In *Table 6.1*, the approximate concentrations of some compounds found in industrial white liquor are given. The positive counter-ions for the compounds below are normally sodium (Na^+), calcium (Ca^{2+}) or potassium (K^+). These counter ions make up the ionic strength of the liquor.

Table 6.1. Composition of white liquor from recovery plant, measured at room temperature. It is diluted when charged to digester. The values given below for OH⁻ and HS⁻ are not the normal concentrations charged.

Compounds in white liquor	[mole/l]
Hydroxide ions OH⁻	2.5
Hydrogen sulphide ions HS⁻	0.5
Sulphate ions SO_4^{2-}	0.04
Carbonate ions CO_3^{2-}	0.3
Sulphite ions SO_3^{2-}	0.03
Tiosulphate ions $S_2O_0^{2-}$	0.007

The carbonate in white liquor originates mainly from incomplete conversion of carbonate to hydroxide in the causticising. The sulphate stems from incomplete reduction of sulphide in the recovery furnace and thio-sulphate is obtained from sulphide oxidised by air. Other compounds can also be present in white liquor such as potassium salts and sodium chloride. These come with the wood or are contaminants in the make-up chemicals and process water. Minor amounts of polysulphide may exist in white liquor as a result of partial oxidation of the smelt from the recovery furnace or of the green liquor. With the exception of the active cooking species, all other compounds are referred to as ballast. The ballast chemicals affect the pulping as an increased ionic strength of the cooking liquor, which is negative for the delignification selectivity. Additionally, high amounts of ballast can cause problems with incrustation in the digester or the recovery plant.

6.3.2 Definitions of Terms Concerning Chemical Charges

The white liquor is made up of sodium hydroxide, NaOH, and sodium sulphide, Na_2S and the active cooking chemicals are hydroxide and hydrogen sulphide ions. The white liquors concentration of hydroxide ions can, however not be expressed as the amount of NaOH charged, since Na_2S is hydrolysed to OH^- and HS^-. The hydroxide ions derive from both NaOH and Na_2S and therefore the term effective alkali is used to define the total amount of hydroxide Equation (6.5).

$$EA(mol) = n_{NaOH} + n_{Na_2S}$$ (6.5)

$$EA(mole/l) = [OH]$$ (6.6)

The sulphide charge can be given as the sulphidity of the liquor. The amounts of Na_2S and NaOH are related to each other and determined by the recovery system. Changes in alkali charge inevitably changes the sulphide charge and vice versa. This has lead to the use of the term sulphidity, which relates the alkali charge to the sulphide charge. The definition of sulphidity is, when expressing the amounts in mole, shown by Equation (6.7).

$$Sulphidity(\%) = 100 \times \frac{2n_{Na_2S}}{2n_{Na_2S} + n_{NaOH}}$$ (6.7)

When using the ions involved, the expression becomes as shown by Equation (6.8).

$$Sulphidity(\%) = 100 \times \frac{2[HS^-]}{[HS^-] + [OH^-]}$$ (6.8)

A solution based on only sodium sulphide thus has a sulphidity of 100 %, Equation (6.9)

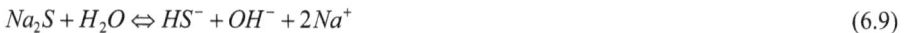

$$Na_2S + H_2O \Leftrightarrow HS^- + OH^- + 2Na^+$$ (6.9)

Effective alkali can also be given in percentage, meaning the weight percentage on wood. The expression is then according to Equation (6.10) with the amounts in weight units.

$$EA(\%) = 100 \times \frac{m_{NaOH} + \frac{1}{2}m_{Na_2S^*}}{m_{wood}}$$ (6.10)

Na_2S^* stands for "the amount of Na_2S in equivalent amounts of NaOH", Equation (6.11).

$$m_{Na_2S^*} = \frac{M_{Na_2S}}{2 \times M_{NaOH}} \times m_{Na_2S} = \frac{78}{2 \times 40} \times m_{Na_2S}$$ (6.11)

To relate Na_2S to equivalent amounts of NaOH is the standard in Scandinavia, whereas in North America Na_2S is given in equivalent amounts of Na_2O.

6.3.3 Liquor-to-Wood Ratio

The liquor-to-wood ratio gives the total amount of liquor, including chip moisture, to weight unit dry wood. The correct way to express it would be ton liquor/ton dry wood, but since the density of liquor is about 1 ton/m^3, it often is given as l/kg. Normal liquor-to-wood ratios range between 3–5 l/kg. Higher liquor-to-wood ratio improves the homogeneity of the cooking process. A larger liquid phase makes it easier for the ions to diffuse and work against concentration gradients.

6.3.4 H-factor – Time and Temperature

The delignification process is strongly temperature dependent. At low temperatures, below 140 °C, the delignification reactions are slow but increase fast in rate as the temperature rises. An increase in temperature with 10 °C results in a more than twofold increase in delignification rate. For that reason, it is hard to judge how far the delignification has proceeded by the cooking time if the temperature has fluctuated. The time and temperature have therefore been combined in a single expression, the H-factor. This has been accomplished with the use of the Arrhenius equation (6.12).

$$k = Ae^{-\frac{E}{RT}} dt \tag{6.12}$$

The rate constant, k, is integrated within the time interval in question and thus the expression for H-factor is formed Equation (6.13).

$$H = \int_{t_0}^{t} Ae^{-\frac{E}{RT}} dt \tag{6.13}$$

It assumed that the activation energy, E (=134 kJ/mole), is the same for the whole delignification process, which is not true but simplifies the calculation. The delignification rate has been determined at different temperatures and the relative delignification rate has been put to 1 at 100 °C. This gives the H-factor integral the following form Equation (6.14).

$$H = \int_{t_0}^{t} e^{(43.2 - \frac{16113}{T})} dt \tag{6.14}$$

The relative reaction rates at different temperatures are given in *Table 6.2*.

Table 6.2. The value of the relative delignification rate, k_{rel}, at some temperatures.

Temperature, °C	k_{rel}/h
100	1
110	3
130	25
150	165
170	927
180	2042

Figure 6.21 shows the temperature profile of a cook with heating from 80 °C to the cooking temperature 170 °C followed by 2 hours cooking time. The graph below shows the relative reaction rate in the cook. The H-factor is the area under the graph of reaction rate.

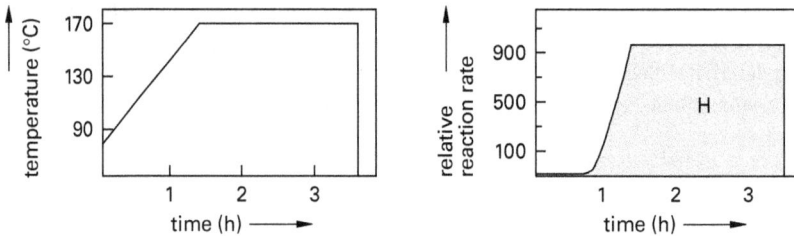

Figure 6.21. Left: temperature profile during a cook with heating up period and cooking at 170 °C. Right: The relative reaction rate during the cook. The H-factor is the area under the relative reaction rate curve.

The heating up period to reach cooking temperature usually contributes with 150–200 H-factor units. The total H-factor requirement depends on the pulp to be produced. The automatic control system for digester registers the temperature and calculates the H-factor. When keeping the chemical charges constant, the H-factor of a cook can predict the kappa number of the pulp, regardless of how temperature and time has been combined to reach the H-factor.

6.3.5 Yield and Rejects

In the delignification process, organic substances in the wood are degraded and dissolved. The yield, in general terms, is defined as the ratio of oven dry pulp to oven dry wood. However, high kappa numbers, inhomogeneous cooking conditions, or occurrence of knots and oversize chips result in knots and shives in the pulp. In chemical pulps, the shives are defined as fibre bundles at least 1 mm long and not able to pass through a 0.15 mm slot. Two yield definitions are used, total yield and screened yield expressed in percentage, Equations (15) and (16). The oven dry weight of pulp, rejects and wood are used.

$$\text{Total yield (\%)} = 100 \times \frac{pulp + rejects}{wood} \qquad (6.15)$$

$$\text{Screened yield (\%)} = 100 \times \frac{pulp}{wood} \qquad (6.16)$$

In *Figure 6.22* a schematic illustration of the yield for kraft cooking is shown. The total yield decreases as the delignification proceeds and more organic material is dissolved. The screened yield goes through a maximum and the position of the maximum is the defibration point. Continuing the cook to lower kappa numbers, the content of shives in the pulp decreases steadily.

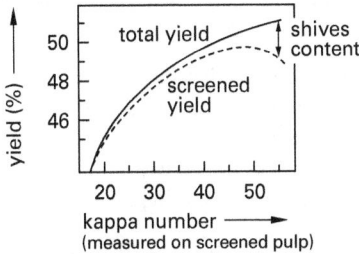

Figure 6.22. A typical relationship between yield and degree of delignification for kraft softwood pulp.

6.4 Impregnation of Wood Chips

6.4.1 Importance of Good Impregnation

In average, a softwood chip consists of half a million fibres, longitudinally connected by the lumen openings and transversely by pores in the fibre cell wall. In order to receive a homogeneously delignified pulp, the ideal would be that all the fibres would get the same amount of chemicals during the same period of time at the same temperature. If chemicals are unevenly distributed in the chip or between the chips, it leads to overcooked fibres and/or bundles of uncooked fibres, so-called shives. The fibres readily accessible to chemicals at the chip surfaces risk being overcooked and the cores of chips may be totally deprived of chemicals and end up as shives. The impregnation step is essential to attempt to reach the ideal conditions stated above. *Figure 6.23,* illustrates an inhomogeneous cook (scenario A) and two more homogeneous cooks (C and D). The narrower the peak, the more homogeneously delignified fibres in the pulp. In scenario A, the impregnation of cooking chemicals is incomplete. This results in fibres with a broad distribution of lignin content. In scenarios C and D, impregnation is improved by presteaming of the chips, which eliminates the high-kappa number tail of scenario A.

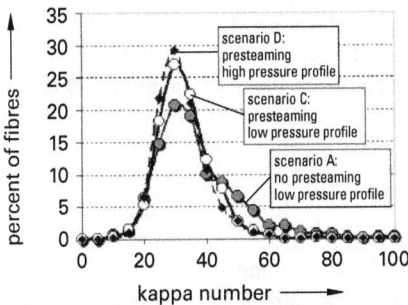

Figure 6.23. Distribution of kappa number of fibres. The better the impregnation, the more homogeneous the cook and the narrower the kappa number distribution. (Malkov, Tikka 2002).

6.4.2 Steaming

Air entrained within the lumen prevents liquor from penetrating into the chip. Furthermore, if liquor would be added, the air would end up captured inside the lumen. It is therefore necessary to remove the air from within the chips, usually by steaming the chips. The air is removed by steaming through several mechanisms.

1. Thermal expansion. The steam heats the air so it expands and is displaced.
2. Increase in vapour pressure. The wood moisture is heated leading to increased vapour pressure within the chips. The increased vapour pressure forces the air out.
3. Partial pressure gradient. Outside the chips, the environment is saturated with vapour, thus creating a partial pressure gradient causing air to diffuse from within the fibres.

Steam entering the lumen condenses and thereby the lumen will be filled with water. The distribution of cooking chemicals within the chips is achieved by liquor penetration and diffusion between liquors of different concentration.

6.4.3 Liquor Penetration

Liquor transport longitudinally occurs through lumen openings. Transport of liquid radially and tangentially, from one fibre to the neighbouring fibres, occurs at different manners depending on type of fibre. Softwood tracheids have bordered pits in the cell wall, providing openings for liquor transport. Hardwood libriform cells are also equipped with pits, whereas the vessels have several different types of pores. The liquor penetration is nevertheless much slower in the tangential and radial direction compared to longitudinal direction. Liquor transport longitudinally is some 50 to 200 times faster than in the latitudinal directions. The lumen openings are much bigger than the pores in the fibre cell wall. Additionally, the transport radially and tangentially can be obstructed. For example, in heartwood and dried wood, the pits of the tracheids may be closed as the torus of the pit membrane has been pushed to one side and clogs the pit opening.

Steaming, and thereby removing the air within the fibres, is essential for good penetration. Higher pressure, higher temperature and longer steaming time improve the air removal and thereby penetration.

Sapwood is much easily penetrated than heartwood. The heartwood has usually lower moisture content compared to the sapwood, and consequently more air, causing the lower degree of penetration.

Water and white liquor penetrate more easily into wood than black liquor. In black liquor, large organic molecules are present which can plug the pores in fibres and thereby making the piths less permeable.

6.4.4 Diffusion of Cooking Chemicals

As the chips are impregnated with cooking liquors, the liquor soaking the chips can be divided into two categories -free liquid and liquid incorporated in fibre lumen, pores and cell wall. The transport mechanism of chemicals to and reaction products from the confined liquid is through diffusion. According to Ficks law, the transportation rate is determined by the diffusion coefficients for the participating ions, the surface available for diffusion and the concentration gradients.

As in the case of penetration, and for the same reason, the rate of diffusion is faster in the longitudinal direction compared to the latitudinal directions. The surface available for diffusion is simply much larger in the longitudinal direction. However, in the course of delignification, material is removed from the cell wall making it more porous. The difference in surface available for diffusion in different directions will diminish with the degree of delignification. The pH of the liquor also affects the surface available for diffusion. The higher the pH, the higher the degree of dissociation of the functional groups in wood components, mainly hydroxide groups. This leads to higher repulsive forces between fibre wall layers and thereby a more porous cell wall. At the pH levels of kraft cooks, the surface available for diffusion is thus quite similar in the different directions.

6.4.5 Importance of Chip Thickness

When the hydroxide ions enter the chip, they are neutralised in reactions with the wood components. Should this result in a pH decrease below 12,5 it would lead to a decrease in delignification rate. The surface available for diffusion decreases in the latitudinal directions below this pH value and the transport rate of fresh alkali to the reaction site is too slow to keep up the delignification rate. At the end of the cook, the delignification of the centre of the chips may still not have reached the defibration point. The insufficiently delignified fibres are not separated from each other and end up as shives in the pulp. This is likely to happen if the chip thickness is too big, *Figure 6.24*.

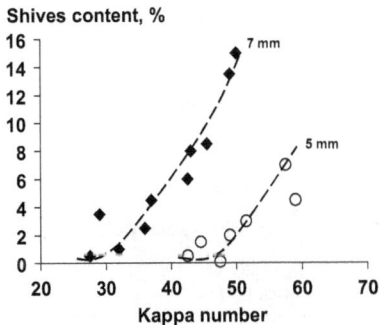

Figure 6.24. Thicker chips give pulp with higher amount of shives. (Hartler and Onisko, 1952).

In general, industrially produced chips are thinner than 5 mm. In the chipping procedure, cracks are introduced in the chips, decreasing the actual chip thickness and improving impregnation.

High liquor-to-wood ratio improves the homogeneity of the cook since the decrease in the concentration of hydroxide ions inside the chips is not as pronounced.

6.5 Pulping Reactions in Kraft Cooking

The main constituents of wood are the polymers cellulose, hemicellulose and lignin. The primary aim of chemical pulping is to remove the lignin and liberate the fibres. However, no cooking chemicals are entirely selective towards lignin. In the course of cooking, also the carbohydrates are degraded and dissolved.

6.5.1 Carbohydrate Reactions

The carbohydrates are attacked already at low temperatures. At the very beginning of the cook, the acetyl groups, found in galactoglucomannan of softwood and xylan of hardwood, are hydrolysed and split off.

A temperature of 100 °C is enough for the peeling reaction to take speed. It attacks the reducing end group in the carbohydrate chain and chews off the sugar unit. The sugar unit now at the end of the chain is transformed into a reducing end group and becomes susceptible to the peeling reaction. The peeling reaction keeps peeling off unit by unit of the carbohydrate chain *Figure 6.25*, resulting in decreased carbohydrate yield. The dissolution of the reducing end groups continues until the end unit manages to convert itself to a more alkali stable structure, in the so-called stopping reaction.

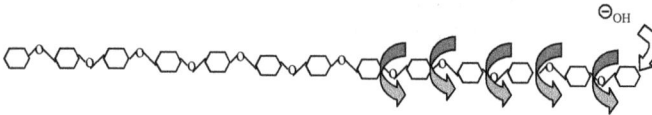

Figure 6.25. The peeling reaction decreases the yield of the carbohydrates by splitting off and dissolving the end units of the carbohydrate chains.

Usually in a cook to produce a pulp meant for bleaching, some 20 % of the carbohydrates are lost, mainly the hemicelluloses, *Table 6.3*.

Table 6.3. Percentage of wood components in wood and pulp (bleachable grade).

	Pine kraft pulp		Birch kraft pulpat	
	originally in wood [%]	yield on wood, [%]	originally in wood, [%]	yield on wood, [%]
Cellulose	41	38	40	34
Glucomannan	17	4	3	1
Xylan	8	5	30	16
Other carbohydrates	5	–	4	–
Tot amount carbohydrates	71	47	74	51
Lignin	27	3	20	2
Extractives	2	0.5	3	0.5
Σ [%]	100	50	100	53

The alkaline hydrolysis starts to have an impact at temperatures above 130 °C, the speed of the reaction increasing with increasing temperature. The alkaline hydrolysis attacks a carbohydrate chain at a randomly chosen bond between two sugar units and breaks it, *Figure 6.26.* The result is two shorter carbohydrate chains. Additionally, the reaction creates a new reducing end group in the carbohydrate chain, which in turn can be the starting point of a peeling reaction. Peeling taking place at sites created by alkaline hydrolysis are called secondary peeling.

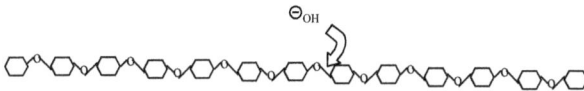

Figure 6.26. A schematic description of alkaline hydrolysis. It attacks a random bond in the carbohydrate chain creating two shorter chains.

The proceedings of alkaline hydrolysis are seen as a reduction in pulp viscosity, as the viscosity value is a measure of the average carbohydrate chain length. The selectivity of a pulping process can be monitored by evaluating the viscosity decrease as a function of the kappa number, *Figure 6.27.*

Since the cellulose chains are so much longer than the hemicellulose chains, the viscosity value gives the degree of polymerisation of the cellulose. However, the mechanism behind carbohydrate degradation is similar for cellulose and hemicellulose and the decrease in viscosity thereby reflects the degree of carbohydrate degradation.

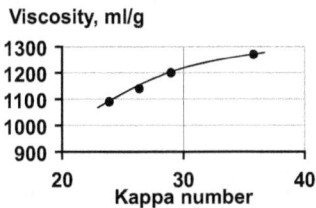

Figure 6.27. In the delignification the process, not only lignin is degraded and dissolved. As can be seen by the viscosity decrease, also the carbohydrates are attacked.

6.5.2 Delignification

The desired reaction is of course delignification of the wood, by degradation and dissolution of the lignin in the raw material. One way to describe the delignification process is by dividing it into three phases with different delignification rates in each phase. By plotting the decrease in lignin to reaction time, the three phases appear, *Figure 6.28.*

1. In the initial phase, the dissolution of lignin is very fast and takes place at temperatures below 140 °C. The dissolved lignin in this phase is also called "extractable lignin". The extractable lignin amounts to some 20 % of softwood lignin and 30 % of hardwood lignin.
2. The bulk delignification is the most selective delignification phase. The major part of the lignin is degraded and dissolved in this phase whereas the carbohydrate dissolution is relatively small.

3. The delignification in the residual phase is very slow. *Residual phase lignin* is defined as the lignin reacting according to the mechanisms of the residual phase. It is desirable that the amount of residual phase lignin is as low as possible, as the selectivity towards degradation and dissolution of this lignin is very poor.

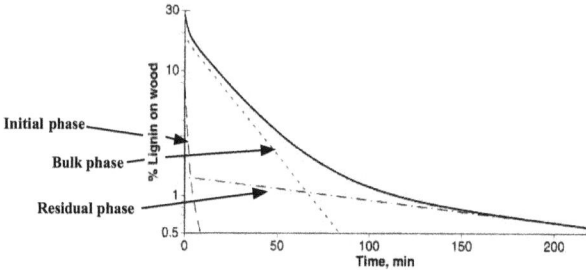

Figure 6.28. The delignification phases. (Lindgren and Lindström 1996).

Residual lignin is the lignin remaining in pulp at the end of the cook. Care should be taken not to confuse these two definitions. For bleachable grades, the delignification has usually been extended until most bulk phase lignin has been dissolved. In reality, this means that the residual lignin in the pulp also can be regarded as residual phase lignin. In comparison, the residual lignin in sack paper grades and liner grades consists of substantial amounts of bulk phase lignin as well as residual phase lignin.

6.6 Influence of Cooking Parameters

The main variables in a kraft cook are concentration of hydrogen sulphide and hydroxide ions, and the temperature. Hydroxide ions are consumed during pulping. They neutralise acidic groups in the carbohydrates and keep degraded lignin in solution.

The hydrogen sulphide ions provide the major part of lignin degradation. However, there is only a slight decrease in the hydrogen sulphide ion concentration between the initial white liquor charge and the black liquor recovered at the end of the cook.

The temperature is usually increased gradually, starting with pre-heating the chips by steaming and then progressively increased temperature until cooking temperature is reached.

The ion strength in the cooking liquor is determined by the sodium ions, present in white liquor as co-ions to hydroxide and hydrogen sulphide.

Dissolved organic matter, mainly degraded and dissolved lignin, naturally increases in the cooking liquor as the delignification proceeds.

6.6.1 Hydroxide Ion Concentration

The delignification rate is strongly affected by the alkali concentration in the cooking liquor. Increased alkali charge increases the delignification rate, *Figure 6.29*.

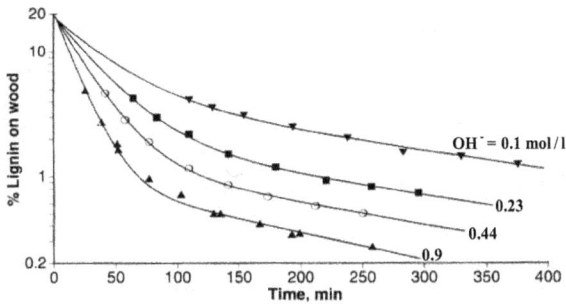

Figure 6.29. The bulk and residual delignification in cooks conducted at different concentration of hydroxide ions. $[HS^-] = 0.28$ mol/l, temp $= 170$ °C. (Lindgren and Lindström 1996).

As can be seen, it is the rate of the bulk delignification that is boosted by higher alkali concentration. The rate of the residual phase delignification is not affected although the amount of residual phase lignin decreases. Extrapolating the residual phase curves concludes this. The amount of residual phase lignin can be deducted from the point where the extrapolation reaches the x-axis. A higher alkali concentration in the cook decreases the amount of residual phase lignin.

The rate of dissolution of carbohydrates increases with increasing alkali concentration. The loss of carbohydrates is made up only to a minor degree of cellulose, since the hemicelluloses are more susceptible to alkaline degradation.

Some 70 % of the glucomannan initially present in wood is dissolved and lost during pulping. In *Figure 6.30,* the dissolution of glucomannan during kraft cooking is depicted. At temperatures below 100 °C soluble glucomannan is lost. Between 100 °C and 130 °C, the losses of glucomannan due to peeling are considerable. At 130 °C, the dissolution seems to come to a halt and the remaining glucomannan is fairly resistant to alkali. An increased alkali charge increases the rate of dissolution but the total loss is not dependent on the alkali concentration.

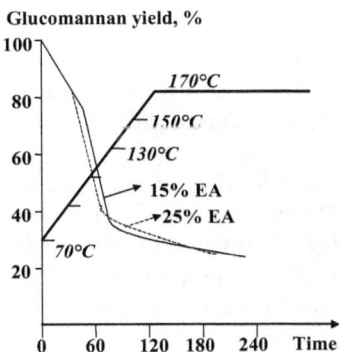

Figure 6.30. The dissolution of glucomannan in a kraft cook of pine at two alkali charges (15 and 25 % EA). (Data from Aurell and Hartler, 1965).

The dissolution of xylan resembles the lignin dissolution, *Figure 6.31.* At temperatures below 130° C, the losses are small and are mostly made up of soluble xylan. The losses at higher temperatures are considerable and due mostly to degradation by the peeling reaction. Some xylan, however, is dissolved without being degraded. This dissolved xylan may later in the cook, when the alkali concentration has decreased, re-adsorb on the fibres. Higher alkali charge gives a significant increase in xylan dissolution. The higher loss of xylan at higher alkali charge is partly attributed to a more extensive dissolution of xylan and partly to less xylan being re-adsorbed.

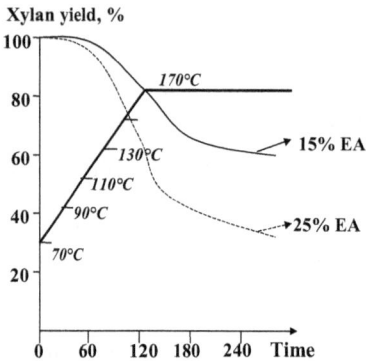

Figure 6.31. Dissolution of xylan in a kraft cook of pine at two alkali charges. (Data from Aurell and Hartler, 1965).

Although only minor amounts of cellulose are lost in a kraft cook, the alkaline hydrolyses does not leave cellulose unaffected. The action of alkaline hydrolyses is to cleave random bonds in the carbohydrate chains and the result is two shorter chains. This can be seen as a decrease in pulp viscosity. An increased alkali charge thus gives pulp of lower viscosity.

6.6.2 Hydrogen Sulphide Ion Concentration

The hydrogen sulphide ion is the main delignifying agent in a kraft cook. Compared to a soda cook, where only hydroxide ions are present, the delignification is much faster in kraft cooking, *Figure 6.32.*

Increased hydrogen sulphide ion concentration increases the delignification rate in the bulk phase, whereas the rate in the residual phase is unaffected. The amount of slowly and unselectively reacting residual phase lignin decreases with increasing concentration of hydrogen sulphide ions. Although the hydrogen sulphide ions do not react with carbohydrates, their presence has a favourable effect on carbohydrate preservation. The increased delignification rate enables the cook to be terminated earlier, leaving less time for attacks on carbohydrates. The viscosity at a given kappa number is thus higher at higher hydrogen sulphide charge.

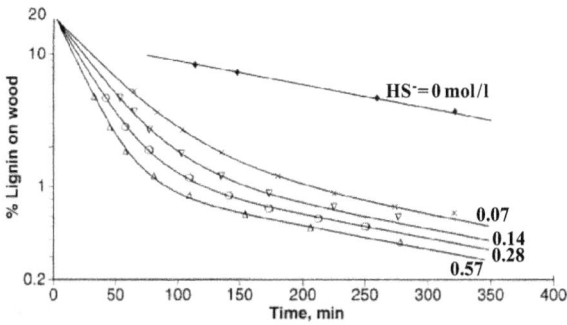

Figure 6.32. The effect of hydrogen sulphide ion concentration on the delignification rate. $[OH^-] = 0.44$ mol/l, temp $= 170°C$. (Lindgren and Lindström 1996).

6.6.3 Temperature

The delignification rate is dependent on the temperature. An increase of the cooking temperature by 10 °C gives a two-fold increase of the delignification rate. In *Figure 6.33*, the cooking temperature has been varied between 150 and 180 °C. Both bulk phase and residual phase delignification rates are affected by the temperature. However, the extrapolation of the residual phase to x-axis reveals that temperature has no influence on the amount of residual phase lignin.

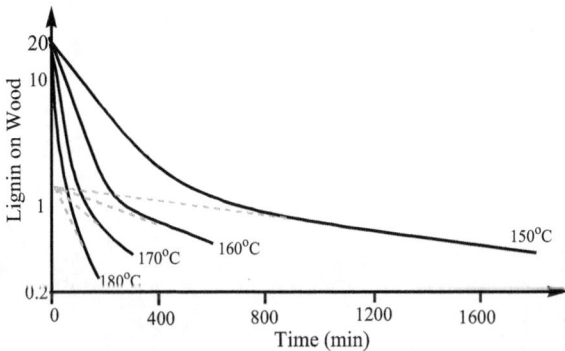

Figure 6.33. Delignification rate affected by varied cooking temperature. $[OH^-]=0.44$ mol/l, $[HS^-] = 0.44$ mol/l (Lindgren and Lindström 1996).

The selectivity of the cook is favoured by a decrease in temperature, due to the difference in activating energy for the bulk delignification, (127 kJ/mol) and cellulose degradation, (180 kJ/mol). A lower temperature gives higher viscosity at a given kappa number.

7 Kinetics of Chemical Pulping and Adaptation to Modified Processes

Ants Teder
Department of Fibre and Polymer Technology, Royal Institute of Technology, KTH

7.1 Introduction

Kinetics is the science that describes the rates of processes and the effects of different variables on the rates. In our case, the chemical pulping of wood, there are in principle three different types of reactions in chemical pulping:

- Delignification (removal of lignin from the wood residue)
- Degradation and dissolution of carbohydrates
- Cellulose depolymerization (hydrolytic cutting of the carbohydrate chains)

Delignification, the degradation and dissolution of lignin, is the desired reaction. Losses of carbohydrate yield and cellulose depolymerization are unwanted but unavoidable consequence of alkaline delignification. Yield is of interest because this is a measure of the production of useful product from a given amount of raw material. The strength properties of the pulp may be compromised if the degree of polymerisation of the cellulose, i.e. the chain length, becomes too low. The degree of polymerisation of the cellulose is often measured as pulp viscosity, *Figure*

7.1. Observe that it is the viscosity (chain length) in the final product, e.g. fully bleached pulp, which is essential. As there is no known process to repair broken chains, a loss in pulp viscosity early in the process chain (e.g. in the pulping process) that has no direct effect on pulp strength can have a negative effect when the viscosity losses in the subsequent bleaching stages are added.

Dry zero span, kNm/kg

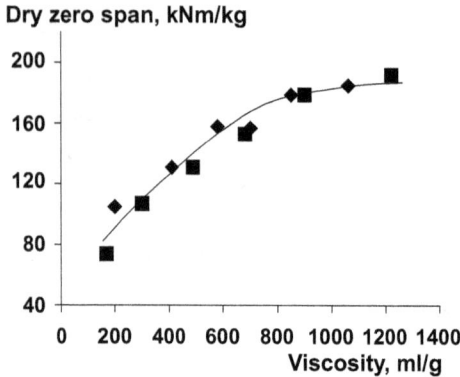

Figure 7.1. The relationship between pulp strength, measured as dry zero-span tensile strength, and pulp viscosity.

Kinetics of pulping processes can be used for process control as well as for process optimisation of an existing cooking system and for identifying new process alternatives. In addition, examination of the parameters that influence the reactions can provide knowledge about which reactions are rate determining. As the first process (delignification) is wanted in pulping and the other two (loss of yield and loss of pulp viscosity) are unwanted, it is common when optimising a pulping process to relate the delignification rate either with the rate of carbohydrate degradation or the rate of viscosity loss. The pulp yield or pulp viscosity at a certain degree of delignification gives a measure of the process selectivity. Observe that it must be always clearly stated which type of selectivity (yield or viscosity) is discussed, as they are not identical.

The most widely used model for process control, Vroom's H-factor model, disregards both the influence of wood species, active chemicals and the reaction mechanisms and is based only on a constant activation energy (134 kJ/mole) for delignification.

The main part of experimental work on process kinetics and its use of generating new optimised processes have been carried out on kraft pulping of softwood (pine and/or spruce) for pulp to be bleached. In the following for simplicity, only this case will be treated, as softwood kraft is the main pulp type in Scandinavia. Analogies can however be drawn to kraft pulping of hardwood.

The wood is assumed to be well penetrated with cooking liquor and we must remember the differences between *kinetics studies*, where the active chemicals are kept constant throughout the cook, and *stoichiometry*, where the pulping chemicals are consumed. The consumption of hydrogen sulphide ions is rapid but small when the wood meets the sulphide containing liquor while the hydroxide ion concentration decreases substantially during the cook.

7.2 The Composition of Kraft Pulping Liquor

The active components of kraft pulping liquor are the hydroxide (OH⁻) and hydrogen sulphide (HS⁻) ions. The presence of sulphide ions (S^{2-}) can be neglected as the equilibrium between sulphide ions and water Equation (7.1) is always displaced to the right.

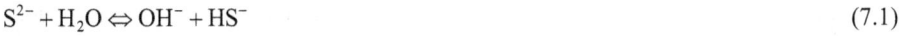

$$S^{2-} + H_2O \Leftrightarrow OH^- + HS^- \tag{7.1}$$

The concept „effective alkali" thus means hydroxide ions and "sulphidity" means the ratio between hydrogen sulphide and the sum of hydrogen sulphide and hydroxide ions multiplied with 200. Thereby, a solution of sodium sulphide in water has a sulphidity of 100 percent.

$$\text{Sulphidity}(\%) = 100 \times \frac{2\left[HS^-\right]}{\left[HS^-\right] + \left[OH^-\right]} \tag{7.2}$$

The concentration of hydroxide ions is essential in kraft pulping while the concentration of hydrogen ions is not. The use of pH ($-\log[H^+]$)for kraft pulping liquors is unsuitable as the ionisation product of water, $K_w = [H^+][OH^-]$, increases with three powers of ten when the temperature is increased from 20 °C to 170 °C.

Kraft pulping liquors as a rule contain some carbonate. When looking on the equilibrium at high alkalinity (high hydroxide concentrations) and high temperatures (conditions at kraft pulping) it is suitable to use the basic constants K_b, Equation (7.3) instead of the more "normal" K_a and the actual process temperature.

$$K_{b1} = \frac{\left[OH^-\right]\left[HCO_3^-\right]}{\left[CO_3^{2-}\right]} \tag{7.3}$$

In liquors of low alkalinity (low concentration of hydroxide ions) and high temperature, carbonate ions are partly hydrolysed into hydrogen carbonate and hydroxide ions, *Figure 7.2*, while at room temperature they are not hydrolysed. The hydrolysis of hydrogen sulphide ions into hydrogen sulphide and hydroxide ions can at most conditions be neglected, irrespective of temperature.

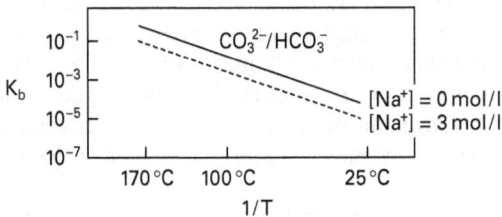

Figure 7.2. The dependence on temperature of the acid-base equilibrium of carbonate.

The positive ions (mainly sodium) have a retarding effect on the bulk delignification rate. This is also true for dissolved lignin in the final part of the cook.

7.3 The Three-phase Model for the Kinetics of Kraft Pulping

The three-phase model, describes the delignification of well-impregnated chips, i.e. it does not treat the essential mass transfer from the free liquor into the chip nor the inhomogeneous in the free liquor during an industrial kraft cook. There are other models available that take care of these aspects.

The delignification during kraft pulping can be divided into three phases: initial, bulk and residual delignification phases. If the liquor composition is kept constant during the reaction, the delignification rate in each phase is directly proportional to the remaining lignin in the wood residue, i.e. the reaction is of first order in residual lignin *Figure 7.3*. This might be astonishing considering the complex structure of the macromolecule lignin but this three-phase model of first order in remaining lignin has been verified in numerous experimental investigations. Observe that this pulp technological delignification rate is related to, but not identical to, the molecular bond breaking reactions that occur in the lignin macromolecule during kraft pulping.

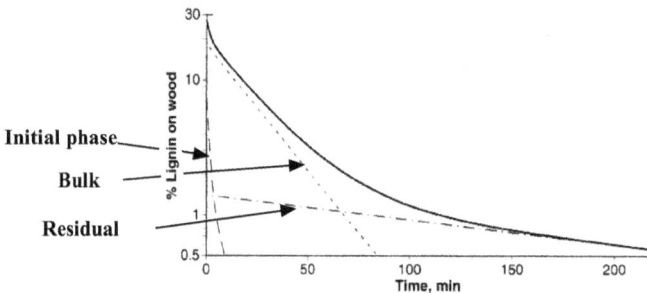

Figure 7.3. The delignification phases. (Lindgren and Lindström 1996).

There are two different views on handling the overlapping of the three different delignification phases. Either they are independent of each other, whereby during each phase only one mechanism is active. The second view assumes all three mechanisms acting simultaneously, but during each phase, one mechanism is dominant. That is to say, during the initial phase the initial delignification is most active until almost all the initial phase lignin is removed and then the bulk delignification mechanism takes over until almost all the bulk lignin is removed etc. The discussion is in principle about the origin of the slowly reacting residual phase lignin: is it formed rather late in the cook or is it present already in the wood (or is formed very early in the cook). The latter model (three parallel reactions) seems more plausible and is used in the following.

The effects of hydroxide ion concentration, hydrogen sulphide ion concentration and temperature (activation energy) on delignification rate are different during the different phases, Equation (7.4). The low activation energy in the initial phase means that this phase is probably governed by diffusion of chemicals. The much higher activation energies of the bulk- and residual phase implies that chemical reactions are rate determining.

Initial $\quad -\dfrac{dL}{dt} = k_1 \left[OH^- \right]^0 \left[HS^- \right]^0 L_i \qquad E_A = 60 \text{ kJ/mol}$

Bulk $\quad -\dfrac{dL}{dt} = k_2 \left[OH^- \right]^{0.8} \left[HS^- \right]^{0.5} L_b \quad E_A = 127 \text{ kJ/mol}$ $\qquad (7.4)$

Residual $\quad -\dfrac{dL}{dt} = k_3 \left[OH^- \right]^{0.5} \left[HS^- \right]^0 L_r \qquad E_A = 146 \text{ kJ/mol}$

The concentration of sodium ions (ionic strength) has a retarding effect on the delignification rate in the bulk phase. No such effect has been observed on the initial phase of delignification. Observe that k_2 and k_3 are affected by conditions in the previous phases (see Section 7.4).

The rate of cellulose depolymerization (chain cutting) can be assumed to be constant during the three phases of delignification, first order in hydroxide ion concentration, zero order in cellulose concentration and with a higher activation energy than for delignification Equation (7.5)

$$-\frac{dC}{dt} = k_4 \left[OH^- \right]^1 \qquad E_A = 180 \text{ kJ/mol} \qquad (7.5)$$

The concentrations of hydrogen sulphide ions, sodium ions and dissolved organic substance can be neglected in regard to the cellulose degradation.

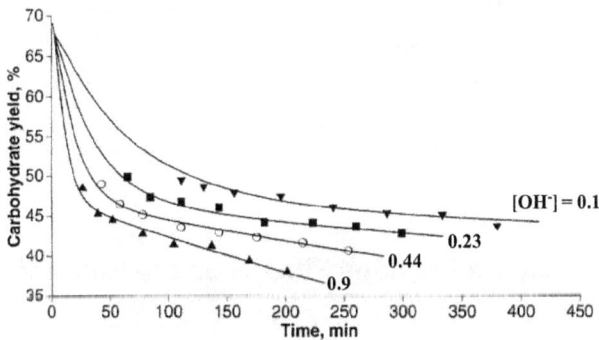

Figure 7.4. The loss of carbohydrate yield at different hydroxide ion concentration. Observe that the liquor composition is assumed constant during the cook. (Lindgren 1997).

The loss of carbohydrate yield can be divided into two phases. The first phase involves a rapid physical dissolution of hemicellulose and primary peeling connected with large alkali consumption, *Figure 7.4*. This rapid carbohydrate degradation and alkali consumption is approximately simultaneous with the initial delignification phase, but it must be remembered that they are two independent reactions

The rapid degradation of carbohydrates in the initial phase can be retarded by oxidation (with polysulphide and/or anthraquinone) or reduction (borohydride or hydrogen sulphide) of the carbonyl end-groups of the carbohydrate chains. The slow loss of yield in the second phase is due to alkaline hydrolysis of glucosidic bonds followed by secondary peeling.

The loss of carbohydrate yield is primarily dependent on the hydroxide ion concentration. The easily dissolved hemicellulose in the first phase is more affected by the hydroxide ion concentration than the second phase.

If the carbohydrate yield is plotted against lignin yield for kraft pulping processes, diagrams are obtained where the delignification selectivity during the three phases of delignification are visualised in principle *Figure 7.5*. The exact form is dependent on temperature and other process conditions.

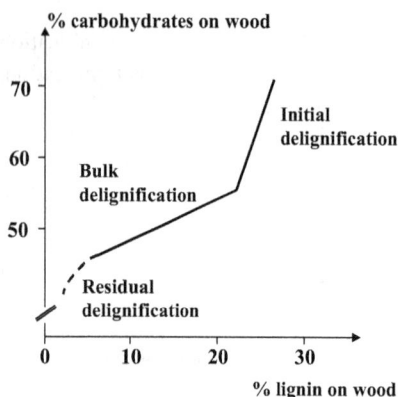

Figure 7.5. Both lignin and carbohydrates (hemicellulose and cellulose) are dissolved during a kraft cook. Principal picture – the exact values depend on process conditions.

That the initial phase shows a low selectivity, regarding carbohydrate yield, depends on the rapid carbohydrate peeling process, especially on the short-chained hemicelluloses. The low selectivity of the residual delignification phase on the other hand originates from that the delignification is considerably retarded.

7.4 Effects of Conditions in the Previous Phases on the Kinetics of Delignification

That there is an effect of the conditions, e.g. hydroxide and hydrogen sulphide ion concentrations, in previous phases on the kinetics of later phases in kraft pulping are rather natural. Chemical reactions (positive and negative) can occur in the lignin molecule without immediately affecting the dissolution of lignin (that is described by the kinetic equations of kraft pulping).

The delignification rate in bulk phase is increased when the concentration of hydrogen sulphide ions in the initial phase is increased from 0 till 0.2 mole/l, but further increase has little effect. Furthermore, by displacing a sulphide containing liquor (e.g. black liquor) through the chip column at the beginning of the cook hydrogen sulphide ions (0.3 mole S/kg wood) can be sorbed by the wood residue, increasing the concentration of hydrogen sulphide ions in the bulk phase. By this sorption hydrogen sulphide ion are carried over to the bulk phase where they increase the delignification rate. The sorption is increased with increased hydrogen sulphide ion concentration and with decreased hydroxide ion concentration, *Figure 7.6*. However, when the hydroxide ion concentration sinks to too low values during the initial phase, especially at low hydrogen sulphide ion concentrations, the delignification in the bulk phase is retarded. There is an additional increase in hydrogen sulphide ions by the carryover of black liquor as a result of the liquor displacement.

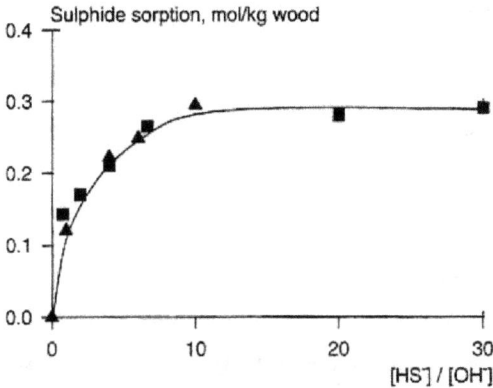

Figure 7.6. Sorption of sulphide into wood as a function of liquor composition. ■=[OH⁻] varied, ▲=[HS⁻] varied.

The amount of residual phase lignin, i.e. lignin to be removed by the slow residual phase process during kraft cook, is dependent on the liquor composition in the bulk phase, but not the temperature.

A combination of low hydrogen sulphide ion and low hydroxide ion concentrations during the bulk phase delignification results in disastrous high amounts of residual lignin *Figure 7.7*.

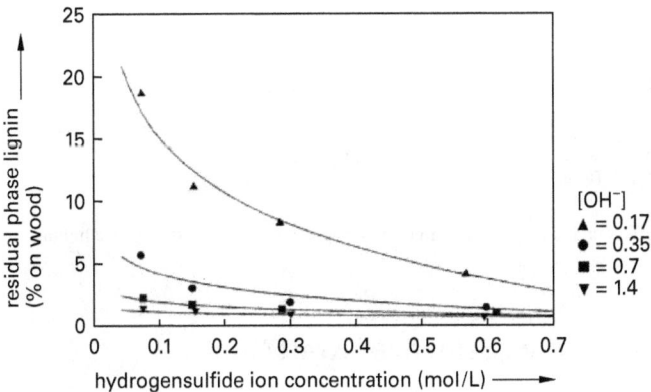

Figure 7.7. Effect of liquor composition in bulk phase on the amount of low-reactive residual phase lignin. (Gustavsson and Lindström 1997).

By increasing the hydroxide ion concentration during the residual phase part of the slow residual phase lignin "is converted" into rapidly reacting bulk phase lignin *Figure 7.8*. This may be due to that, at higher hydroxide ion concentration more α-hydroxyl groups in the non-phenolic β-arylethers are ionised and thereby, more bonds in lignin can be cleaved.

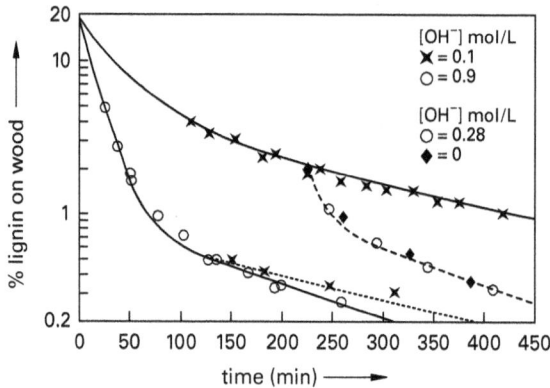

Figure 7.8. The effect of changing the OH⁻ concentration in the residual phase. (Lindgren and Lindström 1996).

The temperature during the previous phases does not affect the amount of residual phase lignin (if the model of three parallel delignification reactions is chosen), *Figure 7.9*.

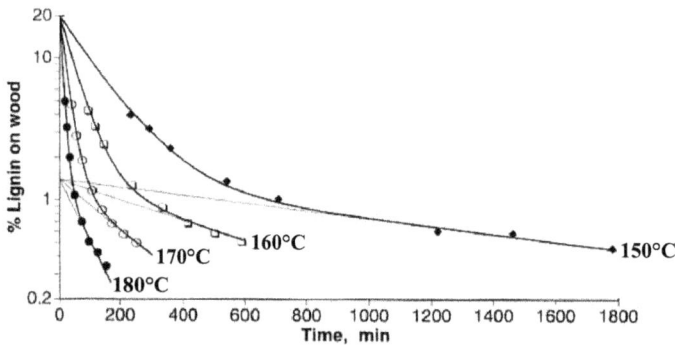

Figure 7.9. Temperature has no effect on the relative amount of bulk phase lignin and residual phase lignin. (Lindgren and Lindström 1996).

7.5 The Four Principles for a „Selective" Kraft Cook

From the above mentioned kinetic knowledge four principles for a selective (high viscosity at a given lignin content or kappa number) kraft cook can be deduced. In principle, the kinetic equations for delignification (2) are divided with the ones describing the rate of cellulose degradation (3). Then the effects mentioned in section 7.4 are taken into account. The essence of this exhibits below and leads to the four principles for a selective cook.

1. Levelled out hydroxide ion concentration during the cook. (In comparison with the conventional cook, decrease at the beginning, increase at the end)
2. High initial hydrogen sulphide ion concentration.
3. Low sodium ion concentrations during the cook, especially during the late part of the cooking.
4. Use as low temperature as possible.

This is a new way of looking on the cook, in terms of concentrations instead of alkali charged on wood. Consequently, adding chemicals throughout the cooking process and carrying part of the cook in counter-current mode can improve the selectivity of the kraft cook.

In a conventional kraft cook, where none of the above four principles are fulfilled, all the chemicals are charged at the beginning of the process. The result is high hydroxide ion concentration initially and less in the end of the cook. Furthermore the black liquor with high concentration of dissolved lignin with corresponding sodium ions and low concentration of hydroxide ions is withdrawn at the end of the cook, which not is desirable according to the principles.

7.6 Adaptation of Continuous Cooking to Modified Kraft Pulping

The first application of the "new " principles to continuous cooking is called MCC (Modified Continuous Cooking) and consists of adding white liquor at three locations along the digester (instead of only at top of impregnation vessel) and carrying the final part of the cook in counter-current mode *Figure 7.10*. In this way the alkali concentration level was decreased in the beginning of the cook and increased at the end. Simultaneously the concentrations of dissolved lignin and sodium ions (not shown in *Figure 7.10*) were decreased at the end of the cook. The simultaneous decrease of hydrogen sulphide ions at the beginning of the cook was a disadvantage, that can be handled if there are two white liquors available (section 7.7) or by using black liquor pre-impregnation/displacement.

In spite of the disadvantage regarding the hydrogen sulphide ion concentration profile an increase in pulp viscosity of 150 dm^3/kg at a given kappa number was obtained. *Figure 7.11* shows the results from the first mill scale experiments 1984 in Varkaus Finland performed by STFI and KAMYR. The scatter in the points is normal for mill scale experiments). With unaltered pulp viscosity it was possible to decrease the kappa number with 9 units (from 32 to 23) but at the expense of 0.5 % loss in bleached pulp yield as the modifications have better effect on the selectivity regarding the hydrolysis of cellulose than preserving the amount of carbohydrates. (At a constant kappa number a yield increase of 0.5 % was observed). The yield loss can be avoided if the higher pulp viscosity is instead used to carry out a more extensive delignification in a subsequent oxygen delignification stage resulting 5 units decrease in kappa number (from 20 to 15).

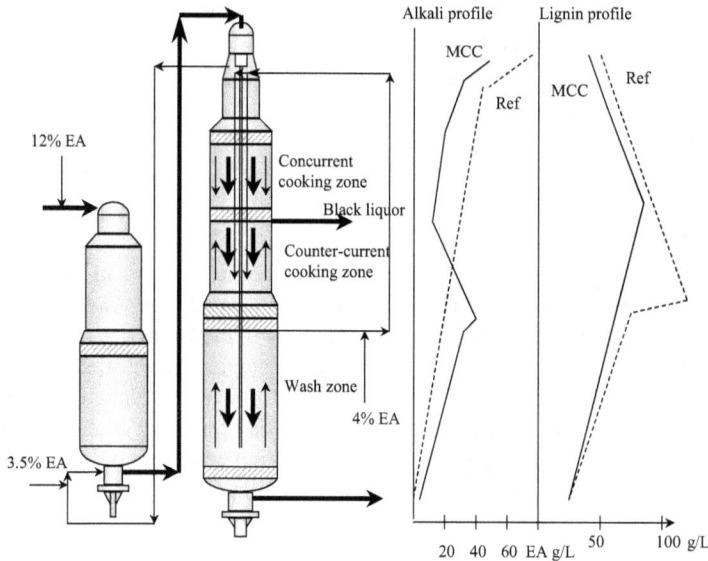

Figure 7.10. MCC –the first of the „new" pulping types based on kinetics.

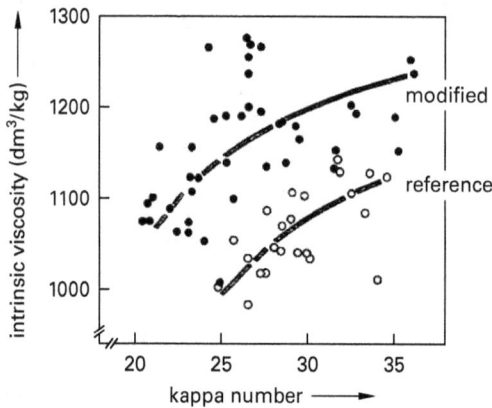

Figure 7.11. Results from mill experiments on the introduction of MCC. The reference is a conventional kraft pulp cooked in the same digester.

The modified kraft pulping process can be further improved by charging some white liquor also in the original hi-heat washing zone thus prolonging the cooking zone, permitting a decrease in cooking temperature, thus improving the selectivity still further *Figure 7.12*. The result is a constant low temperature throughout the cook, which leads to the trade name ITC (Iso Thermal Cooking). The pulp obtained is also easier to bleach, i.e. requires less beaching chemicals per kappa number to obtain certain brightness.

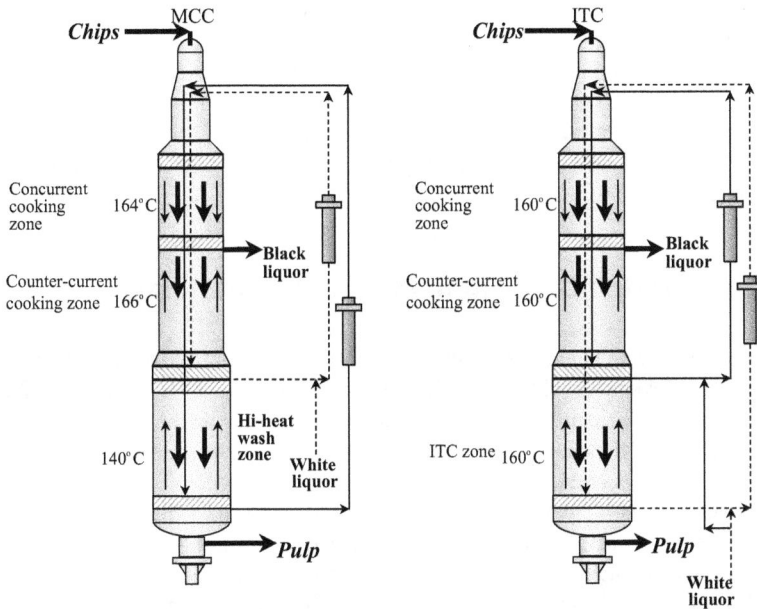

Figure 7.12. Development from MCC to ITC.

A further improvement is introduction of black liquor impregnation (or more correctly black liquor displacement) to continuous cooking (earlier introduced to batch cooking). If all the liquor that is withdrawn from the later cooking phases, is added to the initial phase, also an increased residual hydroxide concentration in the latter parts of the cook can be applied. As this alkali is not wasted. This works, as the excess residual alkali will be consumed when the black liquor comes into contact with wood. This concept is used in the Compact Cooking (Kvaerner

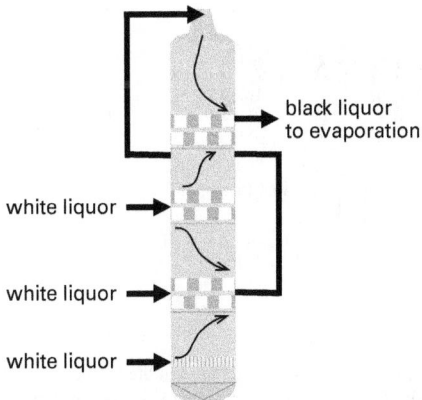

Figure 7.13. Black liquor impregnation is included in the EAPC process.

Pulping) and the EAPC (Enhanced Alkali Profile Cooking by Ahlström- Andritz) concepts, *Figure 7.13*.

The development of modified continuous pulping continues towards better selectivity of delignification (both in respects of carbohydrate yield/composition and pulp viscosity) and towards higher homogeneity and simplicity of equipment. (A short lesson of corporate history of continuous cooking: KAMYR was split into Kvaerner Pulping and Ahlström that was later bought by Andritz).

7.7 Adaptation of Batch Cooking to Modified Kraft Pulping

The primary drive for modifying batch-wise kraft cooking was not to improve selectivity but to improve heat economy in order to avoid the heat loss when the digesters were blown from full (temp 170 degrees) to atmospheric pressure (temp 100 degrees). Displacing the hot black liquor from the digester at the end of the cook with washing liquor of lower temperature and using the black liquor to heat up the chips in the next digester achieved this.

White liquor temperature is increased by heat exchange with hot black liquor before it is charged. It was found out that the displacements should be carried out in two stages, first with warm and then with hot black liquor, for optimal heat economy *Figure 7.14*. The alkalinity is also lower in the "warm" black liquor (than in the "hot" one) that increases the sulphur adsorption (see section 7.4).

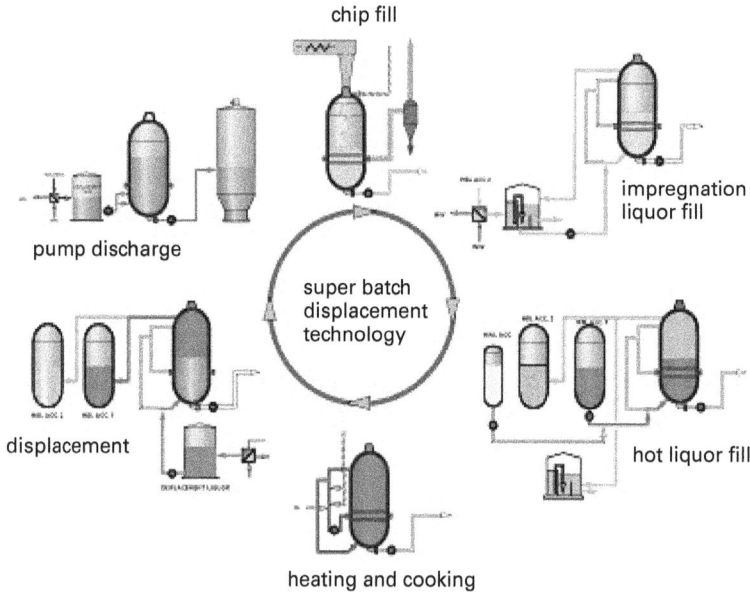

Figure 7.14. The SuperBatch system. The black liquor displaced after the cook is first used in the hot liquor fill. After displacement from that stage it is used in the impregnation. Finally, after displacement from the impregnation stage the black liquor goes to evaporation and recovery furnace. The white liquor is added at the later part of the hot liquor fill. (Metso Paper©).

This procedure requires a couple of new pressurised black liquor tanks and that a system of individual digesters (4–6) is run in exact sequence by supervisory computer control. About half the steam requirement of conventional batch cooking is saved.

It was discovered that in addition the cooks had become more selective mainly as a result of the initial multiple displacements with sulphide containing black liquor, but also the decrease in dissolved lignin and sodium ions during the final displacement of black liquor with washing liquor. The residual alkali in the "hot" black liquor at the end of the cook can be kept rather high as it is utilised in the next cook. The blowing at the end of the cook becomes more gentle to the fibres as high temperatures are avoided, which results in a stronger pulp.

The industrially used batch processes RDH (Rapid Displacement Heating developed by Beloit) and Super Batch (developed by Sunds Defibrator, nowadays Metso Paper) are very similar.

The temperature is as rule not decreased but otherwise the principles of modified cooking are applied to some extent combined with an improved homogeneity.

7.8 Use of Two White Liquors of Different Sulphidities

In a conventional chemical recovery system, based on soda recovery furnace and lime cycle, only one white liquor is obtained with a given proportion of hydroxide and hydrogen sulphide ions. The sulphidity is as rule kept below 40 % in order to avoid too high corrosion and too high emission of malodorous sulphur gases. If in the future other recovery systems based on gasifica-

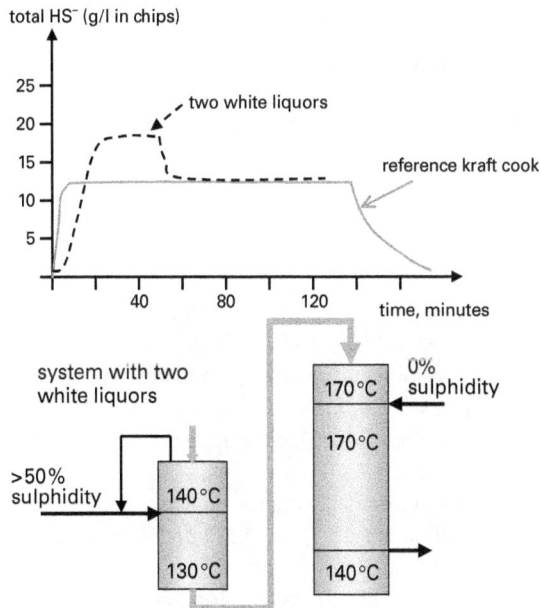

Figure 7.15. Split alkali charge using two white liquors is favourable as a high hydrogen sulphide ion concentration is obtained in the beginning of the cook. Split alkali charge using only one white liquor would result in a lower hydrogen sulphide concentration at the beginning.

tion of black liquor might become available, sodium and sulphur can be obtained in separate streams, thus making preparation of low (or no) sulphidity and high sulphidity white liquors possible. As hydrogen sulphide ions have a large favourable effect in the initial and bulk delignification phases it is advisable to charge the high sulphidity liquor initially and make the later alkali charges with low (or no) sulphidity white liquor, thus avoiding the problems described in section 7.6, *Figure 7.15*. It is an advantage that the white liquor used in the oxygen delignification stages is of low sulphidity as the hydrogen sulphide ions have to be destroyed by oxidation before use.

7.9 Properties of Pulps Obtained with the Different Processes

The unmodified kraft process results in strong pulps. The modifications of kraft process described above, e.g. ITC and Super Batch, result in pulps with still higher strength properties (higher tear strength at a given tensile strength) as a result of higher cellulose yield. The bleachability is improved (lower requirement of bleaching chemicals per unbleached kappa number to reach a given brightness) for ITC pulps.

7.10 Further Reading

7.10.1 Books

Rydholm, S. (1965) *Pulping processes*. Interscience publishers.
Smook, G.A. (1996) *Handbook for pulp and paper technologists*.
Gullichsen J., and Fogelholm C.-J. (eds.) (1999) *Chemical Pulping spec. 6 A. Chapter 3, 5 and 6* Fapet Oy.

7.10.2 Articles

Gustatafsson, L., Teder, A. (1969) Alkalinity in alkaline pulping. *Svensk Papperstidning* 72: 24, 795.
Gustavsson, C, Lindgren C., Lindström, M.E. (1997) Residual phase lignin in kraft cooking related to the conditions in the cook. *Nordic Pulp and Paper Res. J.* 12:4, 225.
Johansson, B., Mjöberg J., Sandström P., Teder A. (1984) Modified continuous kraft pulping - now a reality. *Svensk Papperstidning* 87:10, 30.
Lindgren, C., Lindström, M.E. (1996) The kinetics of residual delignification and factors affecting the amount of residual lignin during kraft pulping. *J. Pulp and Paper Science* 22:8, J 290.
Olm, L., Bäckström, M., Tormund, D. (1996) Treatment of softwood with sulphide-containing liquor prior to a kraft cook. *J. Pulp and Paper Science* 22:7, J 241.

Olm, L., Tormund, D. (2000) Kraft pulping with sulfide pretreatment. Part 2. The influence of pretreatment and cooking conditions on the pulp properties. *Nordic Pulp and Paper Res. J.* 15:1, 70.

Sjöblom, K., Mjöberg, J., Hartler, N. (1983) Extended delignification in kraft cooking through improved selectivity. Part 1.The effects of the inorganic composition of the cooking liquor. *Paperi ja Puu* 65:4, 227.

Sjöholm, K., Mjöberg, J., Hartler, N. (1988) Extended delignification in kraft cooking through improved selectivity. Part 2. The effects of dissolved lignin. *Paperi ja Puu* 70:3, 452.

Teder, A., Olm, L. (1981) Extended delignification by combination of modified kraft pulping and oxygen bleaching. *Paperi ja Puu* 63:4a, 315.

8 Pulp Washing

Lars Miliander
Karlstad University, Department of Chemistry

8.1 Introduction

Washing is an operation that takes place in all kinds of pulp production. In textbooks of chemical engineering the operation is described as solid-liquid extraction and it will be treated as this unit operation in the present context. The design and operation of a washing plant is of great importance especially when considering production of chemical pulp such as kraft pulp and sulphite pulp. Also mechanical pulps need to be washed before further treatment but this chapter deals only with washing of chemical pulp and the reason for this is that the separation of spent cooking liquors and pulp has a large economic impact on the recovery system and the energy situation of the chemical pulp mill. This will be elaborated on further down in this chapter. The washing of unbleached pulp will be considered in detail including the oxygen delignification stage. Washing between bleaching stages will not be treated in this chapter. It is not only a separation process but closely integrated with the bleaching process. The student is therefore recommended to study design of washing in the bleach plant in chapter 10 Bleaching of pulp.

8.1.1 Historical Background

In the beginning of chemical pulp production the main purpose with the washing was to get a clean pulp free from spent liquor. Small production units and an abundant supply of fresh water made it possible to discharge the effluent water from the washing plant directly to the receiving river or lake. Pollution and environmental protection were not discussed those days.

Very soon however it became obvious that pulp washing was important also based on economic aspects. This is especially true for kraft pulp production and the expensive chemicals such as caustic soda and sodium sulphide being used in that process. Sulphite pulp production based on limestone and sulphur had not the economic incentive to improve and close the washing system. Many of these mills discharged the spent liquor all the way until they were either closed or converted to a soluble base such as magnesium or sodium instead of calcium. These changes took place during the 1970's.

The spent liquor accompanying the pulp from the digester contains not only the used cooking chemicals but also dissolved wood substances together named dissolved solids. DS is an often used abbreviation for this parameter. Having in mind that the amount of DS per ton of pulp is in the range of 1.5 ton and 1.3 ton per ton of kraft pulp and sulphite pulp respectively it is obvious that the spent liquor has to be recovered. Today more then 99 % of the spent liquor is washed out of the pulp and taken care of for further treatment in the recovery cycle. The spent chemicals are regenerated to new cooking chemicals and the dissolved wood substance is used as a bio fuel covering a very large part of the energy consumption of the mill. Subsequently less than one percent of the spent liquor is lost in the washing and transferred to the effluent system in later treatment stages of the pulp.

The losses of spent liquor were expressed as losses of sodium sulphate at kraft pulp mills and at sulphite mills the parameter normally used was the recovery rate of dissolved solids. The total amount of Na_2SO_4 in the kraft pulp cooking system is in the range 800 kg per ton of pulp and the change of the losses during the 20th century could best be described in a graph like the following, *Figure 8.1.*

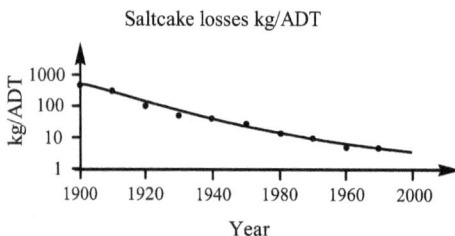

Figure 8.1. Minimizing washing losses during the 20th century.

The washing losses expressed as salt cake (Na_2SO_4) has been reduced from 500 to only about 5 kg per ton of pulp during a period less than 100 years. A lot of hard work lies behind this improvement. In the beginning of the last century the batch diffuser was introduced and the washing loss came down to about 100 kg per ton of pulp. Later on washing on rotating drum filters became more or less a standardised method. The filters were arranged in series and the flow of wash water was counter current to the pulp flow. Two, three and even four filters in series were used. More about this later on.

In the middle of the 20th century continuous cooking was developed and at the same time new washing methods came on stream, like continuous diffusers and hi-heat washing in the digester. After that a number of different washing equipments has been introduced. Among these drum presses, displacement presses, screw presses, pressurized diffusers, horizontal displacement washers and the drum displacer deserve to be mentioned. New equipment arranged in an optimum way resulted finally in the very low washing loss mentioned above.

The amount of water being used for washing has also been reduced to a great extent. Since the spent liquor leaving the washing plant has to be evaporated, in order to increase the concentration of DS from a level 15–17 % (by weight) to at least 65 %, dilution of the liquor with water has to be avoided/limited. After the first dramatic price escalation for fuel oil in the beginning of the seventies, the interest for washing equipment using low amounts of wash water increased a lot.

Today it is the authorities that normally decide how large washing losses can be accepted and these are not anymore given as salt cake losses since the technique has brought us down to a level close to zero. The 5 kg per ton of pulp mentioned above are more or less adsorbed to the surface of the fibre. Instead the amount of dissolved substances remaining in the washed pulp that give rise to oxygen consumption in the recipient is used as a parameter. A standardised method is available today giving the amount of COD per ton of pulp, where COD stands for chemical oxygen demand.

The sulphite pulp industry has during the last 30 years gone through the same kind of development as the kraft pulp industry. Even the very few pulp mills still using limestone as a cooking base have installed modern washing equipment resulting in low losses or described as recovery rate of dissolved solids more than 97 % is achieved.

8.2 Washing Theory

Displacement of one liquid with one other fully miscible liquid out of a pulp pad of a certain thickness is a special case of flow through a porous media. The pulp web is the porous bed and the fibres involved constitute its properties. The liquid phase in such a bed is partly bound to the fibre as a stagnant liquid and partly free moving. The bulk flow during displacement acts on the free part of the liquid and that part will be replaced with certain efficiency due to channelling and variation in bed porosity and other phenomena like particle (fibre) size distribution. The spread out of the displacement front is called dispersion. Introducing a displacing liquid into the pulp bed creates concentration differences between the free bulk flow and the stagnant part of the liquid which in turn starts a diffusion of solute from the stagnant part to the free bulk flow. These transport phenomena are time consuming and depend among other things on the distance for diffusion and the openness of the fibre network. The final cleanliness depends on all these different phenomena and also to the extent of sorption on the fibre surface. When sodium con-

168

stitutes the solute under consideration the sorption has to be evaluated in order to understand the displacement process fully.

The displacement of spent liquor in a pulp bed can be explained by the use of a dispersion model, *Figure 8.2*.

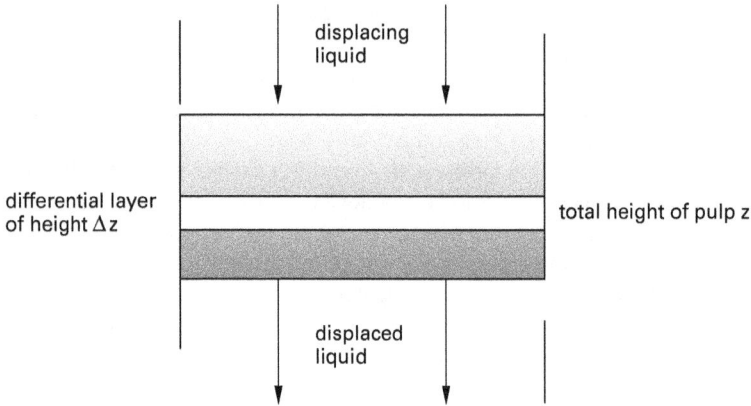

Figure 8.2. A model for the displacement process.

Depending on the dispersion and other phenomena like bed porosity and fibre porosity in the pulp pad the breakthrough curve for the solute concentration will be more or less steep. The process can be mathematically described by a differential material balance over the layer Δz for the solute and the equation will then be given by integration over the total height, z. The equation is a partial differential equation since the term for accumulation in the equation of continuity has to be included in the solution. On the right hand side the equation contains one term for the dispersion and one for the bulk flow of the solute. Numerical solutions are available for this kind of equation but if we introduce certain simplifications like equilibrium between bound/ stagnant and free flowing liquid/bulk an analytical solution shown below can be used.

$$\frac{\partial c}{\partial t} = D_L \cdot \frac{\partial^2 c}{\partial z^2} - u \cdot \frac{\partial c}{\partial z}$$

$$\frac{\partial c}{\partial t} + \frac{(1-\varepsilon)}{\varepsilon} \frac{\partial c_s}{\partial t} = D_L \cdot \frac{\partial^2 c}{\partial z^2} - u \cdot \frac{\partial c}{\partial z}$$

c = concentration of solute
c_0 = original concentration of solute
c_s = concentration of solute in stagnant liquid
u = superficial flow
z = thickness of pulp layer
D_L = Longitudinal dispersion coeff .
t = time
ε = bed porosity

B_0 = Bodensteins number

$c2 = K_0 + K_1 c$

With the assumption above $K0$ is obviously equal to zero

$$\frac{\partial c}{\partial t}\frac{1}{\lambda} = D_L \cdot \frac{\partial^2 c}{\partial z^2} - u \cdot \frac{\partial c}{\partial z}, \text{ where } \quad \lambda = \frac{1}{\left(1 - K_1 + \dfrac{K_1}{\varepsilon}\right)}$$

$$\frac{c}{c_0} = 1 - 0.5\left[erfc\frac{1-R\lambda}{2\sqrt{R\lambda S}} + e^{1/s} \cdot erfc\frac{1+R\lambda}{2\sqrt{R\lambda S}} \right]$$

$R = \dfrac{u \cdot t}{z}$ Number of displacements heights

$S = \dfrac{D_L}{u \cdot z} = \dfrac{1}{B_0}$ Dimensionless dispersion number, dim = 1

The breakthrough curve for the displacement is shown in *Figure 8.3*. A fairly good agreement between experimental data and the model was accomplished in spite of the very simplified relations being used. These studies were carried out on sodium based sulphite pulp at low pH and subsequently the pulp contained a very low amount of sorbed sodium.

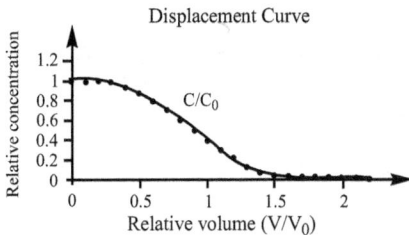

Figure 8.3. Displacement of sodium, experimental data vs model.

A more elaborate model has to be used in order to describe displacement of spent liquor in kraft pulp. Both desorption and diffusion are phenomena of great importance A number of investigations have been made in laboratory and pilot plant scale in order to explain the behaviour of displacement. In a dissertation from 1974 Grähs (Chalmers) presented a very extensive description of the phenomena. In the following text some of these findings will be discussed. The partial differential equation (PDE) for the displacement contains now three different time depending terms. One describes accumulation in the bulk flow, the other diffusion between stagnant and free liquid and the third desorption from fibre surfaces. We have three different concentrations of solute to consider and all three are depending on both time and position in the pulp pad.

$$\varepsilon_d \frac{\partial c}{\partial t} + \varepsilon_s \frac{\partial c_s}{\partial t} + C_F \frac{\partial w}{\partial t} = \varepsilon_d \cdot D_L \frac{\partial^2 c}{\partial z^2} - \varepsilon_d \frac{\partial c}{\partial z} \cdot u$$

ε_s describes the stagnant part of the liquid and ε_{cb} the free part.

$$\varepsilon_s \frac{\partial c_s}{\partial t} = k_1 \cdot a_1 (c - c_s) - C_F \frac{\partial w}{\partial t}$$

C_F, is the pulp consistency, c is concentration of solute in the free part and c_s, is the concentration in the stagnant part. The sorbed amount is given by the parameter w, in kg/ton of pulp.

$$C_F \frac{\partial w}{\partial t} = k_2 \cdot a_2 (c_s - c_s^*) \quad w = f(c_s^*)$$

The suffix * stands for an equilibrium concentration. a_1 and a_2 are specific mass transfer surfaces.

$$w = \frac{A \cdot c_s^*}{1 + B \cdot c_s^*}$$

A and B are the Langmuir constants and w is the sorbed amount of solute per weight unit of fibre.

Boundary conditions

$$u \cdot c_{in} = u \cdot c(0^+) - D_L \frac{\partial c}{\partial z}\Big|_{0^+}$$

Dimensionless variables like the following are normally introduced in order to simplify the boundary conditions.

$$C = \frac{c - c_w}{c_0 - c_w}, \quad C_s = \frac{c_s - c_w}{c_0 - c_w}, \quad C_s^* = \frac{c_s^* - c_s}{c_0 - c_w}$$

$$x = \frac{z}{Z}, \quad T = \frac{t}{\tau}, \quad W = \frac{w - w_w}{w_0 - w_w}$$

τ, is the residence time and c_w, is the concentration in the wash liquor.

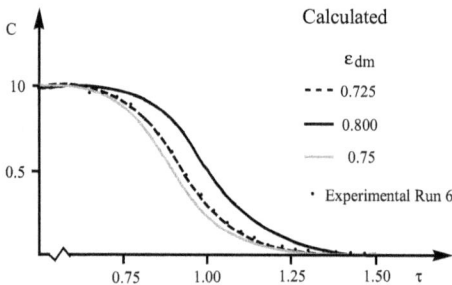

Figure 8.4. A fit of displacement data versus the model by Grähs.

By relative concentration here is meant the actual concentration of solute in the displaced liquid divided by the original concentration in the spent liquor in the pulp before starting the displacement. A very good agreement between calculated displacement and experimental results was presented by Grähs and one example is given below, *Figure 8.4*. The results gave also values for parameters like the dispersion number D_L and the porosity of the bed ε.

The diagram above is a very useful means of explaining the displacement washing operation. It can be used for one single displacement stage and also for a multi-stage counter current operation. In the later case the diagram should be more elaborate and several other functions included.

An infinite number of stages give a breakthrough curve staying on $c/c_0 = 1$ until $v/v_0 = 1$ and after that the relative concentration drops suddenly to zero. The surface under the breakthrough curve is then a measure of how much dissolved solids is recovered during washing. The surface growth under the curve is shown by using the integral of curve. In the ideal case the integral is the same as the diagonal from origo (0/0) to a point (1/1) as it is demonstrated in the following diagrams Figure 8.5–8.6.

ideal displacement

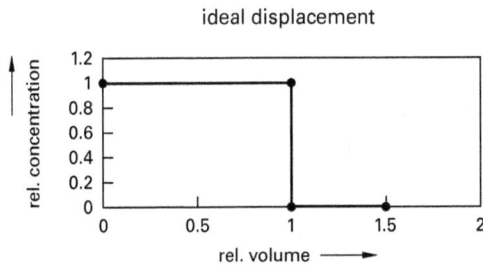

Figure 8.5. Ideal displacement or an infinite number of stages.

infinite number of stage and recovery of COD

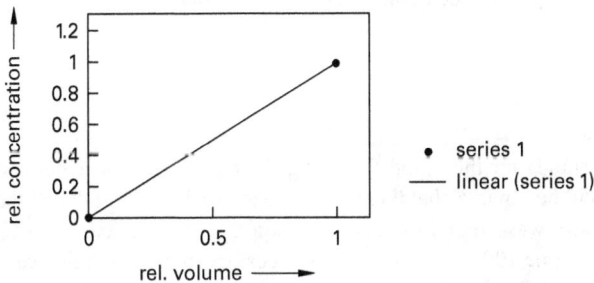

Figure 8.6. The ideal displacement in integrated form as recovery of the solute under consideration.

8.3 Methods for Calculating Results of Pulp Washing

Washing of pulp is a unit operation that can be considered as solid-liquid extraction. Theories developed for this separation process have been applied with great success for describing the washing operation. Standardized methods of chemical engineering have been utilized in order to explain what happens in a multi stage cascade system. Furthermore pulp washing is one of the operations that has been best adopted to the chemical engineering way of handling a process among all the different pulping processes. In the continuation of this chapter COD will be used as a measure of a solute from the cooking of wood.

8.3.1 Displacement Ratio

The parameter being used to describe the efficiency of displacement is the Displacement Ratio (DR). In order to explain this parameter we have to consider a washing stage like the one in *Figure 8.7*. The displacement operation described is based on the same principles irrespective of the equipment that has been chosen for this explanation.

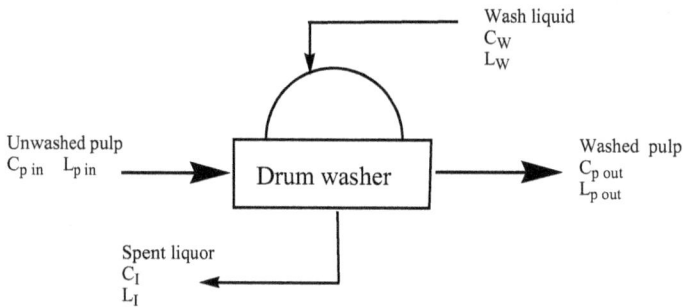

Figure 8.7. A single Washing Stage (Drum Washer).

In the figure C stands for the concentration of solute in the different positions under consideration and it can be either the chemical oxygen demand (COD), dissolved solids (DS) or the concentration of salt cake in the liquid phase. The definition of DR is as follows

$$DR = \frac{\left(C_{pin} - C_{pout}\right)}{\left(C_{pin} - C_w\right)}$$

In the equation above index p stands for the liquid following the pulp flow. DR is defined through this expression and it should be obvious that there is a dependence between DR and the wash liquid amount. By adding more wash liquid the concentration C_{pout} will get closer to C_w and eventually DR will achieve the value 100 %. Therefore it is necessary to give both the wash liquid amount and the corresponding DR value when describing effectiveness. In order to further improve the understanding of DR as a washing parameter it should be stressed that the displacement operation is not supposed to change the consistency of the pulp flow. The dependence of the liquid amount being used can approximately be described with following equation for a drum washer

$$DR = 1 - e^{-L_w/L_{pout}}$$

L is the liquid amount and index w refers to wash liquid and index pout to pulp leaving the system. The equation is the same as the one being used for an ideal mixing stage. A simple material balance around the washing stage shows that wash liquor in excess of the amount in the pulp leaving the stage L_{pout} will end up as dilution of the liquor leaving the same stage L_1.

The displacement of spent liquor out of the pulp pad is a time depending process i.e. an unsteady state operation. The displacement washing taking place in the high–heat zone in a continuous digester is purely counter current washing. Most other displacement washing operations are cross current in radial direction through the pulp web. This is true for drum washers and drum presses of all types and the resulting efficiency depends on the fact that the outer surface of the pulp web, where the wash liquid is added, is better washed than the inner surface adjacent to the wire. During periods with undisturbed production high-heat washing is a real steady state process which is true also for multiple stage drum washers if the process is studied at a stationary point. Mathematical models for the specific operation taking place inside the continuous digester are available and development work is still going on in order to improve these models. In the following we will concentrate the discussion to the unsteady state displacement process being used in continuously working systems.

If the DR value is known as a function of the wash liquor amount it is possible to calculate C_{pin} from the target value C_{pout} and the equally known C_w. When C_{pin} has been calculated material balances both for the washer itself and for the dilution point based on flows, concentration of solute and pulp consistencies is useful in order to find out about C_{pout} from a previous stage. In this way it is possible to work through a complete washing plant which will be fully understood by the reader after taking part of section 8.5 of this chapter.

8.3.2 *E*-factor Method

Due to the limitation of the DR parameter other methods of describing the efficiency of a washing stage have been developed. The most useful of these is probably the Nordén efficiency factor [1] written as N or E. In this chapter E is used as a designation for this parameter. E is the number of theoretical (or ideal) separation stages required to achieve a certain level of cleanliness of the washed pulp. The idea behind this way of treating the washing operation was to find a parameter (E) for the washing efficiency independent of the wash water amount. More about the Nordén efficiency factor will be found in the next section of this chapter. Here we present the theoretical background to the E-factor and we start with a figure showing a multistage counter-current system both as a block diagram and a graphical diagram according to the well known McCabe Thiele–method for calculation of the number of ideal stages. See *Figure 8.8–8.9*. Nordén developed this method of calculating washing results already during the 1960's.

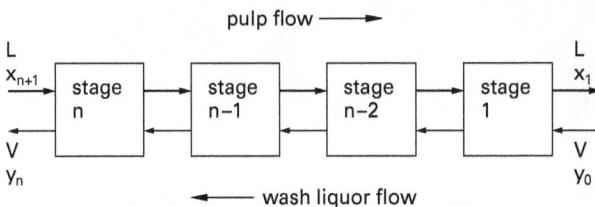

Figure 8.8. Multistage system in counter-current washing.

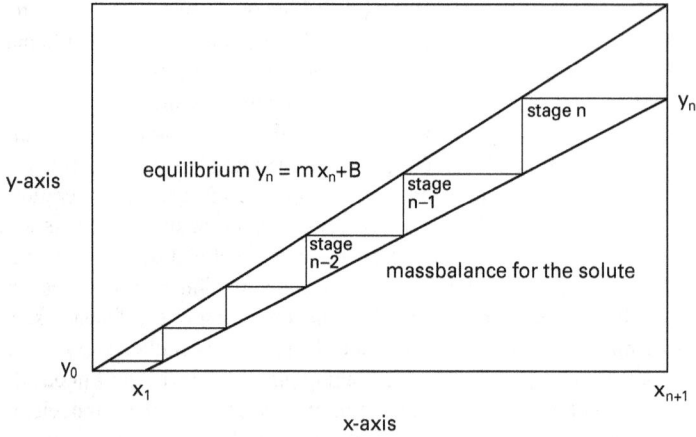

Figure 8.9. McCabe-Thiele diagram for multistage counter-current washing.

If the equilibrium between two miscible liquid phases is supposed to be given by a straight line we get equation (8.1). Index n is used for flows leaving stage n and for the ideal case equilibrium, is supposed between y_n and x_n.

$$y_n = m \cdot x_n + B \tag{8.1}$$

A mass balance for the solute is given by Eq. (8.2)

$$y_n \cdot V + x_1 \cdot L = y_0 \cdot V + x_{n+1} \cdot L \tag{8.2}$$

The flow with index $n + 1$ is one entering stage n.
By eliminating y_n from these two equations we get Eq. (8.3)

$$m \cdot x_n + B = \frac{L}{V} x_{n+1} + y_0 - \frac{L}{V} x_1 \tag{8.3}$$

which is rearranged to the following expression

$$x_{n+1} = \frac{V}{L}(m \cdot x_n + B) - \frac{V}{L} y_0 + x_1$$

$$\begin{cases} A = \dfrac{L}{m \cdot V} \\ x_{n+1} = \dfrac{m \cdot V}{L} x_n + \dfrac{V}{L} B - \dfrac{V}{L} y_0 + x_1 \quad \Rightarrow \quad x_{n+1} = \dfrac{1}{A} x_n - \dfrac{1}{A} x_0 + x_1 \\ y_0 = m \cdot x_0 + B \end{cases} \tag{8.4}$$

When $n = 1$ we get Eq. (8.5)

$$x_2 = \frac{1}{A}x_1 - \frac{1}{A}x_0 + x_1 = x_1\left(\frac{1}{A}+1\right) - x_0\frac{1}{A} \tag{8.5}$$

Further on $n = 2$ gives Eq. (8.6)

$$x_3 = \frac{1}{A}x_2 - \frac{1}{A}x_0 + x_1 \tag{8.6}$$

By eliminating x_2 in Eq. (8.5) and (8.6) we get:

$$x_3 = \left[\left(\frac{1}{A}\right)^2 + \frac{1}{A}\right]x_1 - \left(\frac{1}{A}\right)^2 x_0 - \frac{1}{A}x_0 + x_1 = \ldots$$

$$\ldots = \left[\left(\frac{1}{A}\right)^2 + \frac{1}{A}+1\right]x_1 - \left[\left(\frac{1}{A}\right)^2 + \frac{1}{A}\right]x_0 \tag{8.7}$$

Generalising Eq. (8.7) to n stages gives us Eq. (8.8)

$$x_n = x_1\left[1+\frac{1}{A}+\frac{1}{A^2}+\ldots+\frac{1}{A^{n-1}}\right] - x_0\left[\frac{1}{A}+\frac{1}{A^2}+\ldots+\frac{1}{A^{n-1}}\right]$$

$$x_{n+1} = x_1\left[1+\frac{1}{A}+\frac{1}{A^2}+\ldots+\frac{1}{A^n}\right] - x_0\left[\frac{1}{A}+\frac{1}{A^2}+\ldots+\frac{1}{A^n}\right] \tag{8.8}$$

Now by substituting n to E we can describe the concentration of solute in the liquid following the pulp to the washing plant according to Eq. (8.9)

$$x_{n+1} = x_1\left[1+\frac{1}{A}+\frac{1}{A^2}+\ldots+\frac{1}{A^E}\right] - x_0\left[\frac{1}{A}+\frac{1}{A^2}+\ldots+\frac{1}{A^E}\right] \tag{8.9}$$

The sum above can be written:

$$S_n = \frac{a_1\left(1-r^n\right)}{1-r}$$

u_1 = the first term
r = constant relation between two consecutive terms

$$x_{n+1} = x_1\left[\frac{1-\dfrac{1}{A^{E+1}}}{1-\dfrac{1}{A}}\right] - x_0\frac{1}{A}\left[\frac{1-\dfrac{1}{A^E}}{1-\dfrac{1}{A}}\right] \tag{8.10}$$

By expressing $\dfrac{1}{A^{E+1}}$ explicitly the expression can be simplified to Eq. (8.11)

$$\frac{1}{A^{E+1}}\left(x_1 - x_0\right) = \frac{1}{A}\left(x_{n+1} - x_0\right) - \left(x_{n+1} - x_1\right) \tag{8.11}$$

but according to [4]

$$x_{n+1} - x_1 = \frac{1}{A}(x_n - x_0)$$

which gives Eq. (8.12)

$$\frac{1}{A^{E+1}}(x_1 - x_0) = \frac{1}{A}x_{n+1} - \frac{1}{A}x_0 - \frac{1}{A}x_n + \frac{1}{A}x_0$$

$$\frac{1}{A^E}(x_1 - x_0) = x_{n+1} - x_n \tag{8.12}$$

$$E \cdot \ln A = \ln\left(\frac{x_1 - x_0}{x_{n+1} - x_n}\right)$$

For the most simple case with $m = 1$ and $B = 0$ we get eventually

$$E = \frac{\ln\dfrac{x_{n+1} - y_n}{x_1 - y_0}}{\ln\dfrac{V}{L}} \tag{8.13}$$

This is the so called E-factor in its most simple form. However, Gullichsen et al. [6] have developed the E-factor to take care for situations where the equilibrium cannot be described by a straight line. This is also the prevailing situation when washing kraft pulp to low washing losses, when sorption phenomena have to be considered.

The washing process itself corresponds to a certain E-value, which has to be calculated first. After that different kind of equipment can be connected in series (each with a given E-value) until the total E-value needed/calculated for the process has been exceeded.

8.4 Equipment for Pulp Washing

Many different kinds of equipment are currently being used for the washing operation. In this chapter following equipment or machinery will be described briefly. The interested reader can get much more information from companies developing and delivering this kind of equipment.

- Vacuum filters and pressure filters
- High-heat washing in continuous digesters
- Atmospheric and pressure diffusers
- Horizontal belt washers
- Displacement or wash presses
- Displacement filter washer

There are other types of machinery available, although those mentioned above are the most common.

The E-factor is frequently given for different types of equipment preferably if the units are working at constant consistency, which gives a high E-factor. Efficiency of presses is not normally described by this factor.

8.4.1 Vacuum Filters and Pressure Filters

The washing efficiency of a filter depends on both dilution and thickening of the pulp suspension and the displacement liquid being added through the nozzles to the pulp pad. A low consistency in the filter vat (1 %) and a fairly high consistency on the doctor blade is advantageous for the washing result. The *DR* value is in the range of 0.7 for a drum filter, but as we have stressed earlier, the value is dependent on the amount of wash liquor. *Figures 8.10 to 8.14* contributes to the understanding of drum filters.

Vacuum filters operate with a pressure slightly below atmospheric pressure on the filtrate side created by a fan or a vacuum pump. This facilitates the build up of a pulp web on the drum and the flow of filtrate is due to the pressure deference between the atmosphere pressure and the pressure on the inside of the drum (the filtrate side). The capacity of the filter depends to a great

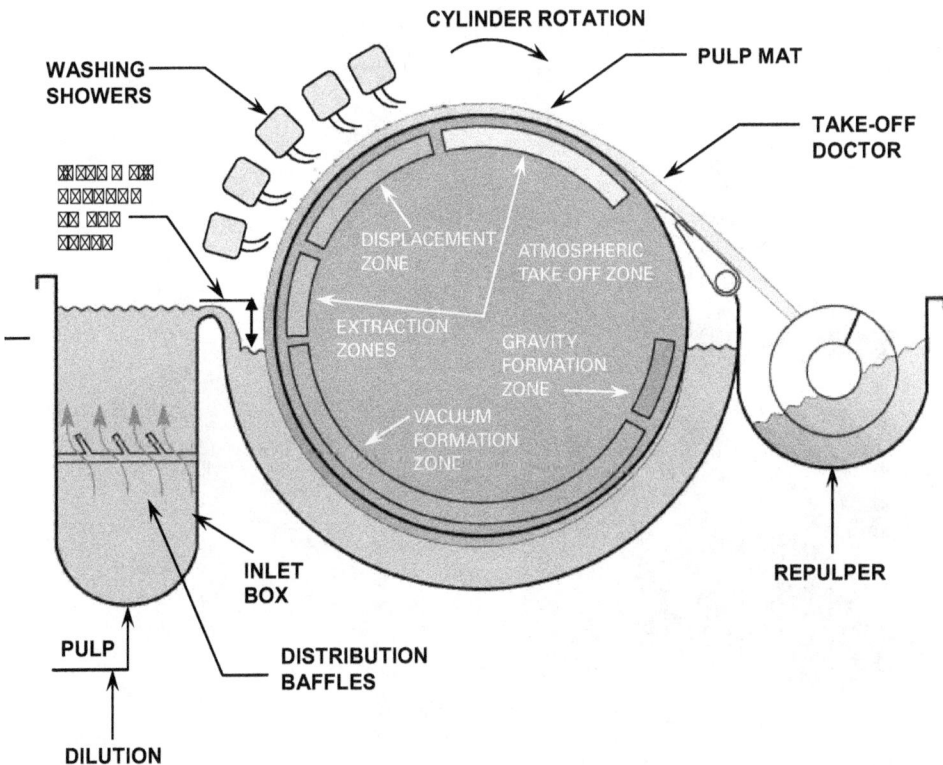

Figure 8.10. Working principle of drum filter. (GL&V).

extent on the pressure difference that can be maintained over the pulp web i.e. the filter cake. Two types of vacuum filters should be compared in this respect. The highest vacuum is achieved with a vacuum pump and an external drop leg down to a filtrate tank. The vacuum is then limited by the steam pressure of water at the actual temperature and therefore the drop leg has to have a length corresponding to the pressure drop. That is the reason why the filters must be located on a level 6–8 meters above the liquid level in the filtrate tank.

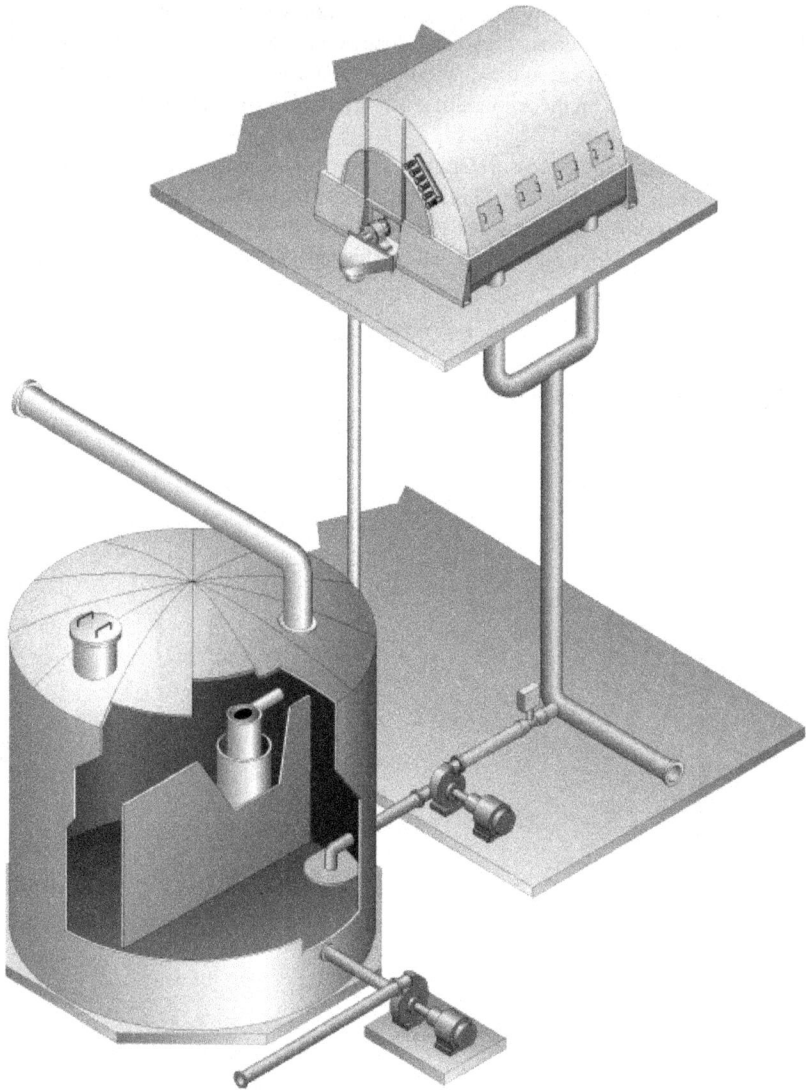

Figure 8.11. Drop leg design for a vacuum filter. (GL&V).

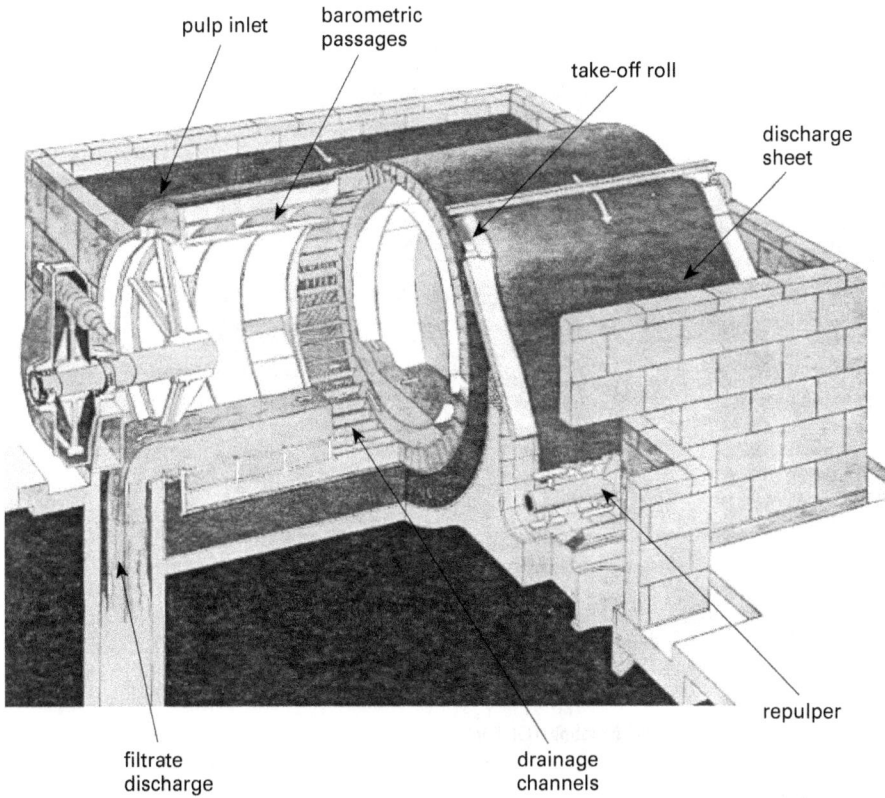

Figure 8.12. Drum filter with a built-in drop leg. (GL&V).

The other type of filter using vacuum has built-in drop legs and then the vacuum achieved depends on the diameter of the filter. Two thirds of the diameter is a practical limit with would be close to three meter water gauge.

The pressure filter has to be surrounded by a completely tight hood and the filter operates with an over pressure on the pulp side. This type of filter has many advantages. On the process side it is possible to arrange with more than one washing stage. The most common type of pressure washer is the two-stage type. Thus the filtrate from the first stage is taken out from the drum with special device and then used again for displacement washing on the drum. The two different filtrates have to be kept apart and utilized in counter current operation. The filter design is shown in *Figure 8.14*. The tight hood of the pressure filter gives a very good working environment.

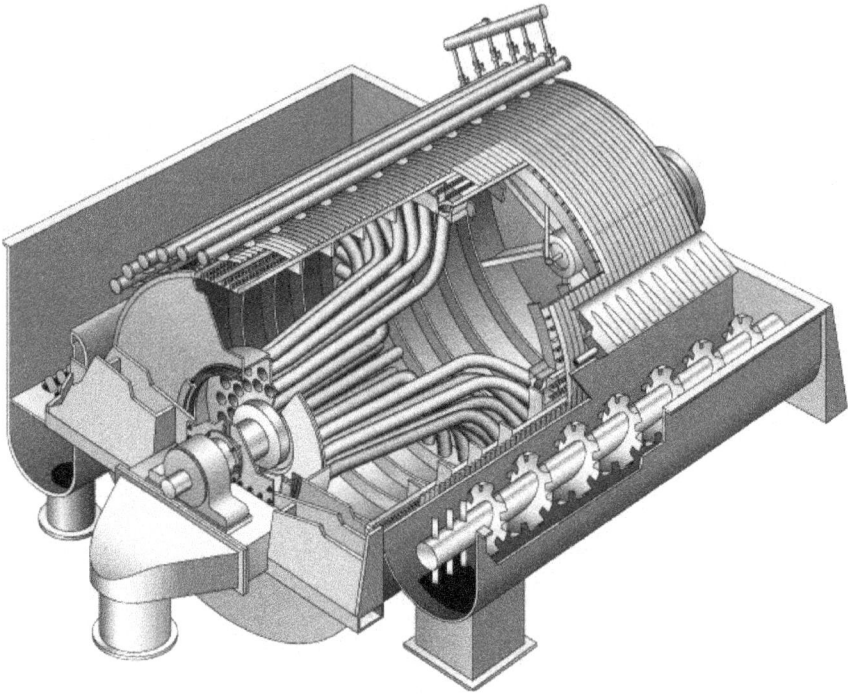

Figure 8.13. Drum filter with filtrate handling. (GL&V).

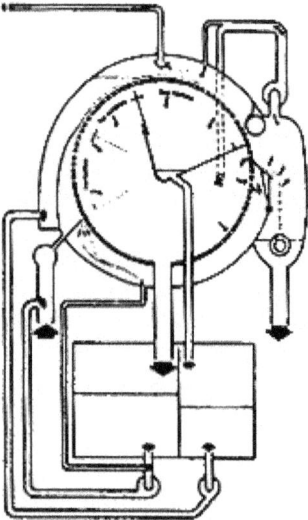

Figure 8.14. The principle for a two-stage pressure washer. (Metso Paper).

The CB (Compaction Baffle) filter, *Figure 8.15*, is a relatively compact pressurized washer that requires much less building space than conventional vacuum or pressure washers.

From the inlet box, stock enters the vat through a deflocculating slice concurrently with the drum rotation. The pulp forms a mat on the rotating drum and is compacted to high consistency by the primary baffle. This compaction expresses strong liquor from the pulp slurry prior to displacement washing.

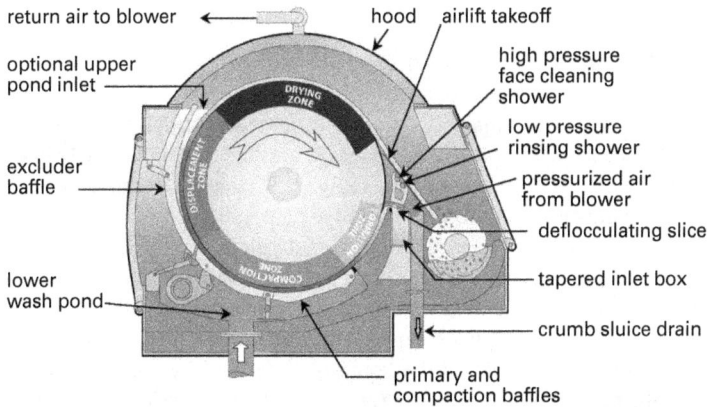

Figure 8.15. CB (Compaction Baffle) filter (GL&V).

A hydrodynamic baffle separates the feed section of the vat from the wash liquor pond. By expressing this liquor prior to application of displacement washing liquor, mixing of the two liquors which would reduce displacement washing efficiency is prevented. An inherent hydrodynamic pressure balance keeps the baffle in contact with the mat but allows the baffle to move freely to accommodate changes in basis weight. The baffle tip is free to travel to the extent that the compaction zone actually diverges.

A wash pond provides plug flow of wash liquor through the compacted mat. Wash liquor is displaced through the pulp mat by the force of the constant pressure in the sealed hood. Washed stock is removed from the washer cylinder by means of an airlift take-off into a pressurized breaker/conveyor.

Hood pressure is maintained by a small blower which re-circulates vapor from the filtrate tank in a closed circuit. Filtrate from the pulp flows horizontally from the washer valve to a filtrate tank. There is no liquor cascading from the washer to the tank to entrain air and generate foam, resulting in much smaller filtrate tanks.

8.4.2 High-heat Washing

This kind of washing is used entirely in the continuous digesters of Kvaerner or Andritz type. High-heat washing is a real counter-current extraction at high temperature. It is a very efficient operation mainly depending on a long retention time. The displacement liquid is flowing upwards and the pulp flow is downwards due to gravitational forces and a higher specific density of the pulp compared to the liquid. The washing zone is placed immediately below the extraction screen where the spent cooking liquor is taken out and brought to the flash cyclones. Very high values of DR have been achieved in high-heat washing, well over 90 %. The retention time in the washing zone has been changed through the years from four hours down to say two hours. This development has resulted in lower digesters and in a few cases already existing digesters has been lowered with as much as 25 meters in order to improve the overall operation. During the last decade the high-heat washing zone has been used also for extended cooking of the pulp simply by adding caustic soda to the displacement liquid entering at the bottom.

8.4.3 Atmospheric and Pressurized Diffusers

Atmospheric diffusers (*Figure 8.16* and *Figure 8.17*) were developed during the 1960's but have been further improved since then. The pressurized one was introduced commercially in the beginning of the 1980's. Both types of diffusers are preferably working at a constant consistency of about 10 %. They require only a very limited space in the pulp mill.

The open type is normally erected on top of the unbleached pulp storage and the pulp is scraped from the top of the diffuser into the storage. No recycling of filtrate takes place. The washing is achieved purely by displacement and the wash liquor is introduced between concentric screen arrangements from which the displaced liquor leave the system. The displacing liquid is flowing in radial direction in the pulp and the liquid is distributed by nozzles rotating/moving in the same way as the top scraper securing an even distribution of wash liquid between the concentric screens.

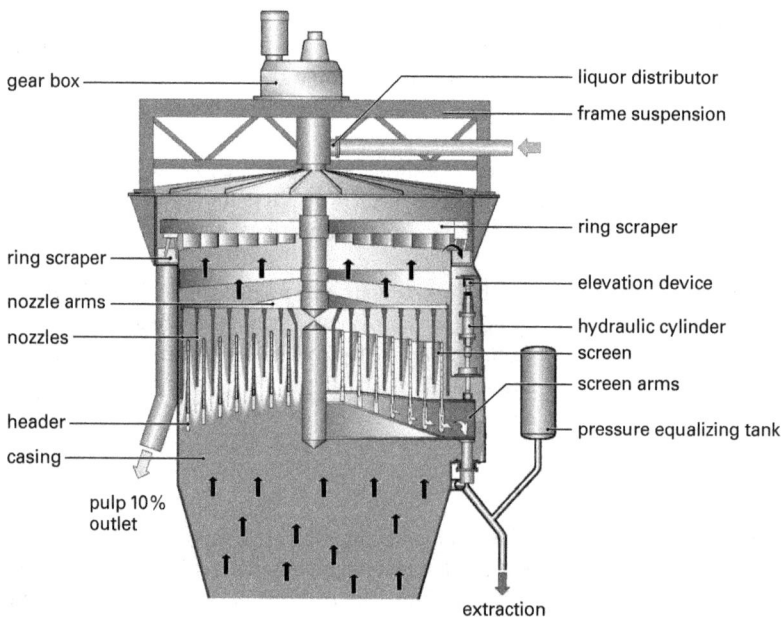

Figure 8.16. Principle of an atmospheric diffuser. (Kvaerner Pulping).

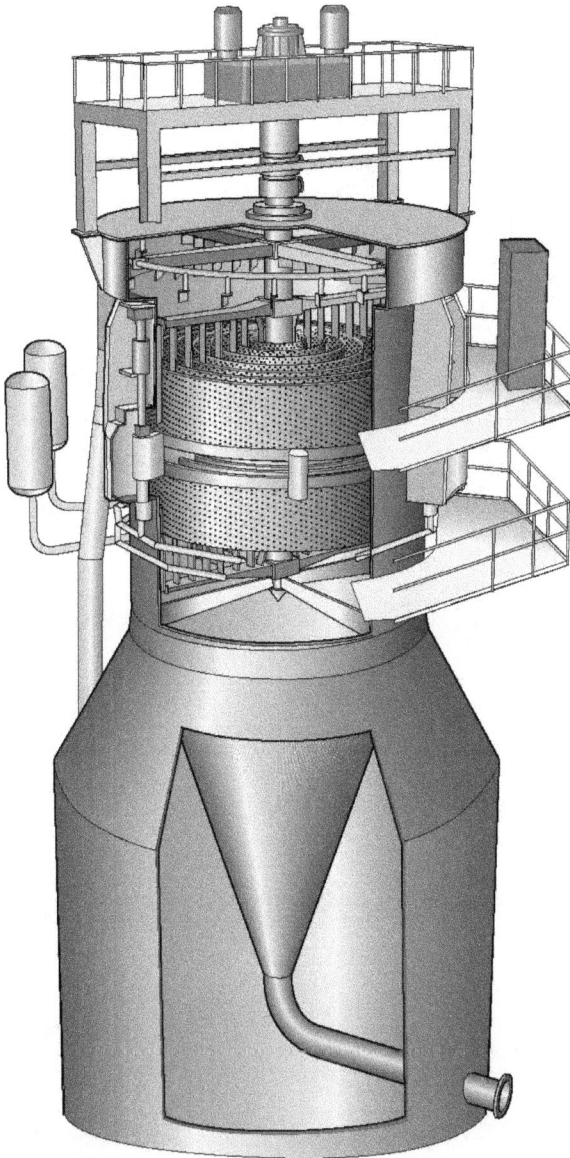

Figure 8.17. Double atmospheric diffuser on top of a pulp storage tank. (Kvaerner Pulping)

The screens are moved upwards at the same speed as the pulp flow for about 15 cm. The movement is then interrupted and the screen is pushed down to the start position by a rugged hydraulic system. This cycle is then repeated again. The flow rate of the pulp suspension is fairly low due to the large diameter (say 5 m) of the washing zone. The number of screen rings and the diameter depends of course on the production level for which the equipment is designed.

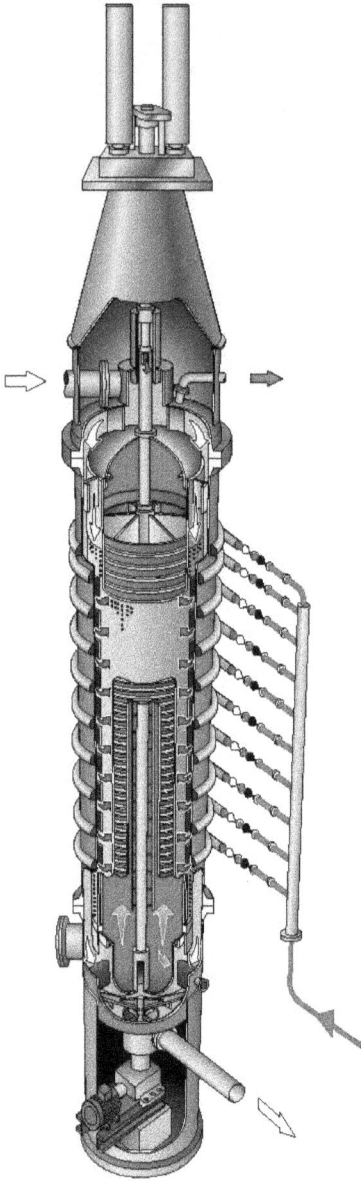

Figure 8.18. Pressure diffuser. (Kvaerner Pulping).

The pressure diffuser is entirely closed and can thus operate at temperatures above 100 °C. The pulp is pumped or blown to the top of the diffuser where the pulp inlet is designed as an annular space between the filtrate screen and the wash liquid distribution slots. *Figure 8.18* should be used for a better understanding of the operation. The washing efficiency is achieved purely by displacement since the unit is working at constant consistency all the way from pulp inlet to the outlet. The displacement ratio is high due to an even distribution of both pulp and wash liquid. Channelling is thus avoided which otherwise should have reduced the efficiency. Problem with entrained air in the pulp suspension is avoided in this type of equipment since it is closed and pressurized and operates at high temperatures.

The filtrate screen is moved up and down with the aid of a hydraulic system. The screen drum is moved downwards a certain distance at the same speed as the pulp and then suddenly brought back up again. The screen drum is slightly conical in order to facilitate the rapid upward movement. These details can be seen in *Figure 8.18*. There is a slightly different pressurized diffuser on the market that has the pulp inlet at the bottom and the outlet on top of the equipment. Both the designs achieve the same efficiency.

8.4.4 Horizontal Belt Washer

This type equipment came on stream in the late 1970's. The design of the machine is very similar to the wire section of a traditional Fourdrinier type paper machine. The pulp inlet is through a head-box at a consistency of about 4 % from which the pulp is evenly distributed on a moving wire. Underneath the wire there are suction boxes maintaining a slight vacuum, which give the necessary pressure drop (driving force) for the displacement operation to take place. As many as five washing stages, can be arranged in one single belt washer. Stage wise the operation is carried out in counter-current flow arrangement. The displacement washing is done at 10 % consistency with small exceptions and the washing efficiency is very high but of course dependent on the number of stages that the equipment has been designed for. The retention time for the whole washing operation is in the range of 30 seconds. *Figure 8.19* explains some of the main features of the machine.

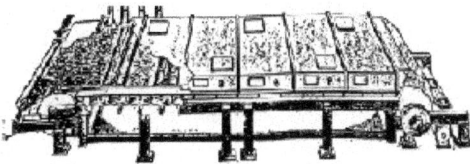

Figure 8.19. Belt washer.

8.4.5 Displacement or Wash Presses

Presses have been used in washing of pulp for a long period of time due to the squeezing action on the pulp. A high outlet consistency of e.g. 33 % means that 22 m^3 of filtrate have been removed from the pulp if the inlet consistency is 4 % (corresponding to 24 m^3 of spent liquor). In the beginning only screw presses and conical disc presses (i.e. DKP) were used. Obviously we have here a very good opportunity to achieve high washing efficiencies simply by pressing out the liquid from the pulp suspension.

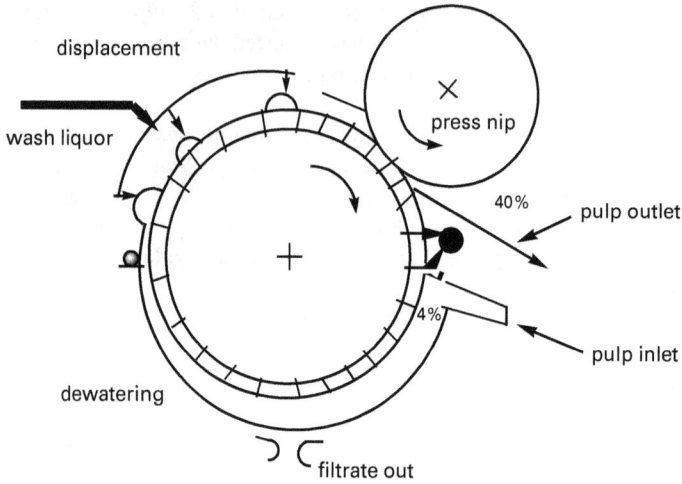

Figure 8.20. The principle of the first wash press with a solid press roll.

Further improved function was a result of the introduction of a displacement zone immediately before the press nip. This kind of displacement could only be achieved on drum presses where the flow of the displacing liquid could be controlled. Drum presses were available at that time but they were not used for washing. When the development to displacement presses started in the early 1970's two different types came on stream. One of them had two perforated drums operating with a press nip between the drums which can be seen in *Figure 8.21*.

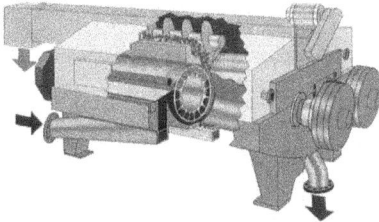

Figure 8.21. A displacement press with perforated drums. (Metso Paper).

The other one had a solid roll running against a perforated drum. In the second case the displacing liquid (introduced under a wash flap with slots) was forced to flow through the pulp pad towards the perforated drum and in the press nip remaining liquor was squeezed out of the pulp pad in one direction. The vat is tapered around the drum and so is the wash flap. The tapering depends on the drainage properties of the pulp. The principle of this displacement press is shown in Figure 8.21.

The press with two perforated drums seems to perform better than the one with a solid press roll. The latest development in this field is therefore a drum press given the commercial name "compact press" with again two perforated drums and with optimised usage of the perforated drum surface. This wash press can be seen in *Figure 8.22*, below.

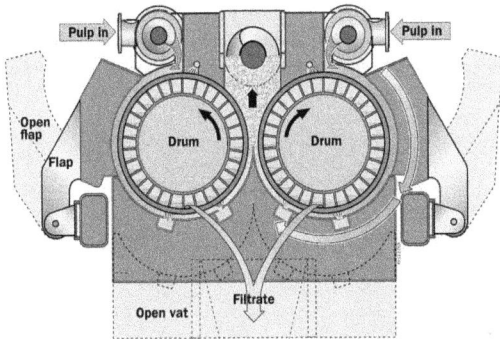

Figure 8.22. The "Compact Press". (Kvaerner Pulping).

8.4.6 Drum Displacement Washer

This type of washer is a modern generation pulp washer and it can be compared with a pressure filter in some respect but in other it is quite different. It is a completely tight unit operating under hydraulic pressure from the wash liquor pumps and as many as four washing stages can be arranged on one single drum. The vat surrounds the drum to a very large extent. There is an opening slot for the pulp discharge after one revolution on the drum.

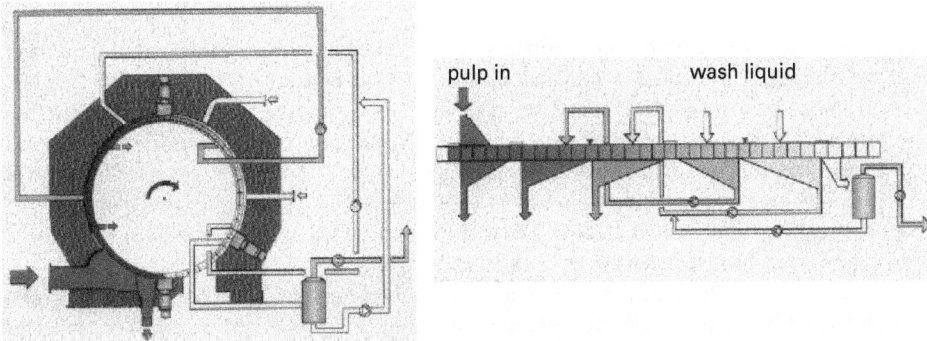

Figure 8.23. Operating principle for a 2-stage DD-washer. (Andritz).

There are no shower pipes as it is on ordinary drum washers but they are replaced by an extended vat including slots for distribution of wash liquid. Inside the drum there are boxes collecting the displaced liquid in separate fractions which is necessary in order to achieve multistage washing.

The DD washer is very efficient and a four stage counter-current unit is compatible with a traditional washing plant based on four drum washers operating in series. *Figure 8.23–8.24* should be used to get a better understanding of the DD filter performance. In Figure 8.23 the counter-current principle is illustrated by inlet flow lines that are colourless for clean wash water, slightly coloured for intermediate filtrate and darkly coloured for spent liquor. The system shown in the figure is a two stage system. The principle of the washing process is best explained by the horizontal presentation of drum displacer. Here we can see the possibility to carry out the counter current washing section wise in order to increase the efficiency of the two stages.

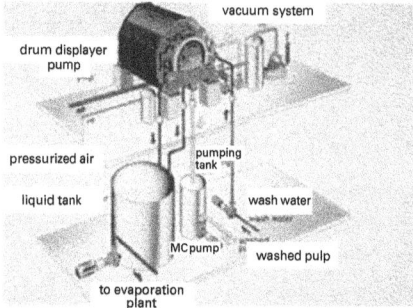

Figure 8.24. A DD-washer and schematic system (Andritz).

Much more could be said about every different kind of washing equipment and the list above is not at all complete. The interested student can easily find much more information about this topic simply by getting in touch with the large supplier companies developing and manufacturing this kind of equipment.

8.5 Commercial Applications

A number of different commercial washing systems will be described in this part of the chapter and the equipment presented in previous part will appear in a great variation in the washing plant. Almost every washing plant is tailor made in order to optimise the performance both from the process and economic point of view. Quite a few washing plants are fairly similar though especially if the digester plant is based on continuous cooking. A more or less standardized system here was high-heat washing followed by an atmospheric diffuser. In the beginning of continuous washing after batch digesters three filters in series became almost a standard in exchange of the obsolete batch system still in use (see *Figure 8.25*).

The washing results from early continuous systems were considered quite good and sufficient for a number of years. The recovery of spent liquor could be as high as 97–98 % for a kraft pulp mill and 90 % for sulphite pulp mills. However with continued concern for environmental protection it became obvious that losses of spent cooking liquor had to be reduced even more. The authorities claimed that 99 and 97 % for kraft and sulphite pulp mills respectively had to be achieved.

Batch cooking systems normally include a blow tank from which the pulp is pumped at a low consistency say 4 % and therefore equipment like drum filters and presses are suitable. Modern continuous digesters blow the pulp at 10 % consistency from the high-heat zone directly to a diffuser, pressurized or not, working at the same consistency as the digester which is advantageous for the system design. This is demonstrated in *Figure 8.26* below with a system starting with digester and diffuser followed by seven DD-washers for post oxygen washing and washing

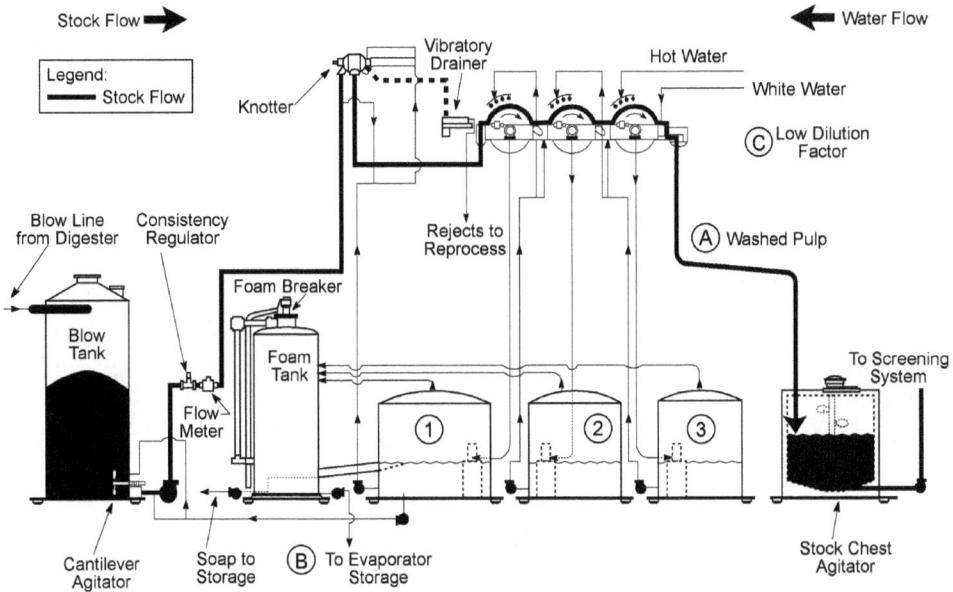

Figure 8.25. Three drum washers in series. (GL&V).

between bleaching stages. More about washing of bleached pulp will be found in the chapter about bleaching technology.

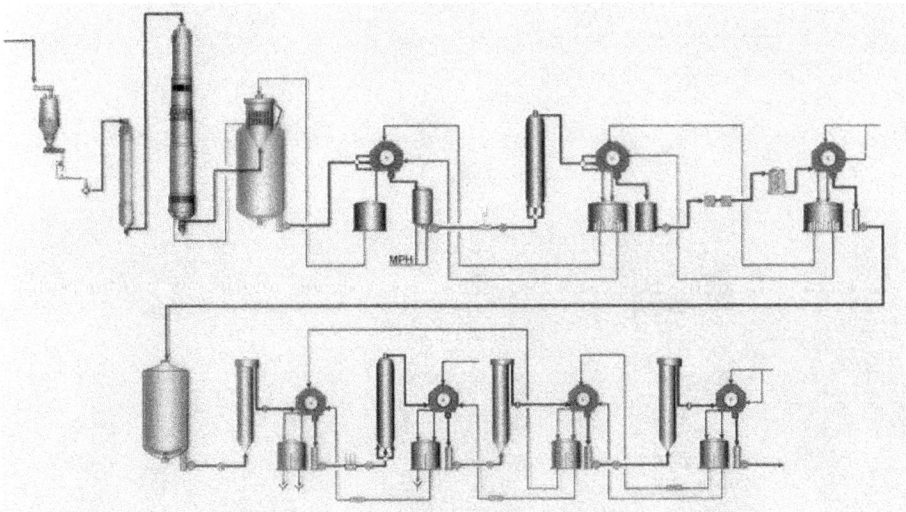

Figure 8.26. A total pulp production system with seven DD-washers. (Andritz).

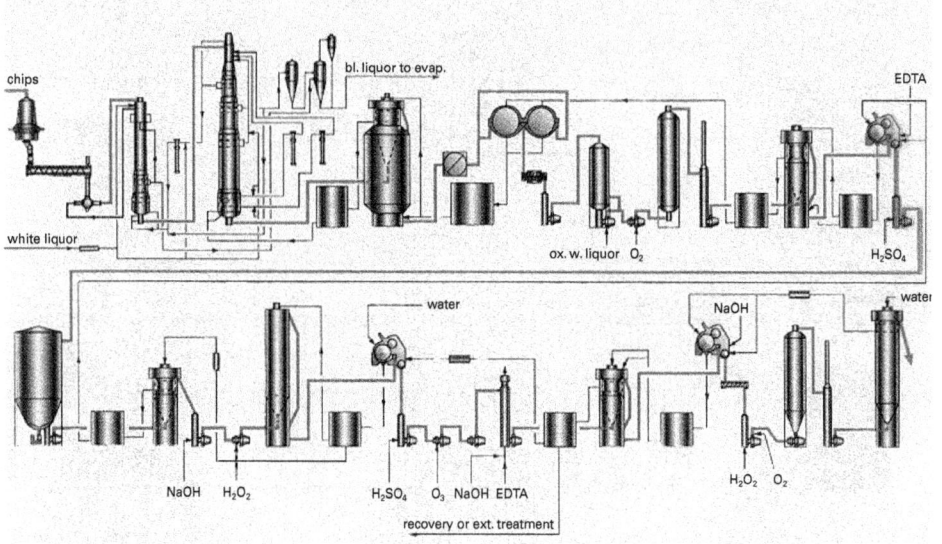

Figure 8.27. A total fibre line based on diffusers and presses (Kvaerner Pulping).

In following *Figure 8.27* we have a different fibre line based on high-heat washing followed by diffusers and wash presses. The whole process is almost completely counter current based on a totally chlorine free bleach plant. An icon showing the old wash press with a solid press roll has been used in the flow diagram.

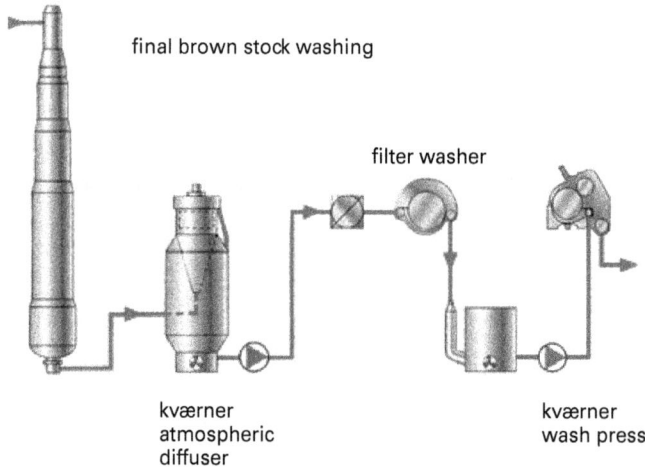

Figure 8.28. Final washing of unbleached pulp with screening involved (Kvaerner Pulping).

post oxygen washing

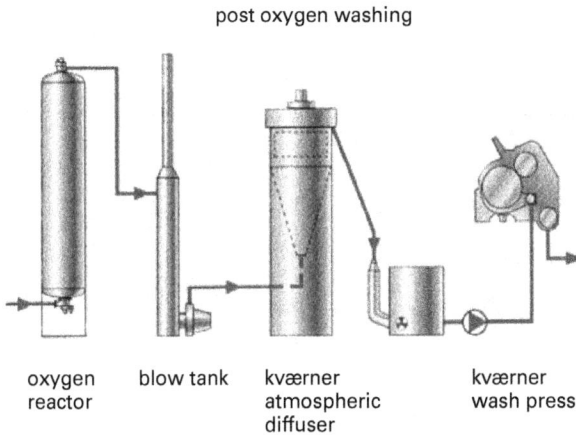

oxygen reactor | blow tank | kværner atmospheric diffuser | kværner wash press

Figure 8.29. Washing after oxygen delignification. (Kvaerner Pulping).

Computer simulations of washing systems have been available for many years. One of the oldest is GEMS. A later development is a simulation called FlowMac which is based on Extend. A lot of system calculations can be made on MS Excel as well.

8.6 Energy Demand and Environmental Impact of Pulp Washing

Both energy demand and environmental impact by the pulp industry have been decreased a lot through the last decades and this trend will continue in the future. The improved pulp washing has contributed a lot to this development. The energy rich black liquor has to be evaporated to a high concentration (normally 65–75 % dissolved solids) before being brought to the recovery boiler. Therefore all dilution of the black liquor during washing has to be limited as much as possible and there are different means of doing this. The old way of reducing dilution was by adding further washing stages often of the same kind as before but a much more efficient method has appeared to be installation of new equipment using less electric power and less washing liquid.

The importance of this is described by *Figure 8.30* where we have on the vertical axes recovery of DS and on the horizontal axes the water usage expressed in relation to the original black liquor amount. There are three straight lines in the diagram and the one with at a slope of 1.0 represents the original black liquor concentration. The line with lower slope corresponds to the concentration of the liquor being sent from washing to evaporation. Finally the line at a slope of 4 to 5 corresponds to the concentration of evaporated liquor. The distance between i) the intersection of the weak liquor line and the recovery curve and ii) the intersection of the thick liquor line and a straight horizontal line on the same level of recovery tells how much water has to e evaporated. The student should as an exercise work through the different expressions needed to verify the situation.

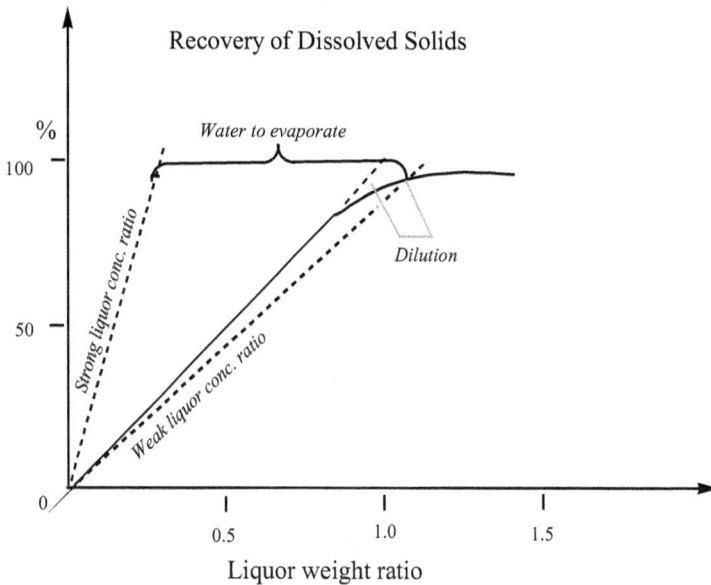

Figure 8.30. How dilution in washing is connected to load on evaporation.

In order to limit dilution in pulp washing different types of presses were introduced. The most efficient ones proved to be the wash presses where a displacement operation took place just before the press nip. The displacement itself was not very efficient but in combination with pressing to fairly high consistencies the final result was very good. One could say that a ‚low dilution technique' was introduced during the 1970's. Measured values for *DR* stayed normally on a level 50–55 % according to the definition. The dilution factor DF was in the range of one m³ per ton of pulp. In order to understand a figure like this it should be compared with what could be achieved on drum washers in series. With 3 washers in series a DF of 3 m³ per ton of pulp was quite normal and 4 washers in series required 2.5 m³ per ton of pulp. If the difference in DF is say 1 m³ per ton of pulp comparing two different washing plants it corresponds to another 1000 m³ to evaporate per day at a production of 1000 ton per day.

Also in the evaporation plant the number of stages has to be optimised against the steam consumption. Capital cost versus operating cost. If an existing mill can reduce the DF with 1 m³ per ton of pulp it would reduce the steam consumption with about 0.25 ton per ton of pulp with 5 evaporation stages. Then it depends on the total energy balance if the lower consumption could be given full credit. Normally the evaporation uses 4 bar steam and the availability of the so called low pressure steam has to be studied in order to understand the economy in a changed situation like this.

The concern about energy usage should not be limited to only steam consumption in the evaporation plant but also include for power usage. The circulation of spent liquor only for transport purposes is power consuming especially for drum washers. A washer requires an inlet fibre consistency of say 1 %. The discharge takes place at 12 % consistency on the doctor table. As a result of this situation we get more than 90 m³ per ton of pulp in an internal circulation on every washing stage. Taking into consideration only the height difference between the level where the washers are placed and the spent liquor tanks, which can be almost 10 meters, it is obvious that the situation requires a lot of pumping energy. Only liquor circulation consumes 3 kWh per ton of pulp and washing stage based on the height difference. If we include for other pressure losses the power consumption for pump circuit will be on the same level as the drive of the drum washer.

When using a press we have to circulate say 22 m³ per ton of pulp going from 33 % to 4 % consistency. If the press comes after a washer only 15 m³ have to be circulated in order to get from 12 % down to 4 % consistency. The distance between the floor where the presses are placed and the tanks below is normally less than half the distance mentioned above for washers. The press itself uses some 10 kWh per ton and the liquor circulation uses less than one kWh per ton.

The environmental impact depends on the final washing loss preferably expressed as kg COD per ton of pulp. However, as a result of the very low losses now achieved it is reasonable to claim that a washing plant has a very limited impact on the environment.

8.7 Washing Results

In this section some washing results collected at operating pulp mills will be discussed. Salt cake losses will be compared with COD and DS losses at variable dilution factors and different types of equipment and the markedly different behaviour.

Saltcake losses have been determined by a standardized method since 1974 (a so called SCAN method). The measurement is based on both a standardized sampling and followed by a specified extraction of all sodium by the usage of a hydrochloric acid solution. The result is then expressed as kg Na_2SO_4 per ton of pulp. With this method also sorbed sodium is included in the result. The sorbed part of the sodium cannot be washed out of the pulp pad in a normal washing operation. Therefore a parallel method of determing saltcake losses has been developed and used through the years. Instead of an acid solution plain water is used for extraction of sodium from the pulp sample. The value obtained in this way is called "easily washable" saltcake losses (in a couple of diagrams below the method is denoted "acc to Kamyr"). The amount of saltcake we get by this method is considerably lower than the amount the acid based extraction gives. The sorbed amount of sodium on the fibre surface varies a lot and a number of investigations have been made on many different pulps. *Figure 8.31* gives an idea about the size of the adsorption.

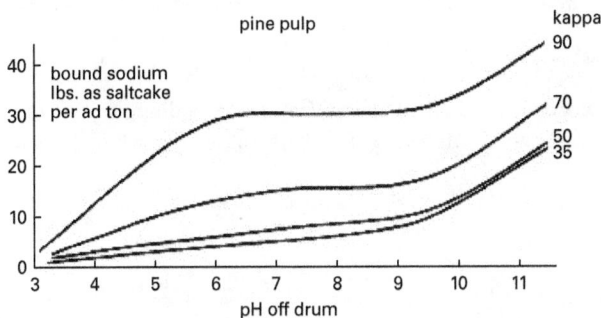

Figure 8.31. Sorbed amount of sodium versus pH [9].

Description of adsorption of sodium on the cellulose fibre can be made by using the adorption isotherm according to Langmuir. The adsorbed amount is expressed by

$$w = \frac{A \cdot B \cdot c}{1 + B \cdot c} \tag{8.14}$$

w = adsorbed sodium [kg/kg pulp]
A = parameter standing for maximum adsorption capacity
B = constant
c = sodium concentration in the liquid phase

The values of the parameters varies when comparing different investigations. However it is fairly safe to say that the remaining sorbed amount of saltcake at the clean end of the washing plant is in the range of 5 kg per ton of pulp.

Standardized methods are being used for determination of COD as well. The problems with sorption is not equally pronounced in this case but sorption phenomena should be remembered and considered when new equipment is to be chosen. There are results available showing that presses are more efficient than other kind of equipment when reducing the amount of dissolved organic material in the pulp. Modern kraft pulp mills have washing losses in the range 5–7 kg of COD per ton of pulp. The lower figure could be valid for unbleached pulp whereas the slightly higher figure could be true after an oxygen bleaching stage. The saltcake losses in a case like this are almost entirely due to sorption phenomena and washable salt cake is close to zero.

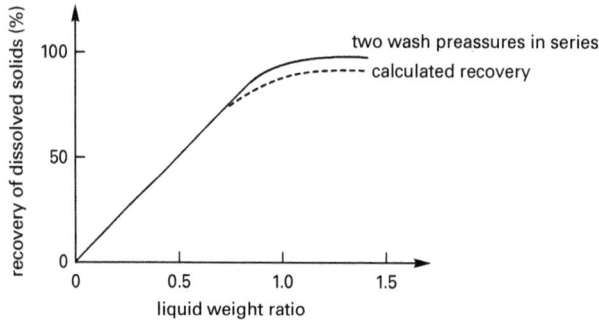

Figure 8.32. Washing result on two Wash presses in series.

The displacement ratio (*DR*), depends a lot on whether COD, *DS* or sodium is used as the solute. Following *Figure 8.33–8.35* show the difference. The diagrams are describing the situation in a post oxygen washing stage.

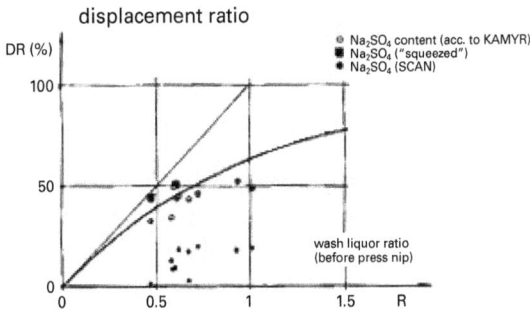

Figure 8.33. Comparing different methods of determination of DR-values.

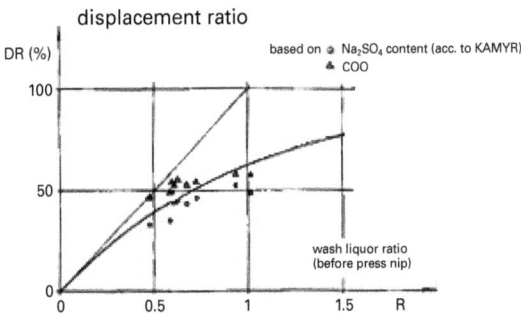

Figure 8.34. Comparing DR-values for different solutes.

The washing efficiency when comparing DS and COD as solutes is almost the same which is explained by the small difference between washable sodium and COD/DS. It should be undersood that the DS value contain inorganic salts like salt cake while the COD value does not.

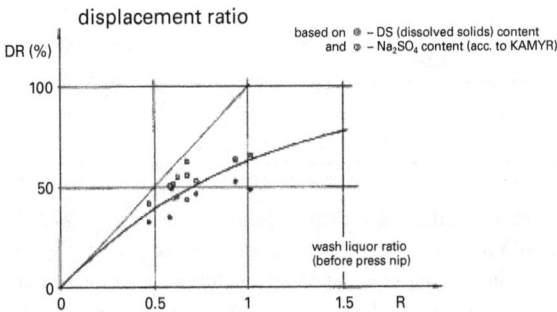

Figure 8.35. Comparing DR-values for washable sodium and DS.

The studies discussed above have been carried out on a wash press and it is obvious that the displacement ratio on this kind of equipment is not very high. If the displacement is done at say

20 % consistency and a dilution of only 1 m³ per ton of pulp is used on the press we have 4 m³ of liquid in the pulp pad and use only 3.3 m³ as wash liquid (based on 33 % outlet consistency). In the figures above the wash liquor ratio (the relation between wash liquor amount and the liquor amount in the pulp pad) is the independent variable and it is kept below one. In the press nip itself it is likely that we have total mixing during the process of increasing consistency.

Improving old washing plants by adding another stage is quite common. In *Figure 8.36* the result of adding a wash press to a 4-stage drum filter washing plant is shown. The recovery of spent liquor went above 99 % and with a lower dilution of the recovered liquor. The low washing loss depended also on the fact that the drum filters could be operated with a higher outlet consistency which is beneficial to the total result.

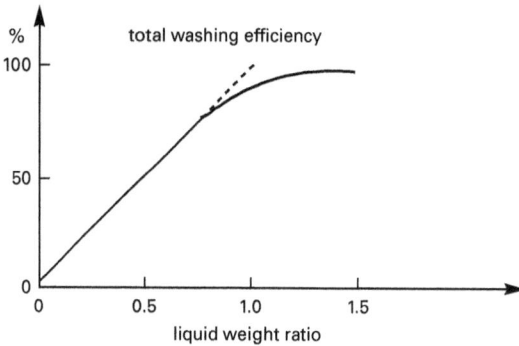

Figure 8.36. Well above 99 % efficiency after expansion with a wash press.

More comprehensive investigations have been made in order to find explanations to the different behaviour when comparing washing of inorganic salt cake versus organics (COD). The data in *Table 8.1* represent a relation between residual COD and saltcake after washing. The study has been carried out on three different types of equipment. According to these results it is easier to get to low levels of COD in relation to salt cake on a wash press compared to drum washers and pure displacement washers.

Table 8.1. Relation between COD and salt cake at post oxygen washing.

Displacement Washer	1.5–2.5
Filter Washer	1.0–1.5
Press Washer	0.5–1.0

The same situation is clearly demonstrated in diagram form in following *Figure 8.38–8.39*. The investigation refers to washing of pulp after oxygen delignification and different washing systems have been compared. Two presses in series give a very low residual COD content but a rather high salt cake content (including sorbed sodium). Two filters followed by a press give the best result when considering both solutes i.e. approximately 5 kg per ton of pulp. The relation between the two solutes is close to one.

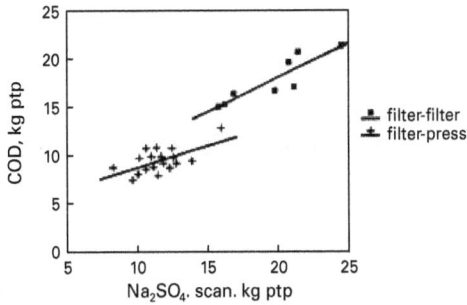

Figure 8.37. COD as a function of saltcake comparing different systems Ref 5.

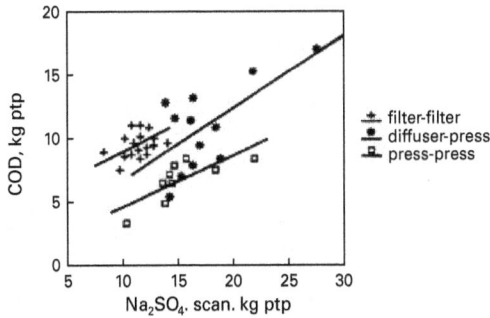

Figure 8.38. COD as a function of saltcake comparing different systems Ref 5.

It is of great importance to achieve low losses in post oxygen washing for two reasons. The first is the fact that all the solute (DS, COD or $NaSO_4$) being washed out until this point is recycled and utilized in the pulp mills recovery system. The other is the fact that a very low washing loss after oxygen delignification gives a low consumption of expensive bleaching chemicals and also lower discharge of pollutants from the bleach plant. However there is an economic limitation to be considered.

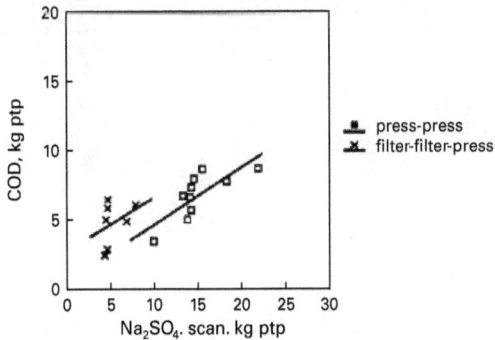

Figure 8.39. COD as a function of salt cake comparing different systems Ref 5.

The control system for a washing plant is not yet based on the true washing result. Instead it is aimed att minimizing variations in the production level and also in the washing result. The rotation (rpm) of drums (filters and presses) is controlled by the inlet flow which is kept at a constant consistency by a CRC loop. CRC stands for Consistency Regulator and Control. The rpm (revolutions per minute) of a filter can also be controlled by the vat level. Level indicators are used in both pulp chests and filtrate tanks. Buffer volumes are important in order to meet production variations in other parts of the mill.

8.8 Process Control

The pulp flow regulator also control the amount of wash liquor being used and the need for dilution. Presses can also be controlled by the load on the press nip to ensure a constant outlet consistency.

The capacity of a washing plant depends on many parameters like e.g. the drainage rate of the pulp and the temperature of both the pulp suspension and the wash liquor being used. The porosity of the pulp web being formed and the specific surface of the fibres are other parameters of interest. The pulp web has not a constant porosity but the situation is even more complicated due to the compressibility of the web. A scientific evaluation or discussion of the phenomena involved is far outside the scope of this chapter.

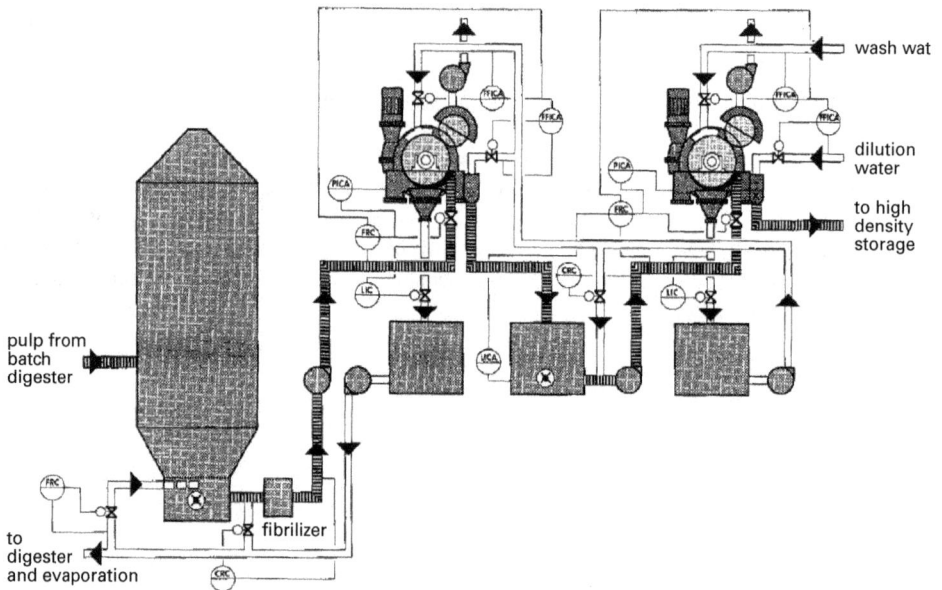

Figure 8.40. Schematic PID for a washing plant.

8.9 Nomenclature

Parameter	Description	Unit
A	$L/m*V$	
A and B	Langmuir constants	
a_1 and a_2	Specific mass transfer surfaces	m^2/m^3
B	Constant	
B_0	$1/S$ Bodensteins number	dim = 1
c	Concentration of solute in bulk flow	kg/m^3 or weight-%
C	Dimensionless concentration	dim = 1
c^*	Equilibrium concentration	kg/m^3 or weight-%
c_0	Original concentration of solute	kg/m^3 or weight-%
C_F	Pulp consistency	weight-%
c_s	Concentration of solute in stagnant liquid	kg/m^3 or weight-%
DF	Dilution factor	m^3/ton of pulp
D_L	Longitudinal dispersion coefficient	m^2/s
DR	Displacement ratio	dim = 1
DS	Dissolved solids	kg/m^3 or weight-%
E	Nordén factor number of ideal separation stages	
K_0 and K_1	Equilibrium constants	
k_1 and k_2	Constants	
L_p	Liquor following the pulp	m^3/ton of pulp
M	Equilibrium constant	
n	Stage number in a counter-current cascade	
p	Subscript = pulp	
r	$u*t/z$ Number of displacements heights	dim = 1
R	Wash liquor ratio – V_w/L_P in diagrams	
S	$D_L/u*z$ Dimensionless dispersion number	dim = 1
s	Index = stagnant part	
t	Time	s
T	t/τ Dimensionless time	dim = 1
U	Superficial velocity	m/s
V	Volume of filtrate	m^3/ton of pulp
V_0	Volume of original liquor	m^3/ton of pulp
w	Sorbed amount of sodium according to Langmuir	kg/ton of pulp

Parameter	Description	Unit
w	Subscript = wash liquor	
W	Dimensionless sorbed amount	dim = 1
x and y	Concentration of solute in L and V respectively	weight ratio
z	Thickness of pulp web	m
ε	Bed porosity	dim = 1
ε_d	Volume of free flowing liquid	dim = 1
ε_s	Volume of stagnant liquid	dim = 1
λ	Constant	
τ	Residence time	s

8.10 Further Reading

Crotogino, R H., et al. (1987) *Tappi Journal* June.

Grähs, L-E. (1974) Washing of Cellulose Fibres. *Dissertation*. Chalmers University of Technology.

Gullichsen, J., Fogelholm C-J. (2000) *Papermaking science and technology, volume 6A, chemical pulping*. Helsinki, Finland.

Gullichsen, J., Östman, H. (1976) *Tappi* 59(6).

Kommonen, F. (1968) *Paperi ja Pu* 50(6)347–357.

Lindström, L-Å., Nordén, S. (1990) *44th Appita General Conference*.

Nordén, H. (1965) *Kemian Teollisuus* 22, p 430.

Nordén, H. (1966) *Kemian Teollisuus* 23, p 343, 586.

Rosen, A. (1975) *Tappi* 58(9).

9 Chemistry of Bleaching of Chemical Pulp

Göran Gellerstedt
Department of Fibre and Polymer Technology, Royal Institute of Technology, KTH

9.1 Introduction

In kraft (or sulfite) pulping, the process does not permit a complete delignification for reasons of pulp yield and quality. The remaining small amounts of lignin, usually in the order of ~2–5 % (on pulp) depending on wood species and process details, give the fibers a brownish color and, therefore, the manufacturing of white fiber products require a further delignification and bleaching of the fibers. This is accomplished by a series of alternating oxidation and extraction treatments ultimately leading to an almost lignin-free fiber. In its simplest form, the chemistry can be written as shown in *Figure 9.1*.

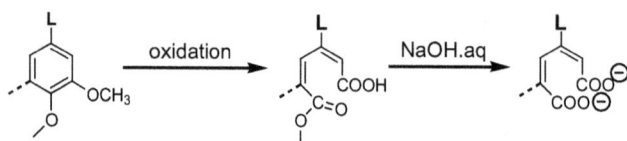

Figure 9.1. General mechanism for the major reaction encountered in bleaching.

In old bleaching technology, a treatment of the pulp with aqueous elemental chlorine (C) followed by alkaline extraction (E) was used as the predominant delignification stage. The final brightening was done with chlorine dioxide (D) and the complete sequence being e.g. CEDED. Already in the 1960s, large efforts were, however, made to substitute at least part of the chlorine with more environmentally friendly alternatives such as oxygen. The discovery that small amounts of magnesium salts, added to an alkaline oxygen stage (O), to a large extent protected the pulp from being degraded paved the way for the technical implementation of oxygen stages with the first commercial installation in the World being done in 1970. Since the effluent from an oxygen stage can be incorporated in the chemical recovery cycle of a kraft mill, this resulted in a much improved environmental situation since the oxygen stage was able to remove more than 50 % of the lignin that remained in the pulp after the cook. Thereby, the amount of dissolved material in the remaining bleaching effluent was reduced to a considerable degree.

The successive installation of oxygen stages together with a gradual replacement of some of the chlorine in the C-stage with chlorine dioxide that started in the 1970s resulted in further improvements of the environmental situation in kraft mills. The gradual and rather slow changes in bleaching technology towards more environmentally friendly sequences were, however, dramatically altered when it was found that the bleaching effluent from a kraft mill contains small amounts of chlorinated dibenzodioxins and dibenzofurans. In addition, a large variety of other low molecular weight chlorinated compounds, aliphatic and aromatic, were also identified in bleaching effluents. Findings such as these resulted in a strong societal pressure on the industry and in a rapidly increased rate of change of bleaching technologies. Elemental chlorine was successively phased out and in the early 1990s, the modern bleaching sequences of today became established with oxygen (O), chlorine dioxide (D) and hydrogen peroxide (P) as the predominant oxidation agents. Some mills have also installed an ozone (Z) stage. The bleaching technologies are usually referred to as elemental chlorine free (ECF) and totally chlorine free (TCF) respectively. Typical sequences are OD(OP)DD and OQ(OP)Q(PO) where Q stands for treatment of the pulp with chelating agent.

9.2 The Structure of Residual Lignin

For bleachable grades of pulp, the kraft cook is carried out to kappa numbers in the order of 25–35 for softwood and 15–20 for hardwood. The kappa number value includes, however, not only the amount of residual lignin but also other oxidizable structures present in the unbleached pulp. The most important of these is hexenuronic acid which is attached to the xylan but other types of reactive structures/functional groups are also present and may, depending on pulping conditions and wood species, contribute several kappa units to the overall kappa number as exemplified in *Table 1*. The desirable brightness increase that is the objective of bleaching is, however, unequivocally related to the elimination of the lignin component as shown in *Figure 9.2*.

Table 9.1. Contributions to the kappa number from lignin, hexenuronic acid and "non-lignin" structures in various unbleached pulps (Li 1999).

Pulp type, Kappa number	Lignin[1]	HexA[1, 2]	Non-lignin[1, 3]
Pine kraft, 18.6	14.3	1.9	2.4
Pine soda/AQ, 18.9	16.8	0.3	1.8
Birch kraft, 13.6	6.5	4.5	2.6
Eucalyptus kraft, 17.2	9.0	7.4	0.8

1) calculated in Kappa number units
2) HexA = Hexenuronic acid
3) Non-lignin = non-specified but oxidizable structures

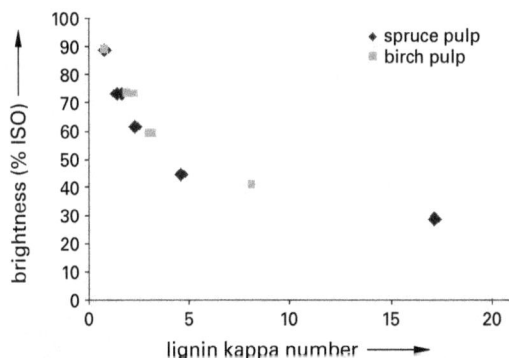

Figure 9.2. Brightness development for industrial kraft pulps as a function of the content of lignin, measured as lignin kappa number. Bleaching sequences: ODEQP (spruce) and OQ(OP)Q(PO) (birch).

The remaining (residual) lignin that is present in the fibers has a chemical structure that apparently is much different from the original lignin in wood. Analytical data show a large decrease in the amount of remaining β-O-4 structures (10–20 % of the original value), an increased amount of free phenolic hydroxyl groups (25–35 units per 100 phenylpropane units) and chemical linkages between lignin and all the major polysaccharides, viz. cellulose, xylan and glucomannan. Furthermore, several new aliphatic methylene groups can be found by ^{13}C NMR analysis of isolated lignin samples indicative of a less reactive chemical structure. Whereas the changes in β-O-4 structures and phenols are expected consequences of the kraft pulping chemistry, the presence of reduced carbon atoms and lignin-carbohydrate linkages (LCC) are not obvious. It has been suggested that the former may be created as a side reaction during the kraft cook but a more likely explanation can be found in the fact that several minor lignin structures with a reduced side-chain are already present in the wood lignin. These, together with fatty acid(s) which seem to be incorporated by chemical linkages to the lignin during pulping, may well explain the rather high amount of methylene groups that are found.

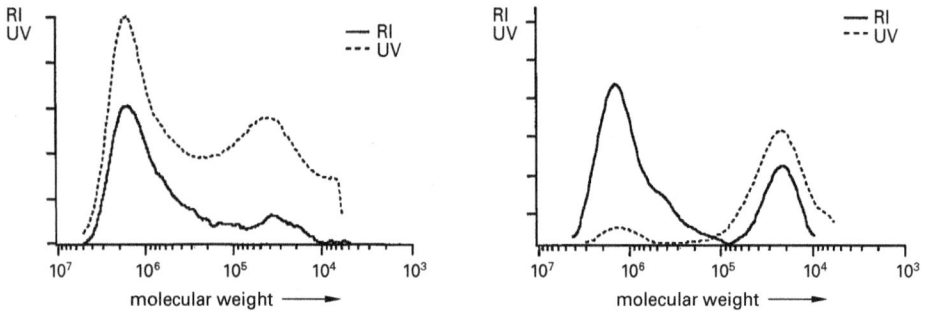

Figure 9.3. GPC-analysis of unbleached kraft pulp from pine (left) and birch (right) after dissolution in dimethylacetamide/lithium chloride. RI = refractive index, UV = ultraviolet detection. Molecular weight calibration with pullulan standards. (Karlsson 1997).

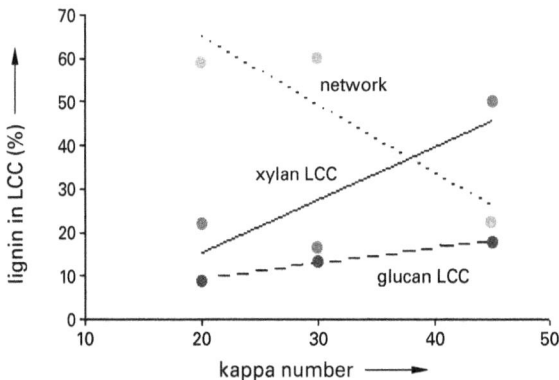

Figure 9.4. Amount of lignin chemically linked to glucan, xylan and glucomannan-xylan (network) respectively in softwood kraft pulps having different kappa numbers. The amount of non-linked lignin is ~10 % of the total (Lawoko 2005).

The presence of chemical linkages between lignin and polysaccharides in wood has been the subject of numerous studies and their presence is fairly well established. For kraft pulps, on the other hand, it is not until very recently that unequivocal proof has been obtained for the presence of such linkages. By GPC-analysis of unbleached kraft pulps from both pine and birch after dissolution in dimethylacetamide/lithium chloride, it was found that the UV-absorbing material (i.e. lignin) is connected to the high as well as the low molecular weight fraction of the polysaccharides (*Figure 9.3*). By a selective degradation of the cellulose portion of kraft pulp by an endoglucanase followed by swelling, alkaline extraction and pH-fractionation, a more detailed picture of the relative amounts of lignin-carbohydrate linkages to the various polysaccharides has also been obtained (*Figure 9.4*). Thus, almost all residual lignin present in a kraft pulp is chemically linked to the polysaccharides with the majority being linked to the hemicelluloses.

9.3 Oxygen Delignification Chemistry

9.3.1 General Aspects

The reactions between kraft pulp and oxygen in alkaline media have been the subject of numerous studies during a long period of time. The desirable oxidation of lignin is, however, always accompanied by a non-selective oxidative degradation of the cellulose and it was not until the beneficial effect of adding a magnesium salt was found that oxygen could be used as a commercial bleaching agent (O-stage). Nevertheless, also in the presence of magnesium(II), a certain degradation of the cellulose takes place during an oxygen stage and, typically, a viscosity loss of some 200 units (dm^3/kg) is observed.

Whereas the first generation of oxygen delignification (oxygen bleaching) stages were used for achieving a very limited degree of the overall delignification required after the cook, the later development has been directed towards more extensive degrees of delignification. The growing environmental awareness, the low cost of oxygen and the much improved performance of modern O-stages (and double O-stages) have been strong driving forces. This development is schematically shown in *Figure 9.5*.

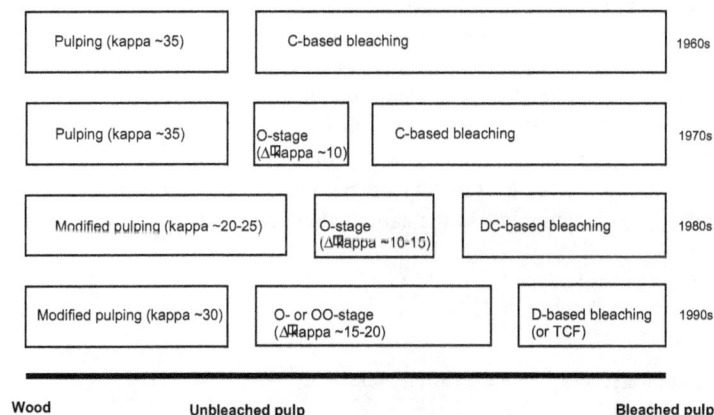

Figure 9.5. Schematic illustration of the development of pulping and bleaching technologies for softwood kraft pulp (Rööst 2004).

At present, technical oxygen delignification is normally carried out at medium consistency with all chemicals, i.e. gaseous oxygen, aqueous sodium hydroxide and magnesium salt, being introduced in a mixer followed by further reaction in a tower during 30–60 min at around

90–100 °C. Shorter oxygen stages as well as more powerful dual stages are, however, also used. The comparatively low rate of oxidation encountered with oxygen (autoxidation), due to the presence of two unpaired electrons in the ground state (triplet), is thus compensated for by reaction in an alkaline medium, using an intensive mixing and having a high temperature. In order to compensate for the low solubility of oxygen in water, a high oxygen pressure (~0.6 MPa) is also applied. Under these conditions, an efficient mass transfer of oxygen to the reacting sites in the fiber wall takes place as schematically illustrated in *Figure 9.6*. Therefore, a comprehensive delignification takes place but without much carbohydrate dissolution. In mill operation the degree of delignification can be around 60–75 % depending on wood species and exact reaction conditions.

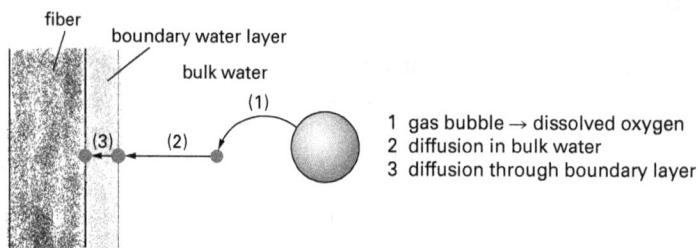

Figure 9.6. Schematic view of the transfer of oxygen from the gaseous phase to the reaction sites in the fiber wall.

Studies with lignin and polysaccharide model compounds have shown that the major reactions of quantitative importance in an oxygen stage are the oxidation of phenolic end groups in lignin and the oxidation of cellulose chains. The former reaction gives rise to delignification whereas the latter induces a chain cleavage and thus a reduction in viscosity which ultimately may lead to a loss of fiber strength. In the lignin reaction, the initial step is an electron transfer from a phenolate anion to oxygen thus giving rise to one phenoxy and one superoxide radical (*Figure 9.7*, reaction 1). Both these species are unstable and will rapidly react further. For the superoxide radical, reaction with a second superoxide radical (dismutation) to form hydrogen peroxide and oxygen (reaction 3) or, alternatively, reaction with the phenoxy radical (in any mesomeric form) to form a new intermediate, an organic hydroperoxide are both viable possibilities. In the presence of transition metal ions in the oxidized state, an oxidation of superoxide into oxygen may also take place (reaction 4). In the carbohydrate reaction, alternative mecha-

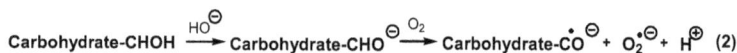

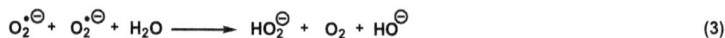

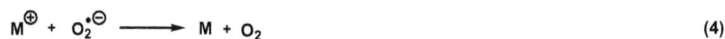

$$\text{Phenol-OH} \xrightarrow{\text{HO}^{\ominus}} \text{Phenol-O}^{\ominus} \xrightarrow{O_2} \text{Phenol-O}^{\bullet} + O_2^{\bullet\ominus} \qquad (1)$$

$$\text{Carbohydrate-CHOH} \xrightarrow{\text{HO}^{\ominus}} \text{Carbohydrate-CHO}^{\ominus} \xrightarrow{O_2} \text{Carbohydrate-}\overset{\bullet}{C}O^{\ominus} + O_2^{\bullet\ominus} + H^{\oplus} \qquad (2)$$

$$O_2^{\bullet\ominus} + O_2^{\bullet\ominus} + H_2O \longrightarrow HO_2^{\ominus} + O_2 + HO^{\ominus} \qquad (3)$$

$$M^{\oplus} + O_2^{\bullet\ominus} \longrightarrow M + O_2 \qquad (4)$$

Figure 9.7. Initial reactions in the alkaline oxygen oxidation of lignin and polysaccharides (reaction 1 and 2) together with major decay reactions of the superoxide radical (reaction 3 and 4).

nisms have been suggested but a reaction proceeding through ionization of an alcohol group and electron transfer to oxygen (reaction 2) is one likely candidate.

The formation of hydrogen peroxide through the dismutation reaction of superoxide and possibly also through further reactions of organic hydroperoxides formed as intermediates means that oxidation reactions typical of hydrogen peroxide may supplement the oxygen reactions in the O-stage. Hydrogen peroxide, in particular in the presence of transition metal ions, may, however, also decompose with formation of hydroxyl and superoxide radicals as shown in *Figure 9.8* (reaction 1). The hydroxyl radical is known as a powerful oxidant with rates of reaction with many organic and inorganic substrates close to diffusion controlled. Thus, any hydroxyl radicals formed can be assumed to rapidly react further by either an addition (reaction 2) or a hydrogen abstraction mechanism (reaction 3) depending on the substrate (lignin or carbohydrate). In addition, hydroxyl radicals can be trapped by reaction with any hydrogen peroxide (reaction 4) or transition metal ions (reaction 5) present in the aqueous phase. The organic radical intermediates formed on oxidation of either lignin or carbohydrates with hydroxyl radicals will rapidly react further and, again, lignin degradation and cellulose oxidation can be expected to occur.

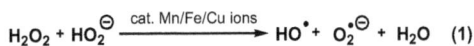

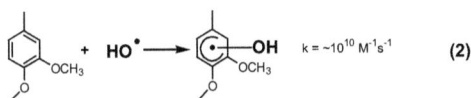

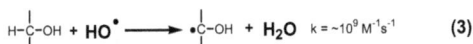

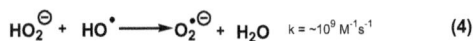

Figure 9.8. Formation of hydroxyl radicals by decomposition of hydrogen peroxide and various decay reactions of hydroxyl radicals (with approximate reaction rates).

9.3.2 The Reactions of Lignin

The alkaline oxygen oxidation of lignin has been the subject of studies with pulp, isolated lignin and with lignin model compounds. These studies clearly reveal that the major reaction is an oxidation of aromatic units in lignin having a free phenolic hydroxyl group. The primary reaction step, formation of a phenoxy radical, is followed by attachment of a superoxide radical or possibly an oxygen molecule to any of the resonance structures and formation of an organic hydroperoxide as shown in *Figure 9.9*.

The hydroperoxide structures are not stable and will undergo different types of secondary reactions. Among these, the oxidative cleavage of the aromatic ring and formation of acidic groups must be regarded as the most important since that reaction will render the lignin more hydrophilic. A certain decrease of the molecular size of the lignin may, however, also play a role since the oxidized lignin must be able to diffuse out of the fiber wall. Analysis by GPC of the dissolved lignin from an O-stage has revealed that the molecular size distribution is rather similar to the lignin dissolved in a kraft cook (*Figure 9.10*).

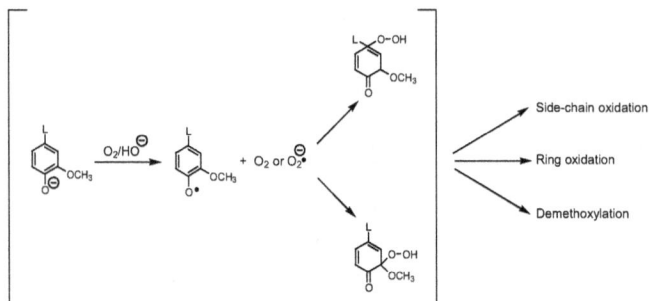

Figure 9.9. The initial reaction step in the oxidation of a phenol with oxygen in alkaline media.

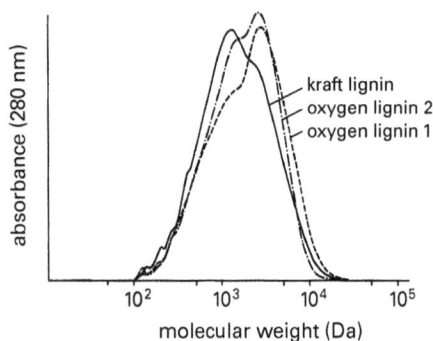

Figure 9.10. Gel permeation chromatography of lignins dissolved from a kraft cook and from a subsequent O-stage respectively. Delignification in the O-stage was around 8 (lignin 1) and 15 (lignin 2) kappa number units respectively. Molecular weight calibration with polystyrene standards.

Experiments with simple lignin model compounds have shown that the formation of acidic reaction products can take place according to several different reaction routes. Thus, by the use of creosol (2-methoxy-4-methylphenol), all the major reaction pathways by which the aromatic ring is degraded have been identified. In addition, oxidation of phenolic lignin structures having a benzylalcohol structure gives the further option of an oxidative cleavage of the side chain (*Figure 9.11*).

In studies with kraft pulp, it has been shown that the kinetics of oxidation of lignin (measured as changes in kappa number) in the O-stage takes place in two distinct phases with the first being less dependent on alkali charge and oxygen pressure than the second. It can be assumed that this behaviour is related to the amount of leachable and/or easily accessible lignin present in the fiber wall after the cook and washing procedure. Consequently, these types of lignin should dissolve in the first delignification phase. For individual lignin structures, on the other hand, the kinetic data reveal a large variability in reaction rates although the presence of highly reactive structures such as stilbenes and enol ethers in lignin is low. Lignin-like phenols show more uniform reactivities with simple mono-phenols being somewhat more reactive as compared to bis-phenols (*Figure 9.12*).

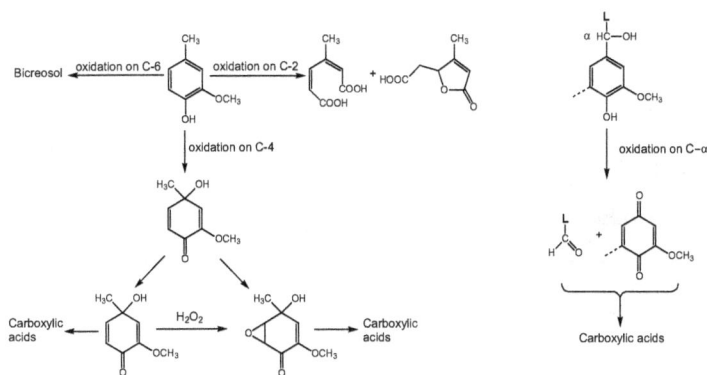

Figure 9.11. Alkaline oxygen oxidation of creosol and formation of acidic products (left hand formula) and oxidative cleavage of a benzylalcohol structure (right hand formula). (In low molecular weight model compounds like creosol, a comprehensive oxidation on C-6 takes place with formation of a radical coupling product, bicreosol. This type of oxidation has much less probability in a polymer material.)

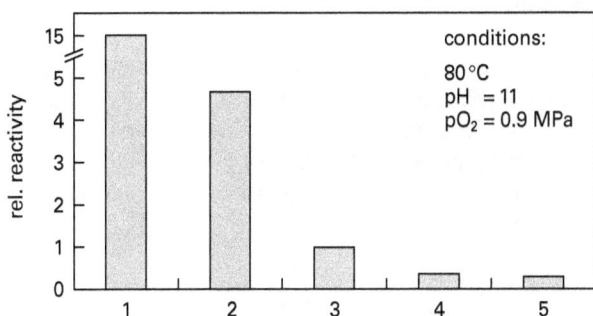

Figure 9.12. Relative reactivities of lignin model compounds on oxidation with oxygen/alkali. 1 = diguaiacylstilbene, 2 = enol ether from β-O-4 structure, 3 = propylguaiacol, 4 = phenolic β-O-4 structure, 5 = 5,5'-bis-propylguaiacol (Johansson 1997).

After pulping, the residual pulp lignin (softwood) contains in the order of 30 phenolic hydroxyl groups per 100 phenylpropane units. After a subsequent O-stage, this number is reduced to values around 10–15 in the fiber lignin. For the corresponding dissolved lignin in the O-stage, a surprisingly high number of phenolic hydroxyl groups can, however, be found *Figure 9.13*). In addition, a large number of carboxyl groups is formed as the result of the oxidation of lignin. This apparent anomaly can be explained by assuming that the analytical value for phenols obtained for unbleached pulp lignin, in fact is a mean value covering a broad range of values with a substantial portion being much higher than the average. A lignin fraction with a high content of phenols can be assumed to be the most reactive and preferentially removed. In addition, a certain formation of new phenolic hydroxyl groups in lignin during the course of an O-stage cannot be excluded, e.g. by hydroxyl radical addition or by cleavage of aryl ether linkages.

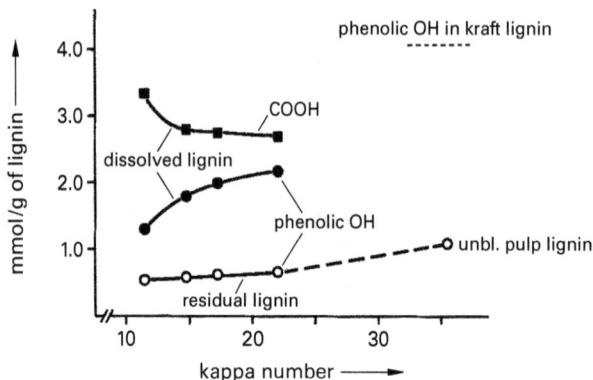

Figure 9.13. Amounts (mmol/g of lignin) of carboxyl groups and phenolic hydroxyl groups in lignins after kraft pulping and oxygen delignification respectively. The O-stage was carried out to four different kappa number values by using different charges of alkali in the stage.

9.3.3 The Reactions of Carbohydrates

The yield of pulp after an O-stage is high demonstrating that lignin is the predominant component being dissolved. Thus, despite the fact that the reaction is carried out in an alkaline medium, the peeling reaction, prevalent during kraft pulping, is of a minor importance. During oxidative conditions, any reducing end group present in the polysaccharides will rapidly become oxidized to the corresponding aldonic acid group thereby preventing the endwise degradation from taking place. The reaction is outlined in *Figure 9.14*.

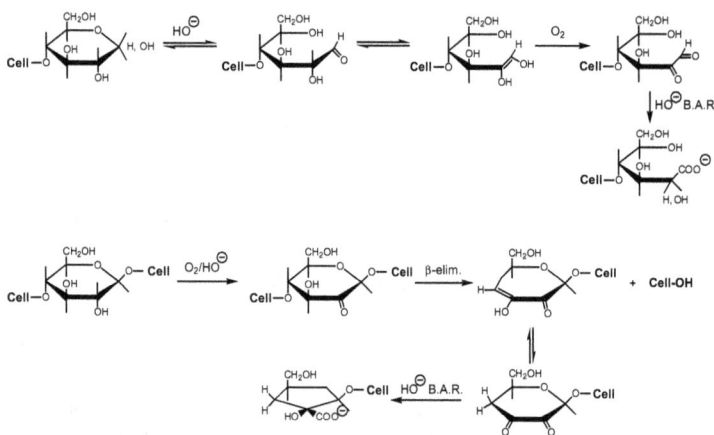

Figure 9.14. Major carbohydrate reactions in the O-stage. Oxidative stabilization of a reducing end group (upper reaction), oxidative cleavage of a polysaccharide chain (lower reaction). B.A.R. = benzilic acid rearrangement.

The fact that a rather large viscosity loss in the pulp takes place during an O-stage has been regarded as a serious drawback and has been a major reason for the slow rate of introduction of oxygen stages in the industry despite the advantage of using an easily available, cost efficient and environmentally friendly oxidant. The reduction of the degree of polymerisation of cellulose (i.e. the viscosity loss) is due to an oxidation of one or more of the hydroxyl groups located along the cellulose chain. Thereby, a carbonyl group is created and, due to the alkaline conditions, an elimination reaction occurs resulting in a cleavage of the chain into two shorter units (*Figure 9.14*). One of these will have a newly formed reducing end group but thanks to the oxidative conditions, the peeling reaction will not take place to any noticeable extent.

The initial mechanism for oxidation of an alcohol group in polysaccharides to a carbonyl group by aqueous alkaline oxygen is not known in detail. Two possibilities are shown in *Figure 9.7* (reaction 2) and *Figure 9.8* (reaction 3). In the former of these, oxygen can promote the final conversion of the carbohydrate radical-anion to a carbonyl group whereas in the latter, the carbonyl group would be formed by addition of a superoxide radical to the carbohydrate radical followed by elimination of hydrogen peroxide (*Figure 9.15*).

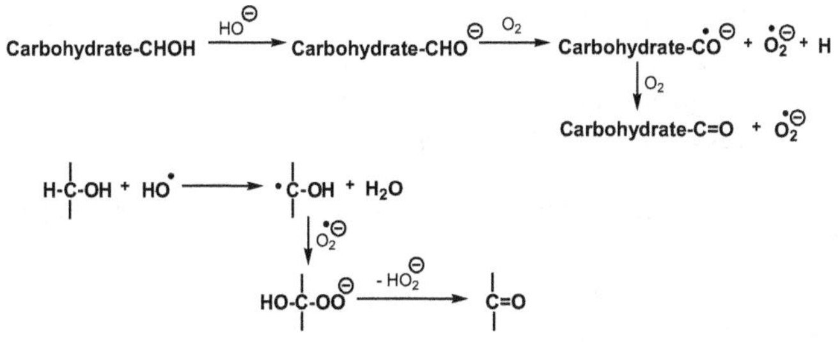

Figure 9.15. Possible modes of formation of carbonyl groups in carbohydrates on oxidation with oxygen in alkaline media.

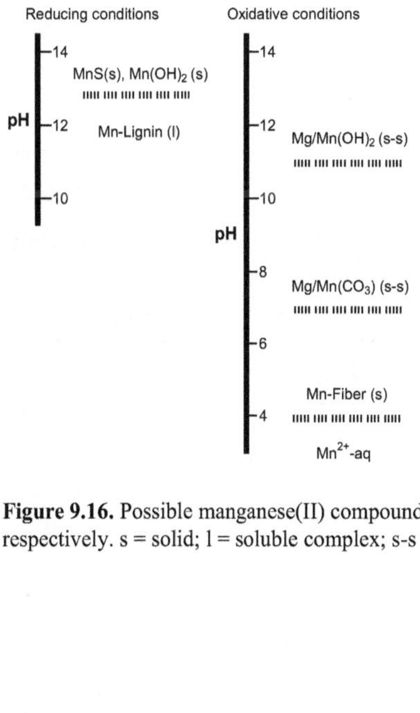

Figure 9.16. Possible manganese(II) compounds under reducing (pulping) and oxidative (O-stage) conditions respectively. s = solid; l = soluble complex; s-s = solid solution (Lidén 1994.)

As noted above, an addition of magnesium salt to the O-stage is essential in order to prevent the extent of degradation of the carbohydrates to reach levels that may affect the strength properties of the pulp. Several different explanations for the "magnesium-effect" have been given but, now, it seems clear that the addition of magnesium(II) results in the formation of a hydroxide or carbonate (depending on pH) which has the possibility of incorporating manganese(II) in the precipitate. Thus, any manganese(II), present in the pulp after the cook will be converted into a solid solution of magnesium-manganese hydroxide (carbonate) as depicted in *Figure 9.16*. Thereby, the catalytic action of manganese ions on hydrogen peroxide giving rise to the rapid formation of hydroxyl radicals can be prevented to a large extent (Figure 9.8, reaction 1).

9.3.4 Dissolved Material in the O-stage

A substantial portion of the residual lignin present in the unbleached kraft pulp, typically ~60–75 %, goes into solution in the O-stage. Except for trace amounts of low molecular weight phenols, such as acetovanillone, this lignin is still polymeric (*Figure 9.10*) albeit containing carboxyl groups as shown in *Figure 9.13*. The total dissolution of carbohydrates, on the other hand, is rather low resulting in a high overall yield of pulp. To some extent, a more comprehensive oxidation of both lignin and carbohydrate fragments takes place. Such reactions give rise to a variety of aliphatic acids, methanol (from methoxyl groups) and carbon dioxide as shown in *Table 9.2*.

Table 9.2. Composition of the bleaching effluent obtained after oxygen delignification of a pine kraft pulp (Pfister and Sjöström 1979).

Compound	Amount, kg/tonne of pulp
Lignin	29
Low M_w lignin products	3.7
Polysaccharides[1]	4.5
Methanol	1.5
Carbon dioxide	7.5
Formic acid	3.0
Oxalic acid	1.1
Acetic acid	1.0
Glycolic acid	1.0
3,4-Dihydroxybutanoic acid	0.9
Minor acids, total	2.5

[1] xylose as predominant sugar moiety

Table 9.3. Changes in the contribution to kappa number from lignin, HexA and other oxidizable structures after an industrial O-stage of spruce and birch kraft pulps respectively.

Pulp	Kappa No	Lignin[1]	HexA[1,2]	Non-lignin[1,3]
Spruce, unbleached	22.5	17.2	1.3	4.0
Spruce, after O-stage	10.7	4.6	1.2	4.9
Birch, unbleached	13.8	8.1	4.7	1.0
Birch, after O-stage	9.6	3.2	4.3	2.1

[1] calculated in Kappa number units
[2] HexA = hexenuronic acid
[3] Non-lignin = non-specified but oxidizable structures

9.3.5 Structural Features of the Kraft Pulp after the O-stage

The comprehensive dissolution of lignin in the O-stage is not accompanied by a similar large decrease in the kappa number of the pulp. The most important reason for this discrepancy can be found in the fact that the O-stage will not affect the amount of hexenuronic acid units, originally present in the pulp after the kraft cook, to any noticeable extent. In addition, a further formation of other oxidizable structures, presumably in the hemicellulose fraction of the pulp, seems to take place. Therefore, the overall decrease in kappa number can be in the order of ~50 % for softwood and ~30 % for hardwood pulps. Two examples of the changes in kappa number taking place when going from unbleached to oxygen delignified pulp are shown in *Table 9.3*.

The lignin remaining in the fibers after the O-stage has a somewhat different chemical structure in comparison to the lignin in wood. Thus, the cleavage of β-O-4 structures that takes place during the kraft cook is reflected in lower values for this type of structure also in the residual oxygen delignified pulp lignin. An apparent increase of such structures after the O-stage in comparison to the value in unbleached pulp can, however, be observed. Therefore, a large inhomogeneity in the structure of the residual lignin after the cook must be assumed and it seems obvious that once the accessible portions of the lignin have been removed by the oxygen oxidation, the remaining part again becomes structurally more like the wood lignin. These features are shown in *Table 9.4*. Other structural changes in lignin in the O-stage seems to include a certain preferential oxidation of simple guaiacyl end-groups. Thereby, and in analogy with kinetic data (*Figure 9.12*), an accumulation of biphenyl (5-5) structures can be found in the residual lignin. As expected, the residual lignin also contains an increased amount of carboxyl groups.

Table 9.4. Analytical data on the frequency of linkage types in (softwood) lignin from wood, unbleached kraft pulp and kraft pulp after an O-stage.

Lignin sample	β-O-4	β-5	b-b	5-5	4-O-5
		NMR-data[1]		Oxidative degr.[2]	
Wood lignin	48	12	3.5	10	5
Kraft pulp lignin, kappa 30–35	10	5	2	12	7
O-stage pulp lignin, kappa 9–11	18	8	2	13	4

[1] NMR-data on isolated lignin samples
[2] Oxidative degradation on wood and pulp samples (includes only phenolic structures)

All residual kraft pulp lignin after the O-stage has been found to be linked to polysaccharides albeit in different proportions as compared to the unbleached pulp. Thus, the xylan-linked lignin which contributes some 20 % of the residual lignin after the cook is further reduced in the O-stage with a decrease that is proportional to the extent of delignification (*Figure 9.17*). For the glucan-linked lignin, a similar trend is found but at lower lignin contents. The major portion of the residual lignin after the cook is linked to both xylan and glucomannan ("network") with the amount of lignin being at least 50 % of the total (*Figure 9.4*). After a subsequent O-stage, this value increases further and 80–90 % of all residual pulp lignin seems to be linked to both of the major hemicellulose components.

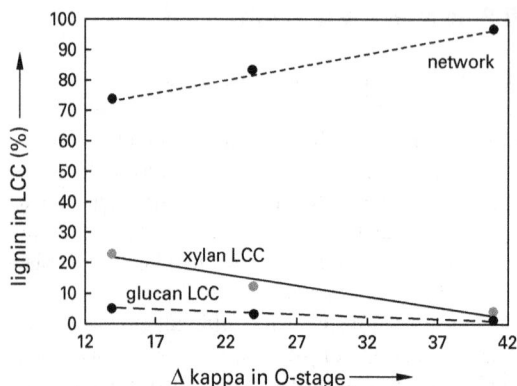

Figure 9.17. Amount of lignin chemically linked to glucan, xylan and xylan-glucomannan (network) respectively in softwood kraft pulp after an O-stage. The conditions in the O-stage was adjusted to obtain a range of pulps having different degrees of delignification. The amount of non-linked lignin was zero (Lawoko 2005).

9.4 Hydrogen Peroxide Bleaching Chemistry

Hydrogen peroxide in alkaline media has been used since a long time for the bleaching of mechanical pulp. Here, it is required that only the chromophoric structures present in the pulp are removed but virtually without any loss of pulp yield and, consequently, rather moderate reaction conditions must be chosen (~60–70 °C). The bleaching liquor must also contain sodium silicate in order to stabilize the peroxide and to provide a buffering capacity to the bleaching system. The same type of bleaching conditions, applied on a kraft pulp, will, however, only give a limited brightness increase and in order to get an efficient bleaching, the lignin must also be removed. Therefore, a temperature around 100 °C must be used and for reasons of chemical recovery, the silicate should be omitted. This type of bleaching requires a pulp in which virtually all types of transition metal ions have been eliminated and it has been found that a pre-treatment stage of the pulp with a chelating agent, EDTA or DTPA, under specified conditions must be carried out (Q-stage). Analysis of pulps which had been subjected to alkaline peroxide bleaching after a Q-stage, performed at different pH-values and temperatures, showed that a high temperature (~90 °C), a pH around 4–6 and one hour reaction time was required in the Q-stage in order to get an efficient utilization of the peroxide and a preserved pulp viscosity. It has also been shown that these conditions result in a very efficient removal of transition metal ions while, at the same time, magnesium ions are retained in the pulp as shown in *Figure 9.18*.

The rate of decomposition of alkaline hydrogen peroxide as well as the rate of oxidation of lignin-like model phenols with alkaline hydrogen peroxide has been studied with and without the presence of catalytic amounts of transition metal ions. In the absence of metal ions, the decomposition of hydrogen peroxide seems to pass through a maximum at around the pK_a-value (~11.6) for hydrogen peroxide. At the same time a maximum in the rate of phenol oxidation can be observed. In the presence of transition metal ions, on the other hand, a strong increase of the rate of both oxidation and peroxide decomposition takes place and the pH becomes less important.

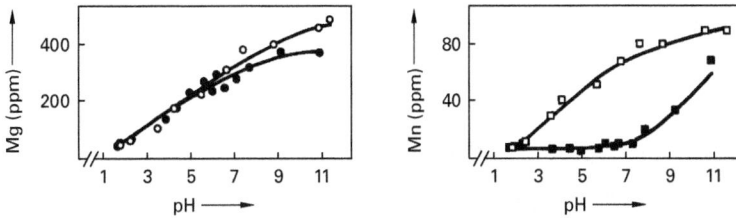

Figure 9.18. Amounts of magnesium and manganese ions in a softwood kraft pulp as a function of pH after treatment with aqueous EDTA at 90 °C and 60 min (Basta et al 1991).

In technical pulp bleaching, the preservation of the bleaching agent, hydrogen peroxide, is of utmost importance since, otherwise, the degradation products, viz. oxygen and water, will provide a very inefficient bleaching system. Despite the removal of almost all of the transition metal ions present in the pulp, a certain decomposition of hydrogen peroxide through spontaneous and/or metal catalysed reactions is unavoidable, however. These reactions will result in the formation of oxygen and water via the intermediate formation of hydroxyl and superoxide radicals according to the reactions shown in *Figure 9.19*.

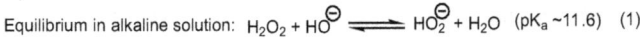

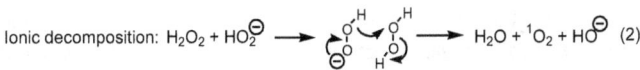

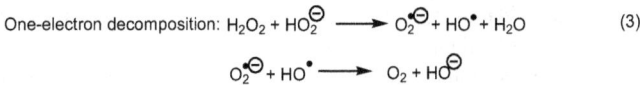

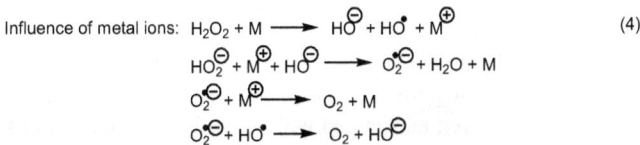

Figure 9.19. Decomposition reactions of hydrogen peroxide in alkaline solution. Influence of transition metal ions (M/M^+ denotes a metal ion). Reaction 2 is of minor importance and will give rise to singlet oxygen.

The radical species present in a P-stage will contribute to the oxidation of lignin but also to a certain oxidation of the polysaccharides according to the reaction shown in *Figure 9.15*. The latter oxidation will give rise to a further drop of pulp viscosity and, for a peroxide-based bleaching sequence, a loss of some 50–200 units (dm^3/kg) can be encountered depending on the conditions in the P-stage(s).

The fact that the residual pulp lignin after an O-stage has "regained" a chemical structure which has a resemblance to the native structure in wood will facilitate the further oxidative degradation by alkaline hydrogen peroxide. The high temperature required (~100 °C) in a P-stage results in formation of a quinone methide from phenolic benzylalcohol structures. Thereby, a nucleophilic addition of a peroxy anion is made possible and, in subsequent reaction steps, a lignin side chain cleavage and fragmentation may occur. The reaction sequence is outlined in *Figure 9.20*.

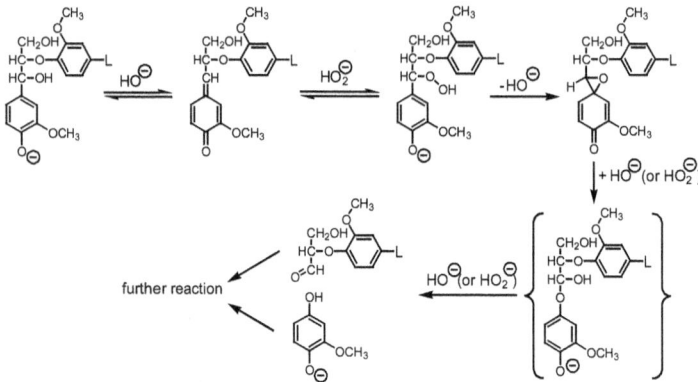

Figure 9.20. Side chain cleavage of a phenolic β-O-4 structure in lignin on oxidation with alkaline hydrogen peroxide. Reaction conditions for L (lignin)=H: 90 °C, 180 min (Heuts 1998).

Thus, the degree of delignification encountered in a P-stage can be substantial and it is accompanied by a considerable brightness increase of the pulp. The latter effect is in part due to the removal of lignin but also to the direct brightening effect exerted by alkaline hydrogen peroxide. The two types of reactions encountered in a P-stage, viz. general lignin oxidation and brightening (chromophore elimination), have been further supported by experiments with isolated lignin samples. Thus, in the alkaline peroxide oxidation of an isolated lignin sample, it has been shown that a gradual (slow) consumption of phenolic lignin structures is accompanied by a formation of carboxyl groups throughout the reaction. Simultaneously, a gradual consumption of hydrogen peroxide takes place. The direct brightening effect on the lignin, on the other hand, is fast but far from complete (*Figure 9.21*). Although the exact structure of the chromophoric groups present in the pulp lignin after pulping and oxygen delignification is virtually unknown, it can be assumed that quinones may contribute to the most reactive portion of these. The rapid oxidation of quinones exerted by alkaline hydrogen peroxide is well known and will proceed by formation of oxirane (epoxide) groups and/or by oxidative opening of the quinone ring (*Figure 9.22*).

In analogy to an O-stage, alkaline hydrogen peroxide is not able to degrade hexenuronic acid groups. Therefore, bleaching sequences only based on oxygen and hydrogen peroxide will result in fully bleached pulps in which the remaining amount of hexenuronic acid may contribute a substantial portion of the final kappa number. One example is shown in *Table 9.5*. The dissolved material from a peroxide stage contains in addition to polymeric lignin a substantial amount of low molecular weight aliphatic acids of undefined origin such as formic acid, glycolic acid, 3-hydroxypropionic acid and 3,4-dihydroxybutanoic acid. In addition, polysaccharides, predominantly xylan, can be found.

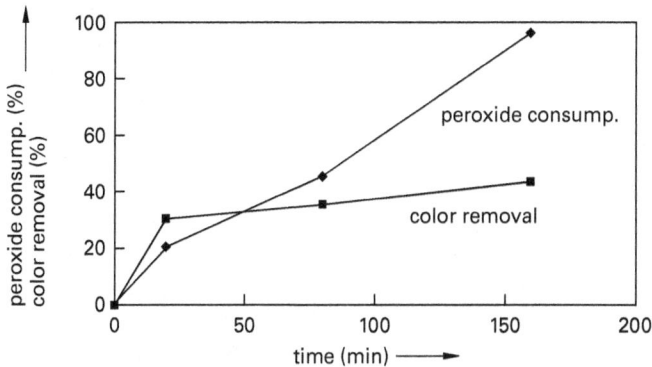

Figure 9.21. Alkaline hydrogen peroxide oxidation of residual lignin isolated from kraft pulp after the O-stage showing the gradual consumption of peroxide and the fast removal of color. Oxidation at 50 °C during 160 min (Gärtner 1998).

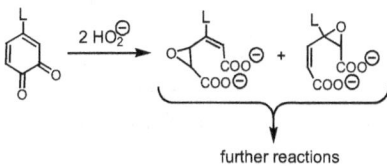

Figure 9.22. Oxidation of a quinone structure by alkaline hydrogen peroxide.

Table 9.5. Changes in brightness, viscosity and kappa number for an industrial birch kraft pulp after an O-stage and subsequent bleaching in a peroxide-based sequence. Contributions to the kappa number from lignin, hexenuronic acid and other oxidizable structures are also shown.

Pulp sample	Brightness, % ISO	Viscosity, dm³/kg	Kappa Number	Lignin[1]	HexA[1, 2]	Non-lignin[1, 3]
after O-stage	59.1	920	9.6	3.2	4.3	2.1
after OQ(OP)	73.3	850	7.3	2.2	3.7	1.4
after OQ(OP)Q(PO)	89.0	710	4.6	0.8	3.0	0.8

[1] calculated in Kappa number units
[2] HexA = hexenuronic acid
[3] Non-lignin = unspecified but oxidizable structures

9.5 Chlorine Dioxide and Chlorine Bleaching Chemistry

9.5.1 General Aspects

Bleaching of kraft (and sulfite) pulp with chlorine dioxide (D) can be done either directly on the unbleached pulp or on the pulp after an O-stage. In that position, the predominant function of the D-stage is delignification with little or no brightening effect on the pulp. Chlorine dioxide is, however, also used as an efficient brightening agent at the end of a bleaching sequence; a development that has been fundamental for the possibility of producing fully bleached kraft pulps. In a "prebleaching" D-stage in combination with an alkaline extraction (E) stage, most of the lignin is removed from the fibers. Typically, the kappa number can be reduced to a value around 4. In a complete chlorine dioxide bleaching sequence, the remaining colored impurities together with some traces of intact aromatic lignin structures are then oxidized at a somewhat higher pH-value (~3.5) as compared to the prebleaching conditions with a pH-value around 1.5–2.5. After such a sequence, the result will be a high purity pulp with a brightness around 90 % ISO.

The rapid development of new bleaching technologies, taking place since the early 1990s, has resulted not only in TCF-bleaching but also in bleaching sequences in which part of the chlorine dioxide charge has been replaced by hydrogen peroxide. In addition, more powerful O-stages have been introduced. The complete substitution of chlorine with chlorine dioxide that has taken place has also resulted in far less acidic bleaching liquors and, consequently, also in lesser risk of getting acid hydrolysis in the polysaccharides during the course of bleaching. At the same time, however, the ability of removing hexenuronic acid groups from the pulp during bleaching has decreased. This is illustrated in *Table 9.6* which shows some typical pulp quality data from bleaching of softwood kraft pulp.

The detrimental effects on pulp quality, e.g. brightness stability, that can be caused by the presence of hexenuronic acid groups, have resulted in a further development of bleaching technology. Thus, the fact that hexenuronic acid is sensitive to acidic conditions has promted some mills to introduce a specific hydrolytic stage which can be done either as an isolated stage or as part of a chlorine dioxide stage. In either case, the objective is to achieve a selective hydrolytic degradation of hexenuronic acid groups as depicted in *Figure 9.23*.

Table 9.6. Changes in brightness, viscosity and kappa number for an industrial softwood kraft pulp after an O-stage and subsequent ECF-bleaching. Contributions to the kappa number from lignin, hexenuronic acid and other oxidizable structures are also shown.

Pulp sample	Brightness, % ISO	Viscosity, dm³/kg	Kappa Number	Lignin[1]	HexA[1,2]	Non-lignin[1,3]
after O-stage	44.7	910	10.7	4.6	1.2	4.9
after ODE	73.3	850	3.3	1.6	0.8	0.9
after ODEQP	88.8	800	1.6	0.8	0.7	0.1

[1] calculated in Kappa number units
[2] HexA = hexenuronic acid
[3] Non-lignin = unspecified but oxidizable structures

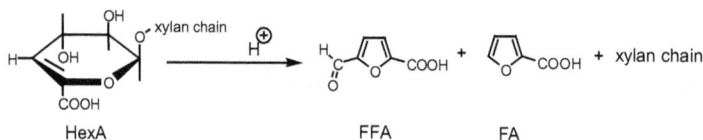

Figure 9.23. Acidic hydrolysis of hexenuronic acid (HexA) resulting in the formation of 5-formylfuroic acid (FFA) and furoic acid (FA).

9.5.2 Reactions in Pre-bleaching

The structure of the residual lignin present in the fibers after either the kraft cook or the subsequent O-stage is still that of a polyphenolic macromolecule as discussed in Section 9.2 and Section 9.3.5. Except for the differences in chemical structure that have been identified, there is also a prominent difference between softwood and hardwood since the latter usually contain a significant amount of hexenuronic acid with a corresponding large contribution to the kappa number (cf *Table 9.5* and *Table 9.6*).

The oxidation of pulp with chlorine dioxide is carried out at a temperature of around 65–75 °C and with a reaction time of 30–45 min. The pH is in the order of 2–3. Normally, the charge of chlorine dioxide, usually expressed as "active chlorine", is based on the incoming kappa number of the pulp. Thus, an oxygen delignified kraft pulp with a kappa number of 12 and a "kappa factor" of 0.2 will use $0.2 \times 12 = 2.4$ % active chlorine or $24/2.63 = 9.1$ kg ClO_2 per ton of pulp (kg/ADt). The initial consumption of chlorine dioxide by the pulp is very fast and within a matter of seconds, the concentration of chlorine dioxide has been reduced to a considerable extent. At the same time, a formation of chloride, chlorite and chlorate ions starts to take place (*Figure 9.24*). Whereas chlorite ions may still participate in the overall reactions of pulp lignin, the formation of chlorate is undesirable since this ion is inert and cannot contribute to any further pulp reactions. In addition, the chlorate ion is biologically active and may cause environmental problems.

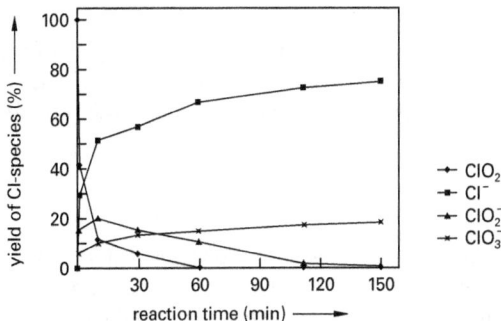

Figure 9.24. Formation of inorganic chlorine-containing species during laboratory bleaching of unbleached softwood kraft pulp with chlorine dioxide (Ni et al 1991).

The mechanism of chlorine dioxide oxidation of lignin has been studied using phenolic as well as non-phenolic model compounds. For both types of structures, the initial reaction is a one-electron transfer reaction giving rise to a phenoxy radical and a radical cation respectively as depicted in *Figure 9.25*. The chlorine dioxide is reduced to chlorite ion. In the next reaction step, a second molecule of chlorine dioxide is added to the radical species with formation of esters of hypochlorous acid. These, in turn, are hydrolysed and, depending on the type of leaving group, give rise to acids of the muconic acid type and to quinones. In these reactions, hypochlorous acid and chlorous acid together with methanol are liberated.

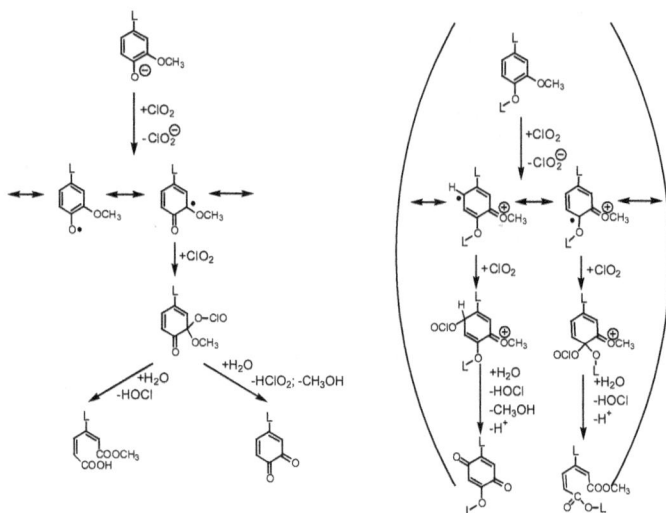

Figure 9.25. Reactions between a phenolic (left) and a non-phenolic (right) lignin structure respectively with chlorine dioxide under acidic conditions (Eriksson 1993). In benzylalcohol structures, a similar reaction will result in an oxidative elimination of the side chain. The reaction of non-phenolic lignin structures is less likely under technical pre-bleaching conditions.

The formation of hypochlorous acid cannot be detected by normal types of analyses of the bleaching liquor due to the high reactivity of this acid (cf. *Figure 9.24*). By addition of a scavenger for chlorine, e.g. sulfamic acid, it is, however, possible to get an indirect analysis according to Reaction (1) shown below.

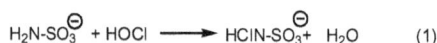

$$H_2N\text{-}SO_3^{\ominus} + HOCl \longrightarrow HClN\text{-}SO_3^{\ominus} + H_2O \qquad (1)$$

Thus, by adding sulfamic acid to the bleaching liquor, it has been shown that chlorine dioxide bleaching of kraft pulp results in the formation of a substantial amount of methanol which occurs in parallel with the dissolution of lignin. Furthermore, by using a kraft pulp having all phenolic hydroxyl groups blocked by methylation prior to reaction with chlorine dioxide, the presence of sulfamic acid was shown to prevent virtually all reactions and the pulp kappa number remained almost unaffected. Pulp bleaching with chlorine dioxide in the presence of sulfamic acid will also give a complete suppression of the formation of chlorate. These observations are in line with the view that i) methanol is formed directly from the oxidation with chlorine dioxide (and subsequent hydrolysis), ii) under the conditions of technical bleaching, non-phenolic lignin structures do not react to any noticeable extent and iii) hypochlorous acid and/or chlorine is involved in the formation of chlorate according to Reaction (2) and (3).

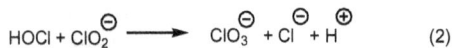

$$HOCl + ClO_2^{\ominus} \longrightarrow ClO_3^{\ominus} + Cl^{\ominus} + H^{\oplus} \qquad (2)$$

$$Cl_2 + ClO_2^{\ominus} + H_2O \longrightarrow ClO_3^{\ominus} + 2Cl^{\ominus} + 2H^{\oplus} \qquad (3)$$

The presence of sulfamic acid in a D-stage has also been shown to decrease the amount of delignification which can be obtained for a given set of reaction conditions. Therefore, from a bleaching point of view, the formation of the highly reactive hypochlorous acid is advantageous since all types of aromatic structures will become oxidized. On the other hand, the presence of the hypochlorous acid/chlorine system in the bleaching liquor will also induce a certain degree of chlorination of the lignin. Of these two species, elemental chlorine is, however, a far more powerful chlorination agent than hypochlorous acid. Since, in aqueous solution, an equilibrium exists between hypochlorous acid, hypochlorite and chlorine as given by Reaction (4) and (5) and further illustrated in *Figure 9.26*, even rather small changes in the pH-value prevailing in the D-stage will affect the amount of chlorinated lignin that is formed.

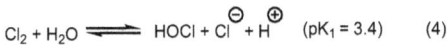

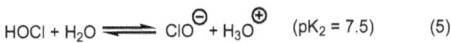

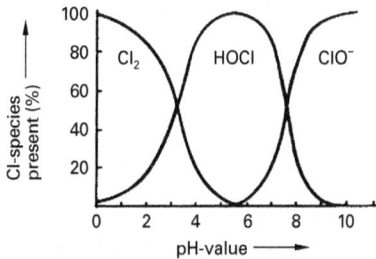

$$Cl_2 + H_2O \rightleftharpoons HOCl + Cl^{\ominus} + H^{\oplus} \quad (pK_1 = 3.4) \qquad (4)$$

$$HOCl + H_2O \rightleftharpoons ClO^{\ominus} + H_3O^{\oplus} \quad (pK_2 = 7.5) \qquad (5)$$

Figure 9.26. Chlorine species present in aqueous solution (room temperature) at different pH-values.

Model studies on the influence of pH for the degree of oxidation and chlorination under conditions similar to those in a D-stage has been carried out on isolated lignin from oxygen delignified softwood kraft pulp. Thereby, it could be shown that a pH around 3 seems to be optimal from an oxidation point of view whereas, at the same time, the degree of chlorination is smaller as compared to reaction at a lower pH-value (*Figure 9.27*). In the figure, the degree of oxidation is expressed as the level of intact methoxyl groups in lignin after the oxidation with chlorine dioxide which for a high degree of oxidation should have a low value. Other ways of expressing the degree of oxidation such as the carbon/oxygen ratio of the lignin after oxidation can also be used and was found to give similar data, i.e. the lowest ratio at pH = 3 for the experiments described in the figure.

In actual pulp bleaching, the degree of oxidation and chlorination in a chlorine dioxide pre-bleaching stage seems to be lower as compared to experiments with isolated lignin as a model compound. Thus, by isolating the remaining pulp lignin after each stage in a complete bleaching sequence and analysis of the methoxyl content, the successive oxidation of the pulp lignin could be followed as depicted in *Figure 9.28*. A corresponding trend in the ratio carbon/oxygen was also found with values ranging from 3.45 (unbleached pulp lignin) to 1.43 (fully bleached pulp lignin). In the first chlorine dioxide stage (D_0-stage), a certain chlorination of the lignin accompanies the oxidation as shown in *Figure 9.28*. Later in the bleaching sequence, a substantial portion of this lignin-bound chlorine is removed and for fully bleached softwood pulps, a residual amount of organically bound chlorine of ~50–100 ppm can be found after ECF-bleaching. The corresponding dissolved high molecular weight material, predominantly lignin, from such a bleaching sequence contains ~10 chlorine/1000 carbon atoms.

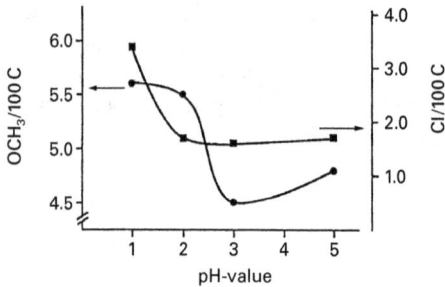

Figure 9.27. Oxidation of isolated residual lignin from oxygen delignified softwood kraft pulp with aqueous chlorine dioxide at different constant pH-values. Number of methoxyl groups and chlorine atoms per 100 carbon atoms.

9.5.3 Reactions in the Final Bleaching Stages

After pre-bleaching with chlorine dioxide, the kappa number (softwood) can be in the order of 3–4. A substantial amount of the oxidizable material present in such a pulp originates from lignin (cf. *Table 9.6*) as revealed by various analytical data. In comparison to native wood lignin, the residual pulp lignin present after the first D-stage contains, however, a large number of carboxyl groups whereas the number of uncondensed guaiacyl groups and methoxyl groups is lower showing that an oxidation of guaiacyl end-groups with formation of acids of the muconic acid type has taken place. In order to remove (most of) the remaining pulp lignin and to

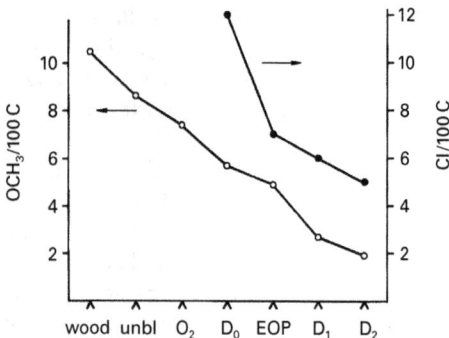

Figure 9.28. Changes in methoxyl and chlorine content in the residual lignin of industrial softwood kraft pulp when going from wood to fully bleached pulp. Bleaching with the sequence OD(EOP)DD. Number of methoxyl groups per 100 carbon atoms and chlorine atoms per 1000 carbon atoms.

increase the brightness to a value around 90 % ISO, either two further D-stages or, alternatively, one D-stage and one P-stage are necessary. In the first of these stages, further delignification takes place under conditions that are typified by longer reaction time (~3 hours), higher temperature (~70–80 °C) and higher pH (end pH at ~3.5–4.0) as compared to the pre-bleaching stage. Similar conditions are also used in a final D-stage. In addition to the oxidation of further phenolic lignin structures, the conditions prevailing in the D_1- and D_2-stages result in oxidation of various non-phenolic aromatic (cf. *Figure 9.25*) and conjugated aromatic structures thus giving rise to both lignin dissolution and non-specified chromophore elimination. The latter reaction is exemplified in *Figure 9.29*.

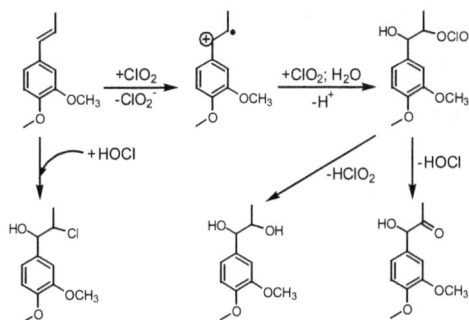

Figure 9.29. Reactions between a conjugated aromatic structure and chlorine dioxide under mild acidic conditions.

9.5.4 Reactions of Carbohydrates

Bleaching of chemical pulp with chlorine dioxide is very selective and unless the pH is very low and/or the temperature very high, the polysaccharide chains are left virtually intact. At the more extreme conditions, acid hydrolysis may take place giving rise to a certain loss of viscosity of the pulp. Furthermore, the hexenuronic acid groups attached to the xylan can be degraded by hydrolysis (cf *Figure 9.23*) and/or by oxidation. The latter reaction may proceed rapidly in the presence of hypochlorous acid/chlorine and more slowly when only chlorite remains as active chlorine species in the bleaching liquor (*Figure 9.30*). In the latter case, the actual oxidizing species may in fact be chlorine dioxide, formed through the slow disproportionation of chlorite according to Reaction (6).

9.5.5 Bleaching with Chlorine

The highly efficient delignification and bleaching of chemical pulps with elemental chlorine was early recognized and chlorine has been the predominant bleaching agent for kraft pulp until the late 1980s. The finding that the reactions between chlorine and pulp constituents give rise to large amounts of chlorinated organic matter of both low and high molecular weight raised, however, a suspicion about negative environmental effects. Therefore, the consumption of chlorine in the world for bleaching of chemical pulp is now rapidly declining and in several countries, chlorine is not used at all.

$$5 \ ClO_2^\ominus + 4 \ H^\oplus \longrightarrow 4 \ ClO_2 + Cl^\ominus + 2 \ H_2O \qquad (6)$$

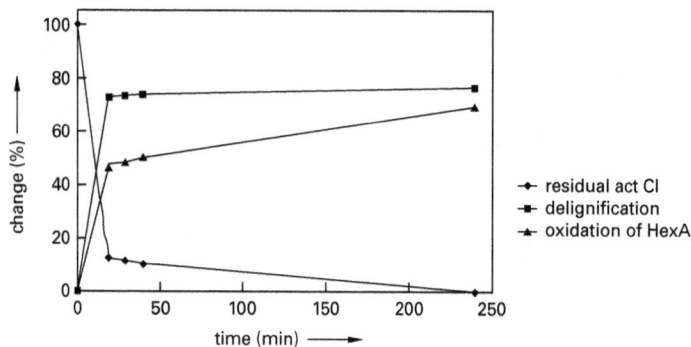

Figure 9.30. Oxidation of an oxygen delignified eucalyptus kraft pulp (kappa number 10.4) with 2.08 % chlorine dioxide (as active chlorine) at pH=3.0 and 60 °C. Separate analyses of lignin oxidation (delignification) and hexenuronic acid (HexA) oxidation.

The high efficiency of chlorine oxidation of lignin can be visualized as shown in *Figure 9.31*. Thus, pre-bleaching of kraft pulp with a mixture of chlorine and chlorine dioxide, followed by alkaline extraction, results in the dissolution of a highly oxidized lignin. A high ratio of chlorine to chlorine dioxide gives, however, a much higher degree of oxidation, measured as remaining amount of aromatic methoxyl groups in the lignin. At the same time, a high degree of chlorination of the lignin takes place.

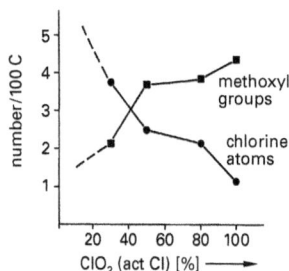

Figure 9.31. Composition of the dissolved lignin (from E-stage) after pre-bleaching of softwood kraft pulp with different ratios of Cl/ClO_2. Degree of oxidation measured as number of methoxyl groups/100 carbon atoms and degree of chlorination as number of $Cl/100$ carbon atoms.

The reactions between pulp lignin and aqueous chlorine may proceed by several different pathways which seem to occur extremely fast and more or less simultaneously. It should be noted that, in addition to chlorine, hypochlorous acid is usually also present in the bleaching liquor as shown in *Figure 9.26*. The major products from these reactions can be summarized as:

- Aromatic substitution, either directly on the aromatic ring or through a side chain displacement reaction
- Hydrolysis of ether linkages
- Oxidation of aromatic rings to quinones and carboxyl groups

The first of these reactions will give rise to a variety of chlorinated aromatic structures with one or more chlorine atoms attached to the same aromatic ring. Furthermore, around 80 % of the available aromatic methoxyl groups have been found to be split off as methanol resulting in a comprehensive formation of new phenolic hydroxyl groups. These, in turn, are sensitive to oxidation, presumably by the hypochlorous acid present and both quinones and various carboxyl groups are formed thus rendering the oxidized lignin soluble in alkali. The various reactions which may proceed on both phenolic and non-phenolic lignin units are summarized in *Figure 9.32*.

The high degree of chlorination of lignin which occurs in a C-stage is reflected in the chlorine/carbon ratio of the dissolved high molecular weight material from CE pre-bleaching with values in the order of 6–7 chlorine per 100 carbon atoms (cf *Figure 9.31*). Furthermore, a large variety of chlorinated low molecular weight compounds are formed.

In the bleaching with chlorine dioxide, a chemistry based on hypochlorous acid/chlorine reactions takes place in parallel to the oxidation caused by chlorine dioxide itself as described above. Consequently, a certain chlorination, hydrolysis and oxidation of lignin will occur also in a D-stage according to the principles described in *Figure 9.32*. Therefore, despite the fact that chlorine dioxide reacts almost exclusively with phenolic lignin structures under the conditions encountered in a D-stage, a comprehensive delignification is still possible. Chlorination of the lignin is, however, unavoidable albeit the degree of chlorination is much lower as compared to a C-stage.

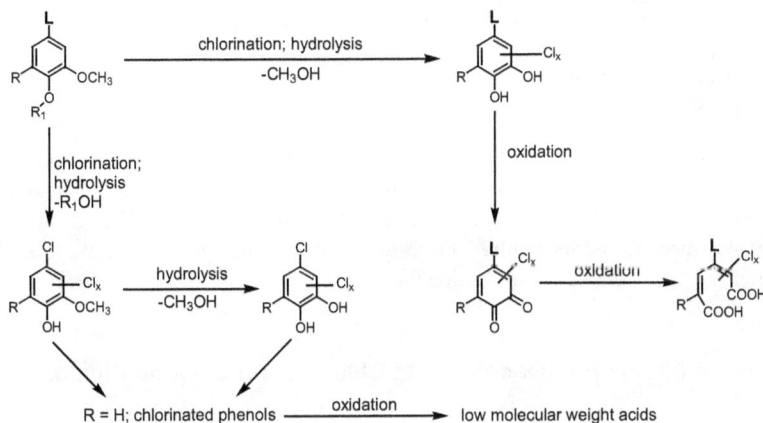

Figure 9.32. Schematic view of the various reaction modes of lignin encountered in pulp bleaching with aqueous chlorine.

9.5.6 Reactions in the Extraction Stage

Bleaching of pulp with chlorine, chlorine dioxide or mixtures of these must be followed by an alkaline extraction stage (E-stage) since the solubility of the oxidized lignin is low under the acidic conditions prevailing in the C- or D-stage. Treatment of the pulp with aqueous sodium hydroxide is usually done at a temperature of 70–95 °C and with a retention time of 1 hour. Normally, an oxidant like oxygen, hydrogen peroxide or both is added in order to further increase the extent of lignin removal and/or to reduce the charge of chlorine dioxide.

The major reaction encountered in an E-stage is the neutralisation of carboxyl groups thereby strongly increasing the water solubility of the oxidized lignin (cf. *Figure 9.1*). In addition, a comprehensive elimination of organically bound chlorine and formation of chloride ion takes place. In the presence of oxygen and/or hydrogen peroxide, an increased formation of quinones and an increased oxidative degradation of such structural units will occur as depicted in *Figure 9.33*.

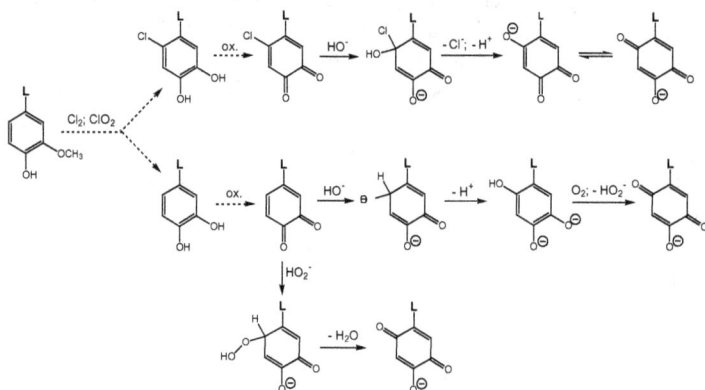

Figure 9.33. Schematic view of important lignin reactions encountered in an E-stage with or without an oxidative reinforcement with oxygen or hydrogen peroxide.

The fact that chlorine dioxide-based bleaching gives rise to a much lower overall formation of hydrochloric acid than bleaching with chlorine results in an efficient elimination of organically bound chlorine, once the D- and E-stage effluents are mixed since the pH of the total bleaching effluent will become rather neutral. From an environmental point of view, this is highly advantageous as further discussed in Section 9.5.7.

9.5.7 Environmental Effects on Bleaching with Chlorine and Chlorine Dioxide

Environmental concern has been a strong driving force for the development of new bleaching technologies that successively have replaced the traditional chlorine-based. Already in the early 1970s, mills in Sweden started to install oxygen stages in order to reduce the amount of discharged organic matter from the bleach plant. Despite some early warnings such as "There is some suspicion that the wood pulp bleaching effluents may be a source of pollution, by reason of their content of chlorinated phenols, derived from lignin." (Sandermann 1974), it was not clearly realized, however, that the bleaching with chlorine constituted a real environmental threat. In 1985, an important finding was made in the US with the detection of highly toxic poly-chlorinated dibenzo-p-dioxins and dibenzofurans being present in river waters downstream from kraft mills (*Figure 9.34*). Although the amounts of these compounds were extremely small, it was quickly confirmed that many pulp mills with production of bleached chemical pulp were also producers of such compounds.

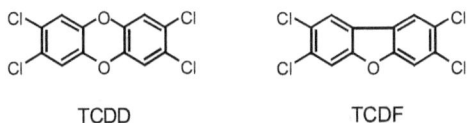

TCDD   TCDF

Figure 9.34. Representative structures for chlorinated dibenzo-p-dioxins (75 isomers) and dibenzofurans (135 isomers). TCDD = 2,3,7,8-tetrachloro-dibenzo-p-dioxin, TCDF = 2,3,7,8-tetrachloro-dibenzofuran.

In Sweden, the since long on-going but long-term changes of bleaching technology in the industry were done with the objective that every new investment should use the best available technology without sacrificing pulp quality characteristics. This work was now challenged by a strongly increased societal and political pressure (*Figure 9.35*) that the industry had to develop new and more "environmentally friendly" technologies at a much faster rate.

Figure 9.35. A provocative example of the public debate on the presence of dioxins in paper products with a text saying that "All chlorine derivatives give dioxins in paper". (Kemisk tidskrift 1988.)

As a result, the transition from chlorine to chlorine dioxide based bleaching (ECF-bleaching) was rapidly introduced at several Swedish mills. In parallel, a development towards bleaching without the use of any chlorine-containing bleaching agents (TCF-bleaching) was initiated. In 1994, this transformation was completed and, today, the major bleaching agents in Sweden are oxygen, chlorine dioxide and hydrogen peroxide. Ozone and peracetic acid are also used in a few mills. The Swedish development of bleaching technologies during the last 25 years can be visualized by plotting the decrease of absorbable organic halogen (AOX) as a function of time. As can be seen in *Figure 9.36*, a steady decrease has taken place but with a sharper slope of the curve during the latter half of the 1980s. Today (2005), the emission of AOX in the bleaching effluent from mills using ECF-bleaching is in the order of 0.1 kg/tonne of pulp, i.e. a value similar to the natural background level in Sweden

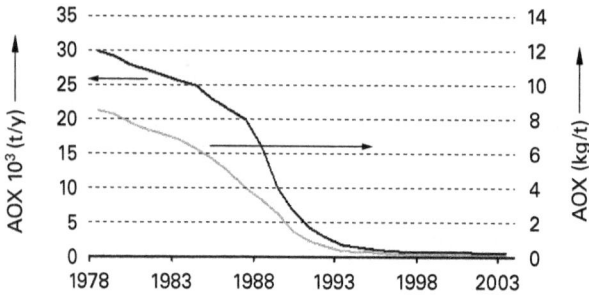

Figure 9.36. The decrease of chlorinated organic substances (AOX) from Swedish pulp mills as a function of time. Total amount from all mills in tonnes/year (left) and average value per tonne (right). Source: Skogsindustrierna.

From an environmental point of view, the most important changes in bleaching technology has been the introduction of oxygen delignification and the use of 100 % chlorine dioxide in the pre-bleaching stage. Since chlorine dioxide predominantly acts as an oxidant, the formation of chlorinated lignin (*Figure 9.31*), low molecular weight phenols and acids can be shown to strongly decrease. At the same time, the formation of chlorinated dioxins and dibenzofurans both in the product and in the effluent is reduced to levels around or below the detection limit (*Figure 9.37*).

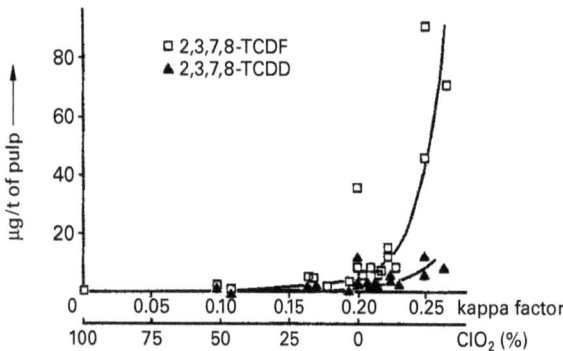

Figure 9.37. Sum of 2,3,7,8-TCDF and 2,3,7,8-TCDD in bleached softwood kraft pulp as a function of the charge of active chlorine in the pre-bleaching stage. (Axegård and Renberg 1989).

The large group of chlorinated phenols, shown to be present in rather high amounts in the bleaching effluent from chlorine bleaching, is strongly reduced in ECF-bleaching. At the same time, the composition is changed such that the amount of di-, tri- and tetra-chlorinated phenols, i.e. the most lipophilic phenols, is reduced to almost zero (*Figure 9.38*).

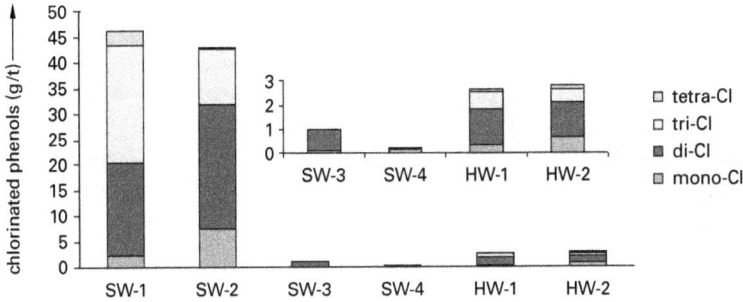

Figure 9.38. Presence of chlorinated phenols (gram/tonne) in bleached kraft mill effluents. SW = softwood, HW = hardwood. Bleaching sequences: SW-1 = O(C92+D8) (EO)DED; SW-2 = O(D25, C70+D5)(EPO)D(EP)D; SW-3 = OD(EPO)D(EP)D; SW-4 = OD(EO)D(EP)D; HW-1 = (D80+C20)(EPO)DED; HW-2 = O(D27, C68+D5)(EPO)D(EP)D. (Strömberg et al 1996).

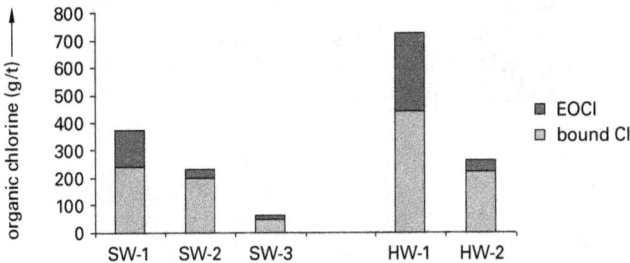

Figure 9.39. Totally bound organic chlorine (Bound Cl) and extractable organic chlorine (EOCl) in softwood and hardwood (birch) kraft pulps. Bleaching sequences: SW 1 = (C85+D15)EDED, SW 2 = O(C85+D15)(E+P)DED, SW 3 = OZEDP, HW 1 = O(C85+D15)(E+P)DD, HW 2 = OD(E+P)DD (Björklund Jansson et al 1991).

9.5.8 Organic Chlorine in Bleached Pulps

In bleaching sequences employing chlorine and/or chlorine dioxide, an introduction of a small amount of organically bound chlorine in the pulp is unavoidable. In analogy to the reactions with lignin, however, the chosen bleaching sequence affects the amount and an O-stage as well as a D-stage will result in a considerable decrease both of the total amount of organically bound chlorine and of the extractable organic chlorine (*Figure 9.39*). For the latter portion, the wood species plays an important role and e.g. in birch kraft pulp the level of remaining extractives after the O-stage is decisive since birch wood has a high content of unsaturated extractives such as squalene and betulaprenols. This type of compounds will easily add chlorine or hypochlorous acid across the double bonds as exemplified in *Figure 9.40*. Consequently, the resulting products may contain considerable amounts of organically bound chlorine and the highly lipophilic character of these can cause problems by the formation of deposits in the mill.

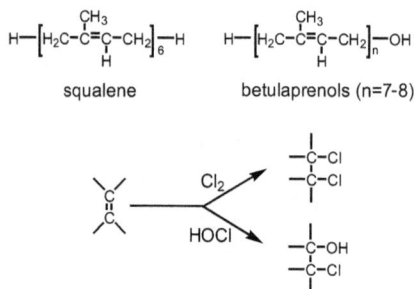

Figure 9.40. Examples of extractives present in birch kraft pulp and the general reaction for chlorination of a double bond in aqueous media.

9.6 Ozone Bleaching Chemistry

Ozone is an extremely powerful oxidant with an oxidation potential of 2.07 V vs NHE at room temperature, i.e. most organic substrates will react with ozone. For this reason, a major use of ozone is as a drinking water purification agent. For bleaching of chemical as well as mechanical pulps, ozone has been tried in laboratory scale since a long time, but for both economic and pulp quality reasons no commercial use has been found until the development of TCF-bleaching in the late 1980s. Today, a few mills in the world are using an ozone stage (Z-stage) either at high or medium consistency, usually as a replacement for a chlorine dioxide stage.

In organic chemistry, ozone has been used since a long time for converting a double bond into two aldehyde end-groups according to the Criegee reaction mechanism. If the reaction is carried out in aqueous solution, carboxyl groups and hydrogen peroxide are formed in addition to aldehydes (ketones). For aromatic rings, the reaction will result in a successive degradation and conversion into aliphatic acids as shown in *Figure 9.41*. The ozonolysis reaction has to be carried out under slightly acidic conditions since, otherwise, a rapid decomposition of ozone into hydroxyl and superoxide radicals will accompany the desirable oxidation reaction (Reaction 1). The presence of these radicals is detrimental and will affect the quality characteristics of the pulp in a negative way as discussed in Section 9.3.3.

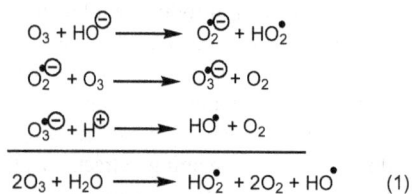

$$O_3 + HO^{\ominus} \longrightarrow O_2^{\cdot\ominus} + HO_2^{\cdot}$$

$$O_2^{\cdot\ominus} + O_3 \longrightarrow O_3^{\cdot\ominus} + O_2$$

$$O_3^{\cdot\ominus} + H^{\oplus} \longrightarrow HO^{\cdot} + O_2$$

$$\overline{2O_3 + H_2O \longrightarrow HO_2^{\cdot} + 2O_2 + HO^{\cdot}} \qquad (1)$$

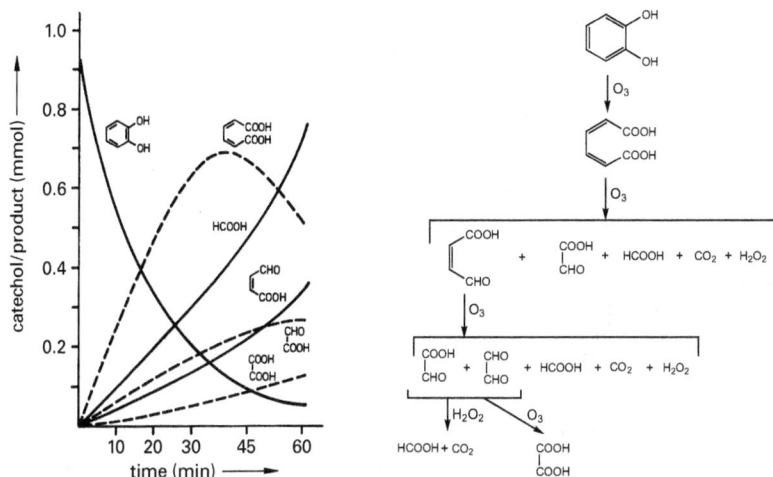

Figure 9.41. Ozonation of catechol in water at 30 °C. Formation of products as a function of ozonation time (left) and suggested mode of formation of products (right). (Yamamoto et al 1979.)

Despite the precaution of carrying out the bleaching reaction under acidic conditions, it has been shown that a certain formation of hydroxyl and hydroperoxy (superoxide) radicals do take place also under these conditions. Thus, in model compound experiments, it was found that ozone degradation of methyl-β-glucoside, a simple model for cellulose, was accelerated if an aromatic compound was present in the solution as illustrated in *Figure 9.42*.

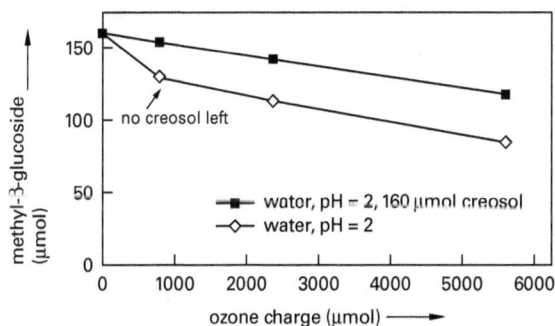

Figure 9.42. Ozonation of methyl-β-glucoside in the presence/absence of an equimolar amount of creosol (2-methoxy-4-methyl-phenol). (Zhang 1994.)

In the figure it can be seen that as long as creosol is present in the reaction medium, a more rapid degradation of the methyl-β-glucoside takes place, thus indicating a more comprehensive formation of radical species. Further confirmation for a reaction between aromatic rings and ozone giving rise to a direct formation of radicals in addition to the ionic reaction with forma-tion of acids (the Criegee reaction) has been obtained in model experiments. Based on these, the initial intermediate in the addition of ozone to the organic substrate, a trioxide, may decompose homolytically instead of forming the desired ozonide. In the latter case, the Criegee reaction, a ring opening reaction will follow whereas in the former case, superoxide radical is formed together with a quinol radical. In the presence of ozone, the superoxide (hydroperoxy) radical will rapidly react further with formation of oxygen and hydroxyl radical. The alternative pathways for the ozone reactions with aromatic rings are shown in *Figure 9.43*.

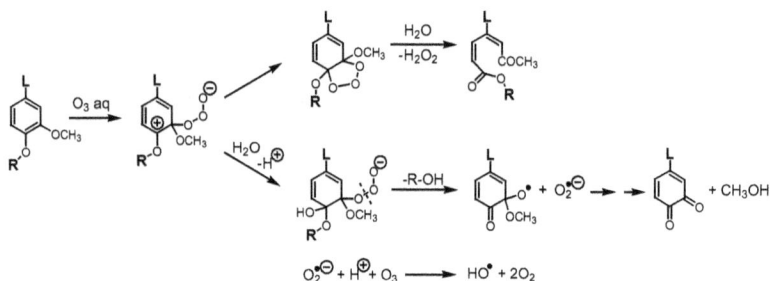

Figure 9.43. Oxidation of aromatic structures according to the Criegee reaction (upper part) and with formation of radical species (lower part). (Ragnar 2000).

The consequences for pulp bleaching of these two modes of reaction is an inevitable oxidation of cellulose which, however, can be kept at a somewhat lower level if the content of lignin is high, i.e. presence of a high content of a radical scavenger (*Figure 9.44*). Thus, after an ozone stage, the cellulose will contain carbonyl groups along the chain which can be shown by treating the cellulose with alkali in the presence and absence of sodium borohydride respectively. In the former case, the carbonyl groups will become reduced to the original alcohol groups and viscosity measurement will not show any substantial difference in comparison to an unoxidized reference sample. If, on the other hand, the oxidized cellulose is directly exposed to alkali and subsequently analysed with respect to viscosity, a noticeable reduction can be found due to the β-elimination reaction in oxidized glucose units (*Figure 9.14*).

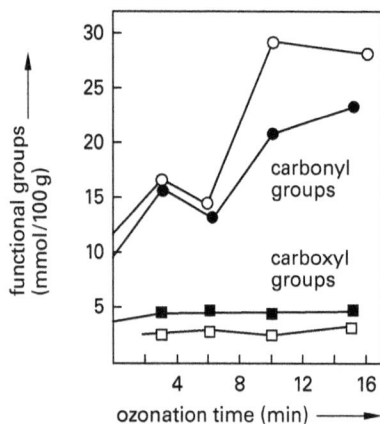

Figure 9.44. Formation of carbonyl and carboxyl groups in unbleached softwood kraft pulp (filled symbols) and in α-cellulose (unfilled symbols) as a function of ozonation time at 23 °C. (Chandra and Gratzl 1985.)

9.7 Other Bleaching Agents

9.7.1 Peracetic Acid Bleaching Chemistry

Peracetic acid has been known since a long time as an efficient and selective bleaching agent for both mechanical and chemical pulps. Thus, at weakly alkaline conditions (pH ~8), an elimination of chromophoric structures in mechanical pulps can be achieved without any substantial loss of yield. For chemical pulps, on the other hand, a comprehensive lignin oxidation and dissolution is obtained at neutral-slightly acidic pH-values. Under such conditions, peracetic acid can also be used for the preparation of holocellulose from wood. Despite the selective reactions of peracetic acid, it has not been used commercially for bleaching purposes due to the high manufacturing costs. The increasing demand for TCF-bleached pulps that started in the early 1990s paved, however, the way for new developments in bleaching technologies and, today, a few mills in the world are using peracetic acid in bleaching sequences such as OQ(OP)PAA+Q(PO). In this position, the pH is around 5–6, the temperature 85 °C and the retention time 1–3 hours.

The major reaction between peracetic acid and lignin is a nucleophilic addition of peracetic acid to electron-rich aromatic systems resulting in a successive oxidation to muconic acid structures via an intermediate formation of quinones. The reaction, outlined in *Figure 9.45*, takes place with phenolic as well as with non-phenolic aromatic structures albeit with large differences in reaction rates. Overall, however, the rate of oxidation with peracetic acid is rather low. In addition to oxidation of aromatic rings, peracetic acid may also induce a side-chain cleavage by oxidation of benzylalcohol structures.

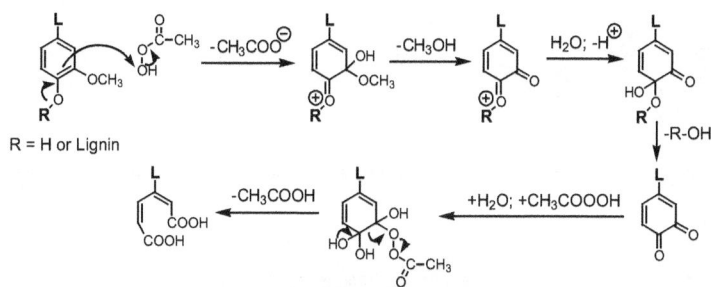

Figure 9.45. Reaction sequence for the oxidation of aromatic lignin structures with peracetic acid. (Johnson and Farrand 1971).

9.7.2 Miscellaneous Oxidation Systems

The bleaching of chemical pulp has attracted considerable research efforts due to the complexity and high costs of the overall process, the difficulties in reaching full brightness (~90 % ISO) in particular for softwood pulps and the environmental problems connected to the bleaching effluents. Thus, in addition to peracetic acid, other peracids have also been tried such as performic acid and peroxymonosulfuric acid (*Figure 9.46*). Neither of these has been able to give any beneficial effects beyond those of peracetic acid, however.

A further group of electrophilic oxidants has been referred to as "activated oxygen" due to their ability to transfer an activated oxygen atom to an aromatic substrate. Dimethyldioxirane is the simplest member of this class and is prepared by reacting acetone with peroxymonosulfate in slightly alkaline solution (*Figure 9.47*). With this oxidant, a selective and comprehensive de-lignification can be achieved but the costs as well as the handling problems are prohibitive for industrial application.

A direct oxidation of lignin with oxygen in neutral or slightly acidic media has been shown to give rise to oxidative side-chain cleavage as well as to products originating from acid hydrolysis (acidolysis). The conditions are rather harsh, however. These problems can be overcome by using a polyoxometalate (POM) with a lower energy barrier to lignin oxidation than oxygen. For re-oxidation of the POM, oxygen can be used, either concurrently or consecutively. In the former case ("the catalytic approach"), a small amount of POM is present together with oxygen thus giving a continous re-oxidation of the reduced form of the POM as it is formed in the lignin oxidation. If a separate oxidation step of the POM is employed ("the stoichiometric approach"), a large quantity of the POM for oxidation of the lignin is required which subsequently must be quantitatively re-oxidized by oxygen. So far, neither of these approaches has found technical application but the possibility of obtaining a comprehensive delignification under mild reaction conditions has attained much interest.

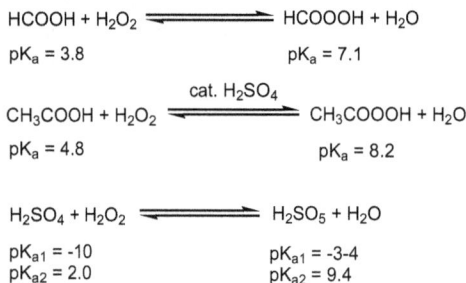

$$HCOOH + H_2O_2 \rightleftharpoons HCOOOH + H_2O$$

$pK_a = 3.8 \qquad\qquad pK_a = 7.1$

$$CH_3COOH + H_2O_2 \overset{cat.\ H_2SO_4}{\rightleftharpoons} CH_3COOOH + H_2O$$

$pK_a = 4.8 \qquad\qquad pK_a = 8.2$

$$H_2SO_4 + H_2O_2 \rightleftharpoons H_2SO_5 + H_2O$$

$pK_{a1} = -10 \qquad\qquad pK_{a1} = -3\text{-}4$
$pK_{a2} = 2.0 \qquad\qquad pK_{a2} = 9.4$

Figure 9.46. Oxidation of formic, acetic and sulfuric acid with hydrogen peroxide to give the corresponding per-acids.

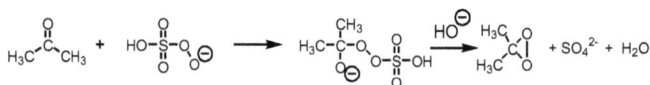

Figure 9.47. Synthesis of dimethyldioxirane from acetone and peroxymonosulfate.

A variety of polyoxometalates has been tried for the oxidation of lignin with one type being represented by $[PMo_{(12-n)}V_nO_{40}]^{(3+n)-}$ and another by $[X^{n+}VW_{11}O_{40}]^{(8-n)-}$ where $X^{n+} = Al^{3+}$, Si^{4+} or P^{5+}. Both of these types are able to react with lignin structures giving oxidative cleavage of side-chain structures and, depending on pH, acidolysis reactions. Further oxidation will, however, also take place and, finally, the dissolved material may be converted into carbon dioxide and water by a complete wet combustion.

Oxidative enzymes such as lignin peroxidases, manganese peroxidases and laccases are able to give a direct oxidative degradation of lignin under mild conditions. In a pulp fiber, however, the size of the enzyme is prohibitive for any reaction. The finding that oxygen and a low molecular weight compound, a "mediator", working together with the enzyme are able to delignify a chemical pulp to a considerable extent has initiated further work in this area (*Figure 9.48*). The major problem encountered so far has been the rather poor stability of the mediator and, consequently, a continuous addition of that compound must be done resulting in a high extra cost. With lignin model compounds, it has been shown that laccase together with HBT and oxygen is able to oxidize aromatic rings to quinones and muconic acid derivatives as well as to give side-chain oxidation and cleavage. Thus, a comprehensive delignification can be expected and the development of cost efficient mediators should constitute a preferred area for further development.

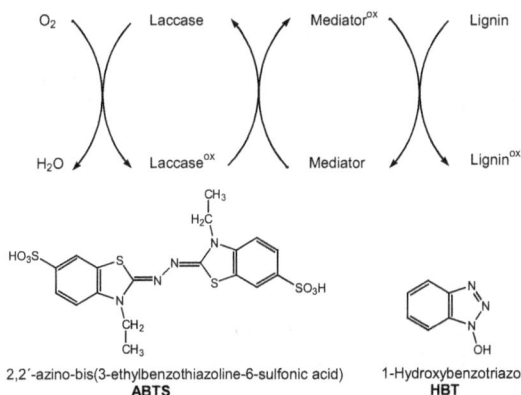

Figure 9.48. Reaction scheme for the oxidation of lignin by laccase in the presence of oxygen and a mediator. In the lower part of the figure, the chemical structure of two different mediators are shown.

9.8 Post-yellowing of Bleached Chemical Pulps

The brightness stability of fully bleached chemical pulps is high and irrespective of the pulp being used at site in integrated mills or stored and transported to a paper mill, the high brightness usually is maintained in the final paper product. Under certain conditions, however, pulp bales, on storage, may give rise to a considerable brightness reversion. It has been demonstrated that this brightness reversion has a maximum at slightly acidic conditions and that the presence of some humidity is required. Although the exact nature of the yellowing reactions still have not been fully identified, a good correlation between the amount of hexenuronic acid in the pulp and the degree of yellowing has been found as shown in *Figure 9.49*.

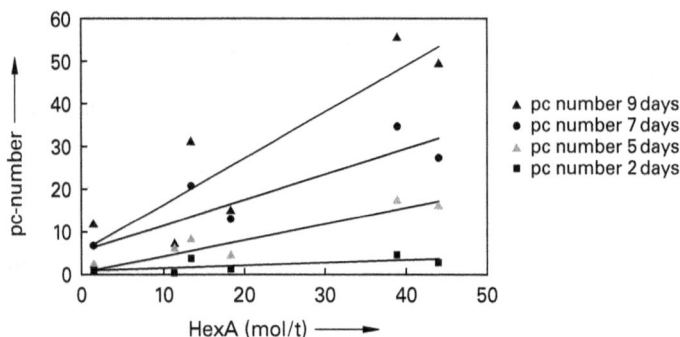

Figure 9.49. Degree of yellowing during 9 days, measured as PC-number, of fully bleached softwood and hardwood kraft pulps as a function of the amount of hexenuronic acid (HexA) in the pulp. (Sevastyanova 2005).

9.9 Further Reading

9.9.1 General Literature

Argyropoulos, D.S. (Ed.) (2001) *Oxidative Delignification Chemistry. Fundamentals and Catalysis. ACS Symposium Series 785.* ACS, Washington DC, USA.

Dence, C.W., and Reeve, D.W. (Eds.) (1996) *Pulp Bleaching. Principles and Practice.* TAPPI PRESS, Atlanta.

Sjöström, E. (1993) *Wood Chemistry. Fundamentals and Applications.* Academic Press Inc, London.

9.9.2 Academic Theses

Eriksson, T. (1993) *Reactions of ozone and chlorine dioxide in bleaching.* Ph.D. Thesis, Stockholm.

Gärtner, A. (1998) *Ligninets reaktivitet vid peroxidblekning. Inverkan av process-parametrar.* Lic. Thesis (in Swedish), Stockholm.

Heuts, L. (1998) *The lignin chemistry of OQP-bleaching. An analytical study.* Ph.D. Thesis, Stockholm.

Johansson, E. (1997) *The effect of oxygen on the degradation of lignin model compounds and residual lignin.* Ph.D. Thesis, Stockholm.

Jääskeläinen, A-S. (1999) *Kraft pulp bleaching with peroxyacetic acid and other peroxy compounds.* Ph.D. Thesis, Helsinki.

Karlsson, O. (1997) *Studies on the extent of bonding between lignin and polysaccharides in pulps.* Ph.D. Thesis, Stockholm.

Lawoko, M. (2005) *Lignin polysaccharide networks in softwood and chemical pulps: Characterization, structure and reactivity.* Ph.D. Thesis, Stockholm.

Li, J. (1999) *Towards an accurate determination of lignin in chemical pulps. The meaning of kappa number as a tool for analysis of oxidizable groups.* Ph.D. Thesis, Stockholm.

Ni, Y. (1992) *A fundamental study of chlorine dioxide bleaching of kraft pulp.* Ph.D. Thesis, Montreal.

Ragnar, M. (2000) *On the importance of radical formation in ozone bleaching.* Ph.D. Thesis, Stockholm.

Rööst, C. (2004) *The impact of extended oxygen delignification on the process chemistry in kraft pulping.* Ph.D. Thesis, Stockholm.

Sevastyanova, O. (2005) *Oxidizable structures in bleached kraft pulps.* Ph.D. Thesis, Stockholm.

Zhang, Y. (1994): *On the selectivity of ozone delignification during pulp bleaching.* Ph.D. Thesis, Stockholm.

9.9.3 References

Axegård, P., and Renberg, L. (1989) The influence of bleaching chemicals and lignin content on the formation of polychlorinated dioxins and dibenzofurans. *Chemosphere* 19: 661–668.

Basta, J., Holtinger, L., and Höök, J. (1991) Controlling the profile of metals in the pulp before hydrogen peroxide treatment. *6th International Symposium on Wood and Pulping Chemistry, Melbourne. Proceedings, Vol 1.* 237–244.

Björklund Jansson, M., Dahlman, O., Månsson, K., Rutquist, A., and Wickholm, J. (1991) On the nature of chlorinated organic matter in bleached kraft pulp. *International Pulp Bleaching Conference, Stockholm. Proceedings, Vol 1.* 123–136.

Chandra, S., and Gratzl, J.S. (1985) Kinetics of carbohydrate and lignin degradation and formation of carbonyl and carboxyl groups in low consistency ozonation of softwood pulp. *International Pulp Bleaching Conference, Quebec. Proceedings.* 27–35.

Johnson, D.C., and Farrand, J.C. (1971) Peroxyacetic acid oxidation of 4-methylphenols and their methyl ethers. *Journal of Organic Chemistry* 36: 3606–3612.

Lidén, J. (1994) The chemistry of manganese in a kraft mill. *3rd European Workshop on Lignocellulosics and Pulp, Stockholm. Proceedings.* 135–138.

Ni, Y., Kubes, G.J., and van Heiningen, A.R.P. (1991) Rate processes of AOX formation and chlorine species distribution during chlorine dioxide prebleaching of kraft pulp. *International Pulp Bleaching Conference, Stockholm. Proceedings, Vol 2.* 195–218.

Pfister, K., and Sjöström, E. (1979) Characterization of spent bleaching liquors. Part 5. Composition of material dissolved during oxygen-alkali delignification. *Paperi ja Puu* 61: 525–528.

Sandermann, W. (1974) Polychlorierte aromatische Verbindungen als Umweltgifte. *Naturwissenschaften* 61: 207–213.

Strömberg, L., Mörck, R., de Sousa, F., and Dahlman, O. (1996) Effects of internal process changes and external treatment on effluent chemistry. In *Environmental Fate and Effects of Pulp and Paper Mill Effluents* (Eds. M.R. Servos, K.R. Munkittrick, J.H. Carey, G.J. van der Kraak). St Lucie Press, Delray Beach, FL, USA, pp. 3–19.

Yamamoto, Y., Niki, E., Shiokawa, H., and Kamiya, Y. (1979) Ozonation of organic compounds. 2. Ozonation of phenol in water. *Journal of Organic Chemistry* 44: 2137–2142.

10 Bleaching of Pulp

Ulf Germgård
Department of Chemistry, Karlstad University

10.1 Introduction

Unbleached chemical pulps, which in most cases are kraft pulps, have a dark brown colour, which is a result of the chromophores in the pulp, and these are mainly found in the residual lignin. However, several paper products like printing and writing paper, food board etc require a bright pulp as the requirements on a good printing paper surface and other related properties are high. Good printing properties mean that a high brightness and a high cleanliness are needed and in the latter case this means a low amount of shives and dirt.

Bleached chemical pulps are found in a number of paper products like fine paper, board, tissue, bleached white-top liner etc. In some cases the bleached chemical pulp is used in combination with unbleached chemical pulp or with bleached or unbleached mechanical pulps. The majority of the chemical pulp that is produced in the world is bleached.

The bleaching process results in a brighter pulp but also in a pulp with higher cleanliness with respect to a low amount of shives, dirt and extractives. Bleaching also results in a pure pulp with respect to bacteria etc which is essential for especially food board and liquid board.

There is a fundamental difference between bleaching of chemical pulps and mechanical pulps. In the bleaching of chemical pulps lignin is oxidized, decomposed and finally eliminated from the pulp fibres. This results in less chromophores in the pulp. In the bleaching of mechanical pulps on the other hand the lignin is not decomposed and eliminated from the fibre. Instead only the chromophores in the lignin are attacked by oxidizing or reducing chemicals. Thus, the major amount of the lignin is still present but it is brighter and the pulp brightness is therefore higher.

As early as in the 18th century it was known that pulp fibres could be bleached with chlorine (Cl_2) and hypochlorite (OCl^- i.e. alkaline chlorine). One stage hypochlorite batch bleaching was the standard bleaching process in pulp mills until the 1930s when bleaching sequences with chlorine-alkali-hypochlorite were developed. Such a simple sequence was sufficient for chemical pulps that were easy to bleach like sulphite pulps but kraft pulps were still not possible to bleach to high brightness with such a sequence without a severe loss of pulp strength.

However, when chlorine dioxide (ClO_2) was found to be an excellent bleaching agent with very little attack on the cellulose in the 1920s it became possible to bleach also kraft pulps to full brightness with only a small strength loss. The first mill installations for chlorine dioxide bleaching were done in 1946 at the Swedish Husum mill. Such bleaching was during the same year also started up in Canada. Chlorine dioxide is today the dominating bleaching chemical used in pulp mills world wide. Bleaching based on chlorine dioxide and where no other chlorine based chemical like chlorine gas or hypochlorite is used is today denoted as ECF (elemental chlorine free bleaching). If chlorine dioxide is not used (and no other chlorine containing bleaching chemicals) the bleaching is instead called TCF (totally chlorine free bleaching). However, very few mills are today producing only TCF bleached kraft pulp and these mills are especially located to Sweden and Finland.

Oxygen (O_2) delignification was developed during the 1950s and 1960s in the Soviet Union as a way to continue the delignification in the cook before the actual bleaching stages. The big challenge was to find ways to avoid the simultaneous carbohydrate degradation during the delignification. The breakthrough was done by French researchers in the 1960s who found that a small addition of magnesium salts could protect the carbohydrates from degradation. The first industrial installation of oxygen delignification was done at the Enstra mill in South Africa in 1970. The start up of oxygen delignification at the Husum and the Gruvön mills in Sweden was done in the beginning of the 1970s. Oxygen delignification is used in all modern pulp mills world wide and all but two Swedish fibre lines producing bleached pulp have installed oxygen delignification thus nearly 100 % of the bleached pulp from Swedish pulp mills is oxygen delignified.

The term totally chlorine free bleaching (TCF) was established during the 1990s after the mill trials carried out by Eka Chemicals at the Swedish Aspa mill in 1989–90. In the mill trials it was shown that it was possible to bleach oxygen delignified softwood kraft pulp to high brightness with only hydrogen peroxide (H_2O_2) if correct conditions were used. This had earlier been impossible to achieve due to the presence of transition metal ions in the unbleached pulp, especially manganese, which resulted in severe decomposition of the added hydrogen peroxide. However, by first treating the pulp with a chelating agent like DTPA or EDTA and then wash the pulp before the hydrogen peroxide was added it was shown to be possible to more or less eliminate the manganese ions prior to the peroxide stage. Earlier hydrogen peroxide had been used but only to a small extent earlier in pulp bleaching but this new discovery made it possible to successfully use even high charges (20–40 kg/t) of hydrogen peroxide in the bleaching sequence. Modern pressurized peroxide bleaching stages uses this finding together with higher temperature to be able to reach high brightness levels with hydrogen peroxide.

Ozone (O_3) bleaching is a relatively new bleaching chemical that was studied in pilot scale in several countries in the 1980s and introduced in pulp mills in the beginning on the 1990s. The most well known installation is the Lenzing sulfite mill in Austria that started an ozone pilot plant of 100 t/d in 1990. In 1992 the Franklin mill in the USA and the Mönsterås mill in Sweden were the first kraft mills to include this bleaching stage on full mill scale. Today about 20 mills in the world use ozone bleaching in their sequences. One of its merits is that softwood kraft pulp of full brightness can be obtained with ozone together with oxygen and hydrogen peroxide stages without the use of any chlorine containing chemicals. Ozone based totally chlorine free bleaching is possible to use also for production of strong fully bleached softwood kraft pulps. Ozone has also become interesting in combination with chlorine dioxide stages as a way to reduce the demand of chlorine dioxide in ECF-sequences.

During the 1980s and 1990s the environmental concern in the pulp and paper industry has increased considerably and the pressure on the industry to become more environmentally friendly has to an increasing degree come from the final customers. Earlier it was the local or federal authorities that tried to push the industry towards reduced pollution. In line with these environmental efforts that especially were concentrated on reduced discharge of chlorinated organic compounds bleaching with elemental chlorine (Cl_2) was phased out and it was replaced with oxygen, chlorine dioxide and hydrogen peroxide.

10.2 Bleaching Chemicals Used

Of practical reasons each type of bleaching stage has got a special letter, *Table 10.1*. By combining these letters according to some special rules it is easy to understand how the bleaching sequence is designed. Thus, a sequence using chlorine in the first stage, alkaline extraction in the second and chlorine dioxide in the third stage the sequence will be: CED. However, if two chemicals are used together in the same stage the two letters from the table are written together within parenthesis and in some cases with a plus between. If for example both chlorine and chlorine dioxide are used in the same stage and if they are charged at the same time the stage is denoted as

(C + D). The chemical that is charged in highest quantity is written as the first one. Thus, in the example above the chlorine charge is larger than the chlorine dioxide charge. If on the other hand one chemical is charged ahead of the other the two symbols are written after each other and in some cases with a comma between. Thus, if chlorine dioxide is charged before chlorine the stage will be denoted as (DC). Although there are strict rules which have been specified by Tappi (The Technical Association of the Pulp and Paper Industry) simplifications are often used. One example is the plus sign that in most cases is eliminated for example in (EOP)- and (PO)-stages thus it can be difficult to know if a sequential or simultaneous charge is used.

Table 10.1. Symbols for bleaching stages.

Bleaching stage	Symbols
Treatment with acid	A
Chlorine	C
Chlorine dioxide	D
Alkaline extraction	E
Hypochlorite	H
Oxygen	O
Hydrogen peroxide	P
Chelating stage	Q
Peracetic acid	T (earlier Paa)
Treatment with water	W
Enzyme stage	X
Dithionite	Y
Ozone	Z

Asterisks have at certain times been used for stages where the same chemical is used but where the bleaching conditions are significantly different. However, in most cases it is better to stick to the nomenclature and try to describe the modifications in another way.

Usually bleaching of pulp is done not only with one but with several bleaching agents and this is done to reduce the bleaching chemical demand and to maintain the pulp strength. In the first stages where the lignin concentration is highest the demands with respect to selectivity are the lowest, as the major amount of the added chemical will be consumed by the lignin. However, the selectivity demand on the bleaching chemicals increases with increasing bleaching.

When the bleaching is done mainly with chlorine containing chemicals i.e. with chlorine, hypochlorite and/or chlorine dioxide, it is common to summarize these charges in one term only and this is called „active chlorine". This means that all charges are recalculated as if only chlorine was used, *Table 10.2*. This means that all chlorine based bleaching chemicals used in such a sequence can be given with only one number.

Table 10.2. Active chlorine content in chlorine containing bleaching chemicals.

Chemical	kg active chlorine/kg
Chlorine (Cl_2)	1
Chlorine dioxide (ClO_2)	2.63
Sodium hypochlorite (NaClO)	0.95

However, the trend is that more and more of the bleaching is done with bleaching chemicals that do not contain chlorine. This means that all oxidizing chemicals used in the sequence are not included in the „active chlorine" number and a new term has therefore been developed, i.e. the oxidizing equivalent or OXE, *Table 10.3*. Thus, by using OXE all oxidizing bleaching chemicals used in a sequence can be included in one number. However, in reality this theoretical number should be used with caution as the same bleaching effect cannot always be expected for different chemicals even if the OXE should be the same.

Table 10.3. Oxidizing equivalents for some bleaching chemicals.

Chemical	e$^-$/mole	Reduced to	OXE/kg
Chlorine (Cl_2)	2	Cl^-	28.2
Chlorine dioxide (ClO_2)	5	Cl^-	74.1*
Hydrogen peroxide (H_2O_2)	2	H_2O	58.8
Sodium hypochlorite (NaOCl)	2	Cl^-	26.9*
Oxygen (O_2)	4	H_2O	125
Ozone (O_3)	6	H_2O	125

* or 28.2 OXE/kg if calculated as active chlorine

10.3 A bleaching Stage

It should be noted that in the very beginning of industrial bleaching, the processes were batch-wise and the pulp was moved to the bleaching vessel by hand force. However, already in the early 1920's the continuous bleaching process was invented.

Bleaching is in most cases done in several stages and each stage is in principal identical but with some smaller differences concerning the type of reactor and the type of washer that is used after the stage. The main differences concerning the reactor are:

1. The pulp flow direction, which can be up-flow, down-flow or upflow-downflow
2. The pulp consistency used in the stage
3. If the reactor is pressurized or not
4. The size of the reactor or tower which indirectly is determined by the reaction time needed.

The final decision of which reactor type and which type of washer that will be used is depending on the bleaching stage and on earlier preferences. In Sweden most stages are of up-flow type, *Figure 10.1*. All modern stages are also of medium consistency type meaning that the pulp consistency is 9–13 %. In older bleach plants there are sometimes stages which are using low consistency i.e. 3–4 % but these stages are now phased out. Ozone bleaching is one exception and such bleaching is sometimes done at 35–40 % consistency also in new systems, see 10.11.

A reactor is denoted atmospheric if there is no overpressure above atmospheric at the top of the reactor. If the reaction on the other hand requires an overpressure for example if a temperature close to or above 100 °C is used a pressurized reactor is needed. A pressurized reactor has a valve at the outlet and the reactor is more rounded at the top as seen in *Figure 10.2*. Inside the bleaching tower either static or mechanical distributors spread the pulp evenly inside the reactor in a way that makes all the pulp flow at the same velocity meaning that a plug flow is obtained.

At the top of the bleaching tower rotating scrapers are forcing the pulp to the outlet in a way that the plug flow in the tower is maintained.

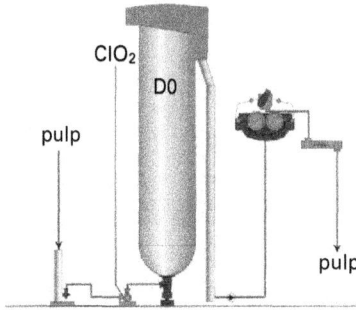

Figure 10.1. A bleaching stage of up-flow type. There is no over pressure at the top of the reactor and the reactor is denoted as atmospheric. The stage is of medium consistency type. (Metso Paper).

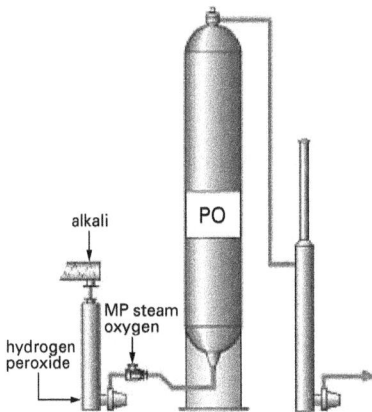

Figure 10.2 A pressurized bleaching stage which in this case is a (PO)-stage. The stage is of up-flow and medium consistency type. (Kvaerner Pulping).

Independently of the bleaching chemical used a medium consistency bleaching stage always uses a medium consistency pump that feeds the pulp through the stage. In Figures 10.1 and 10.2 the pump is mounted on a stand-pipe which always has a certain level of pulp which is needed for the pump to work properly. The pump pushes the pulp through the stage in an even manner and it is important that air is not allowed to mix with the pulp as this can lead to a separation of the pulp suspension into a liquor phase and a gas phase inside the reactor. This is called channelling and it prevents a proper plug flow. Air in the pulp will also negatively influence the possibility to transport and to wash the pulp properly after the stage. From the pump the pulp goes to a mixer or a static distributor where the bleaching chemical is added to the pulp if the bleaching chemical is chlorine dioxide, chlorine, oxygen, ozone (MC) or sometimes hydrogen peroxide. For sodium hydroxide, chelating agents, peracetic acid and sometimes hydrogen peroxide the chemical can also be added to the pulp prior to the pump or in the pump. The requirement on the mixing efficiency is high as poor mixing leads to uneven distribution of bleaching chemicals in the pulp and this results in local over-bleaching in some areas and local under-bleaching in other areas. The final result of poor mixing is that the consumption of bleaching chemicals increase and the pulp strength is reduced compared at a given bleaching result. The requirements are especially high if gas (for example oxygen) is added to the medium consistency pulp slurry or if the bleaching reaction is very fast and the risk for uneven bleaching is obvious. The latter case is particularly valid for bleaching with ozone that reacts extremely fast with the pulp.

When the chemical, or chemicals, are added to the pulp the reaction starts and a certain reaction time is needed. Depending on the chemical used and how far the pulp has been bleached prior to the stage the time requirements vary considerably. The total time requirement is for example less than 1 minute for ozone bleaching and 1–3 hours for final bleaching with chlorine dioxide.

The reactor shown in Figure 10.1 is of up-flow type but down-flow type reactors are sometimes used instead. The benefit by using a down-flow tower is that the pulp level in the tower is not constant and a certain buffer capacity is therefore obtained in each stage meaning that a certain stage can be stopped while the other stages are still in operation. Thus, it is therefore easier to run and control the bleach plant. Another advantage with a mix of up-flow and down-flow reactors is that the energy requirement in the bleach plant is reduced as the pulp does not have to be lifted to the top of the reactor in each stage. A disadvantage with a down-flow reactor is that if the reacting chemical is a gas like oxygen or if the chemical is volatile like chlorine dioxide there is a risk that there will be a phase separation in the top of the down-flow tower. The bleaching chemicals could thus accumulate at the top of the tower and it will not be in good contact with the pulp and the pulp can be under-bleached. Alkaline extraction stages are often of down-flow type and there is no problem with gas separation in such stages as long as oxygen is not used in the stage, i.e. if the stage is of E or (EP) type.

10.4 The Bleaching Reaction in a Given Bleaching Stage

The general reaction pattern in all bleaching stages is the same with an initial fast phase and a slower final phase. One example where the brightness is used as a measure of the bleaching result is shown in *Figure 10.3*. The figure shows that the bleaching proceeds with decreasing efficiency to the so called brightness ceiling where no additional brightness increase is obtained. The figure also shows that the lignin content in the pulp decreases according to a corresponding relationship with an initial efficient phase and a later less efficient phase where the chemical consumption increases considerably for an additional decrease in lignin content. Thus, it is beneficial of economical reasons to concentrate the bleaching to the initial part and to avoid the final part of the reaction.

Parallel to the bleaching effect there is also an attack on the carbohydrates which can be seen by the simultaneous decrease in pulp viscosity, *Figure 10.4*. However, for the carbohydrates the initial part of the bleaching reaction is very selective with a limited attack on the cellulose but with increasing bleaching the attack on the carbohydrates increases. It can therefore be concluded that in the initial part of the bleaching stage the bleaching chemicals charged to the stage are especially consumed in lignin degradation reactions and the selectivity i.e. the lignin degradation rate vs. the carbohydrate degradation rate, is therefore high. The selectivity decreases with increasing bleaching which means that it is recommended of pulp quality reasons to avoid too intense bleaching stages.

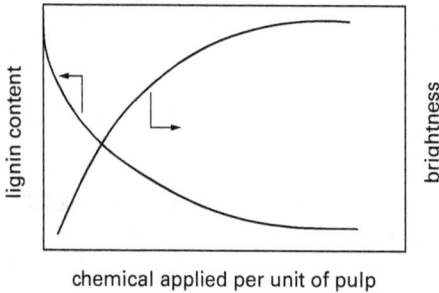

Figure 10.3. The general behaviour of brightness increase and lignin degradation in a given bleaching stage versus the amount of chemicals applied.

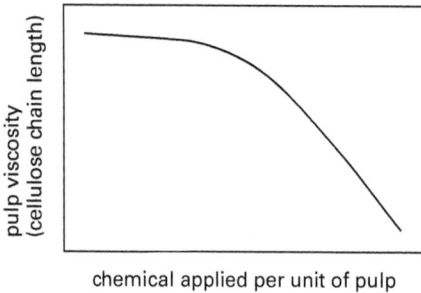

Figure 10.4. The pulp viscosity relationship versus the amount of chemicals applied to a given bleaching stage.

10.5 Bleaching Sequences

As described in section 10.4 the bleaching result that can be obtained in a single bleaching stage is limited and it is usually not sufficient for a pulp mill that produce bleached kraft pulp. Therefore, a sequence of stages are in most cases used and the number of stages is usually 3–5. The maximum brightness that can be obtained is called the brightness ceiling and that brightness is dependent on the wood raw material, the pulping method, the bleaching sequence and the bleaching conditions. The bleaching chemical charge in a given stage is usually aimed for bleaching well below the brightness ceiling of economical reasons. If a higher pulp ceiling is needed more stages, new bleaching chemicals or more intense conditions are needed in the sequence. A chlorine dioxide charge can for example become more efficient if the charge is split up from one to two stages with an alkaline stage between. This split of acidic and alkaline stages results in significantly better bleaching efficiency measured for example as a higher brightness for a given consumption of bleaching chemicals, *Figure 10.5*. The kappa factor on the x-axis in the figure is defined as the active chlorine consumption given in % on the pulp divided by the kappa number of the pulp before bleaching. Thus, if the kappa number of the pulp is 12 and the kappa factor 0.2 the active chlorine charge is 2.4 % or 24 kg per ton of pulp.

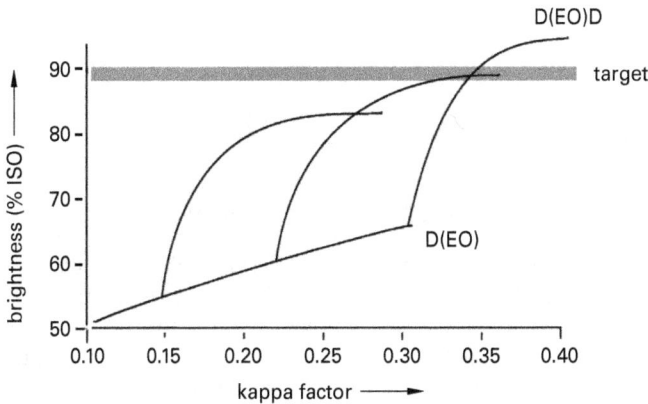

Figure 10.5. Brightness after bleaching with D(EO) or D(EO)D versus the chlorine dioxide consumption denoted as the kappa factor. The latter is calculated as % active chlorine on pulp per kappa number of the pulp prior to bleaching. It is seen that by splitting the chlorine dioxide charge from one to two stages a significantly higher brightness is obtained for a given kappa factor.

The beneficial stoichiometric effect shown in *Figure 10.5* is in a bleach plant obtained by using several bleaching reactors with a washer between each stage, *Figure 10.6*. In mill scale the number of stages in a sequence is usually 3–5 after oxygen delignification if the pulp is bleached to full brightness i.e. at 88 % ISO or higher. Fewer stages are needed if the brightness requirement is lower or if the pulp is easy to bleach. Many hardwood pulps like the most commonly used eucalyptus pulps are easier to bleach than softwood pulps i.e. they require less active chlorine for bleaching to full brightness compared at a given starting kappa number prior to bleaching. However, to also obtain brightness stability comparable to softwood pulps hardwood pulps sometimes need a special treatment which is more discussed below.

In *Figure 10.6* the bleaching towers are of different types. An up-flow tower is used in the first D-stage and up-flow down-flow tower is used in the (EO)-stage where the up-flow part is pressurized as oxygen is added and an overpressure above atmospheric is therefore needed. The final chlorine dioxide stages are of up-flow down-flow type where the major amount of the chlorine dioxide is supposed to be consumed in the up-flow part of the stage. It is not possible to use a down-flow tower without the initial up-flow tube for an (EO) or a D-stage as this would lead to a gas separation at the top of the tower and the bleaching efficiency would therefore be low.

In Figure 10.6 the bleaching towers are of different types. An up-flow tower is used in the first D-stage and up-flow down-flow tower is used in the (EO)-stage where the up-flow part is pressurized as oxygen is added and an overpressure above atmospheric is therefore needed. The final chlorine dioxide stages are of up-flow down-flow type where the major amount of the chlorine dioxide is supposed to be consumed in the up-flow part of the stage. It is not possible to use a down-flow tower without the initial up-flow tube for an (EO) or a D-stage as this would lead to a gas separation at the top of the tower and the bleaching efficiency would therefore be low.

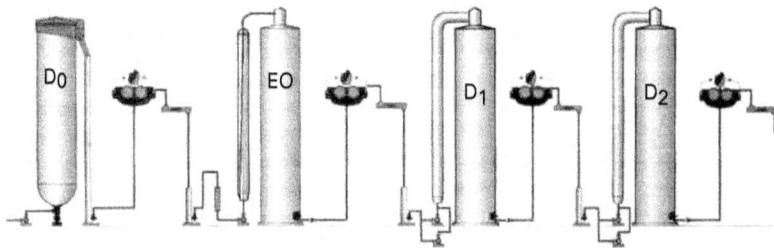

Figure 10.6. A bleaching sequence with four stages according to $D_0(EO)D_1D_2$. (Metso Paper).

Several bleaching sequences are today used in the pulp mills but the D(EO)DD-sequence can be considered as the standard sequence for modern pulp mills all over the world. However, many other sequences are also used and the reason to the large number is due to several reasons. Examples are local considerations concerning pollution, the availability of fresh water, the pulp quality requirements, and the allowed maximum investment and operating cost.

10.6 Bleaching Quality Analysis

There are three bleaching quality analyses that are often used i.e. the kappa number, the pulp viscosity and the pulp brightness. The kappa number is defined in the SCAN-method (Scandinavian pulp, paper and board testing committee) as the number of millilitres of a 0.020 mmol/ dm^3 $KMnO_4$-solution that 1 gram of pulp consumes upon reaction during 10 min at 25 °C. The number is corrected to a consumption of 50 percent of the added permanganate and the amount of pulp given as gram of pulp is 25/the kappa number. The kappa number analysis was once designed to estimate the lignin content of the pulp. However, today a more correct way of explaining the kappa number is to refer to it as an estimate of the amount of structures that can be oxidized in the pulp. The finding that hexenuronic acids that originate from the hemicellulose xylan in the pulp also consume bleaching chemicals has made the kappa number lignin correlation more dubious. For hardwood kraft pulps the kappa number analysis is an especially misleading estimate of the lignin content due to the high content of hexenuronic acids.

The viscosity measured according to the SCAN-method is a measure of the time it takes for a given amount of dissolved pulp in a cupriethylenediamine solution to pass a standardized glass capillary. The value obtained is an estimate of the length of the cellulose chains and thereby a rough estimate of the pulp strength. In practise the relation to the pulp strength is more complicated. Measurements of viscosity after a treatment in acidic oxidising environment, e.g. after ozone or chlorine dioxide treatments, often show contradicting results. It is therefore better to first treat the pulp by alkaline extraction and then measure the viscosity. Another alternative is to use a borohydride treatment to remove the carbonyl groups before the viscosity test. The unit of the viscosity is dm^3/kg. Typically a kraft pulp after cooking has a SCAN-viscosity in the range 1100–1500 dm^3/kg. After oxygen delignification the viscosity is in the range 900–1100 dm^3/kg and at full brightness 700–900 dm^3/kg. In some countries the standardised way of measuring viscosity ends up with a figure in mPs units („milli-Pois"). The correlation with the SCAN viscosity is not linear.

The selectivity of a certain bleaching stage is sometimes calculated as the ratio of the viscosity loss and the kappa number change in a certain bleaching stage i.e. Δ viscosity/Δ kappa number. For oxygen delignification the selectivity is typically in the range of 12–18 (dm³/kg, kappa number) for softwood kraft pulp, where the lower being the more selective and thus more desirable.

The ISO brightness (diffuse intrinsic reflectance factor) is defined as the reflectance measured at 457 nm of an infinite pile of pulp sheets. The brightness is a function of the quota k/s, where k is the absorption coefficient and s the scattering coefficient in the pulp. This correlation was proposed already in 1931 by Kubelka and Munk. During the course of the bleaching the scattering coefficient is normally little affected, while the adsorption coefficient is significantly reduced. The scattering coefficient is reduced by beating and thus beating reduces brightness. Brightness is therefore only measured on unbeaten pulps.

Pulps bleached to 70–85 % ISO are usually referred to as „semibleached". If the brightness is 88 % ISO or higher the pulps are „fully bleached". Pulps having a brightness exceeding 90 % ISO are seldom produced on purpose, but for certain specific applications, e.g. photo paper a brightness of up to 94 % ISO is sometimes requested.

The brightness reversion or yellowing of the pulp is intimately connected to the pulp brightness. The brightness reversion is the decrease in the pulp brightness occurring after the bleaching for example upon storage of the pulp. Factors in the pulp that are known to have an impact of the yellowing are the residual chlorinated extractives, the content of hexenuronic acids and lignin, the two latter being the ones of primary interest in modern mills. External factors that influence the brightness reversion are for example the temperature, UV-light and time. The measurement of the brightness reversion is normally conducted after accelerated ageing carried out by heat treatment or by intense light exposure. The problem with these methods is that the accelerated treatment does not simulate the natural ageing in any exact way. If pulps of different brightness are to be compared with regards to their brightness reversion, the values must be normalised since the brightness scale is not linear. The brightness reversion method must also be stated.

10.7 Oxygen Delignification

Oxygen delignification is used after cooking but before the actual bleaching sequence and its filtrate is in nearly all cases recycled as wash water to the brown stock wash. It is beneficial to include oxygen delignification in a bleaching sequence as it results in reduced pollution, increased pulp strength measured as pulp viscosity if compared at the same kappa number prior to bleaching, reduced kappa number variation to the bleach plant, reduced bleaching chemical consumption and in some cases reduced bleaching cost. Under certain conditions oxygen delignification can also lead to increased pulp yield; however it is not possible to get all the benefits at the same time. Of the benefits above it is the environmental aspect that is the most important one when installing oxygen delignification.

As the filtrates from the brown stock wash are taken to the chemical recovery system this means that the dissolved organic substances from the oxygen stage are recycled and the remaining effluent load from the bleach plant will therefore be reduced compared with the situation before the oxygen stage was introduced. Naturally the kappa number after the cook can be adjusted when oxygen delignification is used. If it is decreased it will further reduce the polluting emissions if the kappa number after the oxygen delignification stage also is decreased to keep the degree of delignification constant but the pulp yield will then also be lower. If it is increased it can improve the pulp yield if the degree of delignification in the oxygen stage is unchanged but the polluting emissions will be higher than in the other alternative.

Bleaching with oxygen is carried out at alkaline pH (final pH 10–11 measured at 25 °C) and at elevated temperature usually about 100 °C. It is usually done in the presence of a small amount of a magnesium salt, like magnesium sulphate, for pulp viscosity protection, *Figure 10.7*. The reason to the beneficial effect of the magnesium ions on the selectivity has been studied by many but according to Lidén et al (2001) the magnesium ions form a solid state where iron and manganese ions are trapped in their +II valance state as carbonates, hydroxides or silicates. These metal ions can therefore not take part in any cellulose degradation reactions. Transition metals and especially manganese ions are active in generating harmful hydroxyl radicals that react with the cellulose and this leads to reduced chain length and thus reduced viscosity.

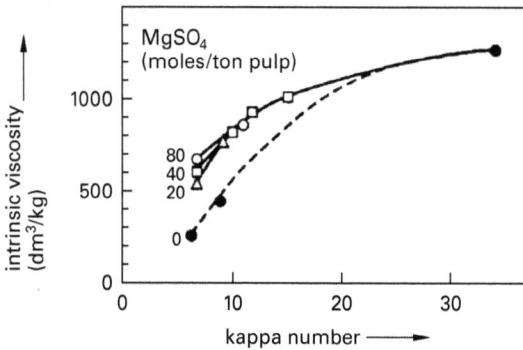

Figure 10.7. The addition of magnesium ions to the oxygen stage slurry increases the pulp viscosity at a given kappa number. A softwood kraft pulp was used. (Gustavsson and Swan, 1974).

The bleaching reaction in the oxygen delignification stage is more selective in its attack on the lignin than prolonged cooking as is shown in *Figure 10.8*. Thus, it is possible to delignify the pulp longer prior to bleaching and still keeping the pulp viscosity constant.

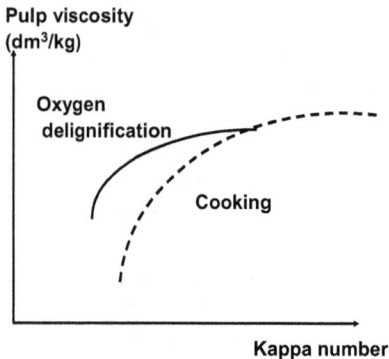

Figure 10.8. The pulp viscosity versus the kappa number for oxygen and non-oxygen delignified pulps.

The positive effect compared with prolonged cooking is also seen on the pulp yield at a certain kappa number, *Figure 10.9*. A further pulp yield increase is obtained if the kappa number prior to the oxygen stage is increased as much as possible. However, the delignification in the oxygen stage cannot be extended too far without becoming poor in selectivity. Thus, there are limits of how high kappa number that can be accepted after cooking if the kappa number after the oxygen stage is to be kept constant.

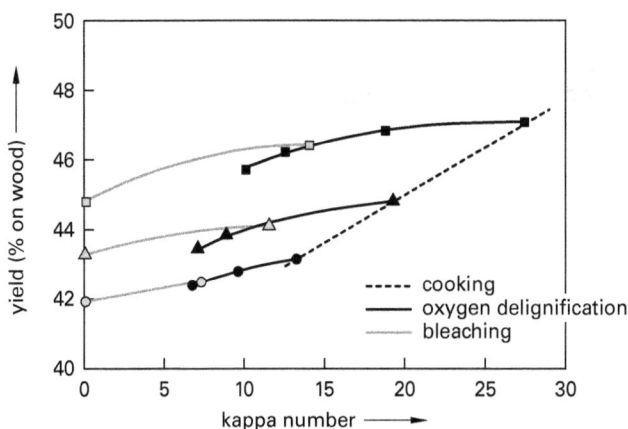

Figure 10.9. The pulp yield after cooking, oxygen delignification and bleaching as a function of the corresponding kappa number of the pulp. (Kvaerner Pulping).

10.7.1 Process Equipment

In oxygen delignification the oxygen molecules have to be transported from the gas phase to the fibre surface via the surrounding liquid film. However, oxygen has a low solubility (0.05 g/l at 20 °C) in water and the solubility decreases with increasing temperature and increasing ionic strength. Thus, the conditions in a pulp mill are not ideal for bleaching with oxygen and the process conditions has to be chosen so the oxygen transport becomes as beneficial as possible.

The solubility of oxygen is increased if the oxygen partial pressure is increased according to Henry's law; $[O_2] = k \times PO_2$ Thus, a high oxygen pressure is used. A relatively dry pulp, i.e. a high pulp consistency, reduces the diffusion resistance in the liquid film that surrounds the fibres. The first oxygen stages were therefore high consistency systems where the high pulp consistency was obtained in a press. In these high consistency (HC) systems the pulp consistency was 25–30 % and the pulp was slowly transported downwards in a reactor where the gas phase was stagnant. The gas phase consisted mainly of oxygen and generated by-product gases as carbon dioxide and carbon monoxide. Oxygen make up was added to the reactor and to avoid a build up of carbon monoxide a certain amount of the gas phase was purged and purified after which it was recirculated back to the reactor. If the gas phase was not properly controlled explosions and/or fires could take place in the reactor.

When new high intensity mixers were developed in the late 1970s it became possible to efficiently distribute the added oxygen gas in a medium pulp consistency slurry, i.e. at 9–13 %, and this simplified the system significantly, *Figure 10.10*. The explosion risks were also eliminated as there was no stagnant gas phase in the reactor. Finally, as the alkali could be better distributed and the bleaching was more homogenous the selectivity was improved. Thus, the medium consistency system quickly took over the market for new systems and many existing high consistency systems have since then been rebuilt to medium consistency systems.

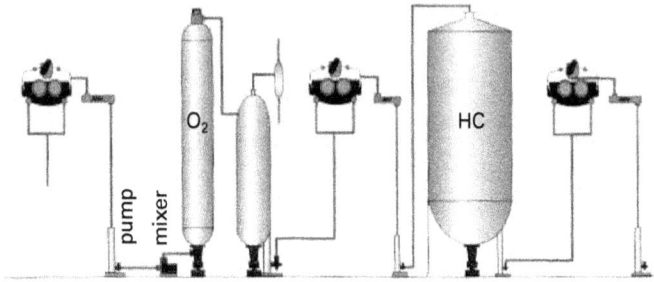

Figure 10.10. A medium consistency oxygen delignification stage consisting of a wash press, a transportation screw, a stand-pipe with a pump, a mixer and a reactor with a following blow tank, a wash press, a HC tower (buffer tank) and a final wash press. (Metso Paper).

The pollution restrictions for pulp mills have continuously increased and the demand on the oxygen stage to become more selective with respect to lignin reactions vs. carbohydrate reactions have therefore increased. More selective reactions mean that the degree of delignification in the oxygen stage can be prolonged without decreasing the pulp viscosity after the oxygen stage. Indirectly this means that less lignin has to be eliminated in the following bleach plant and the pollution load from the bleach plant therefore goes down. The new demands on the oxygen stage led to the findings that splitting the oxygen stage into two reactors where the bleaching conditions were not identical resulted in improved selectivity. By keeping the temperature in the first reactor down and higher in the second reactor the pulp viscosity was found to be best protected.

Although oxygen delignification performed in two reactors was introduced in pulp mills already in the 1980's the modern two reactor processes become a commercial success in the mid 1990's. The conditions used in e.g. the OxyTrac process by Metso Paper are given in *Figure 10.11* Thus, the pressure in the first reactor is 8–10 bar while the temperature is only 80–85 °C. The time used in the first reactor is 20–30 minutes. In the second stage the pressure is decreased to 3–5 bars and the temperature is increased to 90–100 °C to speed up the extraction of the pulp. The reaction time in the second reactor is about one hour.

In the two reactor oxygen delignification process called Dualox and developed by Kvaerner Pulping the first reactor is only a short pipe where the retention time is only a few minutes. By adding an MC-pump after the first reactor, i.e. after the pipe, a high pressure is obtained also in the second reactor thus a high pressure is maintained throughout the whole system. The high pressure during the long retention time is used to speed up the reaction. The pressure in the inlet of both reactors is therefore 8–10 bar, resulting in a reactor top pressure of about 5–6 bar. The temperature in the first reactor is 85–90 °C and 90–105 °C in the second.

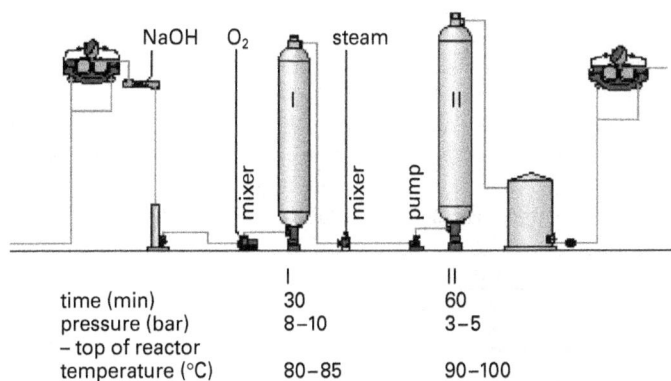

Figure 10.11. A medium consistency oxygen delignification stage with two reactors, i.e. a modification of Figure 10.10. (Metso Paper).

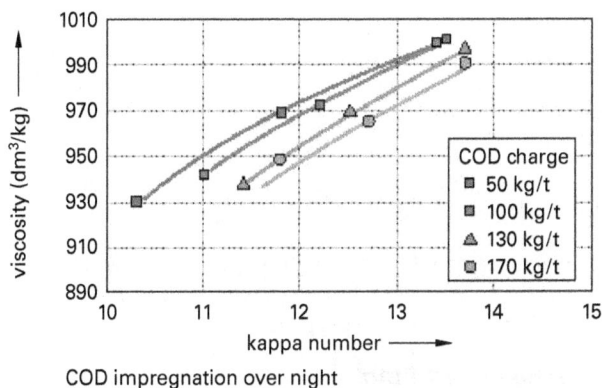

COD impregnation over night

Figure 10.12. The viscosity at a given kappa number after oxygen delignification is decreased if the unbleached pulp contains more unoxidized COD i.e. more dissolved organic material from the digester. Softwood kraft pulp. (Metso Paper).

The efficiency of washing after cooking will influence the amount of organic material that is still in the pulp during oxygen delignification. This remaining amount is often referred to as „carry-over" which is measured as kg COD per ton of pulp or earlier as kg Na_2SO_4 per ton of pulp. It is important that the amount of „carry-over" is on a low level as it negatively affects the selectivity of the oxygen stage. This is shown in *Figure 10.12*. Note that this COD is denoted as „unoxidized" COD. Thus, better washing prior to oxygen delignification reduces the amount of unoxidized COD that goes to the oxygen stage and this makes it possible to increase the degree of delignification in the oxygen stage.

The organic material that is coming to the oxygen delignification reactor directly from the digester is different compared with organic material that has already passed the oxygen reactor and then been recycled back to the wash stage prior to the oxygen stage. The latter type of COD is usually referred to as „oxidized" COD and it has in fact a positive effect on the selectivity,

Figure 10.13. Thus, the organic material that follows the pulp into the oxygen reactor consists of two types of organic material viz. unoxidized and oxidized organic material. The reason to the positive effect of the presence of oxidized organic material is according to Metso Paper that the oxidized organic material is unreactive to alkali and it is therefore not competing with the delignification reactions. Secondly, it is believed that the small amount of peroxide that is formed in the stage reacts more selectively with the pulp as the recycled organic material acts as a chelating agent and therefore protects the peroxide from decomposition reactions. The latter type of decomposition is due to the presence of transition metal ions like manganese.

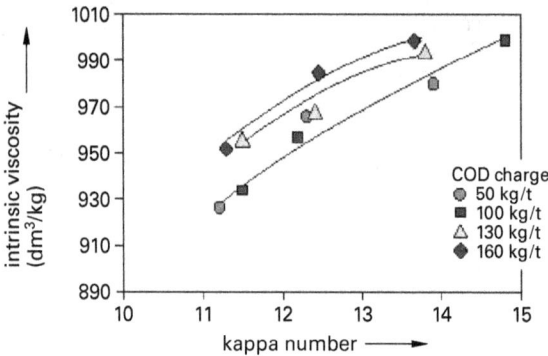

Figure 10.13. The viscosity at a given kappa number after oxygen delignification is increased if the unbleached pulp contains increasing amounts of oxidized COD i.e. dissolved organic material that has already been oxidized in the oxygen stage and then recycled. The same pulp as in Figure 10.12. (Metso Paper).

There is a significant amount of organic material liberated in the oxygen stage and to get the full environmental effect it is important that the pulp after the oxygen stage is well washed. This means that two washing stages are usually needed after the oxygen stage.

World Production of Bleached Kraft Pulp

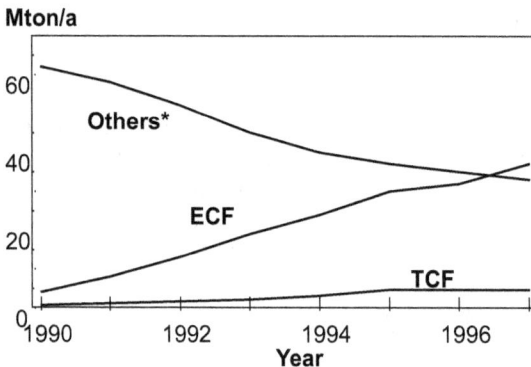

Figure 10.14. The world trend concerning bleaching with sequences including chlorine gas (others), chlorine dioxide but not chlorine gas or hypochlorite (ECF) or neither chlorine gas, chlorine dioxide or hypochlorite (TCF).

10.8 Prebleaching with Chlorine

Chlorine (Cl_2) has for a long time been the preferred bleaching chemical in the first bleaching stage due to its low cost, its high bleaching efficiency and its very high selectivity with respect to lignin reactions. However, during especially the 1980–1990's the chlorine consumption has dropped considerably in the world and it is no longer used in the pulp mills in Sweden or Finland. The reason to this trend was that the environmental problems associated with the discharge of chlorinated organic compounds to the receiving water became obvious and that the consumer pressure against the use of chlorine as a bleaching agent increased considerably. All this together resulted in the rapid development and implementation of ECF and TCF-bleaching, *Figure 10.14*. The figure shows that the dominating bleaching manner today is ECF.

Chlorine (Cl_2) is in water in equilibrium with hypochlorous acid (HOCl) and hypochlorite (OCl⁻) and the mixture between these two species is a function of the pH. Of these Cl_2 and OCl⁻ have been used as bleaching agents in pulp mills. Chlorine was used in the first bleaching stage while hypochlorite was used later in the sequence.

Bleaching with chlorine was originally carried out at a low pulp consistency (3–4 %), at a pH of about 2, at ambient temperature (10–30 °C) and during about 1 hour. The reason to the low pulp consistency was that the mixing technology at that time was still primitive and a low consistency was therefore needed to make it possible to add the large Cl_2-gas volume. As the consistency was low and the bleaching filtrate seldom was recovered large volumes of fresh water was therefore needed to dilute the pulp prior to the chlorine mixer. The temperature was therefore highly influenced by the temperature of the added fresh water and the temperature in the stage could be very low (5–10 °C) during the winter season.

The chemical charge in the chlorine stage was usually high i.e. 2.0–2.5 times the kappa number calculated as kg chlorine per ton of pulp. As the kappa number of the pulp for softwood kraft pulps was usually 30–35, this meant that as much as 60–80 kg chlorine per ton of pulp could be used in a mill. The high charge and the high reactivity of chlorine with lignin meant that the bleaching was very efficient with respect to delignification and cleanliness (= low shive content).

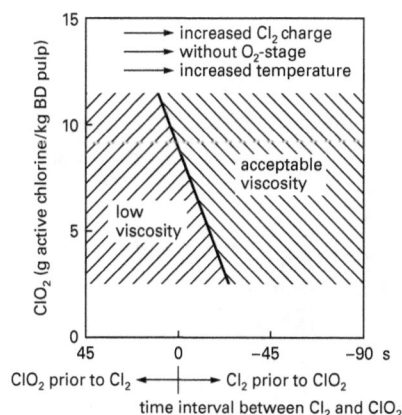

Figure 10.15. The figure shows that if a low charge of chlorine dioxide is used in a chlorine stage to protect the pulp viscosity it is best to charge it after the chlorine. (R Grundelius 1979).

10.9 Prebleaching with Chlorine Dioxide and/or Chlorine

It was discovered in the 1970´s that even a small addition of chorine dioxide to the chlorine stage significantly increased the viscosity of the pulp as chlorine dioxide acts as a radical scavenger. It thus became common to use a mixture of both chlorine and chlorine dioxide in the „chlorine stage". Even as little as 5 % of the total active chlorine charge could positively affect the viscosity especially if the chlorine dioxide was charged in a second mixer after the chlorine charge. This was shown by Grundelius in the classical figure shown below, *Figure 10.15*. The figure shows that if the chlorine dioxide charge was low the chlorine dioxide charge should be done after the chlorine charge i.e. in a second mixer if acceptable pulp viscosity was to be obtained. The arrows at the top of the figure shows that if the charge of chlorine increased, if oxygen delignification was not used or if the temperature was increased the time delay between the chlorine and the chlorine dioxide charges had to increase. The correlation shown in Figure 10.15 indicate that especially the later part of a chlorine stage is detrimental to the viscosity and that part of the stage therefore needs a viscosity protection.

A combination of chlorine and chlorine dioxide will also influence the delignification in the prebleaching as is shown in *Figure 10.16*. The figure shows the kappa number after the prebleaching vs. the fraction of chlorine dioxide used in the stage. Thus, to use the two bleaching chemicals in the best way for delignification the chlorine dioxide charge should be done prior to the chlorine charge which means that two mixers are needed. The poorest result is obtained if chlorine is charged prior to chlorine dioxide. Note also that for a given charge of active chlorine the kappa number after the first bleaching stages is lower if only chlorine is used than if only chlorine dioxide is used in the stage. For non-oxygen delignified pulp the kappa number is instead about the same for these two extreme cases. Thus, the best combination of chlorine and chlorine dioxide with respect to both good delignification and good viscosity protection is to charge the two chemicals according to (D, C + D) and to use a ClO_2-charge of 30–50 %. The reason to the difference between oxygen and non-oxygen delignified pulps is that the oxygen stage eliminates the lignin that is easiest to bleach with chlorine dioxide and the remaining lignin is therefore of chemical reasons less well suited for chlorine dioxide bleaching.

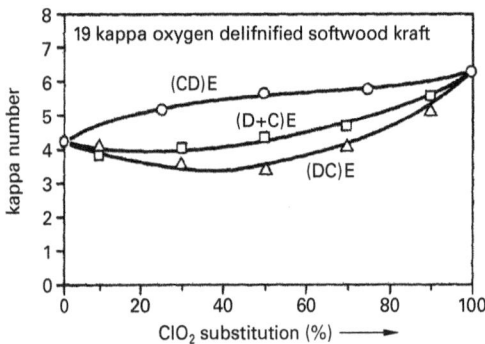

Figure 10.16. The kappa number after (CD)E, (C+D)E and (DC)E prebleaching of an oxygen delignified softwood kraft pulp versus the chlorine dioxide substitution. The same active chlorine charge was used in all cases. (Axegård 1986).

Modern bleaching does not include chlorine gas and the first stage is therefore today usually carried out with only chlorine dioxide. This stage is denoted as D_0 to distinguish it from the first and second D-stages in the final bleaching which are denoted D_1 and D_2. Conditions normally used in such a stage are: pH 2–3, 40–70 °C, 9–13 % pulp consistency, 0.5–1.0 h retention time and a charge of 1.0–2.0 × the kappa number calculated as kg active chlorine /t, see the box below.

Conditions in the D_0-stage

- pH 2–3
- 40–70 °C
- 3–4 % or 9–13 % pulp concistency
- 0.5–1.0 hours
- atmospheric pressure
- 1.0–2.0 x kappa number (aCl/t)

10.10 Final Bleaching with Chlorine Dioxide

Final bleaching to full brightness is often carried out with DD, DED or D(EP)D. The conditions used in these two D-stages, denoted D_1 and D_2 , are different to the conditions used in the D_0-stage. In the D_1-stage the following conditions are usually used; final pH 3.0–4.0, 55–75 °C, 9–13 % pulp consistency and 1–3 h. The influence of the final pH in the D_1-stage on the brightness and the viscosity are shown in *Figure 10.17*. In the D_2-stage the conditions are normally: pH 3.5–4.5, 60–85 °C, 9–13 % pulp consistency and 1–3 h retention time. The corresponding influence of the final pH in the D_2-stage on the brightness and the viscosity are shown in *Figure 10.18*.

Conditions in the $D_{1,2}$ -stages

- pH 3.0–4.0 in D_1 and pH 3.5–4.5 in D_2
- 55–75 °C in D_1 and 60–85 °C in D_2
- 9–13% pulp concistency
- 1–3 h
- atmospheric pressure
- Σ 4–6 x kappa number (E_1)   (a.Cl/t)
- D_1 /D_2-charge ratio usually 2/1–3/1

ECF Bleaching Fundamentals
Optimisation stage by stage - D1 stage

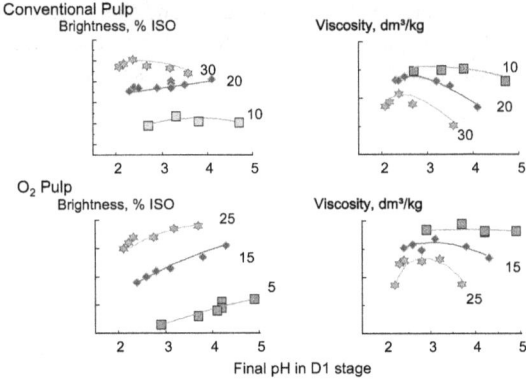

Figure 10.17. The influence of the bleaching conditions on the D1-stage. The chemical charges in kg active chlorine/ton are given in the figures to the left. (Eka Chemicals).

ECF Bleaching Fundamentals
Optimisation stage by stage - D2 stage

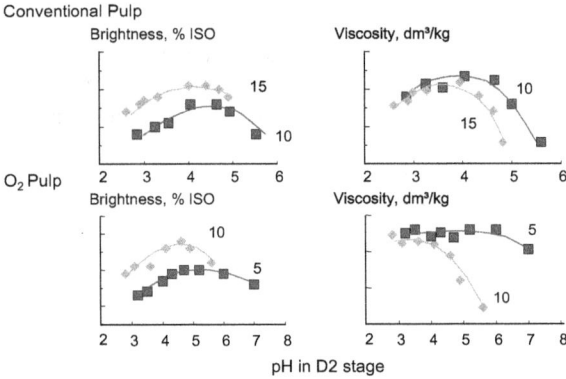

Figure 10.18. The influence of the bleaching conditions on the D_2-stage. The chemical charges are given in the figures to the left. (Eka Chemicals).

The split of chlorine dioxide between the D_1 and the D_2-stages is normally 2/1 to 3/1 to get the best bleaching result for a given charge of chlorine dioxide. This shown in *Figure 10.19*.

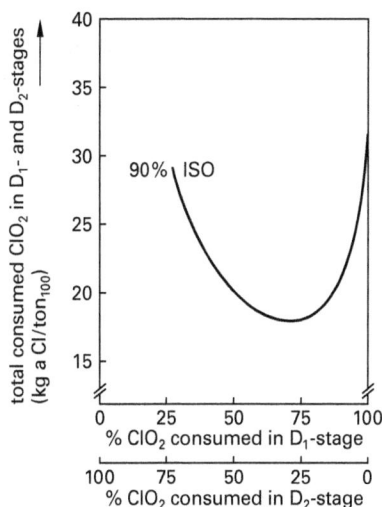

Figure 10.19. The split of chlorine dioxide between the D_1 and the D_2-stages when bleaching to full brightness. (Axegård, Jansson, Teder 1984).

10.11 Ozone Bleaching

Ozone reacts very rapidly with the pulp and the reaction takes place more or less instantly. The importance of a highly efficient mixing of the pulp and the ozone gas is therefore very high. In addition, ozone is sensitive towards high temperatures that cause spontaneous decomposition into oxygen. For this reason ozone stages are run at low temperature, which typically means around 30–60 °C. The stage is carried out at a pH of 2–3 and the retention time is a few seconds up to 1 minute, see the box below. An advantage with ozone is that extractives are removed very efficiently. This decreases the odour and taste in the pulp. Another advantage is its ability to efficiently remove hexenuronic acids in the pulp which improves the brightness stability of the final bleached pulp and reduce the ClO_2–demand in the bleach plant. Removal of hexenuronic acids is especially interesting for hardwood kraft pulps.

> **Conditions in the Z-stage**
> - pH 2–3
> - 30–60 °C
> - 9–13% or 25–45% pulp concistency
> - A few seconds up to 1minute
> - 5–10 bar or atmospheric pressure
> - 3–10 kg/t

In a medium consistency ozone stage the pulp suspension is fed via an MC-pump to two subsequent mixers where the pulp and the ozone gas are mixed and where the reaction takes place,

Figure 10.20. Acid (sulphuric acid) is added for pH adjustment to the pulp suspension prior to the MC-pump. The pressure in the reactor is about 5–10 bar. Upon leaving the second mixer the pressure in the pulp flow is released and the residual ozone is vented to ozone destruction and reuse of the oxygen in the gas phase. The pulp is then fed by a second pump to the subsequent washing stage.

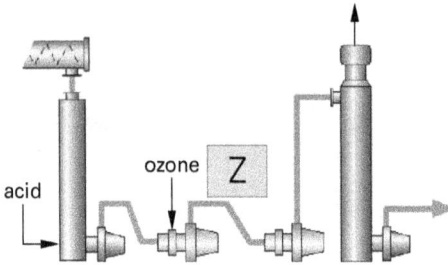

Figure 10.20. A medium consistency ozone stage consisting of a stand-pipe with an MC-mixer, two mixers in series and a final standpipe with separation of the residual gases. (Kvaerner Pulping).

In a high consistency system the pulp consistency used is 35–45 % which significantly facilitates the contact between ozone gas and the pulp fibres. The reaction is then carried out in a reactor with a high gas volume compared with the pulp volume and lower temperature and pressure can therefore be used than in a MC-stage. In the high consistency stage the temperature is usually 30–40 °C. The pressure is atmospheric which is sufficient as the reactor volume is so large compared with the pulp volume. The reactor is also quite different and it is horizontal where the pulp is moved forward by rotating paddles. A slightly longer retention time is utilised than in a MC system but the time is still only about 1 minute. A high consistency ozone reactor consisting of a mixing LC-tank, a press, a fluffer, a horizontal reactor with paddles, a dilution tank and a final wash press is shown in *Figure 10.21*.

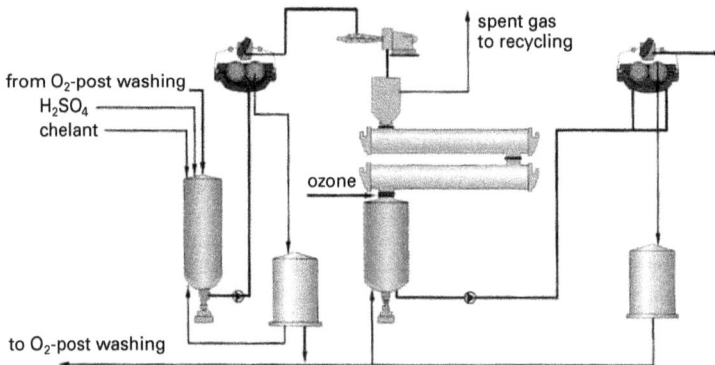

Figure 10.21. A high consistency ozone reactor consisting of a mixing LC-tank, a press, a fluffer, a horizontal reactor with paddles, a dilution tank and a final wash press. (Metso Paper).

A sequential effect can be obtained by combining the ozone stage with chlorine dioxide according to (DZ) or (ZD). A certain amount of chlorine dioxide can then be replaced with the ozone addition and the lower the ozone charge the higher is the ratio of replaced kg chlorine dioxide/ kg ozone. For conventional charges of 3–5 kg ozone per ton of pulp the ratio is 2.5–4 kg replaced chlorine dioxide, calculated as active chlorine, per kg charged ozone.

10.12 Chelating Stages

For hydrogen peroxide bleaching the metal ion content in the pulp is an important variable for successful hydrogen peroxide bleaching. All wood pulps contain metal ions in various concentrations. Some of these may cause decomposition of hydrogen peroxide and this leads to poor bleaching efficiency. The most harmful transition metal ions in hydrogen peroxide systems are manganese, copper and iron ions. Left uncontrolled, trace metal ions can decompose the hydrogen peroxide long before the pulp reaches the desired brightness level. Metal ions can also cause brightness reversion, both during the bleaching process and after the operation is completed. However, there are also metal ions that are beneficial to include in the pulp as these reduce the influence of the harmful ions. Particularly magnesium and calcium ions have this influence. Thus, it is beneficial if the pulp contains as little manganese, cupper and iron ions and as much magnesium and calcium ions as possible when the peroxide is charged to the pulp.

Metal ions can either be removed from the pulp by an acidic treatment at a pH of about 2 and a following wash or by adding a chelating agent to the pulp at a neutral pH and then a wash. In both cases a good metal ion removal is obtained, *Figure 10.22*. The metal ion removal from the pulp is sometimes denoted as Metal Management. Chelating stages are preferentially carried out at as high pH as possible but usually a neutral pH is used to reach the desired low metal ion content.

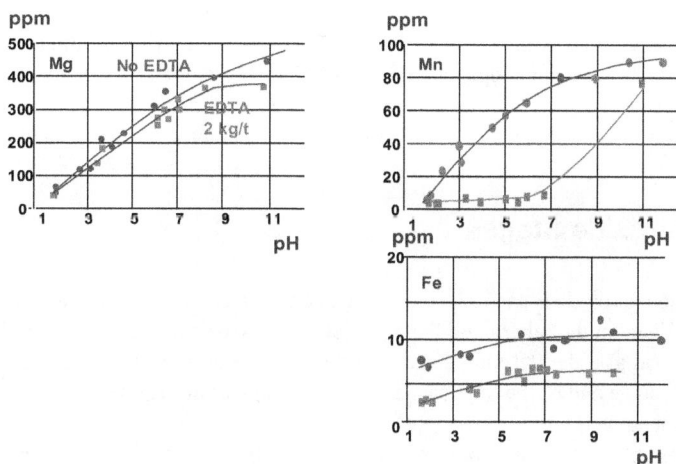

Figure 10.22. The influence of pH on the content of manganese, magnesium and iron ions in softwood kraft pulps without EDTA or with 2 kg EDTA/t. (Eka Chemicals).

The chelating agents form highly stable complexes with many metal ions and the metal ion containing complexes can then be removed in a following wash. An example of the metal removal efficiency after treatment of a pulp sample with EDTA is shown in *Table 10.4*. It is seen that the calcium and magnesium ions are kept in the pulp to a large extent while the unwanted iron, magnesium and cupper ions are mainly removed.

Chelating stages are mainly used prior to peroxide stages. Given below are some examples of bleaching sequences with chelating stages: OQPTP, OQ(PO)DD, O(EOP)Q(PO)Q(PO), (OO)Q(EP)Q(PO) and OQ(PO).

Table 10.4. Metal ion removal from a softwood kraft pulp in a chelating stage (0.2 % EDTA, 90 °C, 1 h, pH 5–7).

	Ca (ppm)	Mg (ppm)	Fe (ppm)	Mn (ppm)	Cu (ppm)
Before chelation	1400	300	11		0.6
After chelation	500–1000	120–280	6–8	<5	0.1–0.2

Conditions in the Q-stage

– metal ion remover prior to P-bleaching

- pH 4–7 (9–10 in alkaline Q)
- pH 5–7
- 50–80 °C
- 9–13%
- 5 min–2h
- amospheric pressure
- 2–4 kg EDTA or DTPA/t

The chelants used primarily in the pulp industry are the sodium salts of ethylenediaminetetraacetic acid (EDTA) and diethylenetriaminepentaacetic acid (DTPA). They remove metal ions by binding them hard to themselves and the metal ions are thus removed from the pulp and washed way in the following washer. Note that no bleaching effect is obtained in such a stage.

10.13 Hydrogen Peroxide Stages

For chemical pulps hydrogen peroxide is used as a bleaching agent when applied as a final stage of a bleaching sequence. When on the other hand the peroxide stage is used in the beginning of the sequence it is mainly used as a delignifying stage. Hydrogen peroxide is also used as reinforcement of the extraction stage and of the oxygen stage. These applications can be denoted as (EP) or (EOP) and (OP) respectively.

In hydrogen peroxide bleaching two reactions are competing with each other. The first one leads to a brightness increase while the second leads to decomposition of hydrogen peroxide into water and oxygen. The reaction is catalyzed by transition metal ions of manganese (Mn), iron (Fe) and cupper (Cu). Hydrogen peroxide (H_2O_2) reacts under alkaline conditions with the hydroxide ion (OH⁻) to yield the perhydroxide ion (HOO⁻). The bleaching is carried out at approximately pH 10.5–11.5 measured as the final pH at 25 °C. A higher pH may result in alkali induced self-decomposition of hydrogen peroxide.

A delignifying peroxide stage is placed at the beginning of the bleaching sequence after an oxygen stage. The stage is then either atmospheric or pressurized. A sequence used for production of fully bleached (>88 % ISO) softwood kraft pulp can for example be the following: OQPD(EOP)D. The corresponding sequence for hardwood pulps can be OQPD. The chemical reactions occurring in a delignifying peroxide stage are closely related to those occurring in an oxygen stage. Magnesium salts are therefore also used in peroxide stages especially the pressurized (PO) stages where the peroxide charge and the temperature are higher than in an atmospheric stage.

The conditions in a delignifying peroxide stage are normally: 90 °C, retention time about 2 hours and a final pH of 11.0–11.5 measured at 25 °C. In a pressurized stage the temperature is higher, 100–110 °C. The benefit of using pressurized peroxide stages compared to an atmospheric stage is that at equal retention time the delignification is faster and the brightness of the pulp is higher, *Figure 10.23*.

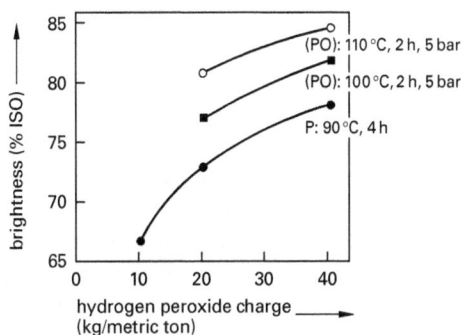

Figure 10.23. Bleaching with P and (PO)-stages shows that a higher brightness is obtained after a (PO)-stage. (Metso Paper).

The correct charge of hydrogen peroxide is depending on the kappa number of the incoming pulp, the target kappa number after the peroxide stage, the degree of decomposition of hydrogen peroxide that occurs and the amount of organic and inorganic substance that are carried over from the Q stage. The charge of hydrogen peroxide is approximately 5–40 kg/t but usually 10–20 kg/t. Residual hydrogen peroxide is zero or very low. The charge of sodium hydroxide is directly proportional to the charge of hydrogen peroxide and is normally 20–30 kg per ton of pulp. Notice that hydrogen peroxide and sodium hydroxide should not be added together. This would lead to too high pH, which in its turn will result in decomposition of hydrogen peroxide.

When the purpose of hydrogen peroxide is mainly bleaching, the peroxide stage is generally a final stage of the bleaching sequence. The kappa number of the pulp influences the brightness obtained with a given charge of hydrogen peroxide. For example will a pulp of kappa number 20 only reach a brightness of about 60 % ISO while the same peroxide charge will give a pulp of kappa number 4 a brightness of 85 % ISO. The retention time also has an influence. If the bleaching tower is too small in proportion to the production an excess of hydrogen peroxide must be charged in order to reach the brightness target. Thus, the residual hydrogen peroxide at the top of the tower is high, sometimes about 15–20 % of charged hydrogen peroxide. This peroxide can be reused if the filtrate from the washer after the peroxide stage is recycled as wash water on the washer prior to the peroxide stage.

A benefit of using peroxide in the final stage of a bleaching sequence is the improved brightness stability. Reversion for a fully bleached pulp (90 % ISO) using a sequence with a final peroxide stage can be limited to 1–2 points compared with 3–4 points for a sequence without a final peroxide stage.

Conditions in the (PO)-stage

- pH 10.5–11.0
- 80–110 °C
- 9–13% pulp concistency
- 1–3 hours
- 3–8 bar
- 5–10 kg O_2/t and 5–40 kg H_2O_2/t

10.14 Other Peracids than Hydrogen Peroxide

A peracid, or a peroxy acid, is an acid that contains a perhydroxide group (OOH^-). The peracids used in pulp bleaching are peracetic acid (CH_3COOOH) and peroxy-monosulfuric acid (H_2SO_5), usually called Caro's acid. The peracids are stronger oxidizing agents than hydrogen peroxide. This means that milder bleaching conditions can be used for the peracids than for hydrogen peroxide. Peracids are usually used as a complement to other bleaching chemicals to reach extra high brightness. In Sweden peracetic acid has only been used in TCF-sequences where full brightness is needed and where ozone bleaching is not used.

The peracetic acid can be used as an equilibrium solution with a mixture of peracetic acid, acetic acid and hydrogen peroxide (Te) or as pure distilled peracetic acid (Td). Of economical reasons only distilled peracetic acid is of interest and its bleaching effect is also equal or slightly better than for the equilibrium type. Due to its relatively high price only a few kg of peracetic acid is normally used per ton of pulp. If used as a final bleaching stage it can raise the brightness by about 1.5–2 % ISO units that can be interesting when high brightness is needed. It is also very selective meaning that the pulp viscosity is marginally affected which naturally is interesting in most bleaching sequences.

The conditions used in a peracetic acid stage are: pH 4–6, 60–80 °C and 1–3 h retention time. It is less sensitive to transition metals than hydrogen peroxide and a previous chelating stage is therefore not as important as before a hydrogen peroxide stage. The effluent emission from a peracetic acid stage contains a relatively high amount of COD as the bleaching chemical itself contribute to the COD emission. The acetic acid formed is however easy to degrade in a conventional external effluent system and the high COD emission is therefore not too problematic.

In *Figure 10.24* the bleaching effect of different peracids versus pH in sequences with the peracid stage followed by QP. As can be seen a neutral pH (pH 6) is recommended and compared at the same OXE charge a better result is obtained with peracetic acid than with Caro's acid. It can also be seen that the distilled peracetic acid results in a higher brightness than if equilibrium peracetic acid is used while the resulting viscosity is the same for the two peracid types.

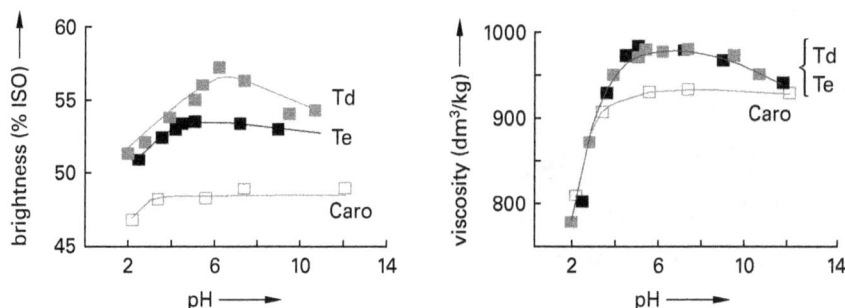

Figure 10.24. Influence of pH in a peracetic acid stage and a Caro's acid stage. T_d denotes distilled peracetic acid and T_e denotes peracetic acid in equilibrium with hydrogen peroxide and acetic acid. (Eka Chemicals).

Conditions in a T (Paa)-stage

- CH3COOOH (decomposes during storage)
- pH 4–6
- 60–80 °C, 1–3 h
- Carried out in existing stages
- 5–10 kg/t

10.15 Bleaching of Shives

After cooking the pulp usually contains a certain amount knots and shives consisting of fibres that have not reached full fibre separation and the particles remaining in the pulp to be bleached should preferentially be removed in the bleaching plant. That the shive bleaching efficiency is high is important as shives in the final bleached pulp reduces the quality of the pulp. However, modern fibre line technology has considerably decreased this problem compared with old technology. Fibre bundles, knots and other particles are more difficult to bleach than individual fibres due to the higher lignin content and to the longer diffusion distance for the bleaching chemicals. One example is shown in *Figure 10.25* which very clearly shows the non disintegrated fibres in the middle of one particle. Both the type of particle that is bleached and its thickness will have an effect on the rate of particle bleaching. There is a competition in the pulp slurry for the bleaching chemicals between the fibres and the particles and the relative reaction rate between these two reactions will determine the final cleanliness of the pulp at any given bleaching result whether this is given as kappa number or brightness.

Figure 10.25. A SEM picture of a bleached shive particle. (Axegård 1979).

It has also been shown that fibre bleaching and particle bleaching are done according to different reaction orders. Fibre bleaching in oxidizing stages follows a reaction order that is 2–5 while particle bleaching is of first order, *Table 10.5*. This means that if the logarithm of the remaining amount of fibres and particles are plotted versus the reaction time the resulting relationships are quite different, *Figure 10.26*. Thus, it can easily be understood that initially only fibres will be bleached while after longer bleaching time fibre bleaching will proceed very slowly and particle bleaching will therefore take over from a kinetic point of view. Particles will react slower with increasing bleaching time but the relative reaction rate versus the fibre bleaching becomes more beneficial.

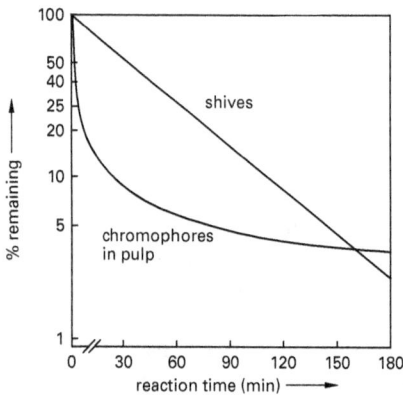

Figure 10.26. The figure shows the remaining amount of shives and chromophores in a chlorine dioxide final bleaching stage where the pH and the chlorine dioxide concentrations were kept constant. The different reaction patterns for shives and pulp fibres are obvious. (Axegård 1979).

Table 10.5. Rate equations for bleaching of fibres and particles (Axegård 1979).

Rate equations and activation energies for elimination of softwood particles and for bleaching softwood kraft pulp.

Bleaching stage	Pulp fibres	Particles
Oxygen Bleaching	$\dfrac{dK}{dt} = -k_1[OH^-]^{0.7} PO_2^{0.5}[K]^{3.2}$ $E_{k1} = 70$ kJ/mol	$\dfrac{d[Particle]}{dt} = -k_5[OH^-]^{0.8} PO_2^{0.5}[Particle]^1$ $E_{k6} = 65$ kJ/mol
Chlorine dioxide in pre-bleaching	$\dfrac{dK}{dt} = -k_2[ClO_2]^{0.5}[H^+]^{-0.2}[CL^-]^{0.3}[K]^5$ $E_{k7} = 60$ kJ/mol	$\dfrac{d[Particles]}{dt} = -k_6[ClO_2]^{0.7}[Particles]^1$ $E_{k7} = 35$ kJ/mol
Chlorine dioxide in final bleaching	$\dfrac{dC_k}{dt} = -k_3[ClO_2]^{0.5}[H^+]^{-0.2}[C_k - Ck]^3$ $E_{k3} = 60$ kJ/mol	$\dfrac{d[Particles]}{dt} = -k_7[ClO_2]^{0.8}[H^+]^{\otimes}[Particles]^1$ $E_{k8} = 30$ kJ/mol where $\otimes = 0$ for shives and knots, $\otimes = -0.3$ for pine bark
First alkaline extraction stage	$\dfrac{dC_k}{dt} = -k_{41}[OH^-]^{0.2}[K] - k_{42}[OH^-]^{-0.5}[K]$ $E_{k41} = 40$ kJ/mol $\quad E_{k42} = 2$ kJ/mol	$\dfrac{d[Particles]}{dt} = 0$
Ozone	$\dfrac{dK}{dt} = -k_5[O_3]^{1.0}[H^+]^0\left([K - K_{00}]\right)^2$ $E_{k5} = 29$ kJ/mol	$\dfrac{d[Particles]}{dt} = -k_{10}[O_3]^{0.8}[H^+]^0[Particles]^1$ $E_{k10} = 28$ kJ/mol

One example showing the cleanliness of a kraft pulp after a D_1-stage where the temperature has been varied between 40 °C and 80 °C is presented in *Figure 10.27*. It is obvious that at a given brightness the cleanliness of the pulp is higher at a lower temperature. The figure also shows that a longer bleaching time is beneficial for the cleanliness.

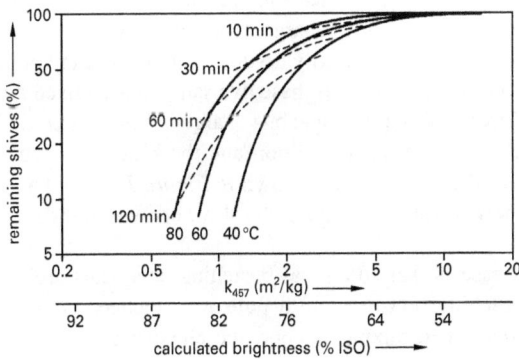

Figure 10.27. Calculated percentage of remaining shives versus calculated chromophore content in the pulp fibres of an O(C + D)E-prebleached pulp after bleaching with chlorine dioxide at different temperatures. The pH and the chlorine dioxide concentrations were kept constant. Reaction times are given as broken lines. (Axegård 1979).

10.16 Bleaching of Sulphite Pulps

The discussion so far has concerned kraft pulps. However, sulphite pulps are slightly different as these are brighter than kraft pulps after the cooking stage and they are easier to bleach to full brightness. This means that fewer bleaching stages or less bleaching chemicals are needed for sulphite pulps. This also means that TCF-bleaching to full brightness can easier be achieved also with sequences that would not be sufficient for kraft pulps. The Swedish sulphite pulp mills are all of TCF-type and sequences used are (EOP)Q(PO), ZEP, (EO)P etc.

The reason for the good bleachability is mainly due to the structure and the amount of residual lignin in the unbleached pulps. Sulphite pulps have during cooking got a lignin that contains sulphonic groups, SO_3H, that makes the lignin hydrophilic and easy to dissolve in the bleaching process. Secondly sulphite pulps have few strong bonds between the phenylpropane building blocs while this type of bond is very common in lignin after kraft pulping. Finally, sulphite pulps do not contain any hexenuronic acids which is beneficial for bleaching and pulp quality reasons.

10.17 The Importance of Good Mixing

Good mixing of pulp and bleaching chemical has always been a critical requirement and many types of mixers have been used over the years. Mixer design has continuously progressed as a result of both practical experience and increasing knowledge of mixing principles and pulp rheology. Modern mixers are therefore small but highly efficient. In the old days mixers for chemical addition could be very large and the largest where probably the Fiscalin mixers of up to more than 2 meters in diameter and these giant mixers were used in chlorine stages. When mixing of chemicals into the pulp suspension is inefficient, some of the pulp receives chemicals in excess of that required, which may be consumed in less desirable reactions such as reactions with carbohydrates. Other regions of the pulp receive insufficient amount of chemicals and the pulp, shives and other bleachable particles may not be adequately bleached. If the pulp locally either gets too high or too low chemical charge the average bleaching result for the whole pulp sample will always be inferior compared with a homogenous charge. One simple way to understand the reason to the poorer bleaching result is to theoretically assume that 50 % of the pulp gets a charge that is too low and 50 % gets a too high charge. If these two samples are mixed the average bleaching result will always be poorer than if a homogenous charge has been used due to the curved relationship between the bleaching chemical addition and the bleaching result. This conclusion is valid for the brightness of the pulp which is shown in *Figure 10.28* but also for the light absorption coefficient (k_{457}), the viscosity of the pulp, the shive content of the pulp etc.

The net effect in the non-homogenous case is that additional bleaching chemicals are required to obtain the desired degree of bleaching. Furthermore a pulp with reduced strength properties will be produced. In other words, when mixing is poor, the chemicals cost is increased and the pulp quality is decreased.

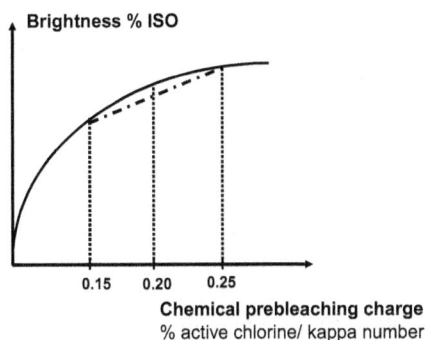

Figure 10.28. The figure shows the effect of poor mixing. If 50 % of the pulp gets a too high chemical charge (0.25) and the remaining pulp gets a too low charge (0.15) the average brightness of these two samples is poorer than if the mixing is good and all the pulp gets a homogeneous charge (0.20).

In modern bleach-plants medium consistency mixers or high shear mixers are used, where the pulp consistency is 9–13 %. In such a mixer the bleaching chemicals and the pulp are mixed during passage through zones of intensive shear. Imposing high rotational speeds across narrow gaps, through which the pulp suspensions flows, creates the high shear, *Figure 10.29.*

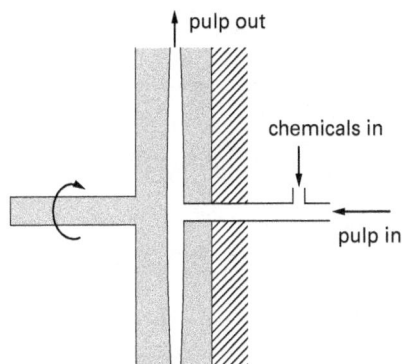

Figure 10.29. The principal design of a high shear mixer.

10.18 Polluting Emissions

Pollution from pulp mills affect both the receiving water and the surrounding air and the magnitude of these emissions are regulated by the authorities in each country. The Swedish pulp and paper mills have always had very strict regulations compared with many other countries and the Swedish industry has therefore done a lot to reduce its pollution emissions. It can therefore be stated that the pulp and paper industry in Sweden is a clean industry with relatively little pollution emissions.

Emissions to the receiving water are easier to detect and also easier to reduce than emissions to the air and the focus has therefore been on water emissions. The parameters mainly analyzed are discharges of organic material (BOD i.e. biological oxygen demand, COD i.e. chemical oxygen demand and suspended solids), discharges of chlorinated organic substances (AOX, chlorinated dioxins) and discharges that are easy to detect (foam, colour and fibres). However, all emissions have been significantly reduced during the last decades although the pulp production has increased and one example is shown in *Figure 10.30*. The figure shows the downward trend for the COD-emissions and the parallel increase in the pulp production.

A lot has been done in the pulp mills to reduce the emissions of polluting substances and examples are: better spill collection and more buffer tanks, better pulp wash, oxygen delignification, replacement of chlorine gas and hypochlorite with chlorine dioxide, prolonged delignification in the digester and in the oxygen delignification stage, better control systems etc. The bleach plant is the dominating source for pollution from the pulp mill. The discharge of COD from the bleaching of pulp is a relatively linear function of the kappa number of the pulp to be bleached, *Figure 10.31*. This means that if the kappa number is reduced prior to bleaching this will directly reduce the COD-emission.

Ways to reduce water pollution

- Prolonged cooking/oxygen delignification
- Better wash equipment   (esp BSW/Post O_2)
- Reduced spill/ more buffer capacity
- Recovery of bleach plant filtrates
- Replace Cl_2 with ClO_2 (AOX)
- Lower multiple in D_0 (AOX)
- Replace ClO_2 with TCF-chemicals (AOX)
- External effluent treatment

COD from swedish pulp mills

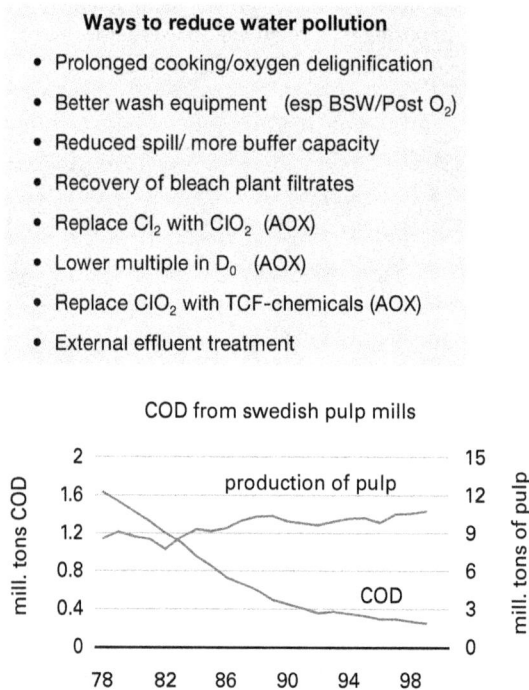

Figure 10.30. The discharge of COD from the Swedish pulp mills has decreased by more than 85 % between 1978 and 2000 while at the same time the pulp production has increased from 9 to 11 million ton/ a. (The Swedish Forest Industries Federation).

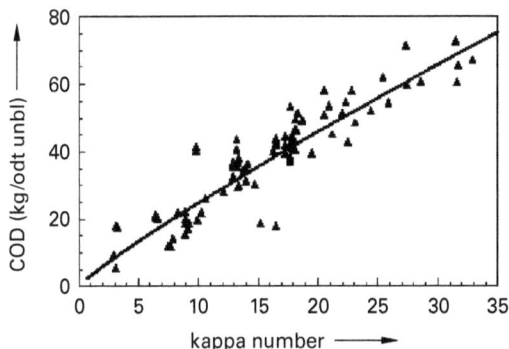

Figure 10.31. The COD-formation from pulp bleaching is directly correlated to the kappa number of the pulp to be bleached. (Metso Paper).

10.19 Bleaching of Mechanical Pulp (Nummerierung geändert)

Bleaching of mechanical pulps is different to bleaching of chemical pulps. For mechanical pulps the intention is only to increase the brightness without loosing pulp yield i.e. without loosing the lignin while for chemical pulps the intention is to increase the brightness by removing the lignin. Bleaching of mechanical pulps is usually performed with hydrogen peroxide or hydrosulphite. These two ways to bleach mechanical pulps are described below.

10.19.1 Bleaching Mechanical Pulp with Hydrogen Peroxide

Hydrogen peroxide (H_2O_2) is the bleaching agent but the active bleaching agent is the hydroxide anion (HOO^-) which is present at high pH. Unfortunately, the hydroxide anion reacts not only with chromophores but also with uncoloured groups. In the presence of metal ions, such as manganese and copper, the peroxide also decomposes into oxygen and water. As a result a silicate addition is often used in the bleaching of mechanical pulps due to its unique ability to decrease the influence of metal ions on the decomposition of peroxide.

The brightness response is directly related to the peroxide charge. In general the pulp is easy to bleach initially and for a given set of bleaching conditions, there is a brightness ceiling where increased peroxide charge has no impact on brightness. This is similar to bleaching of chemical pulps, *Figure 10.32*.

Hydrogen peroxide is only active at high pH and a charge of sodium hydroxide is therefore required in a hydrogen peroxide stage. Unfortunately the use of alkali has also some negative effects. The alkali will neutralize the acidic groups in the pulp and increase the amount of dissolved solids. An excessive alkali charge can also cause yellowing of the pulp, especially at high temperatures. Thus, at a too low alkali charge only a small amount of the peroxide will be active. At a too high alkali charge yellowing of the lignin will counteract the bleaching reactions. Consequently, peroxide bleaching will always have an optimal alkali charge where maxi-

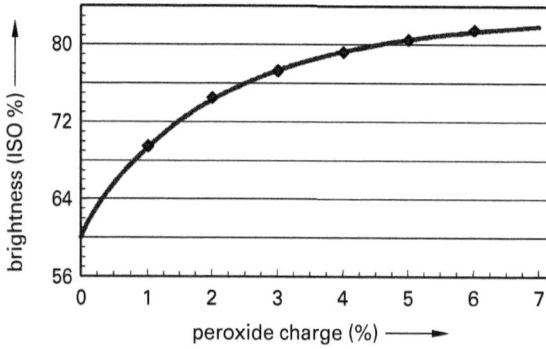

Figure 10.32. The brightness of a TMP pulp versus the peroxide charge. Bleaching conditions used: 3 hours, 70 °C, 3 % silicate, 25% consistency, Spruce TMP, 60% ISO (Metso Paper).

mal brightness is reached, *Figure 10.33.* The optimum alkali charge is influenced by the wood raw material, the metal content in the pulp, the bleaching temperature and the pulp consistency used.

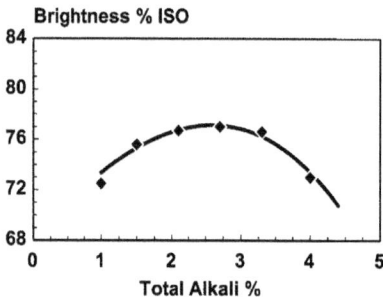

Figure 10.33. The pulp brightness versus the alkali charge. Bleaching conditions used: 3 hours, 70 °C, 3 % silicate, 25 % consistency, peroxide 3%, Spruce TMP 60% ISO (Metso Paper).

The use of sodium silicate (Na_2SiO_3) is an essential part of peroxide bleaching of mechanical pulps. The most important functions of sodium silicate are the buffering capacity and the ability to stabilize hydrogen peroxide solutions. The use of sodium silicate contribute typically to 2–10 % ISO in brightness gain. The charge is typically 1–5 % on pulp depending on the peroxide dose and the metal content of the pulp. A typical bleaching response at different sodium silicate charges is shown in *Figure 10.34.*

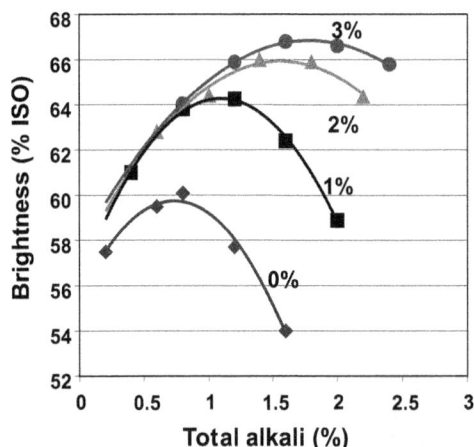

Figure 10.34. Pulp brightness versus the alkali charge at different sodium silicate charges. Bleaching conditions used: 3 hours, 60 °C, 24 % consistency, 2 % hydrogen peroxide. (PQ Sweden).

The hydrogen peroxide bleaching activity is strongly affected by pH as the active bleaching chemical in peroxide bleaching stage is the hydroxide anion (–OOH) which is a dissociation product of hydrogen peroxide. The optimum pH is about 10.5, which provides a sufficient concentration of hydroxide anions while the alkali concentration is still not causing negative effects with respect to selectivity. Silicate acts as a buffer during the bleaching and helps to maintain the desirable pH range longer.

In addition to reactions with pulp fibres hydrogen peroxide also reacts with transition metal ions, which are present in the pulp and in the process waters. These reactions contribute to the loss of peroxide by decomposition and result in formation of hydroxyl and superoxide anion radicals. These radicals react unselectively with the pulp which may contribute to a loss of pulp strength and to a pulp darkening effect unless their concentrations are kept on a low level.

The extent to which the alkali-induced peroxide decomposition occurs can be minimised by chelating the metal ions with a chelating agent such as DTPA, together with sodium silicate or, if the metal content in pulp is low, by using sodium silicate alone. A typical bleaching response for chelated (DTPA) and unchelated TMP pulp is shown in *Figure 10.35*.

The stabilizing effect of sodium silicate is due to the fact that under alkaline conditions it reacts with all multivalent metal ions (including Fe, Mn, and Cu) to form metal silicates, which are non-reactive towards peroxide decomposition. For optimum performance the addition of silicate requires good mixing with the pulp and with other chemicals used. However, a direct mixing with an acidic solution or high solids should be avoided as this might result in precipitation of silica and in extreme cases in scale formation in the process equipment.

Even though peroxide bleaching theoretically can be performed at any consistency, the efficiency is much higher at high consistency, *Figure 10.36*. At high consistency the concentration of peroxide around the fibres will be higher and as a consequence the hydroxide anion will have a better chance to react with the chromophores. Today, bleaching consistencies in the range of 30–35 % are state of the art. When the bleaching is performed at such a high consistency, the pulp must be well washed in order not to increase the rate of peroxide decomposition. Less alkali is also required for a given peroxide charge.

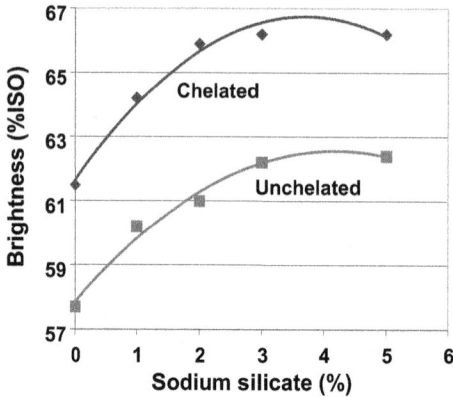

Figure 10.35. The effect of the sodium silicate addition on the pulp brightness at optimum alkalinity for chelated (DTPA) and unchelated TMP pulp. Bleaching conditions used: 4 hours, 65 °C, 21% consistency, 3% hydrogen peroxide. (PQ Sweden).

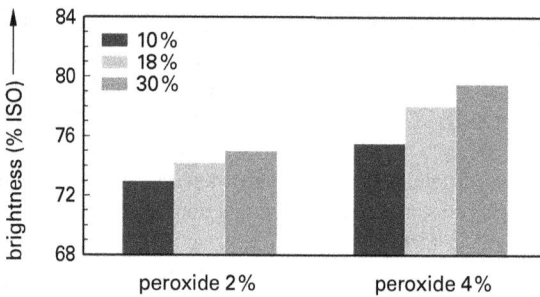

Figure 10.36. Pulp brightness for hydrogen peroxide bleaching with 2 or 4% and at different pulp consistencies. Bleaching conditions used: 3 hours, 70 °C, 3% silicate, Spruce TMP, 60% ISO (Metso Paper).

10.19.2 Modern Hydrogen Peroxide Stages for Bleaching of Mechanical Pulps

A modern high-consistency peroxide bleaching stage can consist of a wire press, a high-consistency tower and a second wire press, *Figure 10.37*. The pulp is bleached at 70 °C for 1–4 hours depending on the peroxide charge. The pulp is discharged at the same consistency and diluted outside the tower to get a plug flow and eliminate plugging of the tower. After dilution, the pulp could go to a paper machine directly or via a following washing stage.

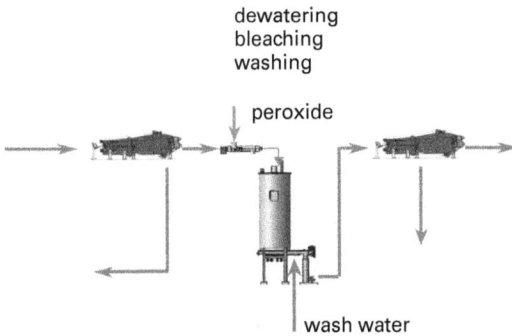

Figure 10.37. A high consistency P-stage comprising a wire press, a mixer, a high consistency bleaching tower and a second wire press. (Metso Paper).

Due to the lower efficiency of peroxide bleaching stages performed at medium pulp consistency (10–15 %) this type of bleaching stage is used only when the brightness requirement is low. In such cases the hydrogen peroxide stage has to compete with hydrosulphite bleaching which also can be used for brightness gains up to 10–12 % ISO. For higher brightness gains, the medium consistency hydrogen peroxide stage has to compete with high-consistency peroxide bleach stages, which will have a much lower peroxide consumption. Consequently, the medium consistency peroxide stage is only recommended as the first-stage in a two-stage peroxide bleach sequence where the second stage is performed at high pulp consistency, *Figure 10.38.* The first medium consistency stage is then performed in a specially designed tower.

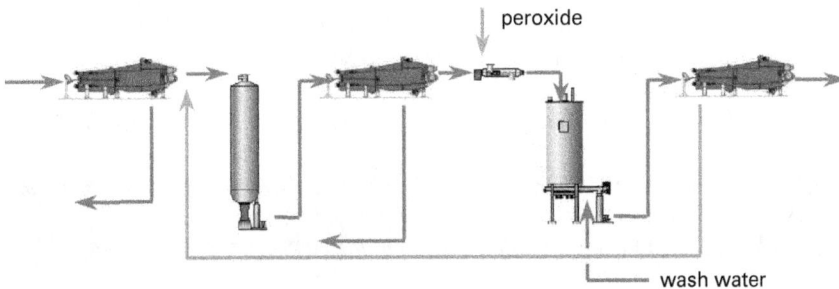

Figure 10.38. A two-stage MC/HC peroxide stage. (Metso Paper).

10.20 Hydrosulphite Bleaching

Hydrosulphite bleaching is usually done in an up-flow tower at 8–10 % consistency. Due to the fact that hydrosulphite reacts with oxygen an efficient deaeration of the pulp before the bleaching stage is necessary in order to achieve a good bleaching result. The brightness increase is limited to 8–12 % ISO and the system is used when only a small brightness gain in the hydro-sulphite stage is required or in combination with hydrogen peroxide where the hydrosulphite stage is used as a final bleaching stage. Even though the additional brightness gain in such a combined system is small, 1–3 % ISO, it could be vital when very high brightness is required, *Figure 10.39*. Note that it is important to place the hydrosulphite tower as close as possible to the paper machine to avoid a brightness loss.

When using a hydrosulphite stage after a peroxide stage, it is important to reduce the pH and eliminate the residual peroxide. The least expensive way to do this is to use sulphur dioxide (SO_2) but since this is quite hazardous, many mills try to avoid this approach. A combination of sulphuric acid and sodium bisulfite is then required in order to both neutralize and eliminate the peroxide.

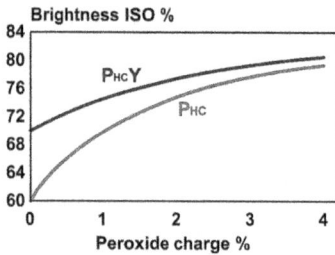

Figure 10.39. The brightness obtained after an HC-peroxide stage and a MC-hydrosulphite stage or after a single HC-peroxide stage. Bleaching conditions used: 0.8% hydrosulphite, Spruce TMP, 60 % ISO (Metso Paper).

10.21 Further Reading

Axegård, P. (1979) *PhD-thesis.* The Royal Institute of Technology (KTH), Stockholm, Sweden.
Axegård, P., Jansson, U., and Teder, A. (1984) *J Pulp Pap Sci* 10(1): J1.
Axegård, P. (1986) *J Pulp Pap Sci* 12(3): J67
Grundelius, R. (1979) *Proceedings, International Pulp Bleaching Conference, Toronto, Canada.*
Gustavsson, R., and Swan, B. (1974) *Proceedings, Tappi Non Sulfur Pulping Symposium, Madison, USA.*
Wiklund, L., Öhman, L-O., Lidén ,J. (2001) *Nordic Pulp Paper Res J.* 16(3): 240–245.

11 Production of Bleaching Chemicals at the Mill

Ulf Germgård
Department of Chemistry, Karlstad University

11.1 Introduction

The chemicals used in the bleach plant are either purchased from chemical suppliers or produced at the mill. The latter alternative is usually called on-site production. The reasons for producing chemicals at the mill instead of buying the chemicals from external chemical suppliers are several and in many cases mill specific. For example will some chemicals decompose during storage like ozone or chlorine dioxide. There is in other cases a significant risk for hazards during transportation or a potential risk for lack of chemicals now and then if chemicals have to be transported to the mill. Transportation of chemicals can in some cases be impossible of economical reasons and one example is chlorine dioxide that is much diluted and the majority of the solution is water. Finally, if the pulp mill is located in a remote area in for example Brazil or

Indonesia it is often impossible of transportation and/or economical reasons to send chemicals to the mill and the chemicals therefore have to be produced at the mill.

Although there are benefits by producing chemicals at the mill there are also some disadvantages. For example is the production relatively small sized compared with the size that chemicals can be produced at for a chemical supplier. Another aspect is that the modern management tend is to focus on the core business and producing chemicals at a pulp mill is not a core business. Therefore, there has been a trend in the industry to move away from on-site production to off-site production of chemicals. However, in some cases a compromise has been reached where a chemical supplier has started up, or taken over an existing, local chemical production unit at a mill. The supplier then sells the chemicals to the pulp mill "over the fence" independently if the chemicals have been produced at the mill or produced by the supplier somewhere else. This trend has started up for chlorine dioxide, ozone and oxygen production at Swedish pulp mills. The benefit for the pulp mill is that they can focus on the pulp production while the chemical supplier can use their skills and know-how in making the chemical production more efficient. Naturally, as part of such a deal the supplier also get a long-term contract with the supplier.

11.2 Oxygen production

In the bleach plant oxygen is used in the oxygen delignification stage, as reinforcement chemical in extraction stages and as a raw material for ozone generation. Oxygen is sometimes also used in the mill in external effluent treatment and in incinerators like the lime kiln or the recovery boiler.

Oxygen is the cheapest bleaching chemical used in the pulp mill and it is produced from air. In air the concentration of oxygen is about 21 % and the rest is mainly nitrogen (78 %) and argon (1 %). The latter two gases are inert i.e. they cannot react with the pulp. Although oxygen is freely available in the air the high content of the other gases in the air makes direct use of air in the bleach plant uninteresting. One reason is that remaining gas in the pulp after the bleaching stage makes washing very problematic as gas bubbles in the pulp significantly influence the possibility for wash water to distribute homogenously in the pulp suspension. Thus, the washing efficiency goes down dramatically if the pulp to be washed contains gas bubbles. Another problem with the use of air is that the mass transfer rate of oxygen molecules to the fibres will be lower as the oxygen partial pressure in the gas also is lower as there is a maximum pressure that cannot be allowed in the reactor. A third problem is that approximately five times larger gas volume has to be charged to the pulp than if pure oxygen is used as only the oxygen part of the air is reactive. Such a large gas volume can make it hard to avoid a gas separation in the bleaching reactor which also is problematic as a homogenous gas-pulp slurry is required. The reactor also has to be much larger if air is used than if only oxygen is used. Thus, of these reasons air is not used in the bleach plant. However, it is not necessary to produce extremely pure oxygen and about 90 % purity is usually sufficient for bleach plant purposes.

The oxygen suppliers produce oxygen by distillation of air at about −190 °C (cryogenic separation) which separates air into oxygen, nitrogen and argon due to the different boiling points. At such a low temperature the gases are liquefied. Liquefied gases are much easier to transport as the volume of the gas is reduced by 600–850 times. The gas produced in this way is very pure, i.e. up to 99.5 %. If such distillation is done at a pulp mill there is no need for liquefaction as the produced oxygen is directly used in the mill and the production cost is therefore significantly reduced. However, on-site oxygen production is usually done with a simpler technology using adsorption/desorption on an adsorption material and such techniques have a lower production cost but the oxygen is also less pure , i.e. less than 93 %. This is however still sufficiently pure for bleaching purposes.

11.2.1 The Cryogenic Air-separation Process

In general, the cryogenic air-separation process based on distillation consists of the following six steps which are also shown in *Figure 11.1*:

1. Air is taken through a filter for dust removal and the purified air is then taken to a compressor where the gas pressure is increased to about 180 bar.
2. The air is then taken to an oil separator and then to a mole sieve for separation of water vapour and carbon dioxide.
3. The purified gas is then cooled down to about -30 °C, i.e. close to the dew point, in a heat exchanger. The cooling media is cold separated gases, i.e. gaseous oxygen (GOX) and gaseous nitrogen (GAN) which come from the distillation column. If a suitably large heat exchanger is used the cold gases can be warmed to a temperature just under that of the incoming air which is interesting to keep the energy consumption down.
4. The air is then taken to an expansion machine where the gas pressure is quickly reduced from 180 bar to 6 bar and the temperature is simultaneously reduced to below -155 °C which means that the gases are liquefied. The gas is then taken to the distillation column where the gas is separated into liquefied oxygen (LOX), liquefied argon (LAR) and liquefied nitrogen (LIN) due to the different boiling points. The cold gases are fed through the heat exchanger and warmed-up while the oxygen, nitrogen and argon liquids are fed into different tanks. The argon needs additional distillation to remove nitrogen before it is taken to the storage tanks.
5. The cooling that is required is produced in a recycle by compressing, cooling and expanding nitrogen to its liquid state. It is a recycle because the gas, which does not condense, is warmed up and fed back to the compressor again.
6. Compressors and other machines need large amounts of cooling water and the heat removal from the cooling water is vented into the air.

Since it's critical from an energy consumption point of view to minimize cold losses, all the critical components like the distillation columns, the heat exchangers and all other cold production equipment are built in a well-insulated tower, consequently often being called a "Coldbox", *Figure 11.1*.

The unique feature of air separation is the great dependency of the different flows. This is because it is a cryogenic process, i.e a low temperature separation, where cooling water and steam cannot be used. The different products or internal flows are instead used for boiling and condensing in the columns, for cooling the incoming air and for sub-cooling the liquid products. Cold nitrogen gas is used in the recycle for liquefaction etc. A disturbance in any of the gas flows gives a disturbance in the pressure and a change of pressure can have a considerable effect on the other flows as the boiling points change with pressure. Changed temperature in a condenser will also change the rate of heat exchange and it will therefore change the gas flow in the column, which will alter the purity of the produced gases.

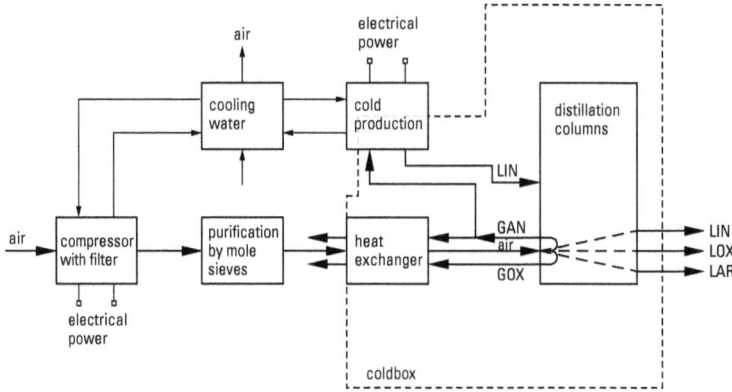

Figure 11.1. An overview of the conventional oxygen production process using air distillation also called cryogenic distillation. (Aga Linde Gas).

11.2.2 On-site Oxygen Production

In some cases an on-site production of oxygen is interesting of economical and/or logistic reasons. For on-site oxygen production the cryogenic process can be used but usually a simplified process based on pressure swings, i.e. gas adsorption and desorption on an adsorbent bed, is instead utilized. The latter processes are denoted PSA (pressure swing adsorption) or VSA (vacuum swing adsorption) The economical savings for production of oxygen on-site can in some cases be considerable and as much as 30 % compared with bulk oxygen produced in a conventional cryogenic production plant has been reported by a supplier. However, even if an on-site production of oxygen takes place usually a back up system with liquefied oxygen usually is available where the liquefied oxygen is taken from an external oxygen supplier.

The pressure swing adsorption (PSA) process was developed in the 1950′s with commercial scale production starting in the 1970′s. In such a process air is passing a molecular sieve that consists of small particles (zeolites) with large amounts of micropores giving a large surface area. The adsorption characteristic is different for different gases meaning here that nitrogen will be adsorbed stronger and to a higher extent to the surface than oxygen, *Figure 11.2*. Thus, the nitrogen molecules will be adsorbed to the column while the oxygen molecules pass the system. Two adsorption columns are usually used where one is in use while the other is regenerated. The pressure swings in a PSA process is between 1 and 7 bar.

Vacuum swing adsorption (VSA) was introduced later where the vacuum is used for regeneration of the absorbent bed although the basic principle is the same. The pressure swings here between 0.3 and 1.5 bar. The term for the present technology is often VPSA.

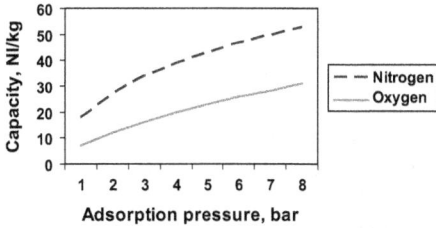

Figure 11.2. The adsorption material in VPSA process adsorbs nitrogen harder than oxygen. Oxygen therefore passes the adsorber while nitrogen is trapped in the adsorbent when air is fed to the adsorbent column. (Based on data from Aga Linde Gas).

In a VSA system the adsorbers are filled with a bottom layer that removes moisture and carbon dioxide from the feed air while the upper part of the adsorber removes nitrogen. Thus, oxygen is passing the adsorber together with a certain amount of argon. The oxygen production stops when the upper layers of the adsorbent is saturated with nitrogen and the purity of the oxygen goes down. When the column is filled with nitrogen molecules it is regenerated with vacuum for removal of the adsorbed gases and during this time usually another column(s) take(s) over so that the total oxygen production process is continuous. The third step is that the absorber is pressurized again by the use of oxygen from another absorber or from a buffer tank before new air is added with a fan to the absorber and the cycle starts over again.

The purity of the oxygen gas can be up to about 94 % but it is usually lower. The capacity of a VSA unit is much smaller than a cryogenic plant at an oxygen supplier but VSA units from 1–200 ton oxygen per day are available i.e. a large VSA unit produces more than enough of oxygen for a pulp mill. The technology has been developed considerably and it is today more energy efficient than the cryogenic process. However, for other types of industries than pulp and paper where the oxygen consumption is much larger cryogenic on-site plants are used and the purity of the produced gas is then also larger. Thus for production of very pure oxygen or for very small or very large production the cryogenic process is the correct one while on-site production is interesting when the demanded purity of the oxygen gas is below about 95 % and when the oxygen demand is not too large, *Figure 11.3*.

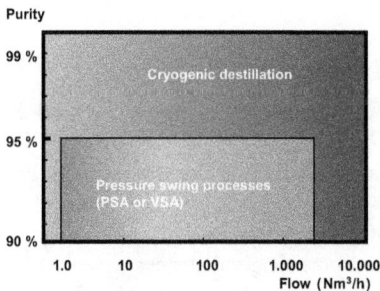

Figure 11.3. The purity of the delivered oxygen versus the production capacity given as m³ oxygen per hour. On-site production with PSA or VSA and off-site cryogenic distillation are shown. Note that the flow scale is logarithmic (Aga Linde Gas).

11.3 Chlorate Production

11.3.1 Off-site Sodium Chlorate Production

Sodium chlorate is produced by electrolysis of sodium chloride and this is in most cases done by a chemical supplier far from the pulp mills. However, in some specific cases is an on-site production also done but this is becoming a rare exception. No Swedish pulp mills are producing chlorate at the pulp mill. The chemical reaction that takes place in the electrolysis cells can be summarized as shown in Reaction 11.1

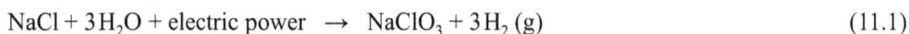

$$NaCl + 3H_2O + \text{electric power} \rightarrow NaClO_3 + 3H_2 \text{ (g)} \tag{11.1}$$

The equation shows that hydrogen gas is produced as a by-product in the electrolysis. It is therefore important to avoid mixing of hydrogen and air as such a mixture is very explosive. Thus, efficient removal of the hydrogen from the electrolysis cells is critical and a possibility to quickly flush the gas system with inert nitrogen is important of safety reasons at production starts and stops. A simplified drawing of a chlorate production plant with electrolysis cells is shown in *Figure 11.4*.

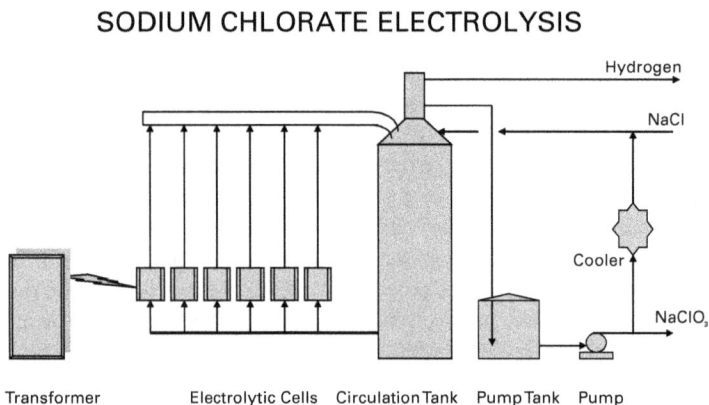

SODIUM CHLORATE ELECTROLYSIS

Figure 11.4. Sodium chlorate production via electrolysis of sodium chloride (Eka Chemicals, Cellchem).

The hydrogen that is produced is utilized for different purposes depending on the local demands. It can for example be used as raw material in the production of hydrogen peroxide if hydrogen peroxide is done on the same place, it can be bottled and sold as hydrogen, it can be used as fuel at the plant etc.

After the electrolysis crystallization is done to separate the chlorate from chloride and other additives and to get a pure product that is possible to transport longer distances at reasonable costs. The transportation to the pulp mill is done with trucks, rail cars or in plastic bags for small consumers. In most cases the sodium chlorate is produced at large production units located in countries where electric power is cheap as the power demand is high, see *Table 11.3*. However, the transportation cost is also important and a compromise has to be reached where the lowest total cost for power and transportation is obtained. Thus, chlorate production is today found in

Sweden (Alby and Stockvik), Finland, France, Canada, USA, Chile, Brazil etc. In some of these countries the power rates are only economically interesting for chlorate production during off peak hours thus a continuous chlorate production is then not possible which naturally is problematic. Examples of countries where the power rates are significantly lower at off peak hours are Japan and Brazil. The necessity to use low cost electric power means that crystallized sodium chlorate is in many cases not even produced in the same country as where the pulp mill is located. Sodium chlorate is thus usually not produced at the pulp mill site. However, in some cases where transportation is problematic like in remote locations in Brazil or Indonesia on-site production can still be interesting. In these cases the production is used directly in the pulp mill and expensive crystallization is therefore not needed.

11.3.2 On-site Chlorate Production

On-site chlorate production is very uncommon and the trend is that the pulp mills prefer to concentrate on the pulp and paper production while production of chemicals is outsourced. However, there are still a few pulp mills that produce their own chlorate. This can be done by producing the chlorate from sodium chloride by electrolysis and then using the product for generation of chlorine dioxide without a further crystallization. This type of on-site production was based on the use of sulphuric acid, giving sodium sulphate as a by-product, has never been a success and has mainly been used in North America and Brazil where transportation to remote pulp mills has been considered to be too costly or too problematic. However, this type of production at a mill site has also been used in Sweden in the 1980's where MoDo and KemaNord (today Eka Chemicals) had a common chlorate plant. The recent trend is to switch from on-site production to purchased crystallized chlorate from a chemical supplier even if the mill is located in a remote area. Another possibility for on-site production is to use a so-called integrated system which combines chlorate production and chlorine dioxide generation, *Figure 11.5*. In the integrated system in the figure the raw material is chlorine, either produced on-site or purchased on the open market, and electric power. Due to the low global demand for chlorine this chemical is today very cheap.

The reducing agent in the integrated chlorine dioxide generator is chloride, i.e. sodium chloride and hydrochloric acid. The full process includes an HCl-burner, which produces hydrochloric acid from chlorine and hydrogen and where the hydrogen is a by-product from the sodium chlorate production. The chlorate is produced in the electrolysis stage and sodium chloride is the raw material. The sodium chloride used is a by-product from the chlorine dioxide generator. Thus, the over all chemical reaction in the integrated process can theoretically be summarized as shown in Reaction 11.2.

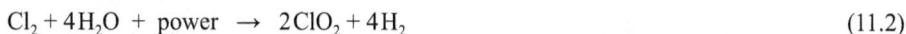

$$Cl_2 + 4H_2O + power \rightarrow 2ClO_2 + 4H_2 \tag{11.2}$$

The equation above is a simplification and in reality there is a simultaneous formation of chlorine in the chlorine dioxide generator as the reducing agent is chloride. The chlorine is an unwanted by-product in the chlorine dioxide solution as it leads to increased formation of chlorinated organic compounds in the bleach plant. The integrated systems of this type have therefore become less interesting as the trend is to use as little chlorine as possible in the pulp mills of environmental care reasons. Another draw back with such an integrated system is that all stages must operate simultaneously and in balance with each other, which require good control and good maintenance of the equipment. No integrated systems of this type have been in use in Sweden.

Integrated process for chlorine dioxide production

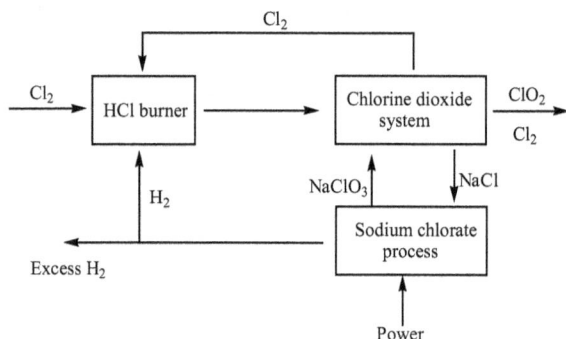

Figure 11.5. An integrated system for production of chlorine dioxide based on chlorine, water and electric power as the only raw material. (Eka Chemicals, Cellchem).

11.4 Chlorine Dioxide Generation

Chlorine dioxide has a boiling point at 11 °C and it has a gas density that is about 2.4 times that of air. Thus, it is a gas at room temperature and it is heavy. It is also explosive above a partial pressure of around 0.1 bar and it is harmful of health reasons with an exposure limit of 0.1 ppm per 8 hours.

Chlorine dioxide has to be produced at the mill as it is only possible to store as a diluted water solution. It also decomposes slowly upon storage and it therefore has to be generated continuously. The chemical is thus always dissolved in water and the concentration in the water solution is low (about 10 g ClO_2/l calculated as ClO_2) which makes transportation very expensive. It is finally a hazardous chemical that is not suitable for transportation.

Chlorine dioxide is produced from sodium chlorate which is reduced to chlorine dioxide in the chlorine dioxide generator. The reaction is carried out at low pH via the simplified chemical reaction below, Reaction 11.3.

$$ClO_3^- + 2H^+ + \text{reducing agent} \rightarrow ClO_2 + H_2O \qquad (11.3)$$

Chlorine dioxide is not stable and it can react further via an exothermic decomposition reaction as shown in Reaction 11.4. This reaction can be initiated by the presence of heavy metals, rust, organic material, sunlight, high temperature, high concentrations in the gas phase and electric sparks. The decomposition leads to an increased pressure that can in the worst case totally destroy for example the chlorine dioxide generator. The modern chlorine dioxide generation process equipment is however designed to withstand such decompositions and it is today very rare that serious damage to the process equipment is caused by decomposition reactions.

$$2ClO_2 \rightarrow Cl_2 + 2O_2 \hspace{4cm} (11.4)$$

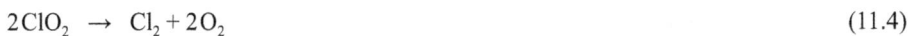

The acid used in the chlorine dioxide generator is usually sulphuric acid or in some cases hydrochloric acid. The reducing agent was earlier either sulphur dioxide or sodium chloride. The name of the first process was the Mathieson process while in the latter case the names were SVP or R-processes. Today, methanol has taken over the role as the dominating reducing chemical but hydrogen peroxide is also becoming an increasing but more expensive alternative. Some examples of acids and reducing agents used in chlorine dioxide generation are given in *Table 11.1*.

The major reason to the change in preference for reducing agents is that the amount and type of by-products produced is directly influenced by the type of reducing agent used in the chlorine dioxide process. The by-products have to be taken care of in the pulp mill and as modern mills have become more and more closed the demand for make-up chemicals has decreased drastically. This is especially the case for make-up sulphate. As the make-up demand in the mills has decreased the additional amount above the need in the mill has therefore little value. Thus, the whole production cost has to be put on the chlorine dioxide production which means that the internal price for chlorine dioxide has increased.

Table 11.1. Acids and reducing agents used in different chlorine dioxide processes.

Acid	By-products generated	
H_2SO_4	Na_2SO_4	
HCl	NaCl	

Reducing agent	By-products salt	By-products in ClO_2-gas
SO_2	Na_2SO_4 and H_2SO_4	Cl_2 (trace)
CH_3OH	$Na_3H(SO_4)_2$	HCOOH
NaCl	Na_2SO_4	Cl_2
HCl	NaCl	Cl_2
H_2O_2	Na_2SO_4	O_2

One example of a chlorine dioxide generator is shown in *Figure 11.6*. This system is atmospheric and it consists of a primary and a secondary reactor plus a stripper. The sodium chlorate solution is passing through these two reactors and the reducing agent is simultaneously added via spargers to the liquor. The generated chlorine dioxide gas is transferred to the absorption tower with the air that is added at the bottom of the reactors and of the stripper. In the absorption tower the chlorine dioxide gas is dissolved in chilled water and then flowing by gravity to the chlorine dioxide storage. After passing the reactors and the stripper the remaining liquid is collected in the spent acid tank from which the acid either is reused in the mill or discharged from the mill if the acid is not needed.

Modern chlorine dioxide generators are of vacuum type where the excess water coming into the system is removed by boiling and where no addition of air to the reactor is used. Thus, the gas phase coming to the absorption tower consists in this case of chlorine dioxide and water vapour while in the atmospheric system the water vapour is replaced with air. The salt in the vacuum system is removed from the reactor as a crystallized salt and this salt is separated on a filter, *Figure 11.7*. Note that only one reactor is used in a modern chlorine dioxide generator and that the process solution is not purged after passing the generator but instead recycled from the generator via the reboiler and back to the generator.

ATMOSPHERIC PROCESS FLOW DIAGRAM

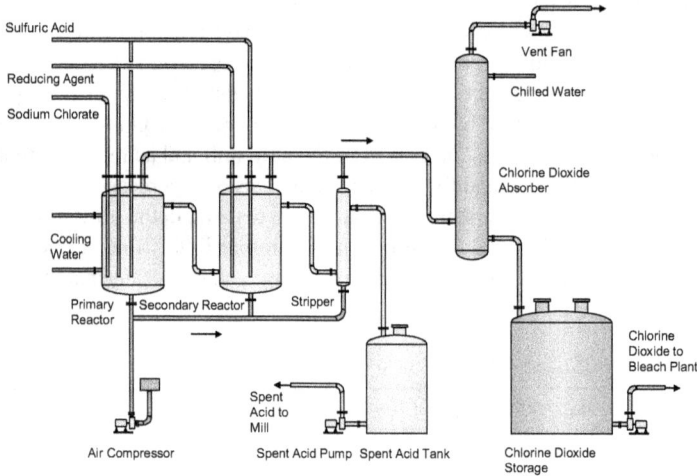

Figure 11.6. An atmospheric chlorine dioxide generator with primary and secondary reactor, stripper, absorption tower and chlorine dioxide storage tank. The process type is HP-A process where hydrogen peroxide is used as the reducing agent. The old Mathieson process is similar if the hydrogen peroxide is replaced with sulphuric dioxide (Eka Chemicals, Cellchem).

VACUUM PROCESS FLOW DIAGRAM

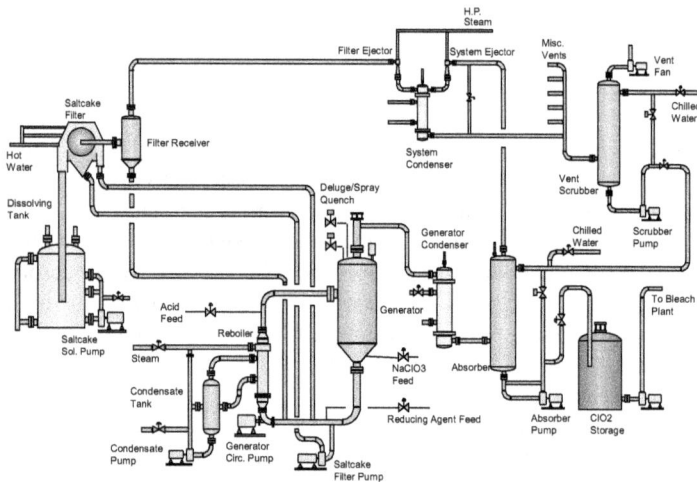

Figure 11.7. A chlorine dioxide system based on the vacuum process principle. All modern systems are of this type. (Eka Chemicals, Cellchem).

The solubility of chlorine dioxide decreases with increasing temperature and the water used in the absorption tower has therefore to be cold (about 10 °C). If the temperature is not low enough the concentration of the chlorine dioxide solution will be too low and/or the absorption efficiency will decrease causing increased emission of chlorine dioxide from the absorption tower and this chlorine dioxide is therefore lost. The solubility of chlorine dioxide in water at 0–15 °C versus the partial pressure is shown in *Figure 11.8*.

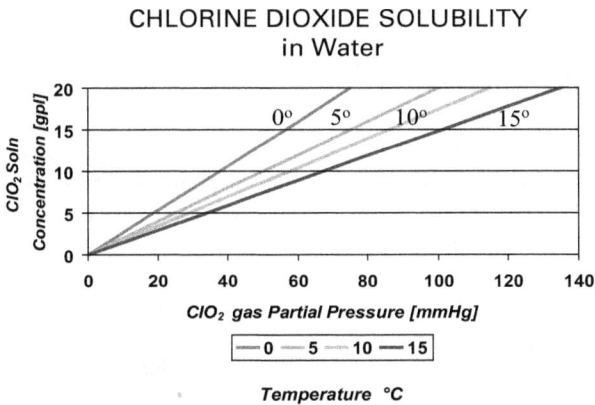

Figure 11.8. The concentration of the chlorine dioxide solution versus the chlorine dioxide partial pressure at 0–15°C. (Eka Chemicals, Cellchem).

Table 11.2. A material balance sheet for some chlorine dioxide processes given as consumption/production in ton/ton ClO2. (Eka Chemicals, Cellchem).

	Mathieson	SVP	SVP LITE	SVP HP
Chemicals in				
$NaClO_3$	1.8	1.68	1.64	1.64
H_2SO_4	1.4	1.57	1.0	0.8
NaCl	–	0.97	–	–
H_2O_2	–	–	–	0.3
SO_2	0.7	–	–	–
CH_3OH	–	–	0.18	–
Power kWh	80	120	100	100
Steam	–	8.7	4.2	5.9
Chemicals out				
Na_2SO_4	1.17	2.30	1.09	1.00
H_2SO_4	1.50	–	0.25	–
Cl_2	–	0.40	–	–
O_2	–	–	–	0.24

The difference between the chlorine dioxide generating processes is particularly evident when comparing amount of input and the output chemicals per ton of chlorine dioxide. Material balances for some typical chlorine dioxide generating processes are given in *Table 11.2*. The two oldest systems are given to the left and the two modern systems to the right. Notice especially the trend for the formation of Na_2SO_4/H_2SO_4 from the old fashioned Mathieson process to the modern SVP HP-process that produce considerably less. This is interesting as modern pulp mills do not need a lot of make-up sulphur due to the more tight closure of the mills.

11.5 Chlorine and Caustic Production

Chlorine has until the mid 1990′s been used in the bleaching of the majority of the pulp in the world and chlorine is a very selective bleaching agent. However, due to the environmental problems associated with the organically bound chlorine that is produced during bleaching with chlorine it has been phased out in all modern pulp mills. Chlorine bleaching has mainly been replaced with oxygen and chlorine dioxide. However, chlorine is still used in many pulp mills where such bleaching is still accepted and/or where chlorine free bleached pulp is not considered as being problematic.

Chlorine is produced via electrolysis of NaCl (brine) according to the following net reactions, Reactions 11.5– 11.7

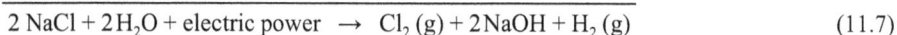

$$2\,NaCl + \text{electric power} \;\rightarrow\; 2\,Na + Cl_2\,(g) \tag{11.5}$$

$$2\,Na + 2\,H_2O \;\rightarrow\; 2\,NaOH + H_2\,(g) \tag{11.6}$$

$$\overline{2\,NaCl + 2\,H_2O + \text{electric power} \;\rightarrow\; Cl_2\,(g) + 2\,NaOH + H_2\,(g)} \tag{11.7}$$

In this process chlorine is produced at the anode and the chlorine gas is taken from the cells, cooled, dried and usually liquefied before transportation to the final consumer if the consumer is not located at the same site. For on-site production chlorine is naturally not liquefied and the cost of chlorine is therefore reduced.

The electrolysis can be done in mercury cell, in diaphragm cells or in membrane cells, *Figure 11.9* The mercury cell process is a very efficient process that produces pure NaOH and it was earlier the dominating process. However, due to unavoidable leaks of mercury to the environment this process has been replaced of environmental care reasons with especially the mem-

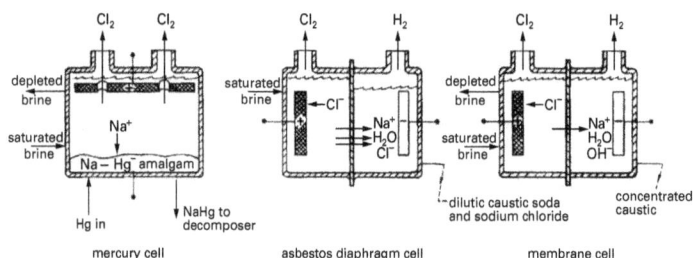

Figure 11.9. Three different types of electrolysis cells for production of chlorine and alkali (Olin Corporation).

brane cell process. The membrane process has a cationic membrane in the electrolysis cell that prevents chloride ions to pass the membrane. Thus, the caustic that is produced is very pure. In the diaphragm process the caustic is contaminated with NaCl and a following purification stage consisting of evaporation and crystallization of NaCl is usually needed.

In Sweden chlorine is still produced at Skoghall, Bohus and Stenungsund but the pulp industry does not consume chlorine since the beginning of the 1990's. However, as late as in the 1970's some Swedish softwood kraft pulp mills consumed as much as 60–70 kg chlorine/ton of pulp in the chlorine stage and 10–15 kg chlorine/ton in the hypochlorite stage.

The chlorine electrolysis is also the main production route for caustic (NaOH) production. Caustic is usually transported to the customers as a 50 % solution but at the pulp mill the solution is diluted to about 5 % before use. As the chlorine and caustic productions are done simultaneously less caustic will be available on the market when the chlorine consumption goes down which results in increased price for caustic. As long as the PVC-production in the world increased, which only consumes chlorine but not caustic, caustic was still in a reasonable supply. PVC is however today slowly replaced with other non-chlorine containing materials and the lower supply of caustic is slowly becoming an economical problem for the pulp industry. Therefore there is an increasing interest in using other caustic sources like oxidized white liquor i.e. internally made caustic, also in the bleach plant. However, oxidized white liquor is not as pure as caustic from electrolysis and it is not possible to use oxidized white liquor directly in for example ozone or hydrogen peroxide stages where even trace amounts of contaminating metal ions can cause decomposition of the chemicals.

11.6 Hydrogen Peroxide Production

Hydrogen peroxide is today used as bleaching chemical for mechanical, chemical and recycled fibres and the demand has increased dramatically during the last 10 years. All modern bleaching sequences use hydrogen peroxide although the actual charge per ton of pulp varies a lot.

Hydrogen peroxide is produced by chemical suppliers in large scale production plants and transported to the pulp mills as a concentrated water solution between 35 and 70 %. Due to the use of stabilizers and the high concentration in the water solution hydrogen peroxide can of safety and economical reasons be transported very long distances for example from Europe to South America. On-site hydrogen peroxide production is not used although such alternatives have been proposed by some manufacturers.

The raw material for hydrogen peroxide production is hydrogen and air or more correctly hydrogen and oxygen, *Figure 11.10*. The actual formation of hydrogen peroxide is done indirectly in a circulating process solution which first is hydrated and then oxidized forming the hydrogen peroxide. The hydrogen peroxide molecules are after an extraction stage with water ending up as a water solution of about 35–40 % concentration which is called raw peroxide. The raw peroxide is not totally pure but contaminated with traces of process solution.

The hydrogen gas is produced in different ways and in many cases is the hydrogen peroxide plant built parallel with another chemical plant where hydrogen is produced as a by-product. Examples of such parallel production plants are chlorine-alkali plants or chlorate plants. However, if such plants are not available the hydrogen can be produced from on-site steam splitting of natural gas or from electrolysis of water.

290

The active agent in the process solution is anthraquinone which continuously is hydrated, oxidized and then dryed after which this cycle is repeated, *Figure 11.11.* The same process is used by all hydrogen peroxide suppliers.

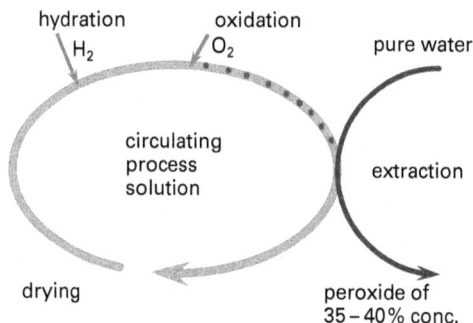

Figure 11.10. A simplified description of the hydrogen peroxide production process. (Kemira Kemi).

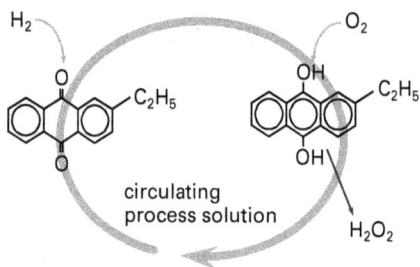

Figure 11.11. Anthraquinone is first reacting with hydrogen and then oxidized with air (oxygen) which forms hydrogen peroxide as a free molecule. (Kemira Kemi).

The total production process for a plant based on hydrogen produced from natural gas is given in more detail in *Figure 11.12.* The natural gas has first been purified from sulphur by addition of ZnO which is reacting to ZnS and separted from the process. Sulphur is a poison for the following hydrogen peroxide process. The purified natural gas is then treated with steam at elevated temperature (890 °C) and in the presence of nickel based catalysts according to Reaction 11.8.

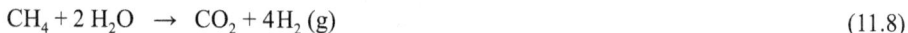

$$CH_4 + 2\,H_2O \quad \rightarrow \quad CO_2 + 4H_2\,(g) \tag{11.8}$$

The hydrogen is then reacting with the anthraquinone in the process solution and the oxidized with air. The raw peroxide produced in the extraction stage has a concentration of the active component of about 35 %. It is then distilled to 50 % concentration before it is taken to storage and later transported to the customers.

Hydrogen peroxide is very sensitive to decomposition according to Reaction 11.9.

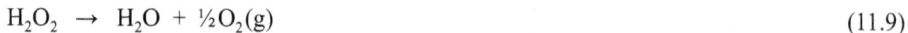

$$H_2O_2 \quad \rightarrow \quad H_2O + \tfrac{1}{2}O_2(g) \tag{11.9}$$

The decomposition results in a considerable formation of gas and thus a considerable gas pressure. As a rule of thumb 1 m³ of 50 % hydrogen peroxide solution produces 1250 m³ if total decomposition takes place. If the hydrogen peroxide concentration instead is 70 % the corresponding gas volume is 2500 m³. Thus, an explosion will occur in such a case. The process solution therefore has to be handled in a proper way. One way is to keep the temperature down as the rate of decomposition increases by 2.3 times for a temperature increase of 10 °C. The decomposition is catalysed by the presence of metal ions like cupper, manganese etc. The process solution also has to be stored at a pH of 4–5 as the rate of decomposition increases at pH-levels below 3 and above 6, *Figure 11.13*.

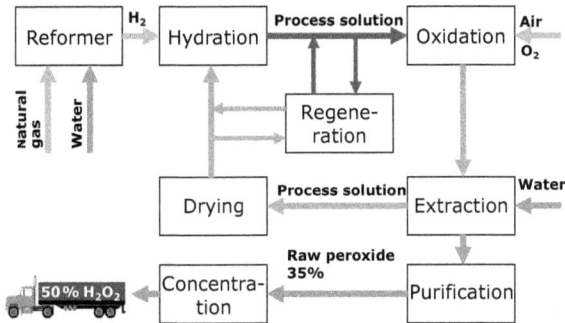

Figure 11.12. A simplified description of the hydrogen peroxide production process. (Kemira Kemi).

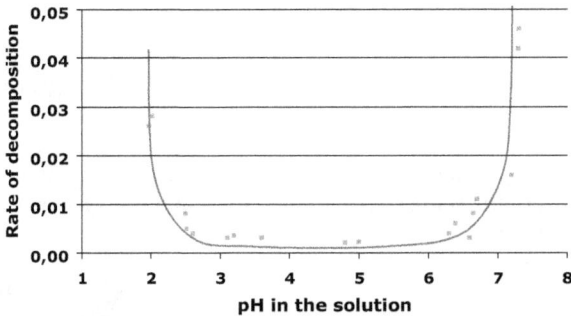

Figure 11.13. The rate of decomposition of a hydrogen peroxide solution is lowest between pH 4 and 5. (Kemira Kemi).

11.7 Peracetic Acid Production

Peracetic acid (CH_3COOOH) is a relatively new oxidizing bleaching agent in the pulp mills. It is a very selective chemical and it is mainly use in TCF-bleaching sequences where high brightness is needed and where ozone is not used. It is produced by mixing acetic acid, hydrogen peroxide and sulphuric acid. However, the resulting concentration of peracetic acid is low and the mixture has to be distilled to make the chemical useful in pulp mills. Such a distillation results in an approximately 40 % solution of peracetic acid of pH 1.5–3 with a few percent of acetic acid and hydrogen peroxide. The residual phase from the distillation contains the unreacted acetic acid and hydrogen peroxide and this solution is recycled and reused, *Figure 11.14*. Peracetic acid is an equilibrium product which means that it has an ability to slowly go back to its raw material, acetic acid and hydrogen peroxide. This reaction is endothermic. To slow the reaction down the distilled peracetic acid is stored and transported at a temperature of about −6 to −10 °C. So far (2001) peracetic acid is not produced on-site at any pulp mill mainly due to the small quantities of the chemical that are used at the mills. The distilled peracetic acid is produced by hydrogen peroxide producers and in Sweden this takes place in Bohus by Eka Chemicals and in Helsingborg by Kemira Kemi.

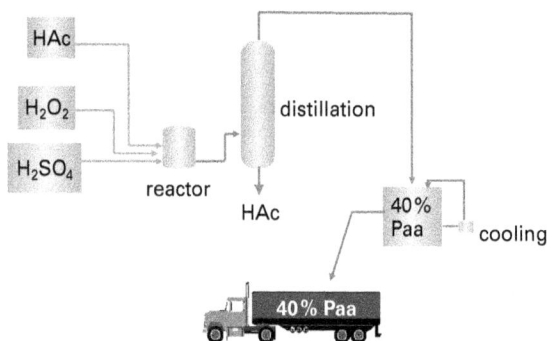

Figure 11.14. Production of distilled peracetic acid from acetic acid, hydrogen peroxide and sulphuric acid. (Kemira Kemi).

11.8 Ozone Generation

Ozone is a gas that is produced from oxygen and electric power in an ozone generator. In the generator the oxygen gas is passing an electric corona discharge and some of the oxygen molecules are exited to ozone according to Reaction 11.10.

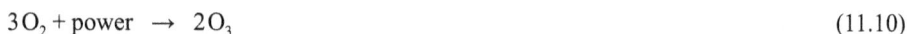

$$3\,O_2 + power \;\rightarrow\; 2\,O_3 \tag{11.10}$$

The resulting ozone gas is in reality mainly a mixture of oxygen and ozone with about 8–14 % ozone depending on the requirement. As the power demand for ozone generation increases considerably with increasing ozone gas concentration the relatively low ozone gas concentration is a compromise between either higher power demand and higher decomposition rate or higher cost for oxygen including higher costs for the increasing gas volume for a given ozone charge.

The correlation between the ozone gas concentration and the power demand in a modern ozone generator is shown in *Figure 11.15*. It is seen that the power demand increases considerably with increasing ozone gas concentration; modern ozone generators are today producing at about 12–13 % concentration. However, the actual ozone gas that is needed in a pulp mill is depending on the ozone bleaching system that is used. In a high consistency ozone bleaching reactor an ozone gas concentration of 6–7 % is sufficient while in a medium ozone bleaching reactor the ozone gas concentration has to be 12–14 % of gas volume reasons.

The ozone generation technology has developed considerably during the last 10 years that has resulted in considerable reductions in power demand per kg ozone. This has made ozone generation cheaper and ozone bleaching economically more competitive. The maximum size of modern (2001) ozone generators has also increased during the last 10 years from about 50 kg/h to 150 kg/h which has further reduced the investment cost and thus reduced the cost for ozone generation.

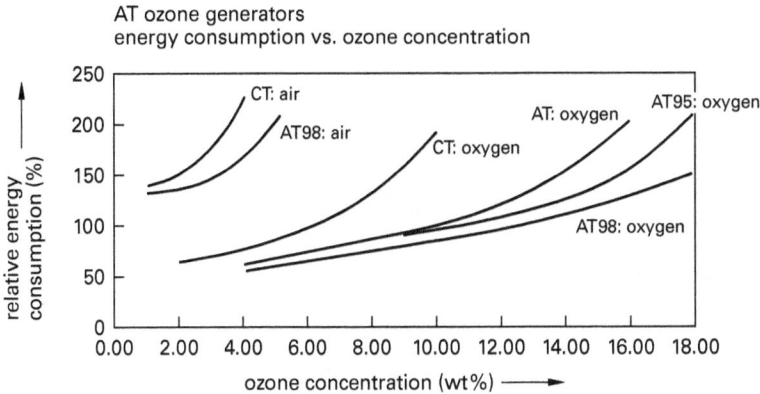

Figure 11.15. The power consumption in ozone generation versus the ozone gas concentration for air and oxygen. The figure shows that the power demand has been decreased during the 1990′s. (Ozonia).

The generated ozone molecules are unstable and they decomposes readily back to oxygen. Therefore, the ozone gas has to be generated in direct connection to the ozone bleaching reactor. The rate of decomposition is also higher at higher temperature and the generation temperature, and the bleaching temperature, should therefore be as low as possible and cooling is therefore needed, *Figure 11.16*. This means in reality that the ozone generation temperature is kept at 20–30 °C.

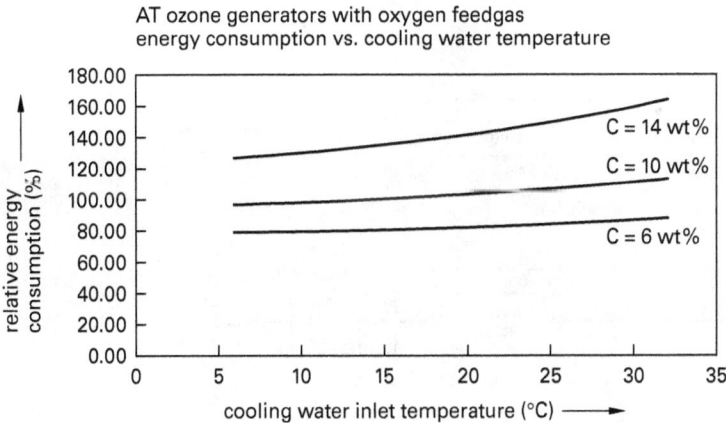

Figure 11.16. The relative ozone consumption versus the cooling water temperature with the ozone gas concentration as parameter. (Ozonia).

294

Finally, to keep the power consumption down it is very important that the gases used in the ozone generator are very dry, *Figure 11.17*. The ozone yield decreases considerably even at very low humidity contents. If recycled oxygen is used as feed gas to the ozone generator it is therefore necessary to completely remove as much of the humidity as possible prior to the ozone generator.

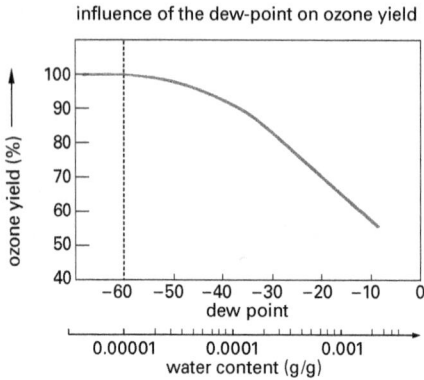

Figure 11.17. The ozone yield versus the influence of the dew-point in the gas. The figure thus shows that the feed oxygen gas should be very dry otherwise the power demand will increase considerably (Ozonia).

In order to keep the cost of ozone down it is important that the by-product oxygen in the "ozone gas" is utilized as completely as possible. Three different methods are then used, *Figure 11.18*. These are denoted:

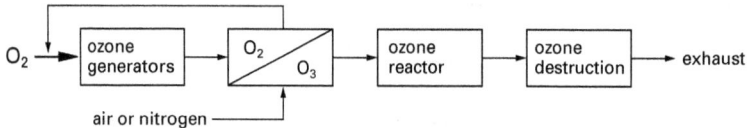

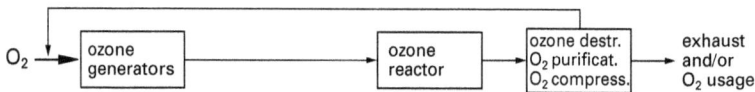

Figure 11.18. Ways to reuse oxygen in connection to ozone bleaching.

1. "Once through" meaning that the residual oxygen gas is directly without any purification used in for example an oxygen stage. This is the cheapest method.
2. "Short loop" meaning that the gas stream is directly after the ozone generator separated into oxygen and ozone and where the oxygen fraction is reused in the ozone generator.

3. "Long loop" meaning that the by product oxygen is taken care of after the ozone bleaching reactor. In this case it is also necessary to purify the gas with respect to humidity as the power consumption of ozone generation increases considerably with the moisture content in the feed gas.

11.9 Oxidation of White Liquor to Alkali

The cooking liquor used in kraft cooking is denoted as "white liquor" and it consists mainly of sodium hydroxide (alkali) and sodium sulphide which both are active in the cook. Chemically white liquor can be described as a mixture of NaOH and $Na_2S/NaHS$. This cooking liquor is internally generated in the chemical recovery system in the pulp mill. However, sodium hydroxide is also needed in the bleach plant for example in the oxygen stage, in the extraction stage(s) and in the hydrogen peroxide stage(s). Thus, one way to get this alkali is to use internally produced white liquor also in the bleaching stages. However, except for cooking purposes white liquor is today only used as alkali in the oxygen stage. In that case however, the hydrogen sulphide ions are first oxidized with air before the addition to the oxygen stage as the cellulose chains in the pulp are significantly degraded during oxygen delignification if hydrogen sulphide ions are present and this will result in reduced viscosity and thus reduced pulp strength. This strength loss is probably a result of the oxidation of the hydrogen sulphide ions during oxygen delignification that generate harmful radicals that attack the cellulose chains and therefore reduce the final pulp strength. Therefore, oxygen delignification stages always uses oxidized white liquor as its alkali. If hydrogen sulphide ions are present in the oxygen delignification stage there will also be an increased oxygen consumption for a given bleaching result which means increasing bleaching chemical cost.

Oxidized white liquor is not used in the other alkaline bleaching stages of two reasons: 1) White liquor is not as pure as purchased hydroxide with respect to metal ions like manganese, cupper, iron etc. Hydrogen peroxide decomposition accelerates considerably in the presence of even trace amounts of such metal ions and oxidized white liquor is therefore not suited for peroxide stages. 2) The sulphides in the white liquor are easily oxidized and if this is done in the bleaching stages a loss of bleaching chemicals will result and the economical loss increases with the price of the lost bleaching chemical. Thus, if white liquor should be possible to use in the bleaching stages the sulphides must be oxidized prior to the bleaching stage and at least the harmful metal ions must be eliminated. Such technologies will become more interesting in the future if the sodium hydroxide price increases further.

The oxidation of white liquor is usually done with air or in some cases with oxygen gas. This is done in the white liquor oxidation stage where the sulphides are oxidized to thiosulphate according to Reaction 11.11.

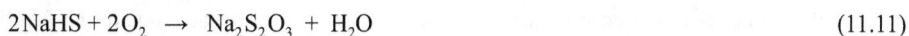

$$2\,NaHS + 2\,O_2 \;\rightarrow\; Na_2S_2O_3 + H_2O \tag{11.11}$$

This reaction is the one that usually takes place in a conventional white liquor oxidation stage. Such a stage usually consists of a reactor where air is sparged into the white liquor at atmospheric pressure. However, further oxidation of the tiosulphate to "totally oxidized white liquor" will result if the residence time and temperature are increased and the reactor is pressurized. This will result in sodium sulphate according to Reaction 11.12.

$$Na_2S_2O_3 + 2O_2 + 2NaOH \rightarrow 2Na_2SO_4 + H_2O \qquad (11.12)$$

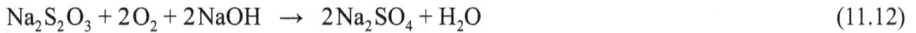

This reaction consumes oxygen, or any other oxidizing agent, and sodium hydroxide. It takes place in the oxygen delignification stage, or in the peroxide stage if oxidized white liquor is used in such a position and the added white liquor is only preoxidized to the thiosulphate state. From the discussion above it can be concluded that oxidized white liquor consists mainly of sodium tiosulphate and sodium sulphate where the final composition depends on the conditions used in the oxidation stage.

11.10 Power Demand for Production of Bleaching Chemicals

The electric power demand for production of bleaching chemicals is given in *Table 11.3*. It is seen that the power demand varies a lot between the bleaching chemicals and that oxygen is lowest in power consumption. Ozone is the highest consumer of electric power calculated per kg of the bleaching chemical. Recalculated per oxidizing equivalent (OXE) oxygen is still lowest but chlorine dioxide is then the highest.

Table 11.3. Power demand for production of bleaching chemicals.

Chemical	KWh/kg	Wh/OXE
O_2	0,5	4
H_2O_2	0,8	13
O_3	10	80
NaOH	1,6	–
Cl_2	1,6	58
ClO_2 (á Cl)	3,7	130

11.11 Future Trends

The market trend is that bleaching chemicals are to an increasing extent produced by chemical suppliers and not by the pulp mills themselves. The bleaching chemicals will in most cases be produced away from the pulp mill and after production transported to the pulp mill. This is the case for chlorate, chlorine, alkali, hydrogen peroxide and peracetic acid. However, some chemicals cannot be transported but has to be produced at the mill site due to the risk for decomposition or for the health hazards that could be the result if leaks. The very high transportation cost that would be the result if transportation was used can also be an obstacle due to the fact that some chemicals are diluted in large volumes of water or other gases. Example in the first case is chlorine dioxide. Examples of the second group are chlorine dioxide, ozone and oxidized white liquor. Finally, oxygen is to an increasing degree produced at the pulp mill and this trend will continue with increasing oxygen demand in the pulp mills. In most cases will the oxygen be produced by a chemical supplier and sold to the pulp mill "over the fence".

12 Recovery of Cooking Chemicals: the Treatment and Burning of Black Liquor

Hans Theliander
Chalmers University of Technology

12.1 Introduction

Black liquor from the pulp washing after the kraft cook and oxygen delignification consists of various organic components, small amounts of inorganic dissolved from the wood, some inorganic compounds originating from the cooking chemicals (the white liquor) and, of course, water. The organic compounds consist mainly of sodium salts of dissolved lignin and dissolved and degraded carbohydrates. The purpose of the recovery plant is to recover and recycle the cooking chemicals and to extract the energy from the organic compounds in the black liquor. The chemical recovery plant may be divided into five different parts:

1. Black liquor evaporation
2. Condensate treatment
3. Black liquor burning
4. White liquor preparation
5. Lime mud reburning

Figure 12.1 is a schematic diagram of the chemical recovery plant.

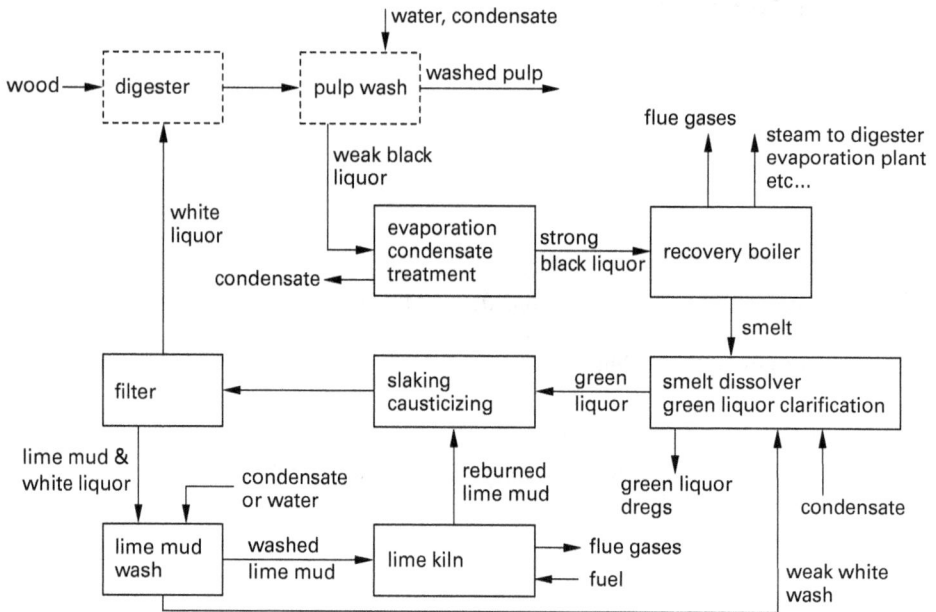

Figure 12.1. A Schematic diagram of the chemical recovery plant.

The weak black liquor is concentrated evaporating the solvent in the black liquor evaporation plant to increase the dry content before burning. It is mainly water that is separated (>95 %), but also small amounts of organic compounds with low molecular weight (e.g. methanol and some sulfur-organic compounds) are separated from the black liquor. The separated mixture of water and organic compounds exit the evaporation plant as a condensate. The organic material is separated from the water in the condensate treatment system and burnt either in the limekiln or in a specially designed furnace. The cleaned condensate is normally used as washing water in the pulp and lime mud washing plants but may also be used in the smelt dissolver. The concentrated black liquor (strong black liquor) is combusted in the recovery boiler and heat is extracted from the hot flue gases formed; the sodium and sulfur compounds (both organic and inorganic) are obtained as sodium carbonate and sodium sulfide. The sodium carbonate and sodium sulfide exit the recovery boiler in melted form and are dissolved in a water solution (weak white wash), thus forming green liquor. The dominating salts in green liquor are sodium carbonate and sodium sulfide, but there are also some sodium hydroxide, sodium sulfate and sodium chloride present. Green liquor also contains small amounts of solid material, e.i. green liquor dregs also known as green liquor sludge. These must be removed before the green liquor is processed in the white liquor preparation plant; here calcium oxide (CaO) is added to the green liquor, forming slaked lime (Ca(OH)$_2$) as a first step. The sodium carbonate is causticized with calcium hydroxide to form sodium hydroxide (dissolved) and solid lime mud (CaCO$_3$), and the white liquor is thereby obtained. The lime mud is separated from the white liquor by filtration or thickening. The white liquor is now ready to be used for cooking wood chips. The lime mud must be cleansed from all white liquor, which is done in the lime mud washing plant. The filtrate from this unit is called „weak white wash" and is used in the smelt dissolver. The calcium carbonate in the lime mud is calcined in the lime mud reburning plant and reburned lime (CaO) is formed. The reburned lime is recycled to the slaking/causticization plant.

It is clear from the above that the chemical recovery plant has two cycles: the sodium cycle, constituted of weak black liquor – strong black liquor – smelt – green liquor – white liquor, and the calcium cycle, constituted of reburned lime – slaked lime – lime mud.

In the chemical recovery system there are a large number of different unit operations, ranging from the huge reactors of the chemical recovery boiler, via equilibrium operations in the evaporation plant and distillation columns (condensate treatment) to solid phase operations such as suspension, filtration and sintering. In order to understand these operations, it is of the utmost importance that knowledge of both chemistry and chemical engineering be combined.

12.2 Properties of Black Liquor

The properties of black liquor, both chemical and physical, are of the utmost importance in the design of the evaporation plant, condensate treatment plant and recovery boiler.

As for as the chemical properties are concerned it is, of course, the composition that is of the greatest importance: it determines, among other things, whether or not there will be scale formation in the evaporation plant, the heat value of the strong black liquor after evaporation and if the heat transfer areas in the recovery boiler will be covered with dust. A rough classification can be performed between organic and inorganic compounds: approx. 2/3 of the solids constituted of organic material and the reminder is inorganic material. *Table 12.1*, gives approximate concentration levels of some common elements. There are also a large number of other elements of lower concentrations (below 1 g/kg dry black liquor): e.g. magnesium, iron, manganese, phosphorus, silicon, aluminum etc…

Table 12.1. Approximate composition of black liquor.

Element	Amount (weight %)	Organic material	Amount (weight %)
Carbon	34–39	Lignin	29–45
Hydrogen	3–5	Hydroxy acids	
Oxygen	33–38		
Sodium	17–25	Extractives	25–35
Sulfur	3–7	Formic acid	3–5
Potassium	0.1–2	Acetic acid	~5
Chlorine	0.2–2	Methanol	~3
Nitrogen	0.05–0.2		~1

The chemical properties are determined by analysis of the different elements and/or of the specific compounds. Standardized methods (TAPPI or SCAN methods) can be found for most of the elements/compounds mentioned above. The chemical properties, together with temperature and pressure, determine the physical and thermodynamic properties of the black liquor as well as the potential for scale formation. Unfortunately, there are not very many general correlations that describe the relationship between the chemical properties, temperature and pressure on the one hand and the physical and thermodynamic properties on the other hand.

12.2.1 Physical and Thermodynamic Properties

Table 12.2, is a summary of the most important physical and chemical properties and the areas in which they are of any importance.

Table 12.2. Properties of black liquor and their importance on the process.

Property	Importance
Density	To obtain correct mass balances. Density is also an indication of amount of solid/dissolved material in the liquor.
Viscosity	Influences the flow conditions and therefore also the heat transfer conditions in the evaporation plant as well as drop formation in the recovery boiler.
Surface tension	Drop formation in the recovery boiler and the formation of a film in the evaporators (particularly falling film evaporators).
Heat conductivity	Heat transfer in the evaporators.
Specific heat capacity	To obtain correct heat balances and it influences the heat transfer in the evaporators.
Specific heating value	The amount of chemical energy that can be converted into heat. Important for the heat balance of the recovery boiler.
Boiling point rise (Vapor pressure depression)	Determines the number of evaporators at a given maximum total temperature difference.
Vapor pressure	The vapor pressure data determines the composition of the evaporated steam and the liquor in the evaporation plant. Is of crucial importance in the design and operation of the columns in the condensate treatment system.

There are no general correlations describing how the physical and thermodynamic properties are influenced by the composition of black liquor. It is therefore necessary to measure these properties so that accurate values may be obtained.

12.3 Evaporation of Black Liquor

12.3.1 Introduction

In the black liquor evaporation plant more than 90 % of the water in the inlet stream (weak black liquor) is removed and strong black liquor is thus obtained. The reason for separating the water is simply to increase the heating value of the black liquor. The incoming weak black liquor has a dry content of 15–20 %, i.e. more than 4/5 of the liquor is water. With that amount of water, the latent heat of the organic material is not sufficient to evaporate all of the water in the black liquor and, thus, the effective heating value is negative. It is, therefore, impossible to combust black liquor containing this amount of water. After evaporation the dry content of the black liquor is 70 to 80 %, which implies that only 1/4 to 1/5 is water. The heating value of the black liquor is substantial when it contains such relatively low levels of water.

Most of the liquid evaporated is water. Some organic compounds, mainly methanol, are also evaporated, but some other organic compounds with a vapor pressure lower, or only slightly higher, than that of water will also be partly evaporated. Water is the main component to be evaporated and, since, the heat of vaporization of water is very high, it is important that the vaporization is performed as energy efficient as possible. Most of the vaporization is, therefore, carried out in a multiple evaporation plant; the final stage in increasing of the dry content may, however, be carried out in a separate evaporation unit. Examples of typical black liquor evaporation plants are shown in, *Figure 12.2.*

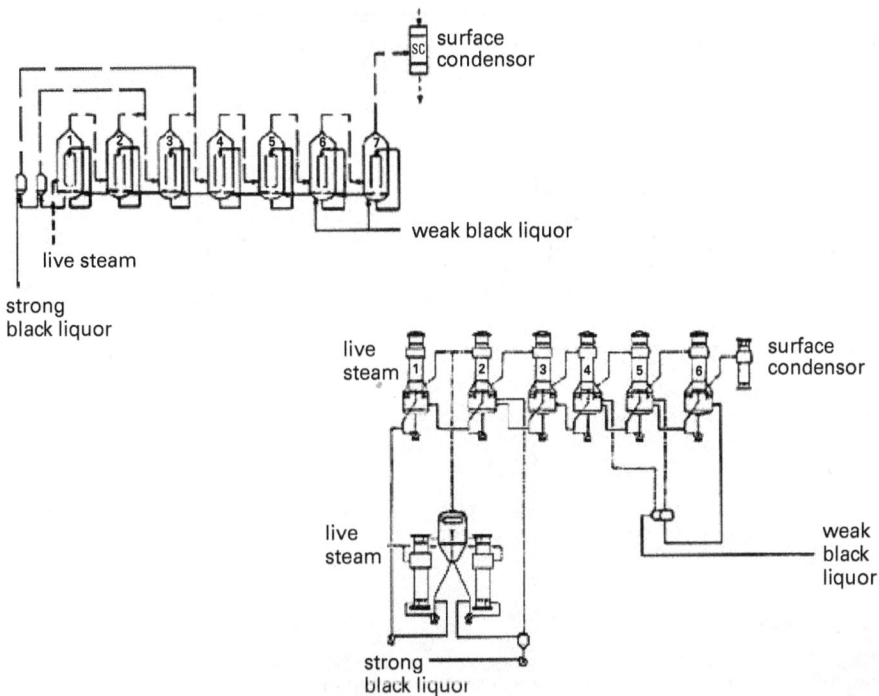

Figure 12.2. Schematic diagram of some typical black liquor evaporation plants.

An evaporation unit consists, in principle, of a heat exchanger and a device for separating the vapor formed from the remaining liquor. The steam formed in the first evaporation unit is used as heating steam in the second unit. The steam formed in the second unit is used as heating steam in the third unit and so on. This means that the condensation energy released by the condensation of the live steam used in the first unit is „reused" in the other subsequent units. This is the key of an energy efficient evaporation plant: the more units there are the greater the energy efficient. A modern evaporation plant consists of six or seven evaporation units: the numbers of units is, in

302

principle, limited by boiling point rises and pressure drops. As can be seen in *Figure 12.3*, there are two common liquor flows in a plant; in a modern system, the flow is almost always counter-current in relation to the steam. In older systems, a mix of co-current and counter-current is often used; an inlet black liquor stream is fed to the 3rd or 4th unit, flows co-current to the last unit, and is thereafter pumped to the first unit before exiting the plant from the 2nd or 3rd unit.

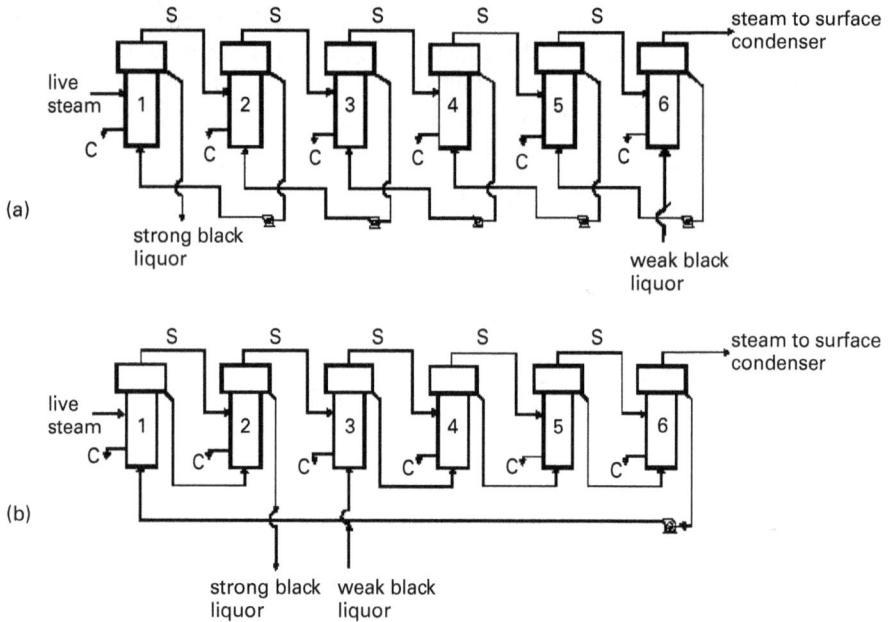

Figure 12.3. Schematic diagram of a counter-current (a), and a mixed co- and counter-current (b), plant respectively. (S is steam and C is condensate).

12.3.2 Equipment

Three different types of evaporators are used in black liquor evaporation:

- The *Kestner apparatus,* shown in *Figure 12.4*, is also known as climbing film evaporator or LTV evaporator (Long Tube Vertical). This type of evaporator is a tube heat exchanger (mounted vertically) with a separation device for the vapor formed and the remaining liquor mounted on top of the heat exchanger. The heating steam condensates on the outside of the tubes, the heat released is transported through the condensate film and the tube wall, and is absorbed by the black liquor. The absorbed heat increases the temperature of the black liquor to boiling point, and the remaining heat evaporates some of the liquor. The liquor enters the heat exchanger from underneath; if the liquor is not at boiling point, the first part of the heat exchanger is used as a preheater, increasing the temperature of the liquor to its boiling point, see *Figure 12.4*. At the level where the liquor has just reached boiling point, the vapor bubbles are rather small and the action is rather gentle. As the liquid rises, however, boiling becomes more intense and the vapor bubbles larger and, as a consequence, the flow pattern becomes more and more turbulent. In the top section of the heat exchanger, there is only a thin film of liquor on the wall of the tube. Tubes are often between 7 and 10 m in length and approx. 0.05 m in diameter. This type of equipment can be used for evaporation up to a dry content of 50 %, above which the viscosity becomes too high for it. It should be kept in mind that this type of equipment is relatively sensitive to the formation of scales, which is also an argument for not using it in the high dry content region. The Kestner apparatus used to be quite common, but almost no new units are installed nowadays for just the reason, i.e. scales are easily formed and the fact that it cannot be used for high dry contents.

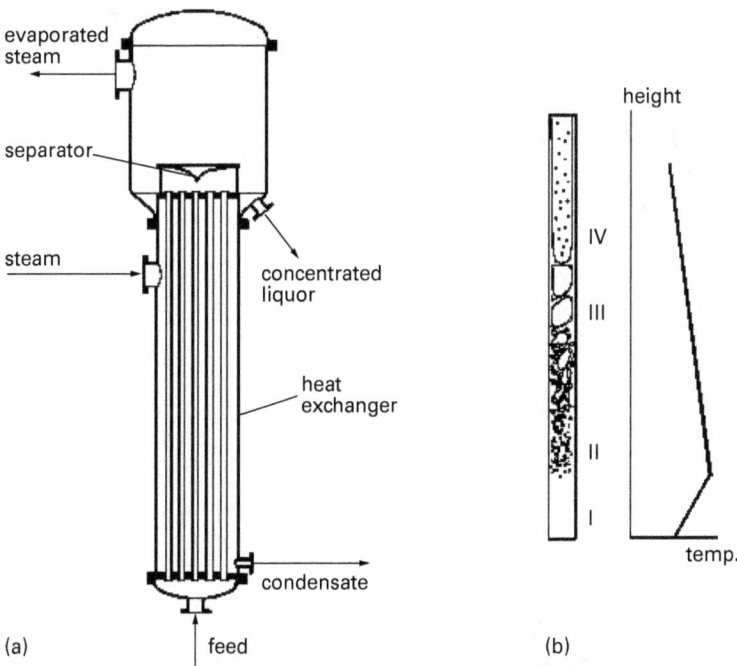

Figure 12.4. Diagram of a Kestner apparatus (A) and the temperature profile (B).

- The *Forced circulation evaporator,* can be used to increase the dry content of black liquor from around 50 % to 70–80 %, i.e. the viscosity of the black liquor is very high. This type of equipment consists of a heat exchanger, a flash chamber and a circulation pump, see *Figures 12.5* and *12.6*. The equipment shown in *Figure 12.6* may also be used as a crystallisator. In order to avoid the formation of scales on its surfaces, the heat exchanger is located well below the flash chamber. This ensures a high pressure in the heat exchanger and, thus, avoids boiling in the tubes. The tubes are normally between 4 and 7 m in length and between 0.03 and 0.05 m in diameter. Compared to the Kestner apparatus, the tubes are shorter and their diameters smaller. The velocity of the liquid in the tubes is normally between 2 and 4 m/s.

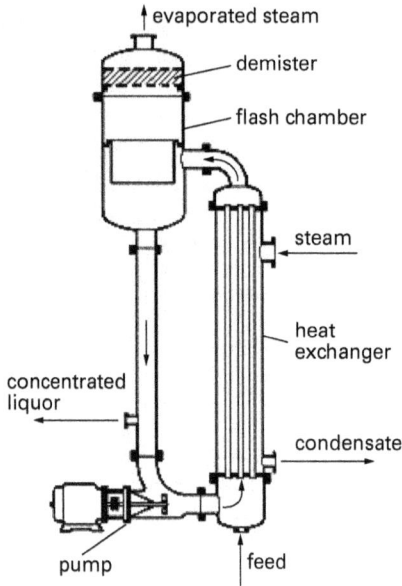

Figure 12.5. Diagram of a forced circulation evaporator.

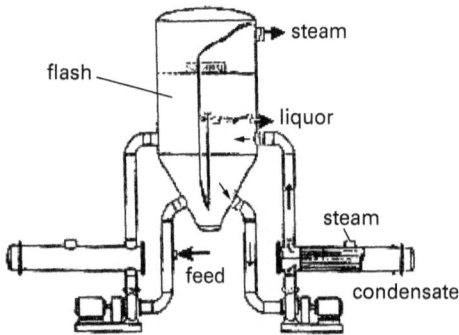

Figure 12.6. Diagram of a forced circulation evaporator designed for crystallisation.

- The *falling-film evaporator,* has become more and more common and is currently the standard type of equipment used in the evaporation of black liquor. This is true for the entire concentration range. A schematic drawing of a falling-film evaporator is shown in *Figure 12.7.* In this type of equipment, the liquor enters the apparatus at the top, where the heat exchanger is located. A liquid film is formed on the inside of the tubes, which moves downwards due to gravitation. In this type of equipment, the separation of the vapor formed and the remaining liquor occurs at the bottom at the apparatus. The heating steam is condensed on the outside of the tubes. The difficult, and also the critical, part is the formation of the liquor film. It is, therefore, not surprising that a couple of different types of equipment for creating the film have been developed, some of which are shown in *Figure 12.8*.

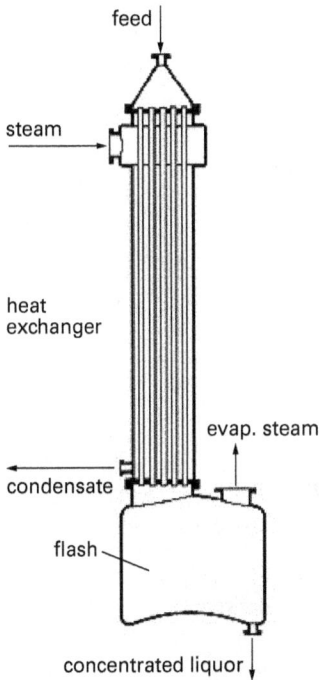

Figure 12.7. Schematic diagram of a fallingfilm evaporator.

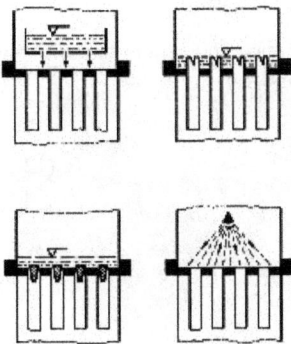

Figure 12.8. Different types of equipment for creating a liquor film.

The heat exchanger in the figures above is of the tube-type, but it can also be in the form of a plate, see *Figure 12.9*. The figures above also indicate that the liquor film is formed on the inside of the tubes. Evaporators where the film forms on the outside of the tubes has also been developed, in which case the heating steam is, of course, condensed on the inside of the tubes.

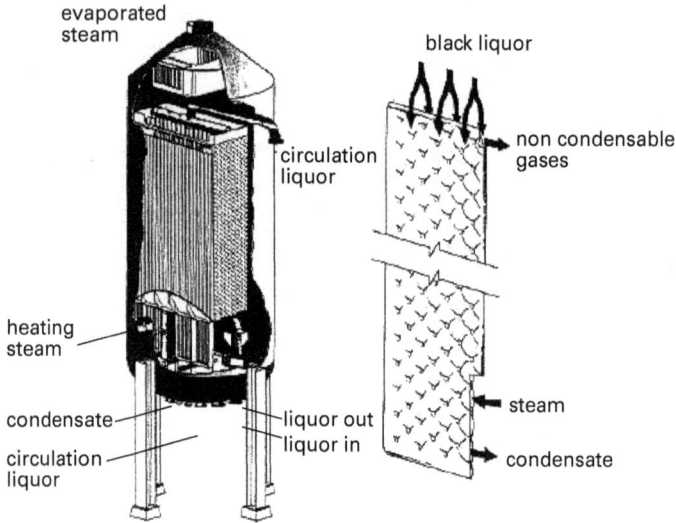

Figure 12.9. Example of an evaporator having plates as the heat exchanger device.

12.3.3 Material and Heat Balances and the „Capacity Equation"

A tremendous amount of energy is converted in a black liquor evaporation plant. In order to obtain high-energy efficiency, it is necessary to have good control of the liquor and steam flows in the evaporation plant. This is achieved with mass and heat balances. The notations normally used in evaporation plant calculations are shown in *Figure 12.10*.

Mass and heat balances can be formulated around each evaporation unit. In the case of a co-current plant, the following balances can be derivated for unit number n:

Total balance:

$$L_{n-1} = V_n + L_n \qquad \text{Eq(12.1)}$$

Component balance:

$$L_{n-1} \cdot x_{n-1} = V_n \cdot y_n + L_n \cdot x_n = \left| y_n = 0 \right| = L_n \cdot x_n \qquad \text{Eq(12.2)}$$

Heat balance:

$$V_{n-1} \cdot H_{V,n-1} - V_{n-1} \cdot H_{L,n-1} + L_{n-1} \cdot H_{L,n-1} = V_n \cdot H_{V,n} + L_n \cdot H_{L,n} \qquad \text{Eq(12.3)}$$

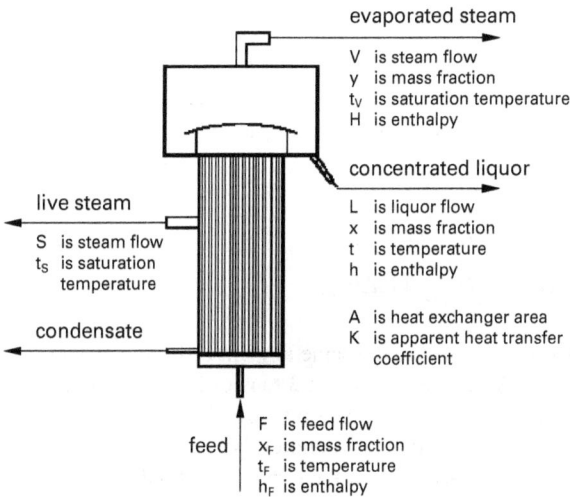

Figure 12.10. Notations used in evaporation calculations.

If it can be assumed that the heating steam and the condensate are saturated, the following equation is valid:

$$V_{n-1} \cdot H_{V,n-1} - V_{n-1} \cdot H_{L,n-1} = \Delta H_{vap,n-1} \cdot V_{n-1} \qquad \text{Eq(12.4)}$$

This implies that the heat balance can be expressed as:

$$V_{n-1} \cdot \Delta H_{vap,n-1} + L_{n-1} \cdot H_{L,n-1} = V_n \cdot H_{V,n} + L_n \cdot H_{L,n} \qquad \text{Eq(12.5)}$$

The capacity equation describes how much energy can be transferred from the heating steam to the liquor:

$$\dot{Q} = V_{n-1} \cdot \Delta H_{vap,n-1} = U_{skb,n} \cdot A_n \Delta T_n \qquad \text{Eq(12.6)}$$

U_n is the apparent heat transfer number and ΔT_n is the temperature difference between the saturation temperature of the heating steam and the exit temperature of the liquid. In a multiple evaporation plant, the sum of the temperature differences is calculated as:

$$\Delta T_{tot} = \sum_{n=1}^{J} \Delta T_n \qquad \text{Eq(12.7)}$$

V_{n-1} is replaced with S if the calculations are for a unit using live steam as the heating medium. A couple of useful definitions describing the energy efficiency of the plant are presented below:

Heat economy = Heat supplied / Amount evaporated   Eq(12.8)

Steam economy = Amount evaporated / Amount of live steam   Eq(12.9)

12.3.4 Heat Economy

Assuming that the heat losses and the boiling point rise are negligible, and that all liquors and vapors are saturated, then all of all liquors and all vapors will have the same numerical value. Under these circumstances the heat balance, i.e. Equation (12.10) can be formulated as:

$$
\begin{aligned}
V_{n-1} \cdot \Delta H_{vap,n-1} &= V_n \cdot H_{V,n} + L_n \cdot H_{L,n} - L_{n-1} \cdot H_{L,n-1} \\
&= \left| H_{L,n} = H_{L,n-1} \right| = V_n \cdot H_{V,n} + H_{L,n} \cdot \left(L_n - L_{n-1} \right) \\
&= \left| V_n = L_{n-1} - L_n \right| V_n \left(H_{V,n} - H_{L,n} \right) = V_n \Delta H_{vap,n}
\end{aligned}
\qquad \text{Eq(12.10)}
$$

The implication here is that it is necessary to use the same amount of heating steam as the amount of vapor produced in each unit. Assuming that about 2 tons of liquid is evaporated in Swedish pulp mills every second, evaporation according to this principle would imply enormous operating costs: this is the reason why multiple evaporation is used where two or more units are coupled together in a „train", see *Figure 12.11*. Live steam (*S*) is used in the first unit. The steam produced in this unit (*V1*) has, of course, a latent heat of evaporation, which can be used for evaporation in subsequent units, i.e. is used as heating steam in the second unit. This means that live steam is not used to evaporate all of the liquid.

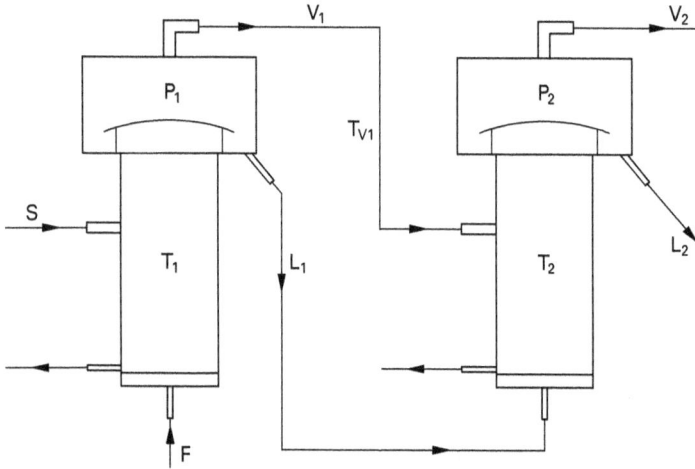

Figure 12.11. A two–unit evaporation plant.

It is necessary to lower the pressure in the units successively in a train of evaporator units in order to obtain a temperature difference between heating steam and liquor. The temperature of the steam from unit 1 (*TV1*) in *Figure 12.11* must, thus, be higher that the saturation temperature (boiling point) of the liquid in unit 2 (*T2*). The pressure of the liquid must be lowered if this temperature difference is to be achieved. *Figure 12.11* shows two units are in series, although a modern evaporation plants tend to have a train of six or seven units in series. From a technical point of view, the number of units are limited by the boiling point elevation, the pressure drops in the pipes and, of course, the saturation temperature of the live steam and the saturation temperature of the exit steam from the last evaporation unit.

12.3.5 Maximum Number of Units

The maximum number of evaporation units that can be used in an evaporation train is determined by the saturation temperature of the live steam, the saturation temperature in the last unit, the boiling point rise and the pressure losses in the pipes, valves etc.. The difference between the saturation temperature of the live steam and that of the last unit is the maximum total temperature difference, which only can be obtained if the boiling point rise and the pressure losses are zero (i.e. a case that never can be obtained in a real system). In a black liquor evaporation plant, the boiling point rise of the liquor is substantial and there are, of course, pressure losses. The consequence of a boiling point rise is that the vapor formed becomes superheated and its properties are thus similar to those of a gas. The heat transfer rate between a gas (superheated vapor) and a tube wall is much lower than that of vapor condensing on a tube wall, as is well known. When a superheated vapor is cooled, the overall heat transfer coefficient is more that two orders of magnitudes lower than for condensation of saturated vapor. Thus, if a superheated vapor is used as a heating medium in an evaporation process a significant part of the heat transfer area available will be used to cool the vapor to saturation temperature. Since the amount of energy delivered when cooling the vapor is very small, it is not feasible, from an economical point of view, to use the heat transfer area for this cooling. The superheated vapor is, therefore, normally cooled to saturation temperature before entering the heat exchanger; this is performed by injecting of saturated condensate into the superheated vapor, see *Figure 12.12*. The heat needed for evaporation of the condensate is taken from the superheated vapor and its temperature then decreases to the saturation temperature.

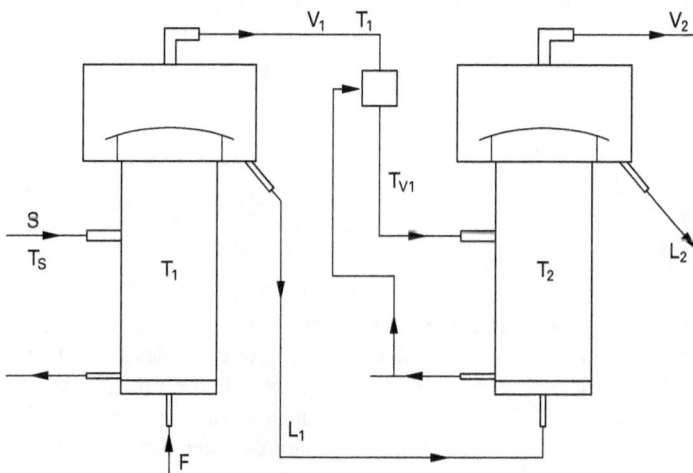

Figure 12.12. Saturating superheated steam by injection with saturated condensate.

Different types of pressure losses give the same result as a boiling point rise, i.e. the total available temperature difference decreases. It is, therefore, important that the design of the steam pipe system minimizes pressure losses. Among other things it is important that control valves are correctly designed. The total available temperature difference can be calculated with the following expression:

$$\Delta T_{tot} = T_S - T_{vn} - \sum_{n=1}^{J} \beta_n - \sum_{n=1}^{J} T_{\Delta P,n}$$

Eq(12.11)

This equation demonstrates that if the boiling point rise and/or the temperature loss due to pressure losses ($T_{\Delta P,n}$) become too large the available temperature difference can, theoretically be negative. This is not a realistic case, as the available temperature difference must, of course, always be positive. However, when the available temperature difference approaches zero the heat transfer area approaches infinity. This shows that the consequence of a boiling point rise and pressure losses is an increase heat transfer area, i.e. the investment cost increases. Since a boiling point rise is due to the presence of ions, there will always be a boiling point rise when black liquor is evaporated. However, minimizing the concentration of ballast salts also minimizes the boiling point rise. It is also important that attempts are made to minimize the pressure losses in the steam pipes.

12.3.6 Scale Formation

A frequently occurring problem in black liquor evaporation is the formation of scales on the heat exchanger surface. Altogether scales may form on the steam side, it is, however, far more common that they are formed on the liquor side; only scale formation on the liquor side will, therefore, be treated here. Scale formation poses a serious problem because the overall heat transfer coefficient decreases very rapidly with increasing thickness of the scales; the heat transferred from the steam side must also pass the scale layer. The overall heat transfer coefficient is, thus, calculated as:

$$U = \frac{1}{\dfrac{1}{h_{steam}} + \dfrac{l_{wall}}{k_{wall}} + \dfrac{l_{scale}}{k_{scale}} + \dfrac{1}{h_{liquor}}}$$

Eq(12.12)

Since the heat conductivity of the scales is much lower than for the steel material in the tube walls (100–1000 times less), the overall heat transfer coefficient is influenced significantly, even when the layers of scales are very thin. If the heat conductivity of the scales is 1000 times less than the material in the tube walls, a layer of a thickness 10 μm will lower the overall heat transfer coefficient by 10 %. *Figure 12.13*, shows a schematic diagram of a pipe with scales. It can, thus, be concluded that it is of the utmost importance that the formation of scales be avoided.

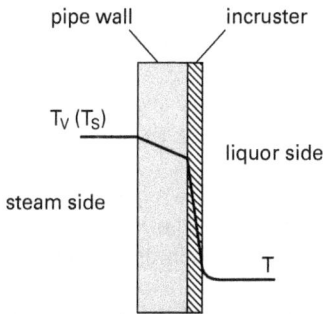

Figure 12.13. Schematic diagram of the temperature profile of a pipe with scales.

It ought to be mentioned that the type and design of the equipment are of great importance in determining the magnitude of the problems caused by scales. The biggest problem is experienced in the Kestner apparatus, due to the fact that the concentration and temperature of the liquor can be very high in the upper part of the heat exchanger. A high concentration always increases the risk of scale formation (c.f. crystallization), whereas a high temperature can be either positive or negative, depending on whether the salts have an increasing or decreasing solubility with increasing temperature. Both the falling-film and forced circulation apparatus are designed to give relatively even temperature and concentration conditions close to the heat transfer area. This reduces the risk of scale formation. It should, however, be kept in mind that scales do actually form in both of these types of apparatus but to a much lesser extent than in the Kestner apparatus.

A first classification of the different kinds of scales formed on the liquor side of the tubes can be made from the mechanisms involved in the formation of scales, see *Table 12.3*.

Table 12.3. Different types of scales on the liquor side of the tubes.

A.	Scales formed from insoluble material in the black liquor	
	1)	Organic material: fibers etc.
	2)	Inorganic material: rust flakes, sand, etc.
B.	**Precipitation of dissolved material in the black liquor**	
	1)	Organic material
		i Tall oil soap (at dry contents above 25–30 %)
		ii Lignin (ph < 11)
	2)	Inorganic material
		i Calcium carbonate (at temperatures above 120 °C)
		ii Burkite, a double salt of sodium sulfate and carbonate (dry contents above 50 %)
		iii Sodium-silicon-aluminum scales (high contents of silicon and/or aluminum)

Whilst scales formed from solid material (rust flakes, fibers etc.) may occur in a Kestner apparatus, they are more seldom in the other two types of apparatus mentioned above. The Kestner apparatus is sensitive to this type of scaling because the black liquor may reach very high dry contents locally in the upper part of the heat exchanger, which may lead to it being solidi-fied (i.e. reaching a dry content of approx. 100 %), with fibers, rust flakes etc. acting as rein-forcement material.

The precipitation of organic material dissolved in the black liquor can be avoided if the tall oil is removed and if the pH is reasonably high, i.e. alkalinity after the cook should not be too low.

It is very difficult to avoid formation of inorganic scales completely: it is a problem uncount-ed frequently in many mills. It is, therefore, not surprising that a number of investigations have been carried out with the aim of understanding the mechanisms involved in the formation of scales. Some of the findings are described briefly below.

It is known that calcium carbonate scales are not formed from solid calcium carbonate parti-cles (i.e. lime mud) in black liquor: in fact it is often calcium ions bound to organic substances that are responsible. The organic compounds starts to decompose at temperatures above 120°C. Calcium ions are, consequently, liberated and, since the liquid contains carbonate ions in a large excess, solid calcium carbonate is formed: this is the reason why calcium carbonate scales are found at temperatures above 120 °C. Since the highest temperature is to be found at the surface of the heat exchanger, it is most likely that the organic compounds decompose there, thus form-ing calcium carbonate scales. These scales can be removed by washing the surface with a solu-tion with a low pH (pH < 4), sulfamin acid is often used.

Burkite scales ($2Na_2SO_4 \cdot Na_2CO_3$) are formed at dry contents higher than 50 %. This type of scale is soluble in black liquor with a low dry content. It is, therefore, common that this type of scale is removed by washing the surface with weak black liquor. Sodium-silicon-aluminum scales (sometimes called silicon-aluminum scales or simply silicon scales, $NaAlSiO_4 \cdot 1/3 Na_2X$, where X may be carbonate, chloride, sulfate etc.) are the most difficult type of scale to handle. It is very difficult to dissolve (solubility is very low), so other methods, such as temper-ature shocks or mechanical methods, are normally necessary. The critical silicon and aluminium concentrations are dependent on the hydroxide ion content and can be described by the follow-ing equation:

$$\log\left[\left(Al^{3+}\right)_{tot} \cdot \left(Si^{4+}\right)_{tot}\right] = -4.6 + 1.4 \cdot \log\left[OH^-\right] \qquad \text{Eq(12.13)}$$

This equation shows that a high silicon concentration can be compensated by a very low alu-minium concentration and vice versa. It should also be remembered that the concentration of hydroxide ions in the black liquor going to the evaporation after the kraft cook is of a great im-portance for the solubility of this scale.

Silicon and aluminum enter the process via the wood, water, make-up chemicals (lime) and biofuel (if used) in the limekiln. In order to avoid the formation of this type of scale, it is of great importance that the wood chips are free from bark, sand etc. and, furthermore, that other sources have as low concentrations of silicon and aluminum as possible. The exit streams for silicon and aluminum are the green liquor sludge, the pulp and, to some extent, the lime grits. A mill with a relatively open system does not normally have a problem with sodium-silicon-alu-minum scales. As more process streams are recirculated, however, there may be a problem if the content of silicon and aluminum concentrations in the inlet process streams are not minimized, and if the concentrations of these substances are not maximized in the exit process streams.

12.3.7 Condensate Treatment

The condensate from the black liquor preparation plant contains organic compounds that must be removed before it can be used for washing water or in the white liquor preparation plant. *Table 12.4,*shows approximate concentrations of some of the most important organic compounds in the condensate.

Table 12.4. Approximate concentrations of organic compounds in the condensate from the black liquor evaporation plant.

Compound	Methanol	Ethanol	Hydrogen sulphide	Methyl-mercaptan	Dimethyl sulphide	Dimethyl disulphide
Conc. (mg/l)	5000–6000	100–200	100–150	400–700	100–200	50–150

Compounds such as methanol and ethanol must be separated in order to keep the amount of dissolved organic material at reasonable levels in the white liquor system. The sulfides are poisonous and very malodorous, and must therefore be separated and destroyed.

All of the compounds shown in the table above have a boiling point lower than water, which implies that these compounds can be separated with distillation. In this case, a column with a long stripper section and no, or a very short rectifying section is used. The heat supply is often live steam added to the bottom tray. Columns used for condensate treatment are often of tray-type (15–30 trays). A schematic drawing of a condensate stripper column is shown in *Figure 12.14.*

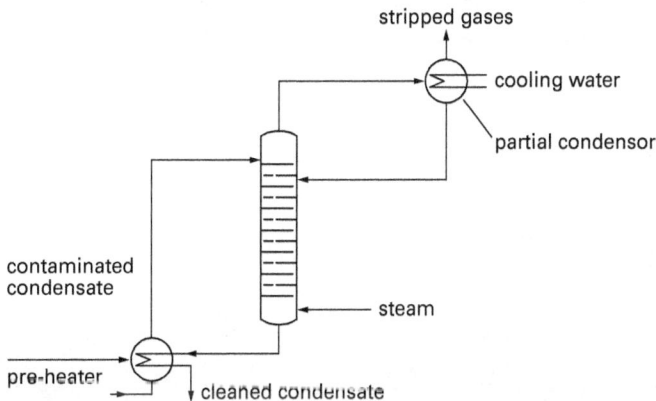

Figure 12.14. Schematic drawing of the condensate column.

The separation efficiency is normally very high: more than 90 % of the compounds mentioned in *Table 12.4* are separated in a well functioning condensate stripper column. Condensate stripping is usually included in the multistage evaporation plant

12.4 The Combustion of Black Liquor

12.4.1 Introduction

The purpose of combusting black liquor is twofold: to extract the latent heat in the organic material and to recover sodium and sulfur in the form of sodium carbonate and sodium sulfide. The method used today is combustion in a recovery boiler of the Tomlinson-type. This type of boiler was developed around 1930, and was a very large improvement on its predecessor when it was introduced: the heat liberated could be recovered as steam and the degree of recovery for sodium/potassium and sulfur increased. Since it was first developed it has, of course, been greatly improved, the basic idea is nonetheless the same as it was 70 years ago. During the 1960s some new ideas about black liquor combustion appeared: One was that black liquor should be gasified in a first stage and, thereafter, the flue gas combusted. After a rather large project was undertaken, development and research efforts almost died out for many years. Since the middle of the 1980s, however, a number of research and development projects dealing with the gasification of black liquor have been carried out and gasification technology is just about to be launched as a commercial product. One of the big advantages of gasification is that it produces burnable gases that can be combusted in a gas turbine and, therefore, the ratio between power production and steam production can be increased tremendously. Even if the introduction of gasification technology becomes very successful, it will take some 30–40 years before all of the old recovery boilers have been replaced by gasifiers. This is partly due to the original investment costs of the recovery boilers as well as the fact that gasifiers are very expensive, around 100 M USD for an average-sized plant. This chapter will focus on the recovery boiler.

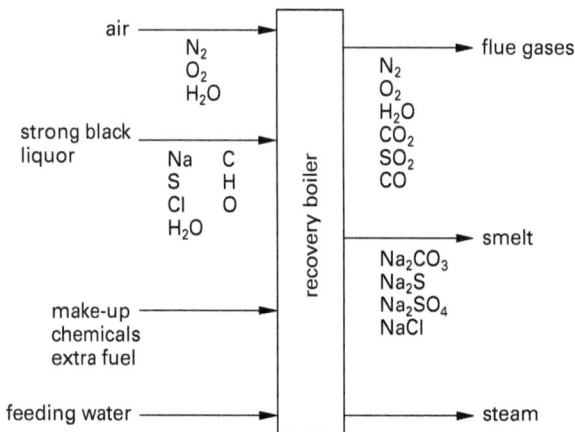

Figure 12.15. Schematic illustration of the chemical processes that take place when black liquor is combusted.

12.4.2 The Recovery Boiler

Strong black liquor is combusted in a recovery boiler. Air is normally used as an oxidant although, in some cases, oxygen-enriched air is used. The resulting products are a salt smelt, flue gases and heat: this liberated heat is used to produce high-pressure steam. *Figure 12.15* is a schematic illustration of the reactants and products to/from a recovery boiler.

The valuable products are chemicals in the form of a salt smelt and high-pressure steam. It should be remembered that sodium is not the only alkali metal: potassium is also present (2–10 %). The levels of sulphur dioxide and carbon monoxide must, of course, be minimized in the flue gases. *Figure 12.16* is a schematic diagram of the recovery boiler; it is an enormous piece of equipment with a height of 60–70 m.

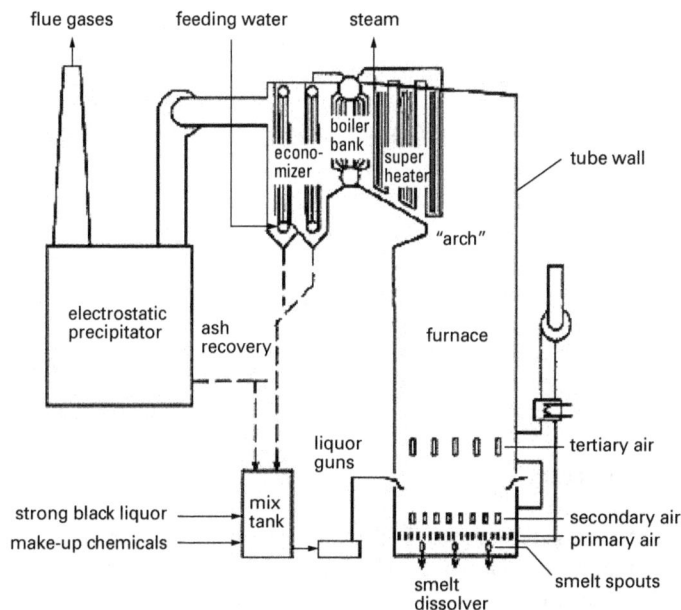

Figure 12.16. A schematic diagram of the recovery boiler.

The recovery boiler consists of a fireplace (the walls are covered with boiler tubes), convection surfaces (super-heater, boilers and economizer), electric precipitation filters and a mixing tank. Strong black liquor enters the mixing tank along with fly ash from the electric precipitator and, if necessary, some make-up chemicals. The resulting mixture is transported via a pre-heater to the liquor guns. These guns spray the black liquor in form of small drops, a few mm in diameter, into the fireplace. As each drop falls to the bottom of the fireplace, it will undergo drying and pyrolysis, and the char formed will be either partly or completely combusted. The inorganic chemicals will form sodium carbonate and sodium sulphide. The rest of the drop (remaining char and inorganic compounds) will fall down to the char bed where the rest of the char is oxidized, and oxidized sulphur compounds will be reduced to sulphides. The carbon is thus used as a reduction agent and exits the char bed as carbon monoxide or dioxide. The salt smelt produced is tapped from the bottom and is then transported via the smelt spouts to the smelt dissolver, where it is dissolved in weak white wash.

The atmosphere in the lower part of the recovery boiler is one of reduction, which implies that combustion of the organic compounds is not complete. It is possible to create a reducing at-

316

mosphere since the air ports are located at different heights: primary air is added close to the bottom, secondary air a little bit higher up but still below the liquor guns and, finally, above the liquor guns, tertiary and eventually quaternary air is added. Above the tertiary air ports, the atmosphere is oxidizing: thus combustion of carbon monoxide, methane, hydrogen, etc. occurs. The walls of the fireplace are covered with heat exchange tubes, which are a part of the heat exchange system for the evaporation of water, so that the temperature outside of these tubes is low enough to solidify the smelt and thus form a solid protective layer. This is necessary in order to protect the construction from the extremely corrosive atmosphere in the recovery boiler: the molten salts in particular are highly corrosive.

When the hot flue gases exit the fireplace, they are first partially cooled when they pass the superheaters (a heat exchanger bank) and the steam inside the tubes is thereby superheated. The partially cooled flue gases pass thereafter the boiler tube bank, where about 50 % of the feed water is evaporated; the remainder is evaporated in the heat exchanger walls in the fireplace. Finally, the flue gases pass an economizer where the feed water is preheated to the boiling temperature. The cooled flue gases pass an electrostatic filter where the fly ash is separated from the gas. Studying the figure above, it is clear that the feed water flows in counter-current relative to the flue gases, except for the fact that some of the water is vaporized in the tubes mounted on the walls of the fireplace. The pressure on the feed water/steam side is normally between 60 and 75 bars (boiling point between 275 and 290 °C) and the steam is superheated to 350–400 °C. There are, however, some systems where the pressure is 100 bars and the steam is superheated to 520–540 °C.

12.4.3 Mechanisms and Chemistry

The chemistry and physics that occur in the recovery boiler are very complicated and are not, in fact, fully understood yet. The most important mechanisms are discussed in this chapter. The discussion will be based on, and start with, a drop of liquor that enters the fireplace in the recovery boiler.

12.4.3.1 The Liquor Drop

Figure 12.17, is a schematic diagram showing the path of a drop of liquor from the liquor gun down to the char bed. During this path, the drop may undergo four different phases: drying, pyrolysis, char combustion and, eventually, combustion of sodium sulphide. It is important to remember that there is a relatively large overlap between the four different operations, i.e. the pyrolysis starts before drying is completed, and so on. This is the result of the huge temperature gradient within the particles: e.g. at the end of the „drying period" the temperature in the centre of the drops can be 160–170 °C but, on the surface the temperature may be 250–350 °C, implying that pyrolysis started at the surface.

The size of the particles is of great importance: if the drops are too small, they will follow the gas flow upwards, and if they are too large, they may not even be dried before they hit the char bed. In the former case, there will be great amounts of fly ash and a big risk of the heat exchanger banks after the fireplace being. Large amounts of fly ash must also be recirculated, resulting in a low efficiency. If the drops are of such a size that they are not dry when they hit the char bed, energy for drying will be taken from the char bed. This may result in the temperature of the char bed being lowered to such a level that combustion ceases. Drop formation is, thus, very important; the size of each drop ought to be between 1 and 4 mm.

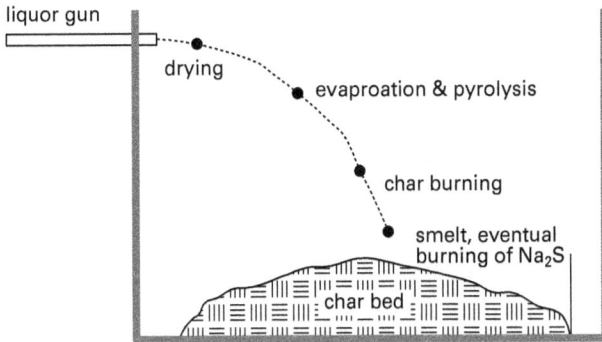

Figure 12.17. Schematic diagram of the route taken by a drop of liquor from the liquor gun to the bed.

It should be observed that the four operations all imply coupled heat and mass transfer, see *Figure 12.18*.

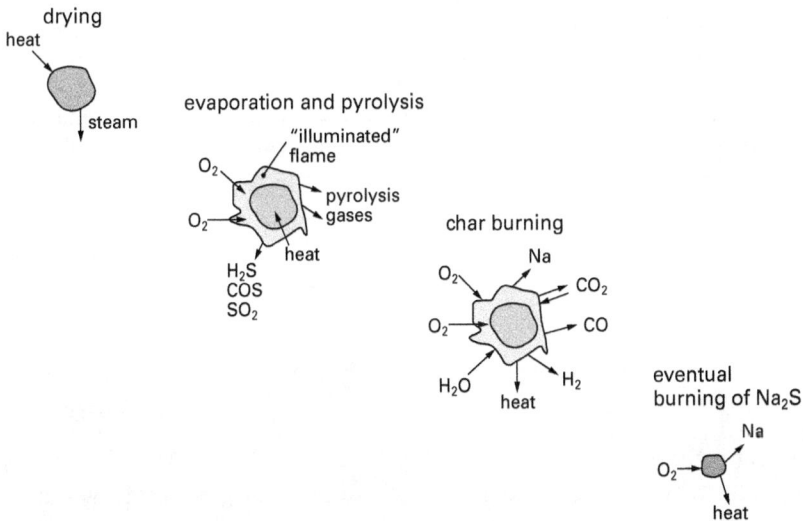

Figure12.18. Schematic diagram of the four operations.

As is indicated in the figure above, the size of the drops changes during the four operations. *Figure 12.19* shows that a drop swells during drying and pyrolysis and during char burning, it gradually decreases as the carbon is burned off. A drop usually swells to about three times its original diameter, but it can swell as much as eight times its original size. Thus, different black liquors have different swelling characteristics. The reason for this is not completely understood, but the pH, amount of inorganic material and the ratio between lignin and carbohydrates all influence the degree of swelling: a higher pH increases swelling, a higher content of inorganics decreases swelling and there seems to be a maximum in swelling if the ratio between lignin and carbohydrates is approx. 1.

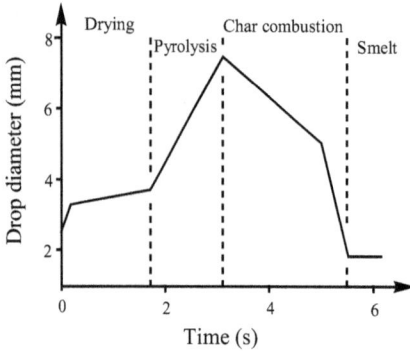

Figure 12.19. An example of swelling and shrinking during the combustion of black liquor.

It has also been observed that the sodium content in the drops decreases 5–20 %during the first few seconds of combustion. This phenomenon cannot be explained by evaporation, and it has been suggested that small micro drops are launched from the drop. There may be several reasons for micro drops being launched, but a reasonable one is that substantial pressure is built up within the drops due to evaporation and/or pyrolysis. When the pressure is leveled out, i.e. the steam/gas flows out of the drops, there will be some liquid entrained with the gas, and small micro drops are formed. An approximate model for the pressure gradient in the drop is obtained if Darcy's equation is used:

$$v = \frac{B}{\mu} \frac{dP}{dr} \qquad \text{Eq(12.14)}$$

where
v is the gas velocity based on the area perpendicular to the flow
B is the permeability
P is the pressure
R is the radius of the drop
The velocity of the gas can be calculated if the amount of gas/steam generated is known:

$$v = \frac{dm}{dt} \cdot \frac{1}{\rho_g \cdot 4 \cdot \pi \cdot r^2} \qquad \text{Eq(12.15)}$$

where ρ_g is the density of the gas at a given pressure. If v in Equation (12.14) is eliminated with Equation (12.15), and the resulting equation is integrated between 0 and ΔP and 0 and r respectively, the following equation is obtained:

$$\Delta P = \int_0^r \frac{\mu \dfrac{dm}{dt}}{B \cdot 4 \cdot \pi \cdot r^2 \cdot \rho_g}$$

Eq(12.16)

The equation above can be used to estimate the pressure difference in the drop. The permeability can be estimated using Kozeny-Carman's equation.

A summary of each of the operations between drying and the formation of a smelt drop is given below.

Drying

During the drying process heat must be supplied to the drop in order to evaporate the water within it. The heat transfer may be divided into two different steps: a. heat transfer from the surrounding to the surface of the drop (convective heat transfer and radiation) and b. heat conduction within the drop. The water vapor evaporated is first transported by means of diffusion and advective transport in the drop and later by convective mass transfer from the surface of the drop to the gas bulk. There are thus simultaneous heat and mass transfer. Although a comprehensive mathematical description is difficult, it is, of course, possible. In this case, water is evaporated and it is a well-known fact that its heat of vaporization is very high. Consequently, a large amount of heat must be transported in relation to the amount of vapor generated and transported. Furthermore, the drops are relatively small in size, which implies that heat transfer probably is the rate-limiting step. Assuming that this is true it is then possible to develop a relatively simple model. The heat transfer to a drop can be described by the following equation:

$$\frac{dQ}{dt} = \frac{A\left(T_g - \overline{T_d}\right)}{\dfrac{1}{h_{\text{eff}}} + \dfrac{1}{6 \cdot k_d}}$$

Eq(12.17)

where
Q is the transferred heat
t is the time
A is the surface area of the drop
T_g is the temperature of the gas bulk
$\overline{T_d}$ is the average temperature of the drop; may be approximated with the saturation temperature.
h_{eff} is the apparent heat transfer coefficient, where both conductive heat transfer and radiation are taken into account
d_d is the diameter of the drop
h_d is the heat conductivity of the drop

The apparent heat transfer coefficient can be calculated as:

$$h_{\text{eff}} = h + \sigma \cdot \varepsilon_{\text{eff}} \frac{\left(T_g^4 - T_s^4\right)}{\left(T_g - T_s\right)}$$

Eq(12.18)

where

h is the conductive heat transfer coefficient
σ is Stefan-Botzman's constant
ε_{eff} is the effective emissivity
T_s is the surface temperature of the drop

Since the black liquor is preheated, only a small amount of energy is needed to heat it to the boiling point if its temperature is below this. Alternatively, if the black liquor is slightly super-heated, only a very small amount of vapor is produced. If these two cases are neglected the coupling equation between mass and heat transfer can be written as:

$$Q = \Delta H_{vap} \cdot m_{steam}$$

Eq(12.19)

where

ΔH_{vap} is the heat of vaporization
m_{steam} is the amount of vapor produced.

If Equation (12.19) is differentiated with respect to time, and the resulting equation is combined with Equation (12.17), the following result is obtained:

$$\frac{dm_{steam}}{dt} = \frac{1}{\Delta H} \frac{A\left(T_g - \overline{T_d}\right)}{\dfrac{1}{h_{eff}} + \dfrac{d_d}{6 \cdot k_d}}$$

Eq(12.20)

This equation can be used to calculate the drying rate. For drops of normal size, the drying is rather rapid and ends after 1–3 seconds.

Pyrolysis

In the pyrolysis stage the organic material (lignin and carbohydrates) is broken down into components with a lower molecular weight, e.g. methane, carbon dioxide, carbon monoxide, hydrogen and water vapor. Sulfur is also partially released in the form of hydrogen sulfide, mercaptanes carbon disulfide, etc.; in fact, the largest release of sulfur is made in the pyrolysis stage. This stage has a reducing, or slightly oxidizing, environment. Should this release occur in an oxidizing environment, the gases released would burn just outside the particle with an illuminating flame. The total process is endothermic or slightly exothermic: pyrolysis is endothermic but, if external combustion is extensive (high content of oxygen in the gas bulk) the total process can be slightly exothermic. The final combustion of the pyrolysis gases occurs above the secondary air level.

Pyrolysis is normally completed after about 2 seconds; during this time, the temperature of the drop has increased from about 200 °C to above 600 °C. It has been mentioned that the volume of the drop increases dramatically during the pyrolysis stage (on average, it increases by 300 %) earlier in this chapter. The internal surface area also increases and, consequently, the inorganic material is spread out over the large internal surface area.

The mechanisms of pyrolysis are even more complicated than those of drying. In a detailed model, it is not only the rate of the transport mechanisms that must be considered; the rate of the different reactions that occur when the organic material is broken down must also be taken into account. Unfortunately, there is no detailed kinetic data of these reactions. An approximate model can, however, be formulated if a rule of thumb is used: the temperature of the particle is 80 % of that of the surroundings when pyrolysis has ended. In other words:

$$T_{end} = T_{Sur} \cdot 0.8 \qquad\qquad Eq(12.21)$$

where
T_{end} is the temperature when the pyrolysis has ended
T_{sur} is the temperature of the surrounding

Equation (12.17) may be used in this case, too, to calculate the amount of heat transferred from the bulk to the particle. In a reducing environment, the surrounding temperature is equal to the temperature of the bulk gas. In an oxidizing environment, the temperature will be higher due to the combustion of the pyrolysis gases released. In the latter case, the surrounding temperature can be calculated with the following equation:

$$T_{g,f} = T_g + \frac{x_{O_2^{fluegas}} \cdot \Delta H_{comb.}}{x_{O_2^{air}} \cdot c_{p,g}} \qquad\qquad Eq(12.22)$$

where
$T_{g,f}$ is the temperature of the combustion zone
$c_{p,g}$ is the heat capacity of the gas
$x_{O_2^{fluegases}}$ is the molar fraction of oxygen in the flue gases
$x_{O_2^{air}}$ is the molar fraction of oxygen in the air
ΔH_{comb} is the heat of combustion

The total amount of energy necessary for pyrolysis may be computed by the following equation:

$$Q = m_{sbl} \cdot c_{p,sbl} \cdot (T_{End} - T_{Dry}) + \Delta H_{pyrolysis} \cdot (m_{sbl} - m_{Char}) \qquad\qquad Eq(12.23)$$

where
m_{sbl} is the mass of the dry black liquor
m_{Char} is the mass of the char
$c_{p,sbl}$ is the heat capacity of the dry black liquor
T_{Dry} is the temperature of the dry drop
$\Delta H_{pyrolysis}$ is the enthalpy of the pyrolysis process

Char Combustion

The reactions that take place during char combustion are mostly heterogeneous. This implies that combustion occurs without a flame. The overall process is exothermic, but both endothermic and exothermic reactions are involved. Char burning is not as rapid as the drying and pyrolysis stages and far from all of the carbon is combusted whilst the drop is falling down to the char bed. The fraction of carbon combusted before the drop hits the char bed depends on i.a. the size of the drop, temperature and oxygen content. The most important char burning reactions are:

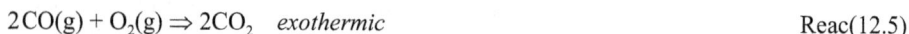

$$2C(s) + O_2(g) \Rightarrow 2CO(g) \quad \textit{exothermic} \qquad\qquad Reac(12.1)$$

$$C(s) + H_2O(g) \Rightarrow CO(g) + H_2(g) \quad \textit{endothermic} \qquad\qquad Reac(12.2)$$

$$C(s) + CO_2(g) \Rightarrow 2CO(g) \quad \textit{endothermic} \qquad\qquad Reac(12.3)$$

$$CO(g) + H_2O(g) \Leftrightarrow CO_2(g) + H_2(g) \quad \textit{exothermic} \qquad\qquad Reac(12.4)$$

$$2CO(g) + O_2(g) \Rightarrow 2CO_2 \quad \textit{exothermic} \qquad\qquad Reac(12.5)$$

The char burning process, when black liquor is combusted, is relatively rapid compared with, e.g. combustion of wood, as the alkali salts catalyse the reactions and the inner surface area is very large. The reactions are faster than the mass transport, which implies that the mass transfer of oxygen to the reaction sites is the normal rate-determining step. It also implies that the atmosphere inside the particles is one of reduction, so the inorganic compounds can then undergo reduction reactions with carbon:

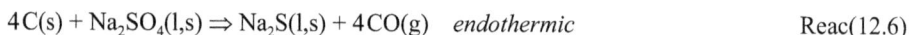

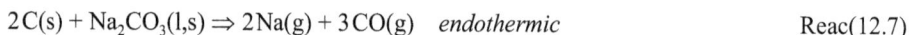

$$4C(s) + Na_2SO_4(l,s) \Rightarrow Na_2S(l,s) + 4CO(g) \quad \textit{endothermic} \qquad\qquad Reac(12.6)$$

$$2C(s) + Na_2CO_3(l,s) \Rightarrow 2Na(g) + 3CO(g) \quad \textit{endothermic} \qquad\qquad Reac(12.7)$$

The inorganic reactions are much slower then the „carbon reactions", implying that only about 10 % of the oxygen originates from reactions with sulfate and carbonate ions. It should also be observed that sodium sulfide may react with oxygen during char burning:

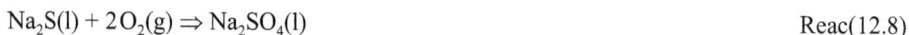

$$Na_2S(l) + 2O_2(g) \Rightarrow Na_2SO_4(l) \qquad\qquad Reac(12.8)$$

The combination of Reactions (12.6) and (12.8) results in a catalytic cycle, the so-called „sulfate/sulfide cycle", see *Figure 12.20*. This mechanism is important, since it increases the overall reaction rate during char burning if the temperature is enough high (>980 °C), which is, in fact, the case.

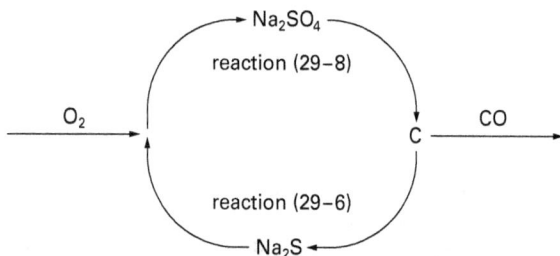

Figure 12.20. The sulfate/sulfide cycle.

The surface temperature of the particles during char burning is much higher than the temperature of the surrounding bulk gas. If it is assumed that the reactions involved are much faster than the mass transfer (which is a reasonable assumption in normal conditions), it is possible to describe the overall rate with mass and heat balances. It has been shown that the overall reaction rate is influenced strongly by the oxygen content, which is an indication that the mass transfer of oxygen from the gas bulk to the surface of the particle is the rate-limiting step. The model is complicated by the fact that the size of the particles decreases as the carbon is consumed. This type of model is not within the scope of this chapter.

Reaction with a Drop Composed of an Inorganic Smelt

It is only for the smallest black liquor drops that char burning is complete before they hit the char bed; only a small fraction of the drops are thus considered is this chapter. The main reaction is the oxidation of sodium sulfide:

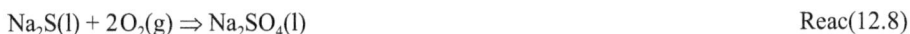

$$Na_2S(l) + 2O_2(g) \Rightarrow Na_2SO_4(l) \hspace{3cm} \text{Reac}(12.8)$$

12.4.3.2 Mechanisms/reactions in the Char Bed

Most of the char particles reaching the bottom of the furnace containing a high level of carbon, so, a char bed will be formed, *Figure 12.21.*

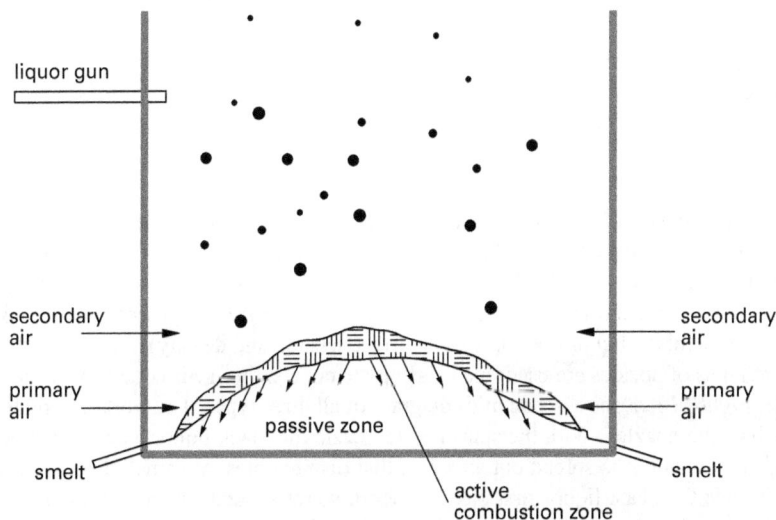

Figure 12.21. A schematic figure of the lower part of the furnace.

The char bed consists of an active combustion zone where combustion (oxidation) reactions occur. It should, however, be remembered that the reduction of sulfur also occurs in this zone. The active area is a relatively small part of the char bed, i.e., it is only a 5–10 cm thick layer on the top. A large part, thus, comprises a passive zone where no oxygen is present, the atmosphere one of reduction and no exothermic oxidation reactions occur. This means that the temperature will be much lower in this zone and the rate of the reducing reactions will also be very low. *Figure 12.22* is a summary of the most important reactions that take place in the char bed.

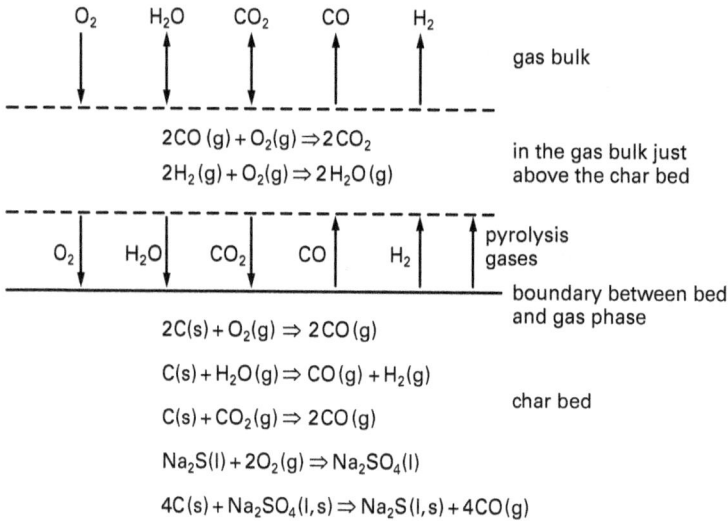

Figure 12.22. Summary of the most important reactions in the char bed, and mass transfer between the bed and the gas bulk.

12.4.4 Equipment

12.4.4.1 Nozzles

It has been shown that the size of the black liquor drops is a very important parameter as far as the combustion process is concerned. It is therefore hardly surprising that the formation of the drops is of the outmost importance in the design and control of the recovery boiler. Drop formation is influenced by several parameters: design of the equipment (nozzles), process conditions and the properties of the black liquor (i.e. viscosity, surface tension and density).

Three different types of nozzles are used: the splash plate nozzle, the „swirlcone" nozzle and the Vtype nozzle, *Figure 12.23* show schematics diagram of all three types. The working principles vary for each of the nozzles. With the splash plate nozzle the black liquor leaves the pipe and hits the angled plate and it is spread out as a disc that disintegrates into small drops. With the „swirlcone" nozzle the black liquor first passes a short, spiral-shaped, channel. It obtains a rotational movement when it leaves the nozzle via a circular orifice, forming a cone with a very thin wall. The cone disintegrates a short distance from the nozzle and drops of black liquor are formed. In the V-type nozzle black liquor is forced through an oval-shaped orifice, spreading it out as a liquor disc, which disintegrates and forms drops.

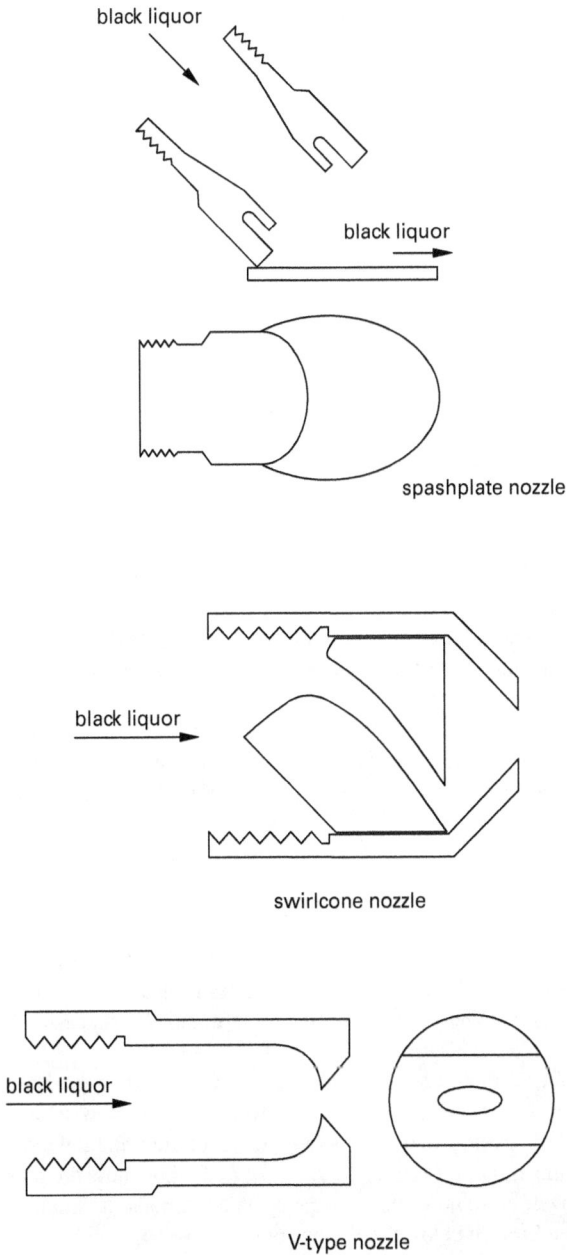

black liquor

black liquor

spashplate nozzle

black liquor

swirlcone nozzle

black liquor

V-type nozzle

Figure 12.23. Examples of the three nozzles used for the formation of drops.

The basic mechanism for the formation of drops is the same for the three different nozzles: a thin, liquid sheet is formed which is transported from the nozzle at high speed. The sheet expands as it is transported away from the nozzle and, finally, when the surface tension forces cannot prevent it from disintegrating, small drops are formed, see *Figure 12.24*.

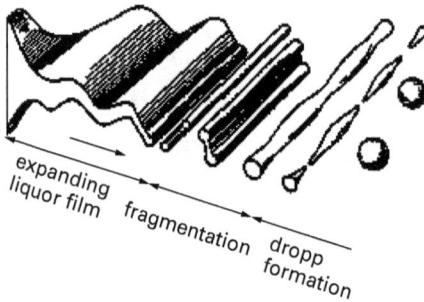

Figure 12.24. Schematic description of the drop formation mechanism.

There are a number of empirical and semi-empirical models describing the diameter of the drop formed as a function of various variables. These models are, however, too specific to be discussed in any detail in this chapter. It has, however, been shown that the most important variables are: the size of the nozzle, the velocity of the liquor and the density, viscosity and surface tension of the liquor. The size of the nozzle and the velocity of the liquor are by far the most important variables; the properties of the liquor are therefore not that important within normal variations.

Nowadays, the nozzles are mounted on stationary liquor guns on two or four of the walls. It is, of course, important that the liquor guns are mounted in such a way that an even spreading of the black liquor is obtained. They are normally mounted 5–6 meters from the bottom of the furnace.

12.4.4.2 Air System

Air supplies the boiler with oxygen, of course, but it also determines the flow pattern in the boiler as well as being the carrier of energy from the furnace to the batteries of heat transfer areas (superheater, boiler bank and economizer). The primary air also determines the shape of the char bed.

Air enters the furnace at three or four different levels: primary, secondary, tertiary and quaternary level. Three levels were used for many years but a fourth level is now used in modern, large boilers. The primary and secondary levels are located below the liquor guns and the tertiary and quaternary levels are located above them, *Figure 12.25*. *Table 12.5*, show how the air is distributed at the different levels as well as some important operation conditions. It should be observed that most of the air enters the furnace below the liquor guns.

Table 12.5. Distribution of air at the different levels, approximate location of these levels and the velocity and temperature of the air.

Air level	Distribution of air (%)	Location (m above the bottom)	Velocity in the nozzles (m/s)	Temperature (°C)
Primary	29–50	0.6–1	29–50	125–200
Secondary	20–50	1.8–3	40–70	125–200
Tertiary	15–40	7–12	50–80	20–300
Quaternary	10–20	(15–30)	60–90	20–200

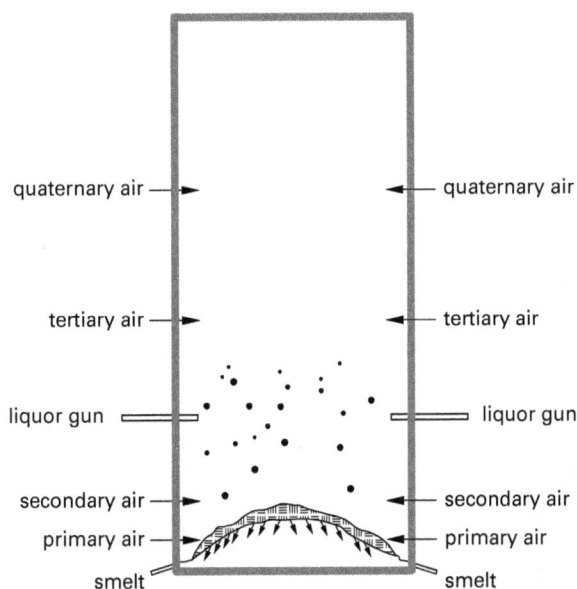

Figure 12.25. Location of the air levels.

The primary air enters the furnace via a large number of nozzles mounted around the furnace; the nozzles are normally 20–40 cm apart. This makes it possible to control the shape and size of the char bed. *Figure 12.26* shows some examples of the results of different air flow rates.

If the flow rate of the primary air is too high, channeling may occur: i.e. the air flows in a narrow passage in the middle of the furnace. The volumes closest to the walls are recirculation zones, see *Figure 12.27*. Since the air flows in a narrow passage, the velocity is much greater than the case where all of the surface area is used; this implies that the residence time is much shorter when channeling occurs, and that material not completely combusted may be transported to the top of the furnace. Consequently, the combustion efficiency decreases and the heat exchanger in the convection part is exposed to much more dust which, in turn, increases the risk of fouling in the heat exchanger areas. The excess of air must also be increased in order to combust the gases completely, which decreases the thermal efficiency of the process.

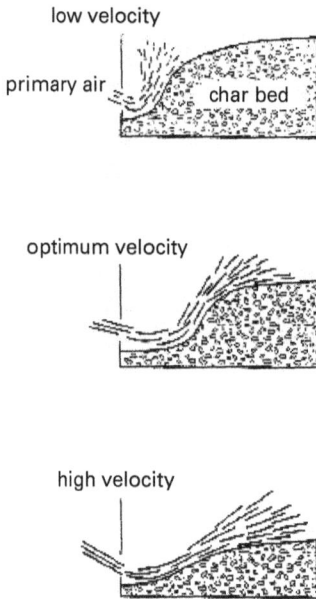

Figure 12.26. Examples of the results obtained when the primary air has different airflow rates.

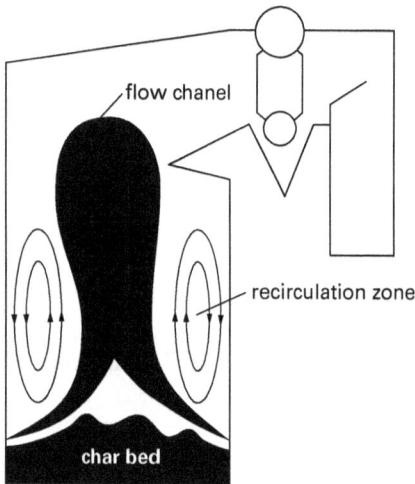

Figure 12.27. Channeling caused by a high flow rate of the primary air.

The secondary air has a big influence of the flow pattern in the furnace, especially if a plug-flow like pattern is to be obtained or not. There are a number of different designs regarding the supply of secondary air. *Figure 12.28* shows two different types: one (a) has the supply of air positioned on two opposite walls, with every second nozzle having a high flow rate and the re-minder a low flow rate. This design decreases the risk of channelling, as the total flow rate of secondary air is sufficiently high. In the second design (b), air is supplied from all four walls in such a way that a rotating flow is achieved: channelling may be prevented here, too, if the total flow of secondary air is high enough.

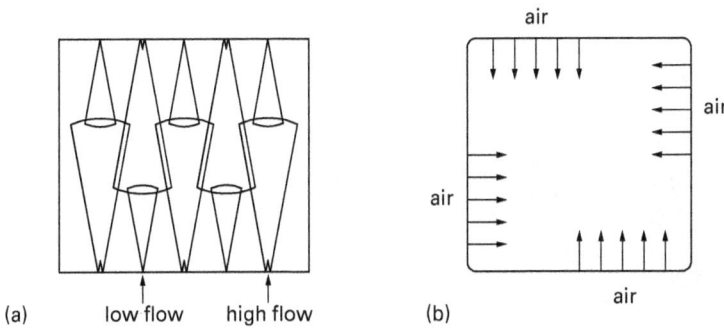

Figure 12.28. Two types of designs for the secondary air supply.

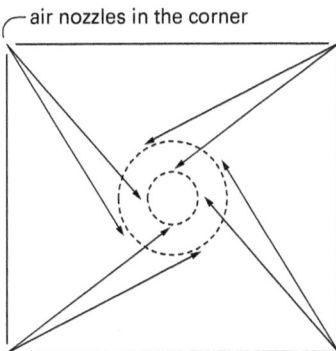

Figure 12.29. An alternative way of mounting the nozzles at the tertiary air level.

The air supply at the tertiary level can be designed in the same way as in *Figure 12.28b*. An alternative design is, however, shown in *Figure 12.29*, where the tertiary air enters the furnace via nozzles mounted in the corners of the furnace. This arrangement may provide more even temperature conditions for the super-heaters.

A fourth air level is common in new large boilers, often being introduced when an old boiler is retrofitted. If the dry content of the black liquor is high (> 75 %) there are some advantages to be gained in having a fourth air level. When the dry content of the black liquor is high, it not only becomes much easier to maintain a high temperature in the char bed, but also becomes possible to redistribute some of the air from the primary/secondary levels to the quaternary level. This reduces the formation of NO_x, and the emission of alkali from the char bed decreases.

12.4.4.3 Heat Exchange areas

The heat exchange areas can be divided into areas in the furnace and convections areas after the furnace. About 40 % of the total amount of heat is transferred in the furnace and the remaining 60 % via the convection areas after the furnace. *Figure 12.30* shows a modern, single-dome boiler.

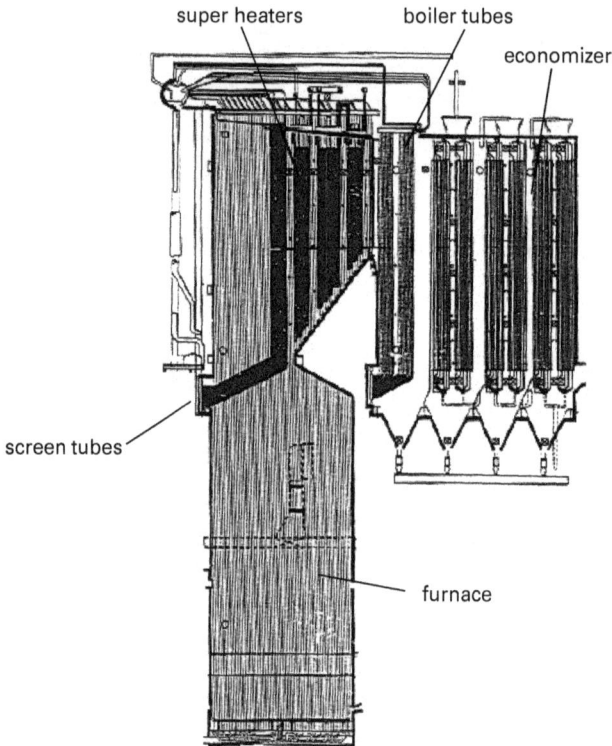

Figure 12.30. Schematic diagram of the heat transfer areas in a recovery boiler.

The heat transfer area in a furnace consists of vertical pipes with a diameter of 0.06–0.075 m. These pipes are welded together to form a gas-tight wall, see *Figure 12.31a*. In the lower part of the furnace, the pipes are often of the compound type, which means that the outer part of the tube consists of stainless steel and the inner part of carbon steel, *Figure 12.31b*. Stainless steel resists the very corrosive environment much better than the carbon steel. In the upper part of the furnace, the tubes are made of carbon steel.

After the furnace passes the flue gases first passes the „screen tubes". The temperature of the flue gases is adjusted (lowered) somewhat before they enter the super-heater tubes: This adjustment is made in order to meet the specification for the temperature of the exit superheated steam. The screen tubes also act as a radiation shield for the super-heater tubes, and are a part of the evaporation heat exchangers. The screen tubes in modern recovery boilers are often replaced by a large „arch", which acts as a radiation shield for the super-heater tubes.

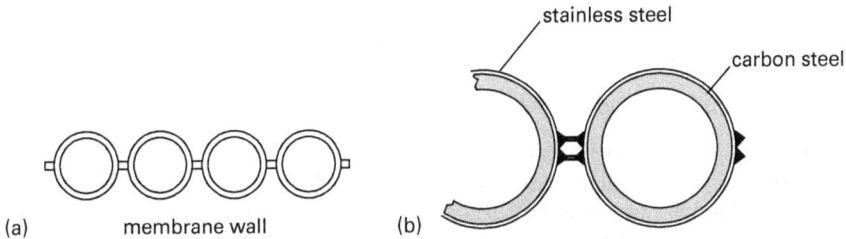

Figure 12.31. Tubes in the furnace.

The saturated steam is superheated in the super-heater tubes. Heat transfer to saturated/superheated steam is the only way by which the super-heater tubes are cooled. Since the heat transfer to steam/gas is much lower than the heat transfer to a liquid or when boiling a liquid, the temperature of the tube walls will be very high, and may cause extensive corrosion. This is why radiation shields are necessary to protect the super-heater tubes, and why the flue gases and steam are arranged in cocurrent mode. These two measures ensures that the temperature of the tube walls is lowered.

The boiler tube bank may be of the one-dome type (modern) or the two-dome type (elder), see *Figure 12.32*. The lower dome in the two-dome type is simply a water buffer that ensures water is always present in the tube bank. In the one-dome type the system is controlled much better, and there is simply no need for a lower dome. In the upper dome, water and vapor are separated. The vapor is transported to the super-heater tubes and the water is recirculated to the boiler tubes. *Figure 12.32* shows that the gas flow is different in the one-dome and two-dome types.

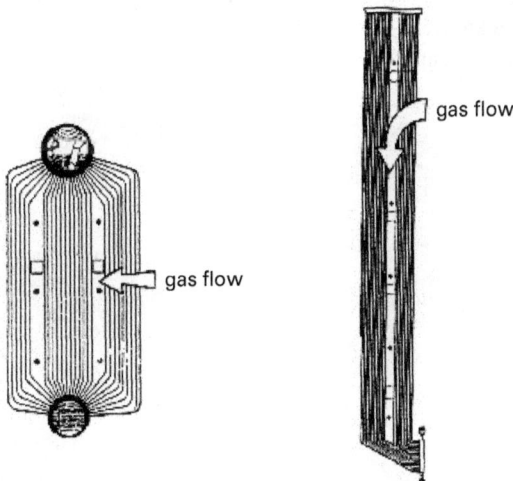

Figure 12.32. Boiler tube banks of the one and two dome types.

The water entering the boiler is pre-heated to a temperature just below the saturation temperature in the economizer, and the temperature of the flue gases is lowered to the exit temperature. The exit temperature should be as low as possible in order to reach as high a level of efficiency as possible. The temperature must not, however, be below 150 °C in order to avoid condensation of low pH condensate, which may cause severe corrosion in the electrostatic precipitation unit.

Soot-blowing of the Heat Exchanger Areas

Naturally, the heat exchange areas will, in time, be covered with dust, and heat transfer will deteriorate considerably. The surfaces must therefore be cleaned at regular intervals. This can be carried out by soot-blowing system or shot cleaning. The steam soot-blowing system, in principle consists of a lance equipped with steam nozzles. The lance is moved into the tube banks and, when the steam is turned on, it blows the detachable dust off the tubes. This type of system is used in the super-heaters and boiler bank. In the economizer, both steam soot-blowing and shot cleaning may be used. The latter method uses metal spheres (made of aluminum or steel) that is falls from the top of the apparatus in question. When the spheres hit the tubes, the dust is removed. The spheres and the dust are collected in the lower part and are separated from each other. The dust is recirculated to the process via strong black liquor and the spheres are used once again to remove more dust.

12.4.5 The Electrostatic Precipitation System

The dust in the flue gases is separated off in the electrostatic precipitation system (EPS). This dust consists mainly of sodium carbonate and sulfate. Since the separated dust is recirculated to the strong black liquor, the recovery of both sodium and sulfur is maximized, whilst the discharge of sulfur is minimized.

Figure 12.33, shows a schematic diagram of an EPS. It consists, in principles, of a charge electrode that charge the dust particles. The flue gases and the charged particles are forced into an electric field where the particles move towards collector electrodes (plates with a charge opposite to that of the particles). These plates are cleaned at regular intervals, when they are shaken so that the dust falls to the bottom and is transported out of the EPS.

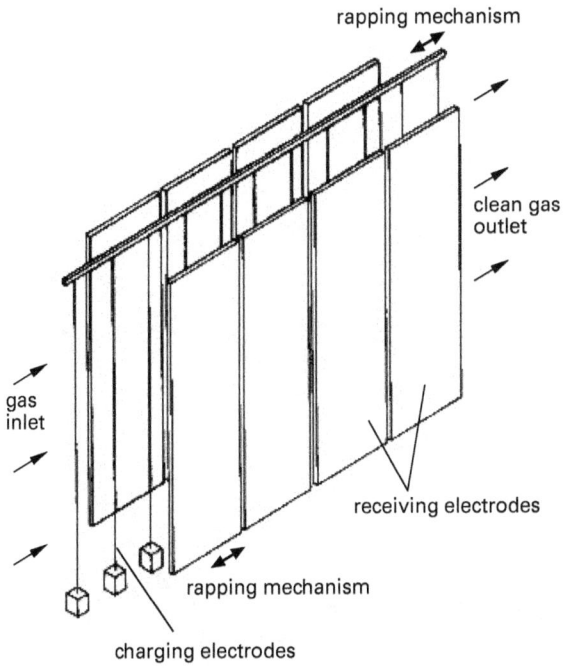

Figure 12.33. Schematic diagram of an electrostatic precipitation system (EPS).

13 The Recovery of Cooking Chemicals: the White Liquor Preparation Plant

Hans Theliander
Chalmers University of Technology

13.1 Introduction

The white liquor preparation plant comprises the part of the recovery cycle where the smelt from the recovery boiler is converted into white liquor that can be used in the pulp digester; *Figure 13.1* is a schematic diagram of this process.

The smelt from the recovery boiler is transported to the so-called „smelt dissolver" where it is dissolved in weak white wash to form green liquor. Weak white wash is, in principle, diluted white liquor; it is the exit liquor stream from the lime mud wash. The main cat ions in green liquor are sodium (~90 %) and potassium (~10 %), and the main anions are carbonate, sulphide and hydroxide ions; small amounts of chloride, sulphate and tiosulphate are also present. In addition, small amounts of solid material, known as „green liquor dreg" (also called „green liquor sludge"), is also present. This dreg/sludge is separated by filtration or sedimentation before the green liquor is transported, via a green liquor storage tank, to the causticising plant. Here burned lime mud (CaO) is added to the green liquor, and the calcium oxide is slaked according to *Reac (13.1)*. As the first calcium hydroxide is formed the causticising reaction starts *Reac (13.2)*, where the calcium hydroxide reacts with the carbonate ions forming calcium carbonate (lime mud) and hydroxide ions. Thus, the white liquor is formed in this step. The hydroxide and the sulphide ions are the active ions in white liquor, which also contains small amounts of carbonate, chloride, sulphate and tiosulphate ions. The cat ions are the same as in green liquor, i.e. sodium and potassium.

$$CaCO_3 \Leftrightarrow CaO + CO_2$$

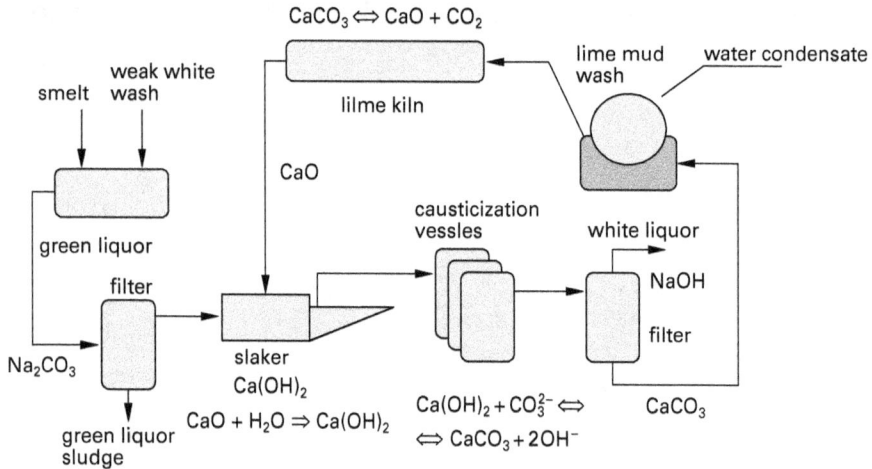

Figure 13.1. A schematic diagram of the process in the white liquor preparation plant.

$$CaO(s) + H_2O \Rightarrow Ca(OH)_2(s) \hspace{4cm} \text{Reac (13.1)}$$

$$Ca(OH)_2(s) + CO_3^{2-}(aq) \Leftrightarrow 2OH^-(aq) + CaCO_3(s) \hspace{2cm} \text{Reac (13.2)}$$

The white liquor and lime mud are separated after the causticizing vessels, normally by filtration, although thickening can also be used. The white liquor is now ready to be used in the pulp digester and is transported to a storage vessel. The separated lime mud is regenerated into new reburned lime mud. However, since the lime mud contains some white liquor, the first step is to separate the remaining white liquor from the lime mud by washing with either water or condensate. The filtrate from the washing step known as weak white wash and is transported to the smelt dissolver. The washed lime mud is transported to the lime mud reburning system where it is dried and heated to above 850 °C; the calcium carbonate decomposes into calcium oxide and carbon dioxide according to *Reaction (13.3)*.

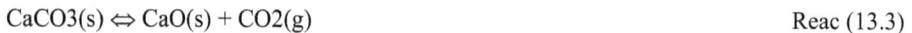

$$CaCO3(s) \Leftrightarrow CaO(s) + CO2(g) \hspace{4cm} \text{Reac (13.3)}$$

Slaking/causticizing, lime mud wash and lime mud reburning is often called the „lime cycle". It can be noted that the calcium participates in three different forms: calcium oxide, calcium hydroxide and calcium carbonate. All three forms are solids but have different crystal structures. *Table 13.1* shows the most important data for the three calcium compounds.

The densities of the industrially reburned lime mud and the lime mud are somewhat lower than the values of the pure compounds given in *Table 13.1*. The density of the reburned lime mud is about 3200 kg/m³, and that of lime mud is normally between 2600 and 2650 kg/m³.

It is also important to bear in mind that *Reactions 13.1* to *13.3* are solid phase reactions and that, consequently, the properties of the solid particles are important for the kinetics of the reactions.

There is also a fourth reaction that must be considered, i.e. the hydrolysis of the sulphide ion:

$$S^{2-} + H_2O \Leftrightarrow HS^- + OH^-$$ Reac (13.4)

This reaction is driven strongly towards the right hand-side at the conditions in green and white liquor. It is, thus, hydrogen sulphide ions, and not sulphide ions, that are present in green and white liquor.

Table 13.1. Some properties of the calcium compounds in the lime cycle.

Name	Crystal structure	Density [kg/m³]	Molar volume [m³/kmol]
Calcium oxide	Cubic 3D structure	~3320	0.017
Calcium hydroxide	Hexagonal, 2D sheets joined together with hydrogen bond⁸.	~2200	0.033
Calcium carbonate	Trigonal 3D structure	~2710	0.037

13.2 The Green Liquor System

The green liquor system consists of the smelt dissolver and a separation system for the green liquor dregs, see *Figure 13.2*. The salt smelt is transported from the recovery boiler to the smelt dissolver via the „smelt spouts". The salt smelt is dissolved in weak white wash in the smelt dissolver and green liquor is formed. There is a small amount of solid contaminates (green liquor dregs) in the green liquor that must be removed. Today, filtration is normally used for separating green liquor dregs, but there are still a large number of mills that use clarifiers for this separation. Some types of primary filters may require a secondary filter for the final dewatering of green liquor dregs. The clean green liquor is transported, via storage tanks, to the causticising plant.

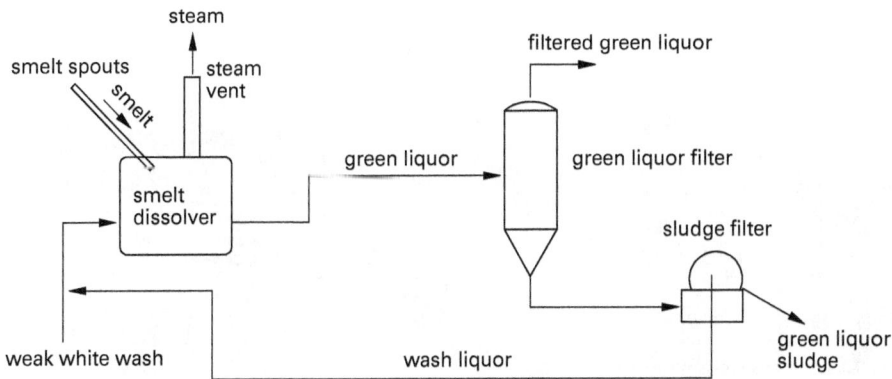

Figure 13.2. Schematic diagram of the green liquor system.

13.2.1 The Chemistry of Green Liquor and its Dregs

Green liquor, as mentioned earlier, is a salt solution. The main ions present are sodium, carbonate, hydrogen sulphide and hydroxide ions, although potassium, sulphate, tiosulphate and chloride ions are also present. About 90 % of the cat ions are sodium and the remaining part is potassium. In most cases, the concentrations of the various anions are presented as sodium salts (e.g. grams of NaOH per kg solution) and it is understood that the amount of potassium is included in this number. This is also made in this chapter. The concentrations are often expressed as g/l. The drawback with this unit is that the volume, and therefore also the concentration, changes with temperature. A more suitable unit for engineers is g/kg, this unit will be used in this chapter. A rule of thumb is that the units based on mass are 15–20 % lower than the units based on volume at room temperature. Table 13.2 shows approximate concentrations of the main salts in „normal" green liquor.

Table 13.2. Approximate concentrations of the main salts in „normal" green liquor.

Salt/ion	Na	K	Na_2CO_3	Na_2S	NaOH	Na_2SO_4	$Na_2S_2O_3$	NaCl
Conc. (g/kg)	70–95	5–15	100–140	30–60	2–25	1–15	1–10	1–10

The green liquor dregs is a mixture of organic and inorganic compounds. This mixture is the most important exit stream („kidney") for a number of „non process elements" (NPE), e.g. magnesium, manganese, iron, aluminium and silicon. The green liquor dregs, however, is compsed of mainly calcium, the content of organic material is normally very low. Table 13.3 shows approximate concentrations of some of the compounds present in green liquor dregs.

Table 13.3. Typical concentrations of some common elements present in the green liquor dregs.

Element	C	Ca	Mg	Mn	Al	Si	Fe	P	Ti
Conc. (mass %)	1–15	10–35	2–15	1–5	0.5–3	0.3–3	0.3–2	0.05–0.1	0–0.05

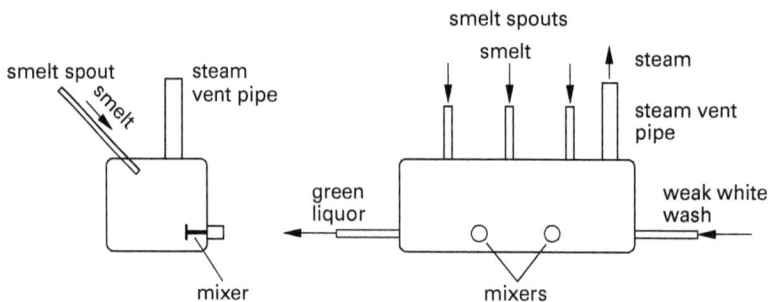

Figure 13.3. The smelt dissolver from two perspectives.

The elements are normally in the form of hydroxides or carbonates, although double salts are not unusual. One double salt that has been discussed and investigated in recent years is hydrotalcite, which is a double salt between magnesium and aluminium ($MgAl_2(OH)_6(CO_3) \cdot H_2O$). This salt is interesting in that it can be used to control the aluminium concentration in the green liquor simply by adding magnesium to the green liquor. The magnesium should not be added in excess since magnesium hydroxide may be formed, which is very difficult to separate (high filtration resistance/low sedimentation velocity). However, if the green liquor sludge has a high concentration of magnesium, aluminium ions may be added in order to improve the separation properties of the green liquor sludge. It is, of course, obvious that there ought to be an optimum ratio between the concentrations of aluminium and magnesium in green liquor dregs. This ratio, however, has not yet been established.

13.2.2 The Smelt Dissolver

Even if very little attention is paid to the smelt dissolver, it is still one of the most important process steps in the white liquor preparation plant: the total ion concentration (the „strength") of the white liquor is controlled in this part of the process. The smelt dissolver is, in principle, a mixing vessel; it is normally rectangular in shape, although it can also be cylindrical, see *Figure 13.3*. The smelt is transported from the recovery boiler to the smelt dissolver via a number of smelt spouts, and weak white wash is pumped into the smelt dissolver. As it is difficult to control the flow of smelt the total concentration of ions must be controlled via the inlet flow of weak white wash. The normal control strategy is to measure the density of the green liquor (which is a measure of the total concentration of ions), which is then used to regulate the flow of weak white wash to the smelt dissolver (i.e. a feedback control system).

In rectangular vessels, the mixers are often mounted from the side, and only occasionally from above. However, investigations have shown that it is difficult to obtain good mixing if side-mounted mixers are used (although it is possible if the energy input is very high), and especially so if the tank is large. It is a good idea to investigate the mixing conditions in the tank if problems relating to the regulation of concentration are experienced.

13.2.3 The Separation of Green Liquor Dregs

Green liquor sludge can be separated from the green liquor either by clarification (sedimentation) or filtration. In earlier systems, a clarification system was used as the primary separation device, and the final dewatering of the sludge was made on a rotary vacuum filter. A schematic diagram of a green liquor clarifier is shown in *Figure 13.4*.

An interesting observation that can be made is that no sedimentation characteristics of green liquor sludge can be found in the literature.

The use of filtration for primary separation has several advantages: the separation efficiency is higher, and the efficiency is not influenced by variations in the density of the green liquor. Four different types of filtration devices are currently in use for the primary filtration of green liquor: tube filter, cassette filter, disc filter and cross-flow filter. They are all pressurized filters, i.e. the pressure on the suspension is above atmospheric pressure. The tube and cassette filters are, in principle, the same type of filter, the filter element being mounted inside a pressure vessel. The difference is rather in the shape of the filter elements: the tube filter has tube-shaped filter elements whereas the cassette filter has leaf-shaped elements. *Figure 13.5* is an example of a tube filter.

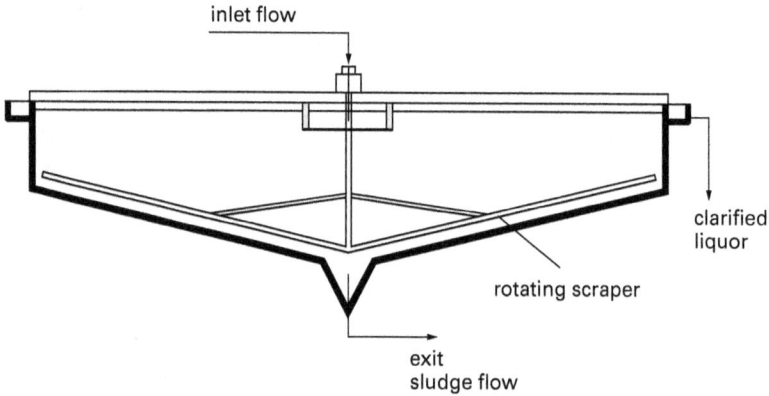

Figure 13.4. A schematic diagram of a green liquor clarifier.

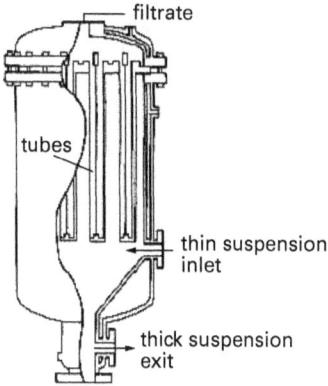

Figure 13.5. An example of a tube filter.

The tube and cassette filters are both batch filters, which imply that when the filter cake reaches a certain thickness (i.e. a certain maximum pressure difference is reached), the inlet flow is closed and a very short period of back flush is applied. This back flush loosens the filter cake from the filter element, which then will sediment to the bottom of the vessel. After a short period of homogenisation, the thick suspension is pumped out from the filter and the filtration cycle starts all over again.

The disc filter is a continuous filter, the basic design of such a filter is shown in *Figure 13.6*. The filter consists of a number of discs that are mounted on a rotating axis. The discs are partly (approximately 50 %) lowered into a vessel filled with the suspension. The discs are normally made in sections, with a filter cloth is mounted on each section. When a section is submerged in the suspension a filter cake is formed; when the section is lifted above the suspension (due to rotation), the cake is dewatered when the liquid is displaced by air/N_2. The filter cake is removed before the section is submerged back into the suspension. It should be remembered that green liquor dregs have a very high filtration resistance. Filterability is improved by the formation of a

„pre-coat" layer of lime mud closest to the filter cloth. Lime mud is much easier to filter, as the flow resistance through a cake of lime mud is much lower than that of a cake of green liquor dregs. When the outmost layer of the filter cake, which consists of green liquor dregs, is removed, a thin slice of the lime mud pre-coat is also removed. This implies that the pre-coat layer of lime mud must be renewed after a certain period of time.

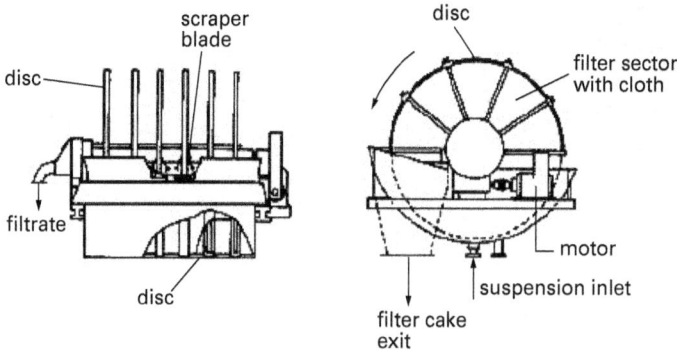

Figure 13.6. Schematic diagram of a rotary filter.

The formation of a filter cake is avoided, on the other hand, in the cross-flow filter, also known as „X-filter". This is achieved by letting the suspension flow downwards through vertical mounted filter cloths, see *Figure 13.7.*

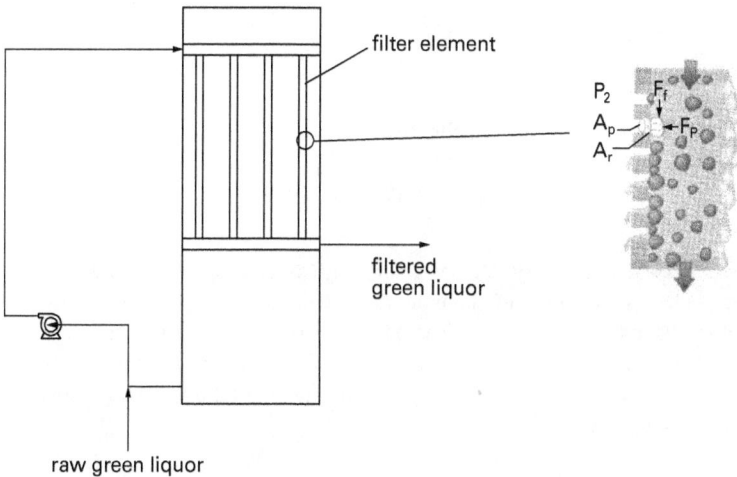

Figure 13.7. Schematic diagram of a so-called „X-filter".

The X-filter is a batch filter and the green liquor suspension must be recirculated a number of times before the suspension has reached a solid concentration high enough for final dewatering on a rotary vacuum drum filter.

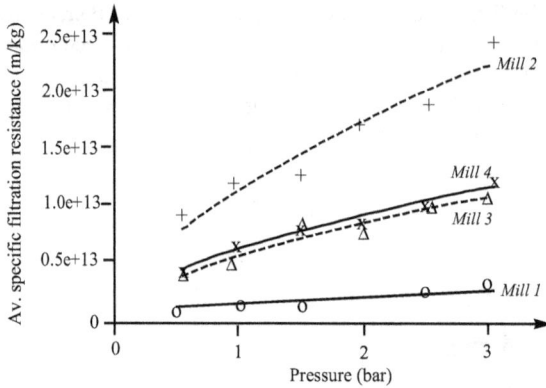

Figure 13.8. Average specific filtration resistance as a function of pressure for four different pulp mills.

There are only a few investigations available describing the filtration characteristics of green liquor sludge. It has, however, been shown that the filtration properties of green liquor dregs vary greatly between different pulp mills, primarily due to the chemical composition of the green liquor dregs. The average specific filtration resistance (αav) versus pressure for dregs from four different mills is shown in *Figure 13.8*. The average specific filtration resistance is defined according to Equation (13.1).

$$\frac{dV}{dt} = \frac{A^2 \cdot \Delta P}{\mu \left(c \cdot \alpha_{av} \cdot V + A \cdot R_m \right)} \qquad \text{Eq(13.1)}$$

where
V is the filtrate volume at the time, t.
A is the filtration area.
ΔP is the total pressure difference across the filter cake and cloth.
μ is the dynamic viscosity.
C is the ratio between the dry weight of the filter cake and the filtrate volume.
R_m is the resistance of the filter cloth.

It can be concluded that green liquor sludge has a very high average specific filtration resistance, and that it is dependent on the pressure applied. A large difference may also be observed between different pulp mills: e.g. the solid dregs from pulp mill 1 had a very high content of calcium. Furthermore, the ratio between the concentration of aluminium and magnesium was highest in pulp mill 2. It has been shown in several investigations that this ratio has a big impact on the filtration properties of green liquor sludge.

13.3 Slaking and Causticising

In the causticising department the carbonate ions in the green liquor are replaced by hydroxide ions, and white liquor is formed. This department consists of a slaker and a number of causticising vessels, see *Figure 13.9*.

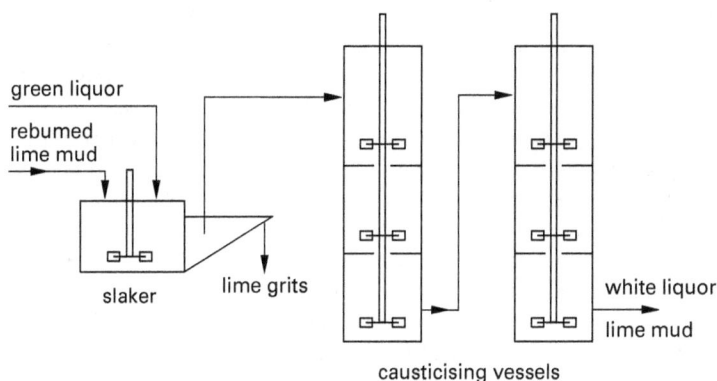

Figure 13.9. A flow sheet of the causticising department.

Green liquor and reburned lime mud (active component: CaO) are added to the slaking vessel (which is basically a mixing vessel) and the lime is slaked according to *Reaction (13.5)*. It is important to remember that as the first calcium hydroxide is formed, the causticising reaction (i.e. *Reaction 13.6*) starts. The causticising reaction is completed in the causticising vessels that follow.

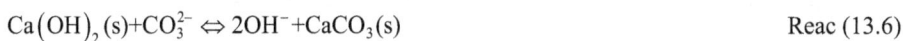

$$CaO(s) + H_2O \Rightarrow Ca(OH)_2(s) \qquad\qquad \text{Reac (13.5)}$$

$$Ca(OH)_2(s) + CO_3^{2-} \Leftrightarrow 2OH^- + CaCO_3(s) \qquad\qquad \text{Reac (13.6)}$$

The composition of the white liquor may vary slightly between different pulp mills; it is in principal the sulphidity (i.e. concentration of hydrogen sulphide) that varies. *Table 13.4* gives an approximate composition of white liquor.

Table 13.4. Approximate composition of white liquor.

Salt/ion	Na	K	NaOH	Na$_2$S	Na$_2$CO$_3$	Na$_2$SO$_4$	Na$_2$S$_2$O$_3$	NaCl
Conc. [g/kg]	70–95	5–15	75–100	30–60	20–45	1–15	1–10	1–10

The rather high concentration of sodium carbonate is due to the equilibrium of the causticising reaction.

13.3.1 Slaking

The slaking reaction is, more or less, completely driven towards calcium hydroxide, and is strongly exothermic; the heat of reaction is 1160 kJ/kg CaO. The reaction is very fast and is completed within 10 minutes and often just in a couple of minutes at a temperature of 100°C. Since calcium hydroxide has a much larger molar volume than calcium oxide (0.033 m³/kmol and 0.017 m³/mol respectively), the solid material swells during the slaking reaction. This creates large mechanical strains in the solid material and, as a consequence, the solid material disintegrates. The average size of the particles after the slaker varies with the quality of the reburned lime mud, but is normally between 15–30 μm. When the suspension leaves the slaker, it passes a classifier; here particles that have not been sufficient disintegrated are separated from the suspension. The separated fraction is called „lime grits" and is reburned lime with a high degree of sintering. They may also contain high concentrations of non-process elements (e.g. silicon).

Two types of vessels are used for the slaking operation in the pulp industry. The most common vessel is equipped with one impeller, which creates a flow high enough to suspend the particles, see *Figure 13.10a*. This type of vessel is normally equipped either with a „high flow impeller" or a pitched blade turbine. The speed of the impeller is normally between 60 and 80 rpm and the average residence time is usually between 10 and 20 minutes. The second type of slaker has two compartments. The impeller used is different in that it resembles a scraper and rotates slowly (~10 rpm) close to the bottom, see Figure 13.10b. This type of impeller obviously creates some mixing, but large particles will not be lifted from the bottom of the compartments. The average residence time in this case is 20–30 minutes.

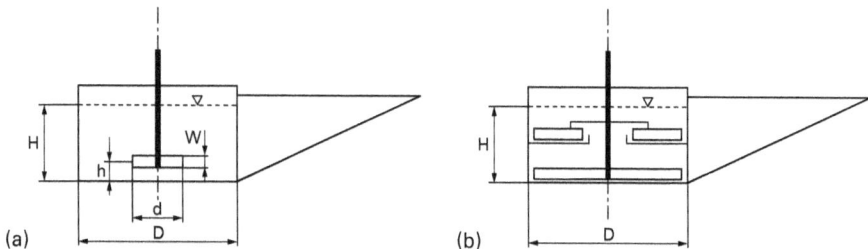

Figure 13.10. Different types of slakers.

13.3.2 Causticising

The causticising reaction occurs in a series of mixing vessels, see *Figure 13.9*. Each vessel is normally divided in 2 or 3 compartments, each compartment being equipped with an impeller. The design of the impeller and the operation conditions are normally such that each compartment can be assumed to be a perfectly mixed vessel. Consequently, the causticising plant can be described as consisting of a number of perfectly mixed reactors, which usually has between 5 and 9 steps. The type of impeller used is normally the high flow impeller type, pitched blade turbine or Ruston turbine; the impeller speed is often between 60 and 80 rpm. The total residence time may vary between 2 and 4 hours.

The causticising reaction is an equilibrium reaction that is weakly exothermic. The equilibrium is influenced by the temperature as well as the composition of the white liquor.

13.3.2.1 The Equilibrium of the Causticising Reaction

The causticising reaction can be described by an equilibrium ratio according to *Equation (13.7)*:

$$K = \frac{\left[OH^- \right]^2}{\left[CO_3^{2-} \right]} \qquad \text{Eq(13.7)}$$

The causticising reaction is one of the „old" inorganic reactions that have been of importance for a long time of period. The equilibrium has been studied for more than 100 years (~1870), but it is only for the past 10–15 years that there has been any consensus between different investigations on the concentration range that is of industrial importance (i.e. ion strength around 4 mol/kg), see *Figure 13.11*.

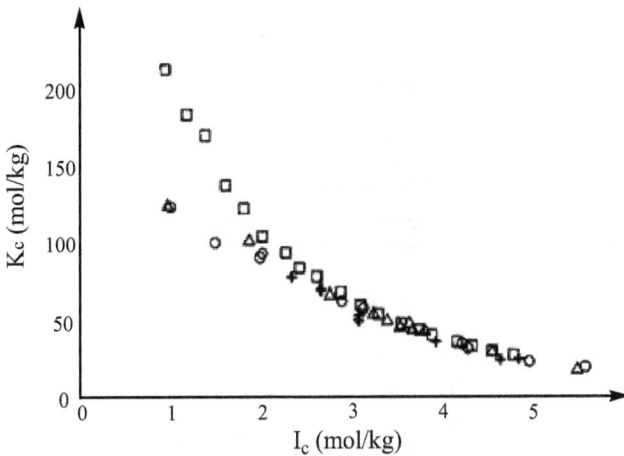

Figure 13.11. The equilibrium of the causticising reaction versus ion strength. Results from three different investigations.

There are no particularly good models available at the present time that describe the influence of both the temperature and the ion strength. *Equation (13.7)* is a simplified Pitzer-model and is, from a theoretical point of view, the most satisfactory equation that describes the influence of concentration. *Equations (13.8)* and *(13.9)* are, however, the best for describing the influence of temperature (based on the Gibbs-Helmholtz Equation).

$$\ln K_m = \ln K(T) + 2f^{\gamma} - \beta_{Na^+}^{*(0)} \cdot m_{Na^+} + \beta_{Na^+,HS^-}^{*(1)} \cdot \left(2m_{Na^+} \cdot m_{HS^-} g(I) \right) \qquad \text{Eq(13.7)}$$

where

K_m is the equilibrium ratio based on molar ratio.
$K(T)$ is the thermodynamic equilibrium constant.
F^γ, $g(I)$ are the parameters influenced by the ion strength.
$\beta_{Na^+}^{*(0)}$, $\beta_{Na^+,HS^-}^{*(1)}$ are the interaction parameters (empirical).
m_{Na^+} is the molality of the sodium ion
m_{HS^-} is the molality of the hydrogen sulphide ion.

$$\ln K_m = \left[A_m + \frac{B_m}{T} \right] + \left[C_m + \frac{D_m}{T} \right] I_m^{n_m} \qquad \text{(unit: mol/mol)} \qquad \text{Eq(13.8)}$$

$$\ln K_C = \left[A_C + \frac{B_C}{T} \right] + \left[C_C + \frac{D_C}{T} \right] I_C^{n_C} \qquad \text{(unit: mol/kg)} \qquad \text{Eq(13.9)}$$

where
K_C is the equilibrium ratio based on the unit mol/kg.
K_m is the equilibrium ratio based on molar ratio.
$A–D$ are empirical constants.
n is an empirical constant.
T is the temperature in K.
I is the ion strength.
The numerical values of the constants in *Equations (13.8)* and *(13.9)* are given in *Table 13.5*.

Table 13.5. Numerical values of the constants in Equations (13.8) and (13.9).

Unit mol/mol		Unit mol/kg solution	
Parameter	Value	Parameter	Value
n_m	0.981	n_C	1.15
A_m	−1.25	A_C	2.67
B_m	909	B_C	915
C_m	−2.82	C_C	−0.074
D_m	−5235	D_C	−82.4

Equation (13.8) describes the influence of concentration reasonably well and can, as a matter of fact, be used to calculate the heat of reaction as it is influenced by the concentration. If *Equation (13.8)* is derivated and combined with the Gibbs-Helmoltz equation the following Equation is obtained:

$$\Delta H^0 = -R\left(B_m + D_m \cdot I_m^{n_m} \right) \qquad \text{Eq(13.10)}$$

Using this equation, it can be shown that the heat of reaction is about −5 kJ/mol at the ion strengths used in pulp mills.

13.3.2.2 The Kinetics of the Causticising Reaction

The causticising reaction is considerably slower than the slaking reaction; a residence time greater that 3 hours is normally required in order to obtain a high conversion at a temperature of 100 °C. It has been proposed in a number of investigations that it is possible to describe this reaction by using a pseudo homogenous assumption. This assumption is, however, not relevant since it has been shown that the reaction occurs in the solid phase, and that it is the mass transfer that controls the overall causticising rate. It has been shown that two different models can be used to describe the kinetics: a progressive model and a shrinking-core model, see *Equations (13.11)* and *(13.12)*. The influence of temperature on the effective diffusivities can be described using an Arrhenius expression, *Equation (13.13)*. The numerical values of the coefficients A and E in the Arrhenius expression are shown in *Table 13.6*. These Equations can be used in „dynamic simulation programme" in design and optimisation studies of the causticising plant.

$$\frac{M_t}{M_\infty} = \sum_{i=1}^{31} x_i \left[1 - \frac{6}{\pi^2} \sum_{n=1}^{20} \left(\frac{1}{n^2} \right) \exp\left(-\frac{D_{eff} \cdot n^2 \cdot \pi^2 \cdot t}{R_i^2} \right) \right] \qquad \text{Eq(13.11)}$$

$$-\frac{dr_c}{dt} = \frac{D_{eff}}{(R - r_c)\frac{r_c}{R}} \qquad \text{Eq(13.12)}$$

$$D_{eff} = A \cdot \exp\left(-\frac{E}{R \cdot T} \right) \qquad \text{Eq(13.13)}$$

Table 13.6. Numerical values of the Arrhenius expression.

Parameter	Progressive model, Eq (13.11)	„Shrinking-core" model, Eq (13.12)
A [m²/s]	$2.74 \cdot 10^{-7}$	$7.45 \cdot 10^{-4}$
E [J/mol]	56000	78800

13.3.2.3 The Size and Form of Lime Mud Particles

Lime mud particles are agglomerates, i.e. small particles gathered together to form larger aggregates. The size of these aggregates can be from a couple of µm up to about 100 µm. *Figure 13.12* shows some examples of lime mud agglomerates; it is obvious that the lime mud particles may have very different structures.

The lime mud agglomerates will be influenced by the mechanical forces due to the mixing in the slaking and causticising vessels and, as a consequence, will disintegrate, see *Figure 13.13*.

It has been shown that this disintegration may be described as a grinding process, which implies that the size of the agglomerates decreases during slaking and causticising. The final agglomerate size is also influenced by the temperature and quality of the reburned lime used. There are a few models that can be used to describe disintegration: one of them is based on classical grinding theory, and assumes that this disintegration of the agglomerates corresponds to a

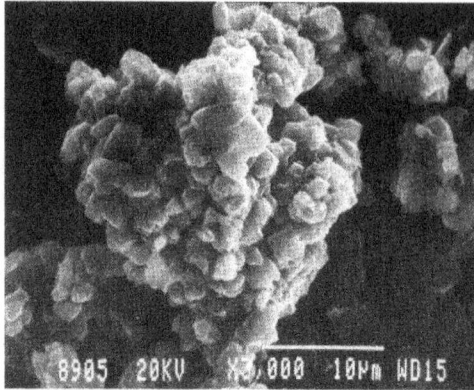

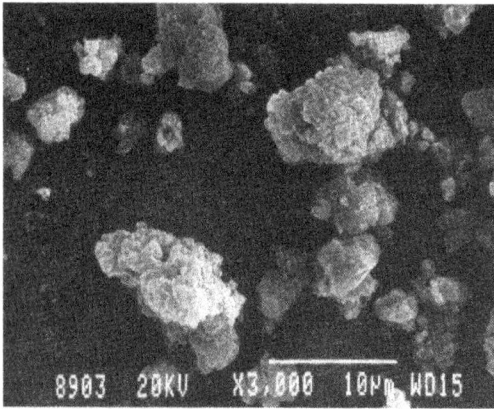

Figure 13.12. Examples of different types of lime mud agglomerates.

certain amount of energy. The energy can be related to the mechanical energy produced by the impellers. *Equations (13.14–17)* may be used to estimate the size of the lime mud agglomerates.

$$R_w = 100\left(1 - \exp\left(-\left(a \cdot d_p + b \cdot d_p^c\right)\right)\right)$$

Eq(13.14)

where

$$d_p = \frac{d}{d_{36.8}}$$

Eq(13.15)

$$E = \frac{m' \cdot N_F}{n-1} \cdot \left(\frac{1}{d_{36.8}^{n-1}} - \frac{1}{d_0^{n-1}}\right)$$

Eq(13.16)

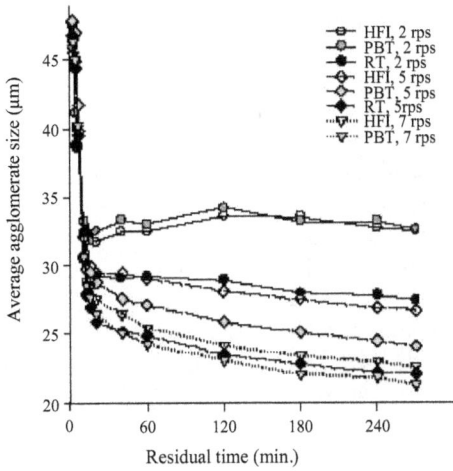

Figure 13.13. Example of the disintegration of lime mud agglomerates. Average agglomerate size versus the residence time, impeller type (HFI= high flow impeller, PBT = "Pitched-blade turbine" and RT = "Rushton turbine") and impeller speed.

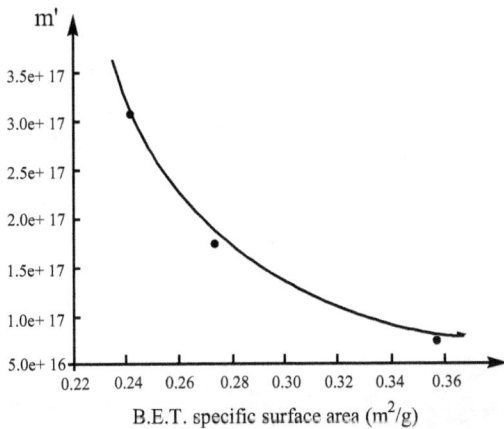

Figure 13.14. The parameter m' versus the specific surface area of the reburned lime mud.

$$N_F = \frac{Average\ value\ of\ the\ turbulent\ fluctuation}{\pi \cdot N \cdot D}$$

Eq(13.17)

where
R_w is the share of the agglomerates smaller than d in size.
a, b are empirical constants (−0.14, 0.6 and 2.1 respectively).
$d_{36.8}$ is the particle size when R_w is 36.8.
d_0 is the particle size when time is equal to zero ($E = 0$).

E is the amount of energy generated by the impellers.

m' is a parameter describing the quality of the reburned lime mud, see *Figure 13.14*.

n is a parameter describing the type of equipment parameter (can, in this case, be approximated to 11.5).

N is the speed of the impeller.

D is the diameter of the impeller.

It has been shown on a laboratory scale that the parameters in the Equation above are valid for a spectrum of different lime qualities. It should, however, be observed that the B.E.T. specific surface area is used as a measure of the quality of the reburned lime mud. The type of equipment also influences the results, and the scale-up to industrial scale is not investigated. Caution must, thus, be taken when these equations are used.

The other model is based on material balances; the probability of a particle breaking and how it is broken (e.g. small fragments being successively loosened from the agglomerate, or the agglomerate being broken into a few large particles). This model is summarized by Equations (13.18–25).

$$R_w = 1 - \sum_j^i W_j(t) \qquad\qquad\qquad \text{Eq(13.18)}$$

$$\frac{dW(t)}{dt} = \sum_{j=1}^i b_{i,j} \cdot S_{Dj} \cdot W_j(t) - S_{Di} \cdot W_i(t) \qquad\qquad\qquad \text{Eq(13.19)}$$

$$S_{Di} = a* \cdot \left(\frac{d_i}{d_{\max}} \right)^\alpha \qquad\qquad\qquad \text{Eq(13.20)}$$

$$b_{i,j} = B_{i,j} - B_{i+1,j} \qquad\qquad\qquad \text{Eq(13.21)}$$

$$B_{i,j} = g\left(f_{i,j} \right) \qquad\qquad\qquad \text{Eq(13.22)}$$

$$f_{i,j} = \frac{1}{s\sqrt{2\pi}} \exp\left(-0.5 \left(\frac{d - \bar{d}}{s} \right)^2 \right) \qquad\qquad\qquad \text{Eq(13.23)}$$

$$\bar{d} = \frac{d_j}{k_1} \qquad\qquad\qquad \text{Eq(13.24)}$$

$$s = \frac{d_j}{k_2} \qquad\qquad\qquad \text{Eq(13.25)}$$

where

$W_i(t)$ is the weight fraction of particles in the size interval i at time t.

$b_{i,j}$ is the fraction of material in fraction j broken down to fraction i.

S_D is the probability that a particle will break.

$a*, \alpha$ are empirical parameters.

B is the cumulative weight fraction of particles smaller than a certain size.

f is the number of particles in one size fraction.

$k1, k2$ are empirical parameters.

Using this model, it has been shown that lime mud particles are broken down into a few, relatively large fragments, which has been confirmed in microscope studies, see *Figure 13.15*. Since this model takes into account how the agglomerates disintegrate, it is more satisfactory than the first model described. How the parameters are influenced by the type of equipment and the quality of the reburned lime mud have not, however, been investigated. The starting point for each new case must, therefore, be the establishment of new values for the parameters.

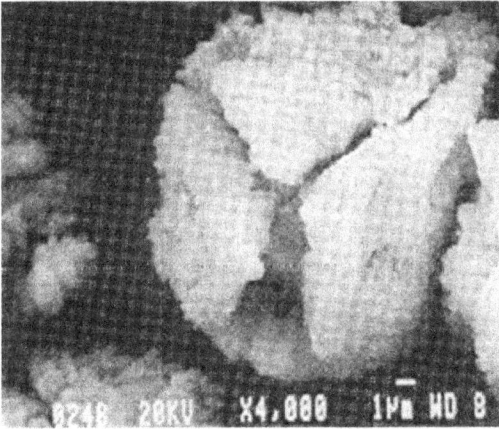

Figure 13.15. Micrograph of a lime mud agglomeration the verge of falling apart.

13.4 The Separation of Lime Mud from White Liquor

The lime mud formed is separated from white liquor after the causticising vessels; in industrial applications, filtration or thickening may be used. Thickening was very common earlier, but nowadays is filtration normally used; all of the pulp mills in Sweden use filtration technique. Particle size plays the most important role in deciding the size of the equipment in both cases.

Filtration is the only technique that will be considered in this chapter.

The filters that are normally used in industry are tube and disc filters of approximately the same type as those used for green liquor filtration, see *Figures 13.4* and *13.5*. Disc filters have the big advantage of having much greater separation efficiency than tube filter (i.e. the dry content of the filter cake is higher on the disc filter). This implies that the washing process of the lime mud that follows filtration can be simplified.

Investigations have shown that the specific filtration resistance, defined according to *Equation (13.26)*, varies significantly between different mills, see *Figure 13.16*.

$$\frac{\mathrm{d}V}{\mathrm{d}t} = \frac{A^2 \cdot \Delta P}{\mu \left(c \cdot \alpha_{av} \cdot V + A \cdot R_m \right)} \qquad \text{Eq(13.26)}$$

The figure above shows that the specific filtration resistance (α_{av}) varies with pressure, which implies that the filter cake is compressible; it also implies that the porosity of the filter

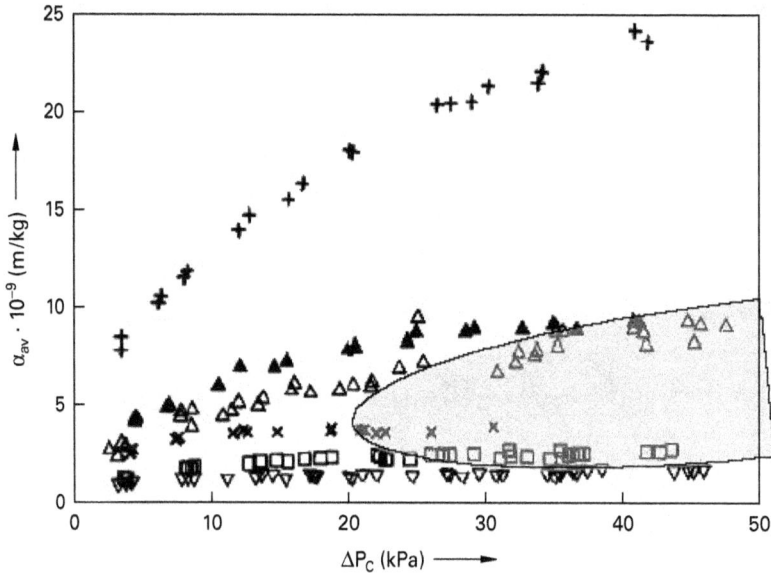

Figure 13.16. Specific filtration resistance versus pressure for 8 different mills in Sweden.

cake (ε_{av}) varies. The porosity of a lime mud filter cake, however, varies very little. *Equations (13.29)* and *(13.30)* can be used in order to describe how the specific filtration resistance and porosity vary with pressure. Typical values of the parameters in *Equations (13.29)* and *(13.30)* are shown in *Table 13.7*. It ought to be mentioned that the values presented in *Table 13.7* should not be used to calculate either the local specific filtration resistance or the local porosity.

$$\alpha = \alpha_0 \left(1 + \frac{P_s}{P_A} \right)^n \qquad\qquad \text{Eq(13.27)}$$

$$(1 - \varepsilon) = (1 - \varepsilon_0) \left(1 + \frac{P_s}{P_A} \right)^\beta \qquad\qquad \text{Eq(13.28)}$$

$$\alpha_{av} = \alpha_0 \frac{(1 - n)\left(1 + \frac{\Delta P_c}{P_A} \right)}{\left(1 + \frac{\Delta P_c}{P_A} \right)^{1-n} - 1} \qquad\qquad \text{Eq(13.29)}$$

$$(1 - \varepsilon_{av}) = (1 - \varepsilon_0) \frac{(1 - n - \beta)\left[\left(1 + \frac{\Delta P_c}{P_A} \right)^{1-n} - 1 \right]}{(1 - n)\left[\left(1 + \frac{\Delta P_c}{P_A} \right)^{1-n-\beta} - 1 \right]} \qquad\qquad \text{Eq(13.30)}$$

Table 13.7. Example of values of the parameters in Equations (13.29) and (13.30).

Parameter	Lime mud Easy to filter	Lime mud Average	Lime mud Difficult to filter
$\alpha_0 \cdot 10^{-9}$ [m/kg]	0.6	1.8	3.1
n	0.15	0.37	0.48
P_A	20	133	164
ε_0	0.672	0.705	0.685
β	0.0	0.022	0.03

The large differences in specific filtration resistances found between different mills are mainly due to different particle size and particle size distribution in the lime muds. This is, in turn, due to variations in the quality of the reburned lime mud, and to some extent, due to different conditions in the slakers and causticising vessels. As mentioned earlier, the specific surface area can be used to describe the quality of the reburned lime mud. Specific filtration resistance versus specific surface area of reburned lime mud is plotted in *Figure 13.17*.

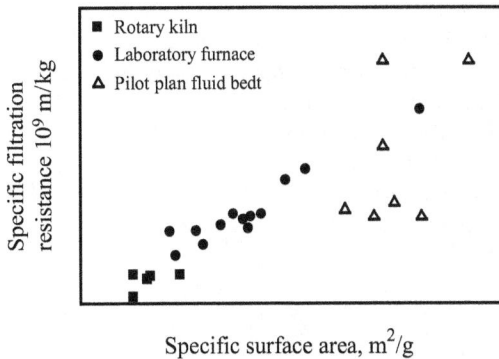

Figure 13.17. Specific filtration resistance of lime mud versus specific surface area of reburned lime mud.

13.5 Washing Lime Mud

After the separation of white liquor and lime mud, the white liquor is then suitable for use in the pulp digester and is transported to a storage tank. The dry content of the lime mud varies between 35 and 70 % (the lower value being valid for a tube filter and the higher for a disc filter), i.e. the lime mud still contains a lot of white liquor. This white liquor must be separated from the lime mud before it can be reburned and is achieved by washing with either water or condensate. One or two washing steps are used: should the dry content be high (i.e. a disc filter is used), one washing step can do the washing, but should the dry content be low (i.e. a tube filter is used), two steps are necessary to achieve a good separation. If two washing steps are used, the first is normally a simple dilution-separation operation, and the second normally uses a rotary drum filter, see *Figure 13.18*. If only one washing step is used, a rotary drum filter is used.

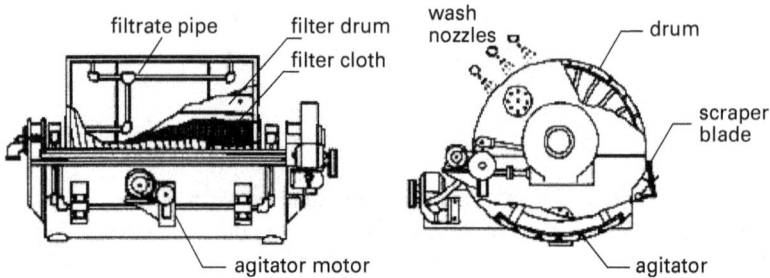

Figure 13.18. Schematic diagram of a rotary drum filter.

On a rotary drum filter, the washing liquid is distributed on to the filter cake by means of spray nozzles. The pressure difference ensures that the washing liquor penetrates the filter cake, thereby displacing the „mother liquor", i.e. the liquor to be separated from the solid material. After washing, air is sucked through the filter cake in order to displace the washing liquor and achieve a high dry content of the filter cake discharged from the filter; the dry content of a discharged cake is normally between 70 and 80 %. The filtrate from the lime mud washing step is called „weak white wash", and is used in the smelt dissolver.

It has previously been shown that the „classical displacement" model can be used to describe the displacement washing of lime mud filter cakes, see Equation (13.31).

$$\frac{c}{c_0} = 1 - \frac{1}{2}\left[erfc\left(\frac{1 - \lambda \cdot W}{2\sqrt{\lambda \cdot W}}\right)\sqrt{\frac{u \cdot L}{D_L}} \; + \; \exp\left(\frac{u \cdot L}{D_L}\right)\left[erfc\left(\frac{1 + \lambda \cdot W}{2\sqrt{\lambda \cdot W}}\right)\sqrt{\frac{u \cdot L}{D_L}}\right]\right] \qquad \text{Eq(13.31)}$$

where
c is the concentration in the exit wash liquor at the washing ratio W.
c_0 is the concentration in the filter cake at time $t = 0$.
λ is a parameter describing the sorption phenomenon (0.85–1 for lime mud).
u is the velocity of the liquor in the filter cake.
L is the thickness of the filter cake.
D_L is the dispersion coefficient.

The Equation above takes into account dispersion phenomenon as well as adsorption. The displacement coefficient of lime mud filter cakes with little or nor defects (cracks etc.) has been determined and is summarized with *Equation (13.32)*, where the dispersion coefficient is a function of the Peclet number. Furthermore, it has been shown that the adsorption on lime mud is rather small.

$$\frac{D_L}{D} = 0.707 + 9.54 \cdot Pe^{1.08}$$ Eq(13.32)

where
Pe is the Peclet number $\dfrac{d_p\, u}{D}$.

D is the diffusivity of the ions.
d_p is the size of the particles in the filter cake.
u is the velocity of the liquid in the filter cake.

The correlations above give results that are slightly too optimistic when applied to the filtration types normally used in pulp mills. This is due to the fact that the filter cakes formed are not free from cracks etc.. The model can, however, be used to calculate the potential of displacement washing

13.6 Reburning of Lime Mud

In the lime mud reburning system calcium carbonate (lime mud) is transformed into calcium oxide (reburned lime mud), which is recycled to the causticising department for causticising of the carbonate ions in the green liquor. The lime mud reburning process can be divided into the following steps:

1. Drying the lime mud.
2. Heating the lime mud to calcinations temperature.
3. Calcinations of the calcium carbonate, Reaction (13.3).
4. Sintering of the reburned lime mud formed.

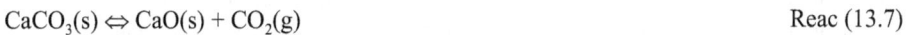

$$CaCO_3(s) \Leftrightarrow CaO(s) + CO_2(g)$$ Reac (13.7)

Both the drying and calcinations processes are endothermic and, consequently, energy must be supplied to the process; this is done via the direct heat exchange of hot flue gases. Burning a suitable fuel directly in the process generates flue gas. The heat demand depends on the type of equipment and its design, but is normally between 6.5 and 9 GJ/tons of reburned lime mud.

Drying lime mud is a relatively uncomplicated and fast operation since it is only surface moisture that is evaporated: it can, therefore, be assumed that the drying rate is determined by the heat transfer. The lime mud is heated to the „calcination temperature", a temperature that varies with the carbon dioxide concentration of the flue gas and the type of equipment used. During calcinations, the calcium carbonate reacts to form calcium oxide and carbon dioxide, according to *Reaction (13.7)*. This is an endothermic equilibrium reaction (~3000 kJ/kg CaO). The relationship between the equilibrium concentration of carbon dioxide and temperature is shown in *Figure 13.19* and *Equation (13.33)*.

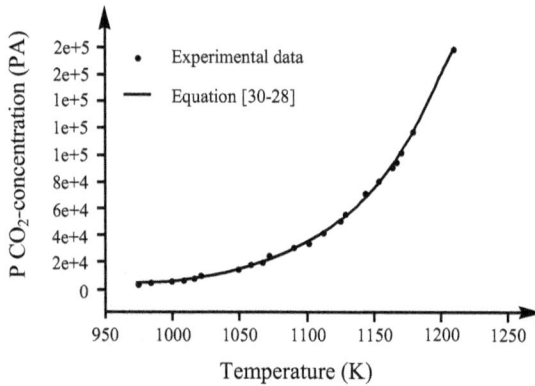

Figure 13.19. Equilibrium concentration of carbon dioxide versus temperature.

$$P_{CO_2} = 2.032 \cdot 10^{-6} \cdot \exp\left(0.0272 \cdot T - 5.2809 \cdot 10^{-6} \cdot T^2\right) \qquad \text{Eq(13.33)}$$

After calcination the lime sinters. Sintering is an operation where the material is rearranged in order to decrease the specific surface area (the driving force is to minimize the surface energy and thus also to minimize the total energy of the system), so small particles will be fused together, forming larger particles. This process occurs on the μm-level, but nodules (i.e. agglomerates consisting of many small particles) on the cm-level are also stabilized. The degree of sintering determines the quality of the reburned lime mud produced. It has been shown that reburned lime that has only sintered to a small extent („light-burned" lime) has a large specific surface area and reacts quickly during slaking and causticising. This type of reburned lime mud also produces lime mud having a small particle size, which implies that the specific filtration resistance becomes high. „Hard-burned" lime mud, i.e. reburned lime mud that has been sintered to a great extent, has a small specific surface area and reacts slowly during slaking and causticising, but the resulting lime mud agglomerates are large and the specific filtration resistance is low. If the sintering process is pushed too far, so-called „deadburned" lime is produced. Dead-burned lime has a very small specific surface area and reacts very slowly during slaking and causticising; it is removed from the process as lime grits. All in all, the degree of sintering can be seen to be very important. There are a number of conditions that influence the sintering rate: the temperature, the composition of the gas (especially the concentration of carbon dioxide) and the solid phase (especially the concentration of sodium and potassium). Unfortunately, it is very difficult to control these conditions (and the residence time) with the equipment used today. *Figure 13.20* shows micrographs of lime mud calcined and sintered during different conditions; it is clearly shown that the particles that have been exposed to higher temperature for a longer time of period are more sphere-shaped and have a larger interface between the particles.

900 °C, 0.5 h

(a)

1000 °C, 1 h

(b)

1300 °C, 1 h

(c)

Figure 13.20. Examples of lime mud calcined under different conditions: a) 900 °C 0.5 h, b) 1100 °C 1h, c) 1300 °C 1h.

The most common equipment used for lime mud reburning is the „rotary lime kiln". As can be seen in *Figure 13.21*, it is, in principle, a long pipe with a length of 50–120 m and a diameter of 2.4–4 m.

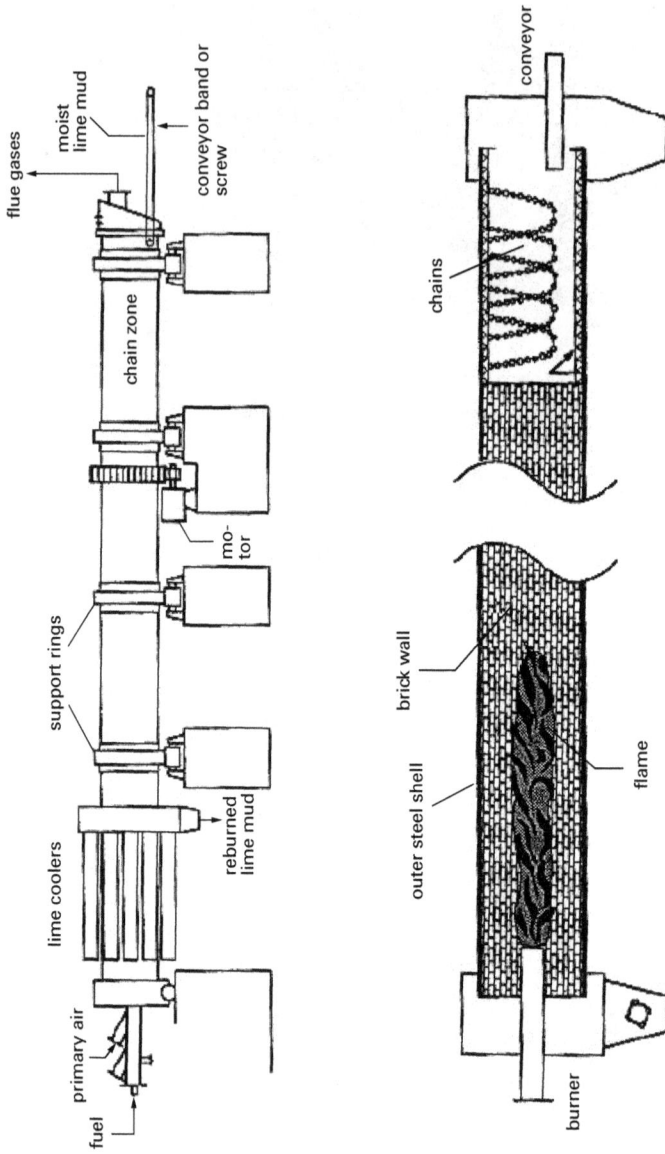

Figure 13.21. Example of a rotary kiln.

The rotary lime kiln is slightly inclined and rotates slowly (1–2 rpm). The moist lime mud enters the lime kiln at the highest end of the kiln. It is transported successively to the lower end as a result of both the inclination and rotation of the kiln. It is, in actual of fact, a bed of solid material that is moved towards the lower end of the kiln. The flue gas is generated at the lower end of the kiln and is moved counter-current with respect to the lime. The flue gas is cooled and heat is transferred to the lime.

The rotary lime kiln is often divided into drying, heating, calcinations and sintering zones, see *Figure 13.22*.

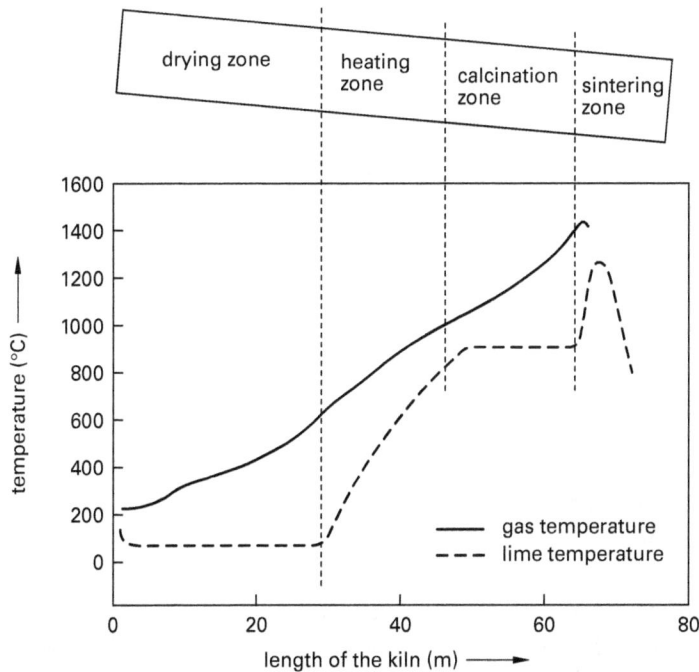

Figure 13.22. Example of a temperature profile in a rotary lime kiln.

The drying zone is often equipped with a chain system to increase heat transport from the flue gas to the lime and to prevent lime from forming very large agglomerates. The chain zone must not be too long and must end before the lime is completely dried in order to prevent heavy dust formation. The trend in recent years has been either to shorten the chain zone or remove it completely and mount so-called „lifters" in the drying zone, which are, in principle, baffles mounted on the inside of the walls. Since the kiln rotates, the baffles will move through the lime bed. There will be a mixing in the bed, and some lime will be lifted up a short distance before falling down onto the bed again. After the drying zone, the lime mud is heated to the calcination temperature (around 850 °C) and calcinations starts. In the calcinations zone, the lime temperature is more or less constant: all of the transferred energy is used for the endothermic calcinations reaction. After the calcinations zone, the lime is heated to a temperature of between 1000 and 1300 °C and the reburned lime mud sinters. The temperature profile in the rotary lime kiln can be calculated using heat balances for the gas and lime mud/reburned lime mud, see *Equations (13.34–38)*:

$$\frac{d}{dx}\left(m_g \cdot C_{pg} \cdot T_g\right) = Q_{g-v} + Q_{g-s} - Q_b - Q_{H_2O,\text{avd.}} - Q_{CO_2,\text{kalc.}} \qquad \text{Eq(13.34)}$$

$$\frac{d}{dx}\left(m_s \cdot C_{ps} \cdot T_s\right) = Q_{v-s} + Q_{vv-s} - Q_{\text{avd.}} - Q_{\text{kalc.}} \qquad \text{Eq(13.35)}$$

$$Q_{g-v} - Q_{v-s} - Q_{v-yv} = 0 \qquad \text{Eq(13.36)}$$

$$Q_{g-s} + Q_{v-s} - Q_{ys-s} = 0 \qquad \text{Eq(13.37)}$$

$$Q_n = h \cdot \Delta T \qquad \text{where } n \text{ can be g–v, g–s, v–s, v–yv, ys–s} \qquad \text{Eq(13.38)}$$

where
x is the length coordinate.
M is the weight.
C_p is the heat capacity.
T is the temperature.
Q is the amount of energy transferred.
H is the heat transfer coefficient.
g is the gas phase.
s is the solid phase.
ys is the surface of the solid phase.
v is the inner wall of the kiln.
yv is the outer wall of the kiln.

In modern lime mud reburning systems, drying is often performed in a separate flash dryer, see *Figure 13.23*. In this type of system, only heating, calcination and sintering occur in the rotary kiln. The introduction of flash drying makes it possible to increase the capacity of the rotary kiln, something that has already been implemented at some mills.

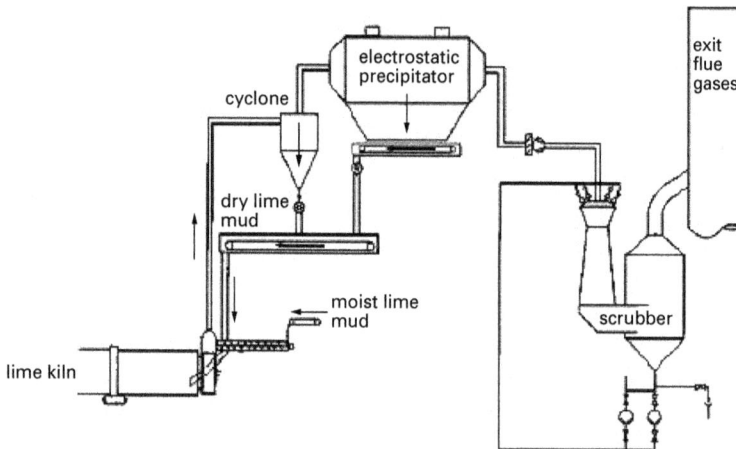

Figure 13.23. Lime kiln with a separate section for drying lime mud.

One very common problem uncounted in the rotary kiln is the formation of „rings" and „balls". „Rings" are thick scales on the inside of the kiln that actually hinder the transport of lime in the kiln, The „balls" are large spheres that can be over 1 m in diameter and can actually make the operation of the kiln impossible. The formation of both is associated with sintering: in this case „intra-particle" sintering occurs. This means that smaller particles/agglomerates are fused together at the contact point, which results in a strong system of particles that can grow to large sizes.

Another type of lime mud reburning system consists of a flash dryer and a fluidised bed reactor, see *Figure 13.24*. The big advantages with this system is that it is much easier to control than the rotary lime kiln. In this case calcination occurs in a fluidised bed in which the temperature is very easily controlled, i.e. it becomes more or less constant. This implies that the degree of sintering becomes even and, as a consequence, the lime quality becomes even. A drawback of this equipment is that the heat demand is higher than for a rotary lime kiln. Only a few of this type have been built, however, since the heat demand is a very important parameter, especially when external fuel is used.

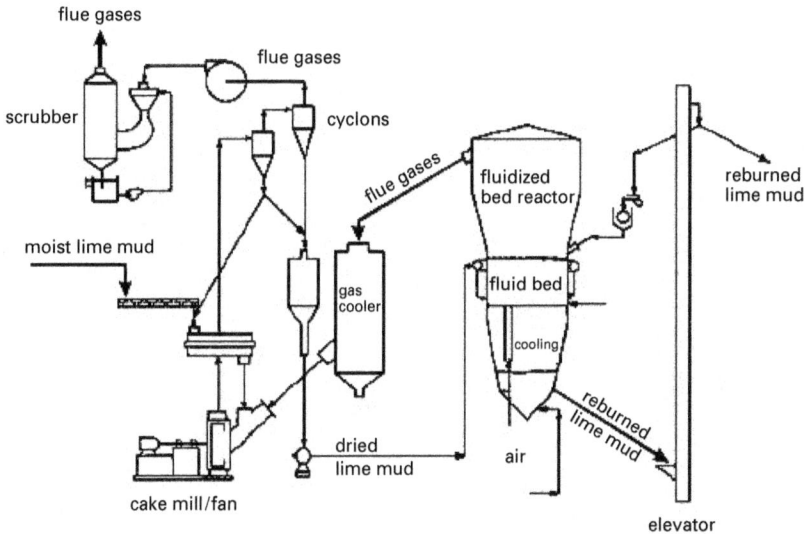

Figure 13.24. An example of a fluidised bed system for reburning lime mud.

If the steam demand is higher than the steam produced in the recovery boiler, an interesting alternative is to combine the fluidised bed reactor with a „steam dryer" and, thus, recover some of the heat used for drying in the form of low pressure steam (4–5 bars), see *Figure 13.25*.

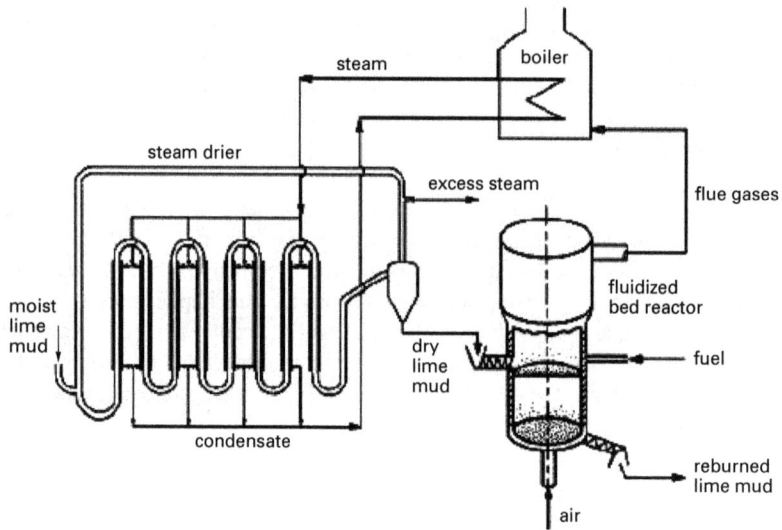

Figure 13.25. A fluidised bed reactor combined with a steam dryer.

14 Towards Increased Closure and Energy Efficiency

Birgit Backlund
Innventia AB

14.1 Introduction

14.1.1 Serious Environmental Problems in the 1960s

Up to the 1960s the forest industry suffered from serious environmental problems. In most cases, chemical recovery at sulphite pulp mills, internal fibre recovery and sedimentation ponds – which are taken for granted today – simply did not exist. Smoking chimneys spread dust and sulphur dioxide in immense quantities. The smell was unpleasant, and the soot settled everywhere in the mills' vicinity.

Watercourses were also severely affected. The effluent from the pulp mills contained large quantities of chemicals and wood residues, including chlorinated dioxins and other polychlorinated organic compounds. Where there was little water circulation, fibre banks started to spread over large areas. In many places the water lacked oxygen. Malodorous hydrogen sulphide built up and the fish died.

14.1.2 The Swedish Model

During the past 20 years, the forest industry has invested some 20 billion kronor in environmental measures. Initially, the work focused on the most glaring environmental problems, in other words, emissions which affected the immediate vicinity of the mills. The first target was to reduce emissions of fibres and oxygen-demanding substances that were the cause of the destruction of all life in the water and at the bottom of the lakes around the mills.

At an early stage, the forest industry opted for a strategy of purifying at source, rather than „end of the pipe" that is to say, they attempted to prevent pollution by making improvements to the processes. Throughout, the main principle has been to reduce the lignin content of the pulp by means of extended cooking and bleaching with oxygen prior to final bleaching. Once the lignin content has been reduced, the pulp is easier to bleach and the need for bleaching chemicals has been steadily reduced. Only when these measures were insufficient was biological and chemical purification of the waste water used. Since the development of new, improved processes represented pioneering work, it was frequently encouraged by The Swedish Environmental Protection Agency. Even though results were not always immediately evident, it is now generally agreed that the Swedish model for environmental work has been very successful.

14.1.3 Oxygen and Chlorine Dioxide Bleaching

In the beginning of the 1970s the first full scale installations of oxygen bleaching were introduced in Sweden. As pre-bleaching with oxygen significantly reduces the lignin content, this almost halved the need to use chlorine gas in the final bleaching stage. Oxygen bleaching has been in use at all Swedish kraft pulp mills.

By the middle of the 1980s, oxygen bleaching had become an established technique. Emissions of chlorinated organic compounds had been reduced to some 5 kg AOX/tonne of pulp compared to around 10 kg in the 1970s. The proportion of chlorine dioxide was successively increased during the 1980s while the use of chlorine gas was further reduced. Chlorine dioxide is a considerably milder bleaching agent from an environmental point of view since it does not create highly chlorinated organic compounds. The need for chlorinated bleaching agents was further reduced during the 1980s with the introduction of reinforced and pressurised alkali stages in which hydrogen peroxide and oxygen could be used. At this point, the major environmental clean-up had been completed. Compared with their competitors in other countries, Swedish companies had low emissions – but high costs for their environmental measures. At that time, the cost of producing one tonne of pulp was 100–150 kronor more than in the main competitor countries and the market was not prepared to pay more for products from mills with low emissions.

14.1.4 A Green Market

The collective pressure from government authorities, environmental opinion and the market would compel the industry to take further measures. 1985 was a tumultuous year with three main events accelerating developments. Firstly, forest industry researchers themselves observed that the biological effects of the emissions from two particular pulp mills were greater than had been expected. The Environmental Protection Agency soon confirmed these results with similar ones from an investigation outside a third mill, where fish were also found to have been harmed. The results received considerable publicity in both Swedish and foreign media. The same year also saw reports that researchers had discovered measurable quantities of chlorinated dioxins, a highly toxic group of substances (*Figure 14.1, Figure 14.2*), in the sludge outside an American pulp mill. Traces of these toxic substances were also found in pulp and paper products. This prompted the Swedish forest industry to begin investigating effluent and products from Swedish pulp mills. Their results showed that the pulp industry was only responsible for a small part of the total emissions throughout the country. Research also showed that the minute traces of dioxins which could be found in paper products did not represent a hazard to health. Despite this, wide-ranging research and development efforts were made to eliminate the formation of dioxins in the pulp process. In 1988, Swedish researchers found the solution: the formation of chlorinated dioxins and dibensofuranes could be eliminated by keeping the ratio of elemental chlorine (Cl_2) to lignin content under a certain level, i.e avoiding high concentrations of chlorine during long time in contact with pulp.

The chlorine debate was now in full swing. Swedish Greenpeace and the Swedish Society for the Conservation of Nature were keenly engaged in the debate. This course of events made a reaction from politicians and environmental authorities inevitable. The licences made for subsequent years involved even tighter restrictions on chlorinated emissions. Sweden became a frontrunner in the „race against chlorine".

The chlorine debate gave rise to fundamental changes in market conditions in many countries. In Germany, Sweden's largest market for pulp, intense demands were made for „low chlorine" and then „chlorine free" products. Other countries followed Germany's example. After some hesitation, the industry reacted vigorously to these new challenges.

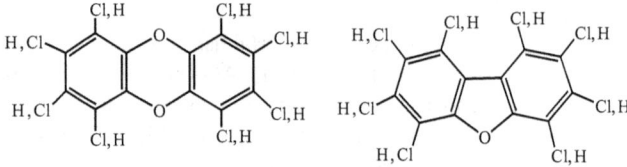

Figure 14.1. Structure of polychlorinated dibenso-p-dioxins (PCDD) and polychlorinated dibensofuranes (PCDF). Two or more of the hydrogen atoms 1-9 are replaced by chlorine atoms.

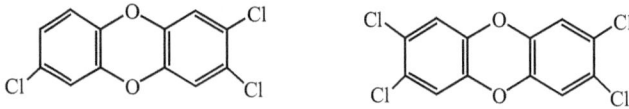

Figure 14.2. 2,3,7-trichloro-dibenso-p-dioxin, to the left, has approximately the toxicity of strychnine. 2,3,7,8-tetrachloro-dibenso-p-dioxin, to the right, has four chlorine atoms instead of three, and is 15 000 times as toxic.

The method that had the greatest impact, a Swedish invention that came to be known as modified or extended cooking, was introduced at the end of the 1980s. At the beginning of the 1990s extended cooking had become an established technique. Super-Batch and ITC cooking are examples of commercial trade names. Extended cooking reduced the need for bleaching further and made way for bleaching with peroxide and ozone, as well as bleaching with extremely small amounts of chlorine dioxide. Chlorine gas was completely removed from the process and is no longer used as a bleaching agent by any pulp mills in western Europe.

The internal measures have been complemented at many mills by various types of biological purification treatment. Emissions of oxygen demanding substances have been greatly reduced. Emissions of chlorinated substances, measured as AOX, have on average been brought down to 0.2–0.3 kg per tonne of pulp. The new processes have also altered the chemical composition of the chlorinated substances. Highly chlorinated substances have been eliminated. The chlorinated pollutants in today's emissions have a far lower environmental impact than those in emissions in the 1970s. The standard bleaching methods today are called ECF (Elemental Chlorine Free), i.e. no Cl_2 added, and TCF (Totally Chlorine Free), i.e. no Cl_2 or other chlorine compounds added. Comprehensive studies, comparing the effects of effluent from mills that bleach with chlorine dioxide and those that do not, have not demonstrated any significant differences between the two methods.

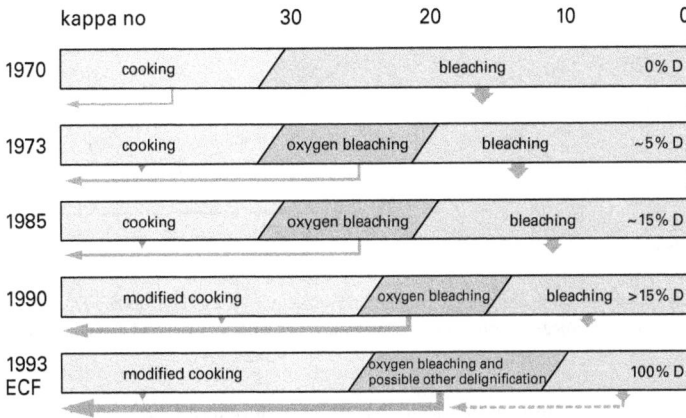

Figure 14.3. The development of methods used for delignification and bleaching of kraft pulp. Oxygen bleaching and modified cooking has made it possible to exclude the use of chlorine gas and, in some cases, also of chlorine dioxide. „% D" is the ClO_2-share of the total amount of oxidation equivalents (ClO_2 and Cl_2) charged in the first bleaching stage.

14.1.5 The Situation Today

The emission situation today is compiled in *Table 14.1*.

Table 14.1. The emission situation today in kraft pulp production

Problem	Current situation
Fibre emissions	Solved
BOD emissions (deoxygenated water)	Solved
Sulphur dioxide emissions	Almost solved
Odours (sulphur compounds)	Unsolved but greatly improved
Oil consumption	Not completely eliminated
COD emissions (organic substances)	Much has been done, some work remains
Emissions of organically bound chlorine, AOX	Much has been done, possibly solved
Chlorate emissions	Solved
„Dioxin" formation in processes	Solved
Nitrogen oxide emissions	Low emissions
Nutrient salt emissions (nitrogen and phosphurus)	Unsolved
Metal emissions	Uncertain whether this is a problem
Chemical additives	Situation unclear, research underway

pulp- and
paper mills 340 000 (8%)
purification plants 50 000 (1%)

other industry 10 000 (1%)

other eg: 3 700 000 (90%) COD to surrounding sea

natural origin, agriculture, forests and forestry, rural settlements, refuse tips etc.

4 100 000 tonnes per year

rentention: 50% of inland emissions

Mill. tonnes COD / Production of pulp / COD / Mill. tonnes of pulp

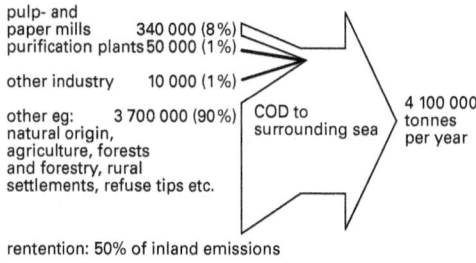

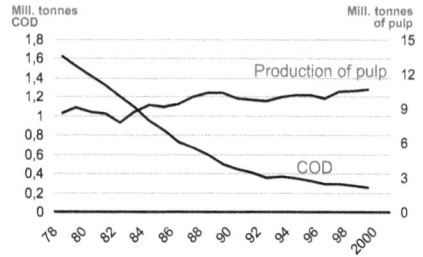

Figure 14.4. The share of the Swedish emissions of oxygen-demanding substances (COD) emanating from the pulp and paper industry has declined as a result of process improvements, reduced water consumption and effluent treatment.

- *COD*, Emissions of dissolved organic substances, expressed in terms of COD, have been reduced by some 80 per cent since the beginning of the 1980s, *Figure 14.4*. In 2000 they amounted to 255,000 tonnes. The substances derive from the wood and consist of a large number of compounds, from simple alcohols to complex, lignin-based compounds. The COD measure states how much oxygen is needed to decompose the organic material. The emissions have mainly been reduced by increased recycling (closed loop systems) in the mill processes and by biological treatment of the effluents.

AOX, Emissions of chlorinated organic substances, expressed as AOX (adsorbable organic halogens), have been reduced by about 97 per cent since the mid-1980s, *Figure 14.5*. The volume in 2000 was about 740 tonnes. This has been achieved by modifying the cooking and bleaching methods used to produce chemical pulp.

- *Sulphur,* Emissions of sulphur compounds amounted to some 5,200 tonnes S in 2000, *Figure 14.5*. Sulphur compounds are formed during the chemical pulp production process and from fuel combustion. Sulphur dioxide has health effects and adds to acidification. Certain reduced sulphur components (e. g. H_2S, methyl mercaptane, dimethyl disulphide) have unpleasant odour that is noticed already at extremely low concentrations. Emissions have been radically reduced since the 1970s, largely due to the large-scale replacement of oil by biofuels such as recovered black liquor (spent cooking liquid), bark and other wood residues. More efficient energy systems and improvements in process methods have also contributed to this development.

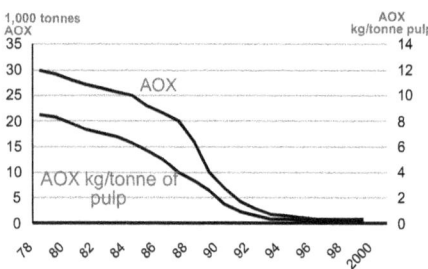

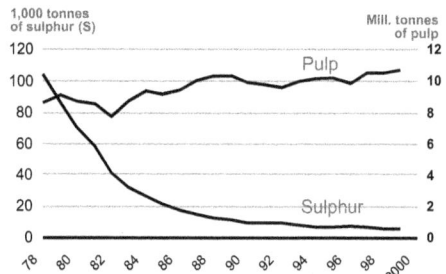

Figure 14.5. The emissions of AOX to water and sulphur to air have been decimated during the last two decades.

If the newest commercial techniques are combined, further environmental improvements can be obtained. In *Table 14.2*, a theoretical reference mill is compared with an average Swedish mill year 2000. The reference mill represents a kraft mill for bleached softwood market pulp in which the most recent technology used in commercial operation in Finland and Sweden year 2000 is implemented in all process departments. The reference mill has dry debarking (bark press water to black liquor evaporation), continuous cooking to kappa number 27, selective slot screening and extensive washing before and after two-stage oxygen delignification to kappa number 10. The pulp is bleached to 89 % ISO-brightness with Q(OP)(DQ)(PO). The bleaching stages have wash presses instead of filter washers. Alkaline bleaching filtrates are recycled to the brown stock washing, while acid bleach filtrates are led to the biological treatment. The black liquor is evaporated further (80 % DS instead of 70 %). This leads to a high temperature in the recovery boiler and this a diminished SO_2 emission. The temperature of the steam from the recovery boiler is raised (490 °C, 80 bar instead of 450°C, 60 bar). Spill liquors are efficiently recovered. The lime kiln uses gasified bark, while surplus bark is sold. Excess steam is used to produce condensing power.

Table 14.2. If the newest commercial techniques are combined, further environmental improvements can be obtained. Adt = air dry tonne.

	Theoretical reference mill 2000	Swedish average 2000
Emissions		
SO_2 to the atmosphere, kg /ADt	0.2	1.1
NOx to the atmosphere, kg NO_2/ADt	1.0	1.3
Total process water effluent flow, m³/ADt	16	58
Process water COD after effluent treatment, kg/ADt	7	26
Process water AOX after effluent treatment, kg/ADt	<0.1	0.21
Emission of „fossil" CO_2 kg/ADt	−260	320
Energy surplus		
Electric energy, kWh/ADt	540	−140
Fuels (bark, tall oil), GJ/ADt	3.1	−0.6

14.2 The Vision of an Ecocyclic Pulp Mill

The popular terms „closed" and „emission-free" mill are not altogether realistic. A pulp mill can never be run without emissions of e. g. water and carbon dioxide. A more accurate definition of an ecocyclic mill is that it creates no environmentally hazardous emissions. All substances leaving the process are re-used or treated so that they cause no harm to nature when they are released into the surroundings.

The largest potential to further adapt a kraft pulp mill to an ecocyclic system is to improve energy utilisation. It is also important to decrease air emissions, in particular NOx, and the organic and inorganic content in wastewater effluents. Solid waste has become an important political issue, leading to increased costs for landfilling. On a long-term scale the elements from wood should be recovered in such a way that they can be returned to the forest or used in another acceptable manner.

The vision of the totally ecocyclic chemical pulp mill is illustrated in *Figure 14.6*. All of the raw materials are renewable and all of the material produced within the mill is useable. Resources are handled in an efficient way with minimal losses. All material flows are parts of closed loops; nothing crosses the borders of the system except energy. Wood fibres are converted to paper products that are recovered after use. The recycled fibres are used again in paper products and finally composted or incinerated with recovery of the heat value. Solar energy is converted to chemically bound energy in the trees. In the pulp mill, the heat value of the wood lignin is converted to steam and electric energy. Excess energy from the pulp mill can be used elsewhere in society. The heat value of the fibres is recovered when the spent paper products are incinerated. Trace elements from the wood are separated through incineration and precipitation reactions and brought back to the forest soil. In addition, the carbon dioxide formed from incineration is assimilated by the trees.

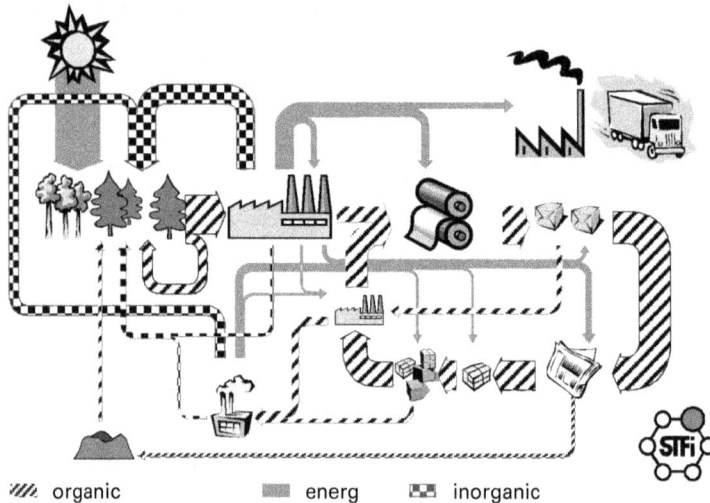

////. organic　　　■■■ energ　■.■ inorganic

Figure 14.6. The vision of an ecocyclic pulp mill.

14.3 System Closure in the Bleach Plant

The remaining emissions to water from bleaching operations contain organic and inorganic substances. The organic substances are mainly degradation products from lignin and extractives. Among the substances that can have health effects on living organisms are chlorinated and unchlorinated phenolic compounds, fatty acids, resin acids and sterols. Inorganic substances – deriving from the wood – that might have environmental effects are nitrogen and phosphorous compounds (nutrients) and heavy metal ions.

The trend goes towards increased recycling of the bleach plant effluents to the unbleached fibre line. The bleaching effluents can then be evaporated and incinerated with the black liquor or separately. However, the accumulated filtrate volume from the bleach plant washers is too large to be evaporated with reasonable energy economy. The volume can be decreased through countercurrent recycling of the filtrates through the bleach plant washing stages, *Figure 14.7*.

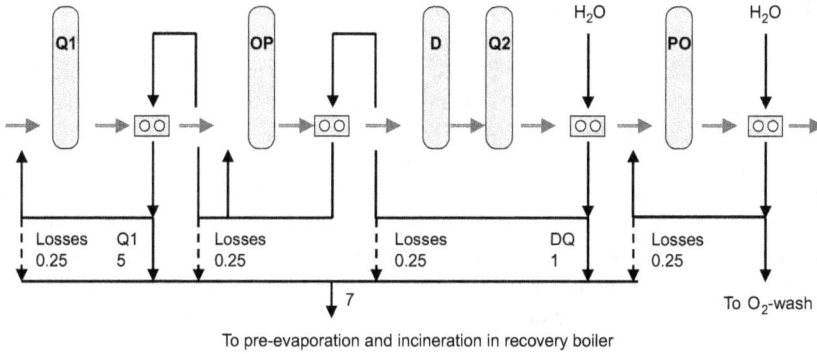

Figure 14.7. A theoretical example of extensive closure of the bleach plant. Filtrate flows [m³/ADt] in the reference mill bleach plant with extensive closure. A total of 7 m³/ADt is pre-evaporated: 5 m³ from the Q_1-stage, 1 m³/ADt from the DQ-stage and 1 m³/ADt from accidental overflows.

A high degree of closure, with evaporation and incineration of the bleach plant effluents, disturbs the chemical balance of the mill, *Figure 14.8*. The main problem is that sodium and sulphur that normally leave the system with the effluents will be enriched in the system. This sodium and sulphur have to be removed by other means. One way is to purge more ash from the electrostatic precipitators of the recovery boiler. In reality, it is not realistic to purge these amounts. Other possible methods to take care of the superfluous Na and S is internal production of a) alkali through oxidation of white liquor and b) sulphuric acid from black liquor sulphur. Internal production of these chemicals decreases the inflow of Na and S to the system.

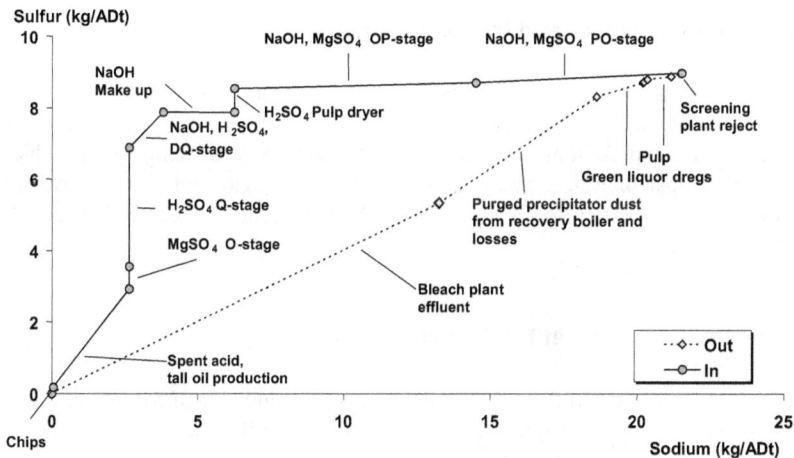

Figure 14.8. The sulphur-sodium balance for the theoretical reference mill (see Table 14.2). The bleaching sequence used is Q(OP)(DQ)(PO).

14.4 Problems in a Kraft Mills with High Degree of Closure

A kraft mill needs to handle several hundred tons per year of various NPEs (non-process elements). Carbon, hydrogen and oxygen are the building blocks of the carbohydrates, lignin and extractives found in wood. In addition, varying amounts of other elements such as Na, K, Mg, Ca, Ba, Al, Si, N, P, S, V, Cr, Mn, Fe, Ni, Cu, Zn, Cd, Pb and As are found in wood. The elements that enter the pulp mill with the wood are to a large extent separated from the fibre flow during the pulping processes, and leave the system as air contaminants, water contaminants or as solid waste. Another important source of NPEs, especially silica and aluminium, is lime (CaO) that is used in the preparation of cooking chemicals.

Increased closure of the mill leads to the build-up of many of these elements, which can cause different types of process disturbances. The main negative effects resulting from an increased concentration of NPEs in the chemical recovery process are deposits, plugging, corrosion, and increased dead load (including extra water), *Table 14.3*. The largest problems in the bleach plant are deposits and the decomposition of hydrogen peroxide, caused by the catalysis of transition metal ions.

Table 14.3. Negative effects caused by non-process elements (NPEs).

Negative effects	Elements
Scaling, deposits	Al, Si, Ca, Ba
Plugging in the recovery boiler	K, Cl
Corrosion	K, Cl
Inert compounds in the lime cycle	P, Mg, Al, (Si)
Disturbances in hydrogen peroxide bleaching	Mn, Fe, Cu
Environment (air and water)	N, P, Cd, Pb and other heavy metals

14.4.1 Deposits in the Digester and the Evaporation Plant

In the digester, calcium carbonate cause deposits. In the black liquor evaporation plant, other types occur as well: sodium aluminium silicates and burkeite, $2Na_2SO_4 \cdot Na_2CO_3(s)$. The solubility of calcium carbonate and sodium aluminium silicate declines with temperature, which means that they often deposit on heat exchanger surfaces. The risk of calcium carbonate scaling in the evaporation plant may be lowered by heat treatment of the liquor, i.e. controlled precipitation, ahead of the evaporators.

14.4.2 Deposits and Corrosion in the Recovery Boiler

Deposits in the upper furnace area of the recovery boiler are often due to high concentrations of chloride and potassium compounds. High concentrations of these NPEs may result in sticky or sintered deposits on screen tubes, super-heater sections or in the generating bank. This is because two- or multi-phase mixtures with a major molten phase fraction (about 20–70 %) are formed within specific temperature ranges, as a result of the melting point depression of the normal fly ash components – sodium sulphate and carbonate. If potassium and chloride levels are increased these temperature ranges will become larger, and in addition will start at lower temperatures. In addition, low melting „acidic sulphates" ($NaHSO_4$, $Na_2S_2O_7$ and their potassium analogues) may form, especially at high SO_2 and O_2 levels. At high chloride levels, corrosion will also become a problem in the recovery boiler.

14.4.3 Inert Compounds in the Lime Cycle – Phosphate and Lime Quality

About one third of the phosphorus entering the mill with the pulpwood will follow the pulp into the bleach plant while the remaining two-thirds will follow the black liquor to the chemical recovery plant. The phosphorus that enters the recovery boiler can be found in the smelt and dissolved into the green liquor. A very small amount of phosphate will be removed with the green liquor dregs because the concentration of phosphate in the green liquor does not normally exceed the solubility limit of calcium phosphate and the precipitation rate of calcium phosphate is low under green liquor process conditions.

The solubility of calcium phosphate is much lower in the white liquor than it is in the green liquor and therefore practically all of the phosphate will precipitate out during the causticising process and accumulate in the lime cycle. This will result in a considerable lowering of lime quality, expressed as free CaO, and the decrease of free CaO will be about 5 weight % per weight % P in the lime, i.e. phosphorus contributes strongly to the inert load in the lime. This is one important reason why it is today necessary to keep the lime cycle somewhat open, adding 5–10 % make-up lime.

14.4.4 Deposits in the Bleach Plant

Scaling can occur in tanks, pipes, wash filters and heat exchangers, when the pH in the bleaching filtrates change between the stages. Deposits of calcium oxalate are predominant at pH values below 8, and for calcium carbonate within the pH range of 8–12. Deposits of barium sulphate may appear throughout the pH range of technical interest, i.e. between 2 and 12. Low temperatures promote the formation of calcium oxalate and barium sulphate deposits because the solubility of these compounds increases with temperature. The opposite applies to calcium carbonate.

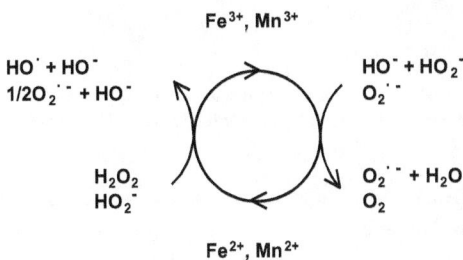

Figure 14.9. Catalytic decomposition of hydrogen peroxide in a bleaching stage by the Fenton cycle.

14.4.5 Catalytic Decomposition of Hydrogen Peroxide

Several transition metals, e.g. Mn^{2+} and Fe^{2+}, are easily oxidized in the presence of e.g. hydrogen peroxide which then forms radical species. These radicals may then reduce the metal ions back to the bivalent stage while forming oxygen and water. This process is described by the so-called Fenton cycle, *Figure 14.9*.

Swedish researchers have recently been discovered that it is possible to stabilise manganese and iron in oxidation state II by enclosing the Mn^{2+} and Fe^{2+} ions with Mg^{2+} in a solid solution with the hydroxide, carbonate or silicate anions, e.g. $(Mg^{2+}, Mn^{2+})(OH)_2(ss)$.

14.5 Separation Methods for Problem Substances – Kidneys

Problems with NPEs can be solved through efficient use of present „kidneys" and, if necessary, to introduce new more efficient ones. One important limiting factor is the cost for further handling of the generated wastes. For many of the NPEs there are well established „purging mediums", such as the removal of transition metals and aluminium by green liquor dregs, and the removal of phosphorus with lime mud, *Figure 14.10* and *Table 14.4*.

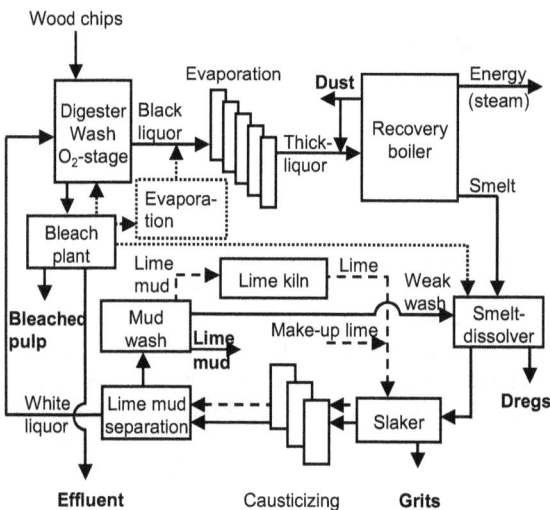

Figure 14.10. Purging media for NPEs in a kraft pulp mill. The liquor cycle is represented with solid lines, and the lime cycle with dashed lines. The red dotted lines represent a proposed concept for the recycle of bleach plant filtrate.

The Champion International BFR concept (Bleached Filtrate Recycle) is an example of a new kidney system. The concept includes a potassium and chloride kidney (purging of recovery boiler ESP ash), and in addition an ion exchanger unit for metal ions (Ca^{2+}, Mg^{2+}, Mn^{2+}, etc.) in the filtrate from the chlorine dioxide stage. The cleaned filtrate is returned to the bleach plant. Some other possible separation methods for mills with a high degree of closure are summarised in *Figure 14.11*.

Table 14.4. Removal of different NPEs from the chemical recovery process. Me = Mg^{2+}, Ca^{2+}, Mn^{2+}, Cd^{2+}, Pb^{2+}, plus others. x = removal, (x) = removal of Si^{4+} (as the concentration in the white liquor is higher than approx. 15 mmol/l), and o = removal of small amounts.

Kidney	Al	Si	P	K	Cl	Me
Green Liquor Dregs	x	o	o			x
Grits	o	o				
Lime mud	x	(x)	x			x
Recovery Boiler/ESP[1]) Ash				x	x	

[1] Electrostatic precipitator

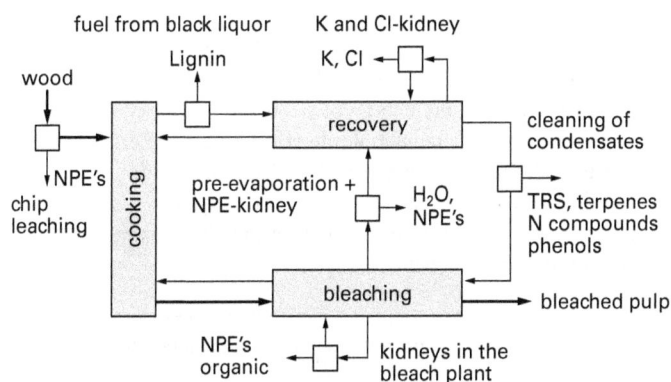

Figure 14.11. Possible purging processes in an extremely closed chemical pulp mill.

14.5.1 Leaching of NPEs from Wood Chips

The majority of the non-process elements (NPE) that exist in the pulp mill have their origin in the wood raw material. If the amount of NPEs in the chips can be reduced, the need for kidneys within the process itself can be eliminated or reduced. A Swedish method to leach metal ions and other NPEs out of the chips, before they enter the digester, is under development. Apart from metal ions, the method removes phosphorus and chloride from the wood. The leaching is performed at around pH 3 and 90 °C. The conditions are a compromise between maximal leaching and conserved pulp strength. One possibility is to use acid bleaching filtrates as leaching liquor. The elements in the leachate can be brought back to the forest soil after reprocessing.

The method does also have energy benefits for the mill: Less Ca means less scaling on the evaporators and less K means less scaling in the recovery boiler.

14.5.2 Membrane Filtration

A pressure-driven membrane kidney can be used both to treat outlet streams prior to biological treatment and to clean internally recycled streams. The economic requirements for the use of membrane technology have, as of yet, not been resolved. The largest potential for membrane filtration as a kidney lies in the separation of high molecular weight compounds and divalent cations from the first alkaline bleaching stage and the chelating stage. Acidic filtrates are less suited for membrane filtration, since they give severe fouling problems, destroying the capacity of the membrane. The reason is that organic acids are protonized in acidic filtrates and adsorb strongly onto the membrane.

A COD retention of approx. 80 % can be achieved with a combination of UF (ultrafiltration) and NF (nanofiltration). The corresponding values for divalent ions is close to 100 % and approx. 80 % for monovalent ions. This treatment results in very clean filtrates, suitable for recycling.

14.5.3 Evaporation of Bleaching Filtrates

The emissions to water from the bleach plant could be eliminated through evaporation of the water content of the effluent, which leaves a concentrate or solid that is more easily taken care of. In order to boil off 1 tonne of water, roughly the energy of 1 tonne of steam is needed. The energy consumption can be strongly reduced by using a multiple effect design, where the effects are arranged by decreasing pressure, such that the steam boiled off from the previous effect can be used as the heating medium in the following effect. In the pulp industry today, six or seven effects are normally used for evaporation of black liquor, which means that no more than one ton of steam is required to boil off approx. 5 m^3 of water.

Table 14.5. The additional equipment that is required and the equipment that can be excluded in two cases of evaporation of bleach plant liquors. The comparison is made against the theoretical reference mill (see chapter 14.1.5).

Case 1 „Direct pre-evaporation and combustion in the recovery boiler"		Case 2 „NPE chip kidney and separate evaporation and incineration"	
Added	Excluded	Added	Excluded
Pre-evaporation of bleach plant filtrate	Effluent treatment	Pre-evaporation of bleach plant filtrate	Effluent treatment
Chloride kidney precipitator ash		NPE chip kidney	
Increased area in remaining evaporation effects	One effect in the black liquor evaporation stage	Increased area in remaining evaporation effects	One effect in the black liquor evaporation stage
Technique to control the sulphur balance		Technique to control the sulphur balance	
Technique to control the sodium balance		Drying of concentrate from separate evaporation	
Ozone treatment of condensate		Ozone treatment of condensate	

The most common evaporation device today is the falling-film evaporator. The liquid is fed to the top of the heating tubes and flows down the walls as a film. The circulation can be adjusted such that good heat transfer is obtained, and dry surfaces do not appear on the heat transfer walls. Regarding bleach plant effluents, research is ongoing on the development of evaporators that are able to use low-grade energy. Lower temperatures make it possible to consider unconventional materials such as plastics.

The changes needed in a mill with evaporation of bleaching effluents are summarised in *Table 14.5* for two cases: with and without NPE leaching of chips (see section 14.5.1). In Case 1, the effluent is pre-evaporated without any pre-treatment. The concentrate is then evaporated together with the black liquor. In addition, chloride is purged by dissolving and crystallizing precipitator ash. The sodium balance is handled by total oxidation to sulphate of white liquor, while surplus sulphur is removed by heat treatment of black liquor. Condensate from the preevaporators is used as wash liquor in the bleach plant.

The condensate from the pre-evaporation stage can be used as wash liquor in the bleach plant. However, in case 2 there is a risk that substances (terpenes, phenols etc.), dissolved from the wood in the „chip" kidney may contaminate the condensate. At present it is not known to what extent this occurs and whether it will substantially affect the bleaching process.

14.5.4 Cleaning of Condensates

Besides water vapor, a number of organic substances are released during the evaporation of black liquor and the cooking stage. The organic compounds originate from the wood or are formed during the process. The implementation of modern internal condensate fractionation and the collection of condensates from each evaporation stage make it possible to obtain fractions with different levels of impurities. The most contaminated condensate has a COD content of 10–15 g/l. The majority of the COD is methanol, but it also contains ammonia, terpenes, TRS reduced sulfan compounds and guaiacol. The most contaminated condensates come from the digester and from the least concentrated black liquor stage, where the most volatile components are released. The second cleanest condensate comes from the following stages in the evaporation plant, i.e. in the direction of the liquor flow. It is common practice today to clean the most polluted condensates by so called steam stripping, while the cleaner condensates are reused in the process as they are or are led to the wastewater treatment plant.

The steam stripper is a distillation column where the rectifying column consists of only a few trays. The extracted steam, which contains organic vapors, is taken to a methanol column. In the condenser methanol is condensed, together with the compounds with a higher boiling point. The non-condensable gases are incinerated in the recovery boiler or in some other combustion unit. The methanol is used as fuel in the lime kiln or recovery boiler.

14.5.5 Electrodialysis for the Separation of Chloride and Metal Ions

Electrodialysis is a separation method that uses electrical potential difference as its driving force. Its capability to separate ions with different charges makes it a potentially interesting kidney for the pulp industry. The separation mechanism is based on the use of ion selective membranes and on an electrical field that is transverse to the membrane. The negative ions move towards the anode and positive ions towards the cathode. It is possible to separate an ionic solution into a clean and concentrated solvent by forming a stack, with electrodes placed at both ends, in which anion and cation selective membranes alternate with spacers placed in between them.

Electrodialysis is capable of separating chloride from recovery boiler ESP dust at a relatively low cost. A potential method to separate chloride and also some potassium from ESP dust with small losses of sodium and sulphate is to combine leaching and electrodialysis. An interesting long-term application is the separation of chloride ions (or sulphate ions) as hydrochloric acid (or sulphuric acid) from ESP dust while recovering sodium as hydroxide. This requires the use of bipolar membranes - a sandwich consisting of an anion and a cation selective membrane with water in between.

14.5.6 Biological Treatment – Decomposition of Organic Substances

Biological purification of wastewater involves the decomposition of organic compounds at a neutral pH with bacteria, or another type of microorganism, so that foremost carbon dioxide, water and new bacteria (bio-sludge) are formed. During anaerobic treatment a bio-gas is formed. Bio-gas is a mixture of high-energy methane and carbon dioxide that includes a small amount of bio-mass.

Some particles, in particular lipophilic substances, adsorb onto the bio-sludge that is generated. Thus, bio-treatment involves a combination of destruction and separation. Different methods are used in order to secure a high content of bacteria in the system. In activated sludge plants („AS" and plants with extended aeration „LAS") the bio-mass is separated by sedimentation and then brought back into the aeration process. It is also possible to use a supporting material (various plastic bodies) on which the bio-mass can grow, thereby retaining this material in the system. This is referred to as AS with suspended supporting material or bio-beds with stationary material. A simpler variation of aerobic treatment is the aerated basin, in which the sludge is not recycled.

Almost all organic material can decompose biologically; the practical applicability of biological methods depends on the reaction rate. The methods are efficient for naturally existing substances with molar weights below 500–1000 Dalton. Biological methods can also be used to convert different nitrogen compounds into nitrogen gas, and to bind phosphorus as polyphosphate.

Biological treatment is normally used to clean external wastewater, but the method can also be used to clean some process streams internally. In most cases, the spreading of bacteria from the biological system into the rest of the process must be prevented. This can be achieved with a microfilter or with an ultrafiltration membrane. The bio-stage and the membrane can work separately, or be built into one unit, referred to as a membrane bio-reactor. This combination implies an internal destruction of the concentrate from the membrane stage, but can also facilitate the filtration.

14.5.7 Wet Oxidation – Decomposition of Organic Substance

Wet oxidation is a method used for the oxidation („combustion") of oxidizable materials in an aqueous phase with air or pure oxygen as the oxidizing agent. This method is today an established process technique. It has been practiced, in a few cases, in the pulp and paper industry during the past decades, both directly in the process and as a sludge treatment method.

In comparison with combustion in a furnace, wet oxidation has the advantage that no water is vaporized. Therefore, the heat of combustion is only dependent on the enthalpy difference between input and output liquid streams. This means that the lowest concentration of oxidizable materials for autogenous combustion is considerably lower with wet oxidation than it is with combustion in a furnace. As a rule of thumb the proportion of oxidizable material that corresponds to about 10 g COD/l may be stated as a lower limit for autogenous wet oxidation. Wet oxidation is normally performed between 200 °C and 300 °C and between 30 and 200 bar.

The main reaction products from organic compounds are carbon dioxide and water. Minor amounts of organic fatty acids are often present, foremost acetic acid, a compound that is oxidized very slowly under wet oxidation conditions. This indicates that a complete reduction of COD will usually not be achieved. The formation of NOx is negligible due to the low reaction temperature, compared to that of combustion. Sulphur is oxidized to sulphate which remains in the water phase, so that the portion of SO_2 in the exhaust gas is negligible.

14.5.8 Adsorption and Ion Exchange – Separation of Chloride and Metal Ions

In adsorption processes, separation is achieved by the attachment of one or more compounds to a solid adsorbent. Changing the process conditions regenerates the adsorbent. In an ion exchange process, a chemical reaction takes place between the ions in solution and the ions in the adsorbent. In general, ion exchange is considered to be a special type of adsorption. A typical example is the softening of water. In the adsorption process the sodium, which is attached to the sorbent, is replaced by calcium ions. To achieve regeneration a concentrated solution of sodium salt is used such that the reverse reaction takes place.

Adsorption used for the final effluent treatment of discharged effluent is a well-proven technology. In the pulp industry, adsorption has been used to reduce the color of bleach plant effluent. Ion exchange is an important part in the BFR concept (Bleach Filtrate Recovery) for closing of the bleach plant, developed in the US. The process includes an ion exchange stage that separates metal ions from the filtrate after the chlorine dioxide stage. Mill results with filtrate from the first chlorine dioxide stage, treated in a cation exchanger, have shown a removal efficiency of Ca^{2+}, Mg^{2+} and Mn^{2+} of ≥ 90 % and for Fe $^{2+} \geq 70$ %. Ion exchange can also be applied for the separation of chloride from recovery boiler dust.

14.5.9 Crystallization – Separation of Chloride and Metal Ions

Crystallization can be seen as a process where solid particles are formed from a homogenous phase (a liquid in this application). Equilibrium is attained when the mother solution is saturated and the solute concentration corresponds to the solubility curve. In practice, crystallization can take place when a liquid is cooled down or with the addition of more of the dissolved substance. Nucleation, crystalline growth and secondary changes of the suspension characterize the process.

There are a number of potential applications where crystallization can be used as a kidney in a pulp mill. One large-scale application is as a part of the BFR (Bleach Filtrate Recovery) concept developed in the US. In addition to the separation of metal ions by ion exchange the concept includes the separation of chloride by crystallization. Re-crystallization is the only method capable of separating chloride that is industrially proven. The separation efficiency of potassium and chloride is at least 90 %. The sodium and sulphate losses are about 20 %.

Another potential method to separate chloride from the system is to use the crystallization method on green liquor. Sodium carbonate precipitates, while sodium chloride and sulphide remain in solution. Sodium carbonate is dissolved and causticised in the usual manner. The solution remaining after the solid phase is separated is evaporated such that sodium chloride precipitates and sodium sulphide remain in the solution. Another possible application is to use crystallization to precipitate metal ions from acidic bleach plant effluent before they are recycled.

14.5.10 Freeze Concentration of Black Liquor and Bleaching Filtrate

A special case of crystallisation is freeze concentration, which is a potential process for the concentration of dilute fluids. The method is based on the fact that when a solution containing dilute dissolved solids is cooled to its freezing point, only the solvent freezes. The crystals tend to form a highly structured pure solid that excludes molecules other than the solvent. The other molecules accumulate at the phase surface. Therefore, these crystals must be washed carefully so as to obtain a solvent that contains a low as possible concentration of impurities after it has melted. The method requires very little energy as a heat pump can be used for freezing and freezing requires less energy than boiling.

The usefulness of this process for the concentration of black liquor was examined in the USA at the end of the seventies but failed due to the complexity of the process. The method has also been studied for the concentration of bleach plant filtrate but once again the operating difficulties were found to be to large.

14.5.11 Absorption, Scrubbing – Gas Cleaning

Absorption refers to the transport of substances from a gaseous phase to a liquid phase. The driving force is the fact that the system is not in equilibrium, i.e., the partial pressure of the substance is higher in the gas phase than in the gas layer close to the surface (that is in equilibrium with the liquid). A gas can be efficiently cleaned by a selective choice of absorption liquid. In addition to absorption, scrubbing also includes the washing of particles from the gas stream. Scrubbers are sometimes used in equipment that recovers heat from moist air, i.e. separates vapor from air.

Absorption has been used to a large extent in the pulp industry, for example for the separation of sulphur dioxide, hydrochloric acid gas and hydrogen sulphide from smoke and vent gases, as well as for the cleaning of odorous non-condensable gases. An alternative method for the latter is incineration.

14.5.12 Mechanical Separation

Common to all mechanical separation methods is that they are used to separate a fluid from a solid bed. These methods are prevalent in the manufacturing of pulp and paper. Important applications are washing, dewatering and pressing of pulp, the separation of lime from white liquor, the washing of lime, the separation of green liquor dregs from green liquor, the pressing of bark and the dewatering of different effluent sludge streams.

The mechanical separation methods are energy efficient. If a dry material is the desired end product it is most energy efficient to first dewater the material mechanically before thermal dewatering. An example is the pressing of market pulp to high dryness prior to the drying.

The black liquor is separated from the pulp by washing with water. In this case it is important to obtain a clean pulp using as little wash liquid as possible. The reason being that the added wash liquid must be removed in the evaporation plant. Efficient washing requires a high fibre concentration into the stage, displacement at near plug flow and a high fibre concentration out of each washing stage.

Another application where it is important to wash efficiently with a high dryness is the last lime washing/dewatering stage prior to the limekiln. It is essential that the green liquor filter filtrate contains a minimum of impurities and, at the same time, that the separated green liquor dregs has a high dryness content. It is also important that the bark and sludge are pressed prior to combustion as a high dryness is advantageous from an energy point of view.

14.5.13 Flotation

Flotation refers to the lifting of flocculated material to the surface of a liquid with the aid of air bubbles that attach themselves to particles whose specific weight is close to that of the carrying liquid. The material that floats to the liquid surface is separated mechanically or by overflow. Flocculation chemicals and acid/caustic solutions, for adjustment of pH of the incoming water, are added before air is injected into the system. In the micro-flotation process air bubbles with very small diameters are formed.

Flotation technique is very robust and therefore it is used, for example, for the separation of soap from black liquor and the separation of printing ink from waste paper. Other possible applications include the separation of NPEs in the bleach plant.

An interesting separation method, which includes precipitation of NPEs from bleach plant filtrate, flocculation and flotation, is the so-called Netfloc process. Precipitation is achieved by the addition of carbonate (Na_2CO_3) and oxidized white liquor at a pH around 11. After a given retention time, flocculation chemicals are added to the flotation tank. There is no need to add air, since there is sufficient air dispersed/dissolved in the filtrate. The process has been tested in full scale for Q-stage filtrate from a QQ(OP)(QPaa)(PO) bleaching sequence. The separation efficiency is 80–90 % for most of the metals present (Mn, Fe, Mg, Ba, Al), and about 70 % for calcium.

14.6 Residual Products

Even if landfilling of solid waste from pulp mills is relatively harmless from an environmental point of view, the issue has become important due to increasing environmental fees on deposition of industrial waste. Another factor is the ecocyclic perspective: nutrients and trace elements from the wood ought to be brought back to the forest soil in order to avoid long-term long-term depletion. However, the conditions vary largely between geographical areas, and some forests may have a natural inflow of trace elements from decomposition of natural minerals.

A kraft mill produces approximately 60 kg solid residual products per tonne of pulp, not including wood room waste (sand and gravel) or building and demolition wastes. The most important process wastes are summarised in *Table 14.6*. The residual products contain elements from the wood. In addition, process chemicals can be found in the lime sludge and the grits ($CaCO_3$), in the green liquor sludge ($Mg(OH)_2$), in the ESP dust (Na_2SO_4/Na_2CO_3) and in the biosludge (Cl-compounds, S-compounds and complex binders).

Table 14.6. Solid residual products from the kraft mill processes.

Residual product	Amount, kg/t	Characteristics
Lime sludge	≈ 20	Mainly $CaCO_3$. Enrichment of P, Mg, Mn and other PFEs. Mainly salts that are difficult to dissolve.
Grits	≈ 3	Similar to lime sludge.
Green liquor sludge	≈ 30	Mainly $CaCO_3$, Na- and K-salts. Enrichment of Mg, Mn, Fe, and other PFEs and certain heavy metals. Mainly salts those are difficult to dissolve.
Recovery furnace ESP dust	≈ 5	Mainly Na_2SO_4, Na_2CO_3, NaCl (ant the corresponding K-salts). Mainly salts that are easily dissolved.
Bark ash	≈ 5	Mainly $CaO/CaCO_3$, K-salts. A mixture of salts with different solubilities.
Sludge from biological treatment of effluents	≈ 3	Residual products from e.g. the bleach plant. Organic material, certain Cl-salts, complex binders. A high content of organic material. After incineration, a mixture of salts with different solubilities.

The main ingredients in the waste streams are Ca, K, Na, Mg, Cl och P, meaning that their economic value is low per kg, even after reprocessing. Since the amounts are considerable, it might nevertheless be possible to find acceptable ways to reprocess the wastes into usable products. Some possibilities:

- Recycling of metal salts to the forest (pH increase and nutrients)
- Filler in road embankments
- Sealing material in landfills
- Raw material for concrete

The only residual product that is not suitable for recycling to the forest is the recovery furnace ESP dust. The salts are very easily dissolvable and contain too much Na, Cl and S. The other residual products should be possible to recycle after reprocessing.

Grits, green liquor sludge and lime sludge might find use as fillers in roadbanks. However, it is uncertain how resistant the materials are under acid conditions. Sintering could increase the stability.

Regarding the use as sealing material in landfills, green liquor sludge is the only residual product to be considered, due to its very low permeability. The limited stability at low pH is a drawback in this application as well. Besides, the need for landfill sealing material is probably decreasing in the future.

Lime sludge and grits are probably suitable as concrete ingredients.

14.7 Energy Potential of the Kraft Mill

14.7.1 The Pulp Mill's Energy Balance

A kraft pulp mill converts considerable amounts of energy, and is to a large extent self-sufficient in bio-energy. The utilization of the biofuels black liquor and bark amounts to approx. 25 GJ/ADt or a yearly consumption of 3–5 TWh (thermal energy) in a modern Swedish mill. Mainly, this energy leaves the mill as low value waste heat after being used for process steam and power. The technical potential to further improve the energy management within the kraft pulp mill is considerable. Effective energy use leads to a surplus of bio-based fuel that can be exported as fuel, power or heat. A lowered intake of water, and thereby a decreased effluent volume, eliminates or strongly decreases the need to heat incoming fresh water, and reduces at the same time the energy requirement of the wastewater treatment plant. As a result of mill closure, the surplus heat will not be absorbed by the incoming fresh water or leave with the outgoing effluent. Therefore, a more closed mill requires a separate cooling system to remove surplus heat.

An overall energy balance over the theoretical reference mill (see section 14.1.5) is shown in *Figure 14.12*. The amount of energy used in the process corresponds to the differences in energy content between incoming wood and outgoing pulp plus tall oil: 25.3 GJ/ADt (21.9 of which comes from the black liquor and 3.4 from the bark),. The largest part of this energy leaves the different processes as low value waste heat (23 GJ/ADt) whereas almost 10 % is „exported" as power. The power and steam requirements of the different process units in the reference mill are shown in *Table 14.7* and *Table 14.8*.

Table 14.7. Power demand of different process units in the reference mill with ECF bleaching.

Process unit	kW/ADt
Wood Yard	45
Digester Plant	44
Brownstock Washing and Screening	60
Oxygen Delignification	60
Bleach Plant	80
Post-Screening	45
Pulp Dryer	120
Evaporation Plant	30
Boilers	90
Causticizing and Lime Reburning (including gasification of bark)	60
Cooling Tower etc.	20
Raw Water Treatment	20
Wastewater Treatment	30
Chemical Plant (including oxygen)	10[1]
Miscellaneous Users and Losses	30
Total	740[2]

[1] TCF: 60 kWh/ADt; [2] TCF: 790 kWh/Adt

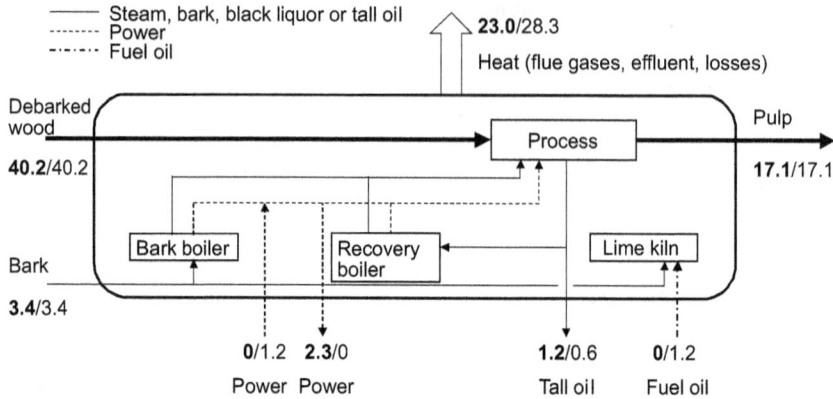

Figure 14.12. Thermal energy balance for the reference mill (figures before slash sign), version with bark boiler instead of bark export. As a comparison (figures after slash sign) the average value for Swedish mills that produced market pulp in 1994 is shown. The energy content in the wood and the pulp is given as combustible dry solids. Unit: GJ/ADt.

Table 14.8. Steam demand of the reference mill and a normal mill for fully bleached kraft pulp (unit: GJ/AD).

	Normal mill	Reference Mill		
		ECF Summer	TCF Summer	ECF Winter
Steam Requirement				
Digester Plant	3.0	1.7	1.7	2.2
Oxygen Delignification	0.6	0.4	0.4	0.4
Bleach Plant	1.3	1.0	1.7	1.0
Evaporation Plant + Stripper	5.1	3.6	3.6	3.7
Pulp Drying Machine	3.0	2.2	2.2	2.2
Soot Blowing Steam + Bottom Blow	1.1	1.0	1.0	1.1
Miscellaneous	1.3	0.5	0.5	0.6
Total Steam Requirement	15.4	10.4	11.1	11.2
Purchased Fuel (lime kiln)	1.8	0	0	0

From an energy and capital cost point of view it is normally more efficient to produce pulp and paper at the same site in an integrated mill, rather than to produce pulp at one site and paper at another. There is a large energy advantage in not needing to dry the pulp before it is transported to the paper mill. Another environmental advantage is that an integrated paper mill can use bio-based energy whereas a non-integrated paper mill normally uses fossil fuel as its energy source.

14.7.2 Increased Heat Efficiency

The reference mill has a large surplus of energy that can potentially be used e.g. for the production of power. Increasing process integration and the incorporation of new technology can increase this surplus energy further. There are several possible uses for this surplus energy, *Figure 14.13*. Energy may be exported as power, heat (steam or hot water) or fuel (wood fuel, bark, lignin or methanol). An addition of natural gas may utilize the systems resources more efficiently since a gas turbine has a limited efficient operating area.

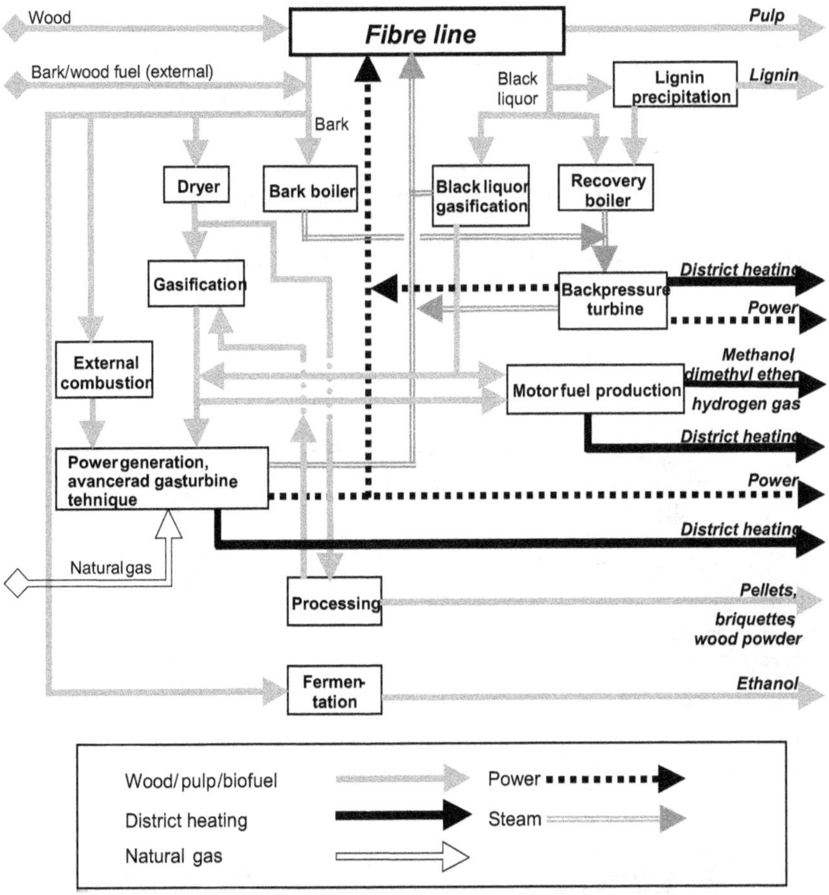

Figure 14.13. Alternative ways to extract surplus energy from a hypothetical pulp mill with an energy surplus.

A large part of the surplus energy in the reference mill is lost with the cooling water from the condensing turbines. Large energy losses can be avoided by either implementing high efficiency power generating processes, or by dispatching surplus energy (fuel) to more efficient external uses. The reference mill steam consumption can be reduced by at least 3.5 GJ/ADt if maximum heat integration is applied. The largest potential saving lies in pulp drying with steam at low temperatures in combination with a heat pump or heat integrated evaporation.

14.7.3 Opportunities to Increase Power Production

The reference mill's power production in a back-pressure turbine is 790 kWh/Adt (version with bark boiler instead of bark export), which is higher than what is normal in present-day Swedish pulp mills: 700–740 kWh/ADt. This difference is a result of higher steam pressure and temperature. The reference mill's steam data point, 79 bar and 485 °C, is equivalent to the standard for modern recovery boilers, whereas today's typical steam data point is 60 bar and 450 °C. It is possible to achieve an even higher power production than that of the Reference Mill. However, this requires more advanced alloy materials in the superheater and in other critical parts of the recovery boiler.

A possible future alternative to the combustion of black liquor in a recovery boiler is combustion in a pressurized gasifier, *Figure 14.14*. In the gasification of a fuel the solid material is converted to an energy-rich gas (primarily hydrogen gas, carbon monoxide and hydrocarbons). This technique has not been implemented for black liquor yet, but full-scale trials are being made.

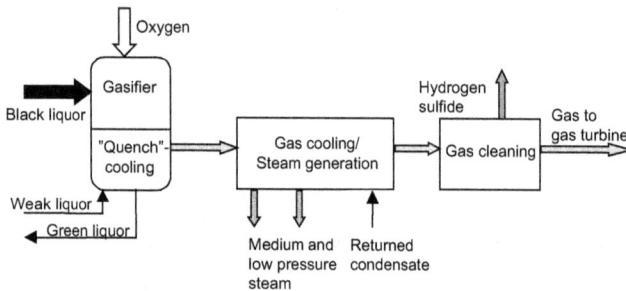

Figure 14.14. Chemical and energy recovery with pressurized black liquor gasification. Generated hydrogen sulphide can be converted to sulphuric acid and polysulphide via Sulphur production in the Claus Process.

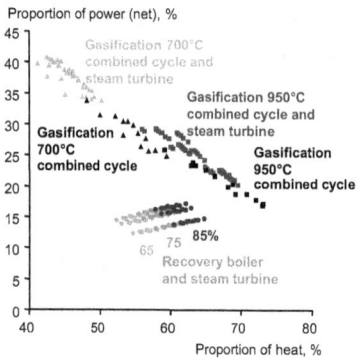

Figure 14.15. The division of the energy content in the black liquor between power and heat for different pressurized black liquor gasification alternatives. Recovery boilers with different steam data points and black liquor solids content are included as a comparison.

The gasifier design and the process chosen for the production of power greatly affect the split between power and heat, *Figure 14.15*. Normally the black liquor gasification process is divided into two groups depending on whether the sodium is in the smelt (ca 950 °C) or is in a solid form (ca 700 °C). The most developed gasification technology is the smelt process. The discussion below refers to the pressurized oxygen-blown smelt process. In *Figure 14.15* this process is represented as the gasification alternative at 950 °C. The released sulphur is absorbed by the green liquor. Another possibility is to convert the hydrogen sulphide to liquid sulphur in a Claus Process for further conversion to sulphuric acid or polysulphide.

The fact that the sulphur is partly separated from the black liquor during gasification makes it easier to control the sulphur and sodium balances. This in turn increases the degrees of freedom to produce cooking liquors with different sulphidity or the generation of polysulphide and sulphuric acid. However, this technology is complicated. A process drawback of black liquor gasification technology is that the green liquor that is formed has a higher sodium bicarbonate content than that from a recovery boiler system. The reason for this is the release and reabsorption of H_2S in the gasification case. As a result of this the lime demand in causticising is higher and therefore also the energy required for lime reburning.

In *Table 14.9*, the net power production from a gasification plant (gas combined cycle process) is compared with that from the reference mill's recovery boiler.

14.7.4 Gasification of Bark and Other Wood Waste Material

In addition to black liquor, bark and other wood waste material can be used for power production in a pulp mill. As an alternative to incineration in the bark boiler, the biofuel can be gasified. After cleaning, the generated gas can be used in a gas turbine process. At present-day conditions the gasification of bark and other wood waste for the production of power is not economically feasible. The efficient utilization of fuel and the large power potential does however make gasification of biomass, combined with power production and residual heat utilization, an interesting possibility in the long term.

14.7.5 Production of Solid Fuel

A more efficient heat use will result in an increase of the kraft mill's energy surplus. Production of more steam in the recovery boiler is often not a solution, due to lack of a nearby paper mill, community or other consumer. An alternative to power production is to sell the surplus as bark and solid lignin fuel. This is especially attractive in the common case that the recovery boiler is a bottleneck for the capacity.Excess lignin can be removed by acid precipitation from the black liquor evaporation, or from a cooking liquor circulation. In the latter case, besides the energy efficiency, improvements in cooking efficiency, chemical consumption and bleachability could be expected. *Table 14.9* indicates that close to 8 GJ/Adt of fuel can theoretically be recovered from the reference mill. However, with today's oil prices, the production of lignin fuel would in most cases not be economically feasible.

Table 14.9. Key data for theoretical green field Model Mills compared with the theoretical year 2000 Reference Mill with ECF bleaching. All Model Mills have a kappa number of 27 after cooking. For comparison, data for the average year 2000 Swedish softwood bleached kraft pulp mill is included.

Case	Sold fuel (GJ/Adt)	Sold power (kWh/Adt)	Net operating costs [1] (%)	Capital investment (%)	Emission of „fossil"CO_2 (kg/Adt)	COD to recipient (kg/Adt)
Reference Mill with ECF-bleaching	3.1	540	100	100	–260	7
Reference Mill with TCF- bleaching	3.1	490	98	99	–270	7
Swedish average 2000	–0.6	–140	120	n.a.	+320	26
Model mill 1. Chip kidney	3.1	500	100	101	250	<1
Model mill 2. Export of excess lignin	7.5	0	99	101	–420	7
Model mill 3. Pressurized black liquor gasification. Conventional lime cycle	2.3	1050	95	108	–390	7

[1] Costs for wood, chemicals and energy, minus sold power, bark and tall oil.
Power: 0.20 SEK/kWh, bark: 0.1 SEK/kWh

14.7.6 Production of Motor Fuels

Instead of power, black liquor gasification could produce motor fuels – methanol for petrol engines or dimethyl ether (DME) for diesel engines, *Figure 14.16*. In comparison with other potential biomass sources for fuel production, black liquor has the great advantage that it is already partially processed and exists in a pumpable, liquid form. Using black liquor as a raw material for methanol/DME production would have the following advantages:

- biomass logistics are extremely simplified as the raw material for fuel making is handled within the ordinary operations of the pulp & paper plant
- the process is easily pressurised, which enhances fuel production efficiency
- the syngas has a low methane content, which optimises fuel yield
- pulp mill economics becomes less sensitive to pulp prices as the economics are diversified with another product
- gasification capital cost is shared between recovery of inorganic chemicals, steam production and syngas production.

The cost would correspond to methanol production from other sources, but today the fuels cannot compete economically with ordinary petrol/diesel from fossil sources.

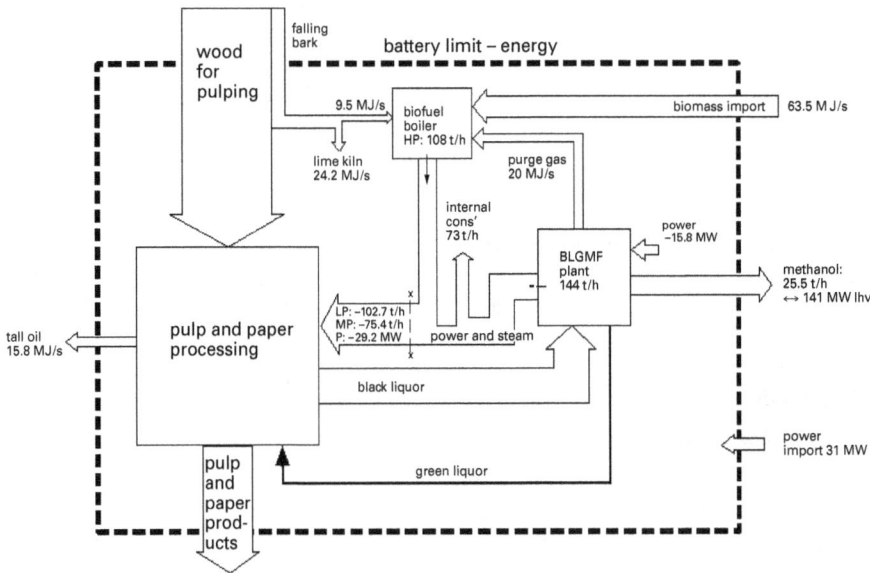

Figure 14.16. Process flow diagram for a possible black liquor gasification process with methanol production.

14.8 Carbon Dioxide Emissions

The bioenergy from the kraft mill can decrease the emissions of CO_2 from fossil sources. Direct and avoided emissions of carbon dioxide from the different phases in the life cycle of the reference mill are shown in *Table 14.10*. The negative sign for avoided emissions of carbon dioxide means that society's net emission of fossil CO_2 is reduced as a result of the pulp production. For comparison, data for the average Swedish mill for bleached softwood kraft pulp in 2000 is presented. The Swedish average in 2000 is based on weighted averages of data from all softwood bleached market kraft pulp mills in Sweden. In contrast to the 2000 Swedish average mill, the reference mill has a negative net emission of fossil CO_2. The main reason is that the reference mill has a surplus of bark and black liquor fuel. The bark is sold, while the surplus black liquor is best suited – based on present conditions in Sweden – for the generation of condensing power at about 540 kWh /ADt. This surplus power is assumed to be used to replace power from a natural gas combined cycle power plant. *Table 14.9* indicates that an even higher energy efficiency, and hence, a larger effect on the emission of greenhouse gases, can be obtained with pressurized black liquor gasification or by removing excess lignin fuel.

Table 14.10. Direct and avoided carbon dioxide emissions for different phases in the life cycle of the pulp production, „from plant seedling to pulp bale", for the year 2000 reference mill with ECF bleaching. The incineration of fibres in the paper products corresponds to an additional amount of 1400 kg avoided CO_2/ADt.

Phase in life cycle	Emissions of fossil based CO_2, kg/ADt	
	Reference Mill	Swedish average 2000
Forest operations	60	60
Transportation of wood	40	40
Production and transportation of chemicals	100	120
Pulp mill, mineral oil	0[*)	50
Pulp mill, power	0	50
Sum direct emissions	200	320
Avoided mineral oil	–260	0
Avoided power (natural gas combined cycle)	–200	0
Sum avoided emissions	–460	–100
Sum direct minus avoided emissions	–260	320

[*) $< 5\ CO_2$

14.9 Further Reading

Skogsindustrierna (Swedish Forest Industries Federation), 2002.

www.skogsindustrierna.org

Ecocyclic Pulp Mill. Final report KAM 1 1996–1999, STFI, Stockholm 2000.

15 Paper Recycling

Per Engstrand and Björn Johansson
Holmen Paper

15.1 Introduction

Recycling of paper has been a neglected part of the pulp and paper literature. This is a big mistake because it is indeed a substantial part of the pulp and paper industry. Paper based on recycled fibre can be made with high quality. The recycled fibre is mainly used in newsprint, tissue and board production but the usage in higher grades of paper is increasing.

The production of paper using recovered fibres is an old technique. The first type of recovered fibres used where old textile fibres. The deinking in terms of recovering printed products was started with an invention in 1774 by Claproth, which concerned removal of printing ink from the fibres. In spite of this invention the real „boom" in recycling of paper started first in the middle of the 20th century. The recovered paper usage has increased by approximately 6 % annually between 1980 and 2000.

The principles of recycling of paper have not changed during this time period. The stock preparation process for recovered paper production consists of three main processes:

1. Repulping of the paper
2. Removal of contaminants
3. Bleaching of the fibres

Recycled paper of a certain grade (packaging, printing, office papers) is disintegrated in a pulper. After pulping contaminants are removed by means of separation steps. Examples of contaminant are: metal objects, glues and adhesives, printing ink and garbage (non-paper components). The technical principles used to separate these contaminants from the fibres are based on physical differences between the materials as: particle size, density and hydrophobicity. The corresponding separation techniques used are screening, cleaning and flotation. Depending on the paper grade that is produced bleaching is applied to various extents in order to obtain the right optical properties of the pulp. Although this description of the paper recycling process may sound simple the process lay-outs of recycling mills are far from simple and the process in itself complex. The aim of this chapter is to give an over-view of the technical and chemical processes involved in production of recycled fibres for various types of paper production.

15.2 Recycle Fibre Statistics – Background and Use

The largest part of the recycled fibre is as already mentioned board, newsprint and tissue. However, there may be some interest for some basic statistics for recovered paper. A figure from 2000 shows that 46.5 % of the fibres used for paper production in Western Europe were based on recovered fibres. It should also be noted that the annual growth for recovered fibre is higher than for chemical and mechanical fibre.

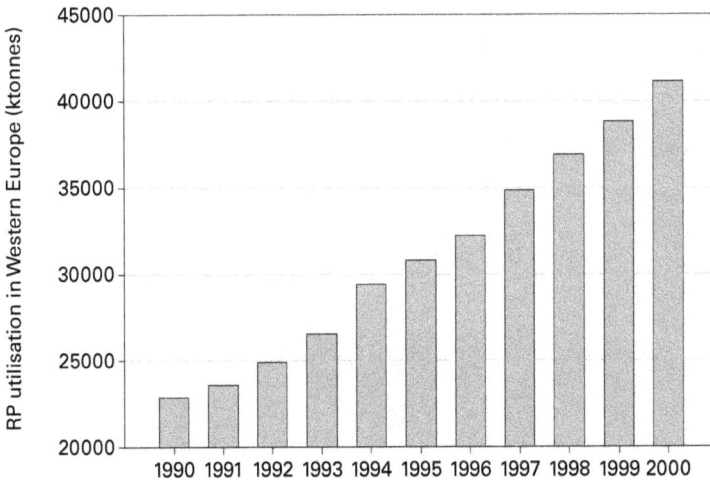

Figure 15.1. The annual growth in recovered paper utilisation in Western Europe (2000). The growth rate has been approximately 6 % annually during the last ten years. Observe that the total paper production for this geographic area was 90542 ktonnes (Confederation of European Paper Industries (CEPI)).

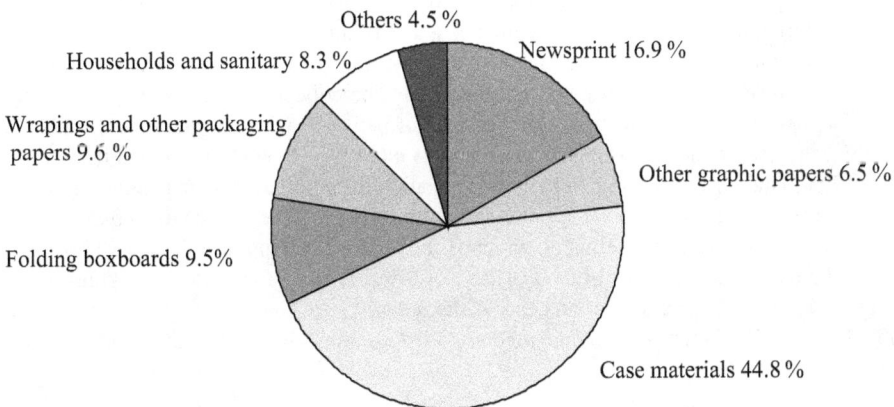

Figure 15.2. The utilisation of recovered paper in various paper production sectors. The larger part, almost 2/3, of the recovered paper is used for production of various types of packaging materials (case materials, folding box boards, wrapping and other packaging papers).

The larger portion of the recovered paper is used for packaging grades. A figure from Western Europe shows that almost 2/3 of the recovered paper is used for production of various types of packaging materials. The high usage of recovered fibres for these types of grades is further emphasised by the utilisation rate[1] for these grades that is close to 90 %. The two other main areas for use of recovered paper are newsprint and hygiene paper production. Newsprint stands for 16.9 % of the recovered fibre usage and an utilisation rate of almost 66 % while the corresponding figures for hygiene paper are 8.3 % and 64 % respectively.

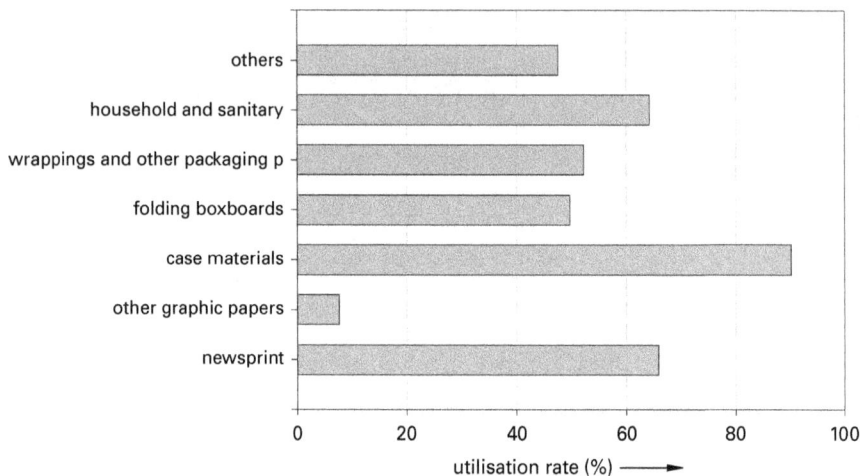

Figure 15.3. Utilisation rates in Western Europe for various paper grades. Utilisation rate = the percentage of recovered paper utilisation compared to the total paper consumption. Observe the high usage of recovered paper for production of packaging, newsprint and hygiene paper production. Also note the low utilisation rate for fine paper production (other graphic papers).

Another important aspect when discussing recovered paper is the rate of collection. To discuss the rate of collection two definitions are needed. The apparent collection, which is the utilisation of RP plus the export minus imports and the collection rate, which is the apparent collection, compared to the total paper consumption. The collection rate for Western Europe was 52 % year 2000. The collection rate varies a lot between different countries, from 18 to 70 %. Sweden has a collection rate of 63 %, which is interesting to note since the utilisation rate only is 17 %. This is of course due to the low population in combination with a high virgin fibre usage. Sweden is together with some other countries a net importer of recovered paper. These numbers can be compared to Germany, which is the largest net exporter of recovered paper in Europe who has a collection rate of 69.8 % and an utilisation rate of 60.1 %. The countries with the highest recovered paper utilisation being in the order of size: Germany (10921 ktonnes, year 2000), France (5778 ktonnes), United Kingdom (4882 ktonnes), Italy (4620 ktonnes) and Spain

---

[1] The utilisation rate is the percentage of recovered paper utilisation compared to the total paper production.

(3829 ktonnes). This can be compared to the utilisation of RP in Sweden and Finland, which is 1816 and 685 ktonnes/year respectively.

15.3 Raw Material Aspects

The deinking of paper is very much dependent on the raw material. The recovered paper can very roughly be divided into 4 groups:

- Old newsprint (ONP)
- Old magazines (OMG),
- Mixed office waste (MOW) and other high grade papers
- Old corrugated containers (OCC).

There are numerous recovered paper qualities but the four mentioned above give a good description of the recycling process. The dominant raw material used for newsprint production is a mixture of ONP and OMG. CEPI statistics from 2000 show that more than 90 % of the recycled fibre for newsprint production is based on an ONP/OMG furnish. The ratio between these two components is very important and is very dependent on the regional usage of various types of paper. A high ratio of magazines are usually considered as favourable for the deinking results (better optical properties) but also results in a lower yield due to a higher ash content[1]. The MOW mainly consists of computer printouts and xerography printed papers. MOW is considered as a high quality raw material and is usually used for production of fine paper or tissue. As seen in *Figure 15.4*, 39 % of the fine paper and 60 % of the tissue production of recovered fibre are based on high-grade raw material.

OCC on the other hand is mainly used for packaging paper production and is usually considered as a contaminant in a printing paper or tissue furnishes. 57 % of the raw materials for packaging material are based on OCC. The OCC furnish is usually not deinked since the optical properties of the paper produced from it is not a primary concern. There are also a number of higher quality raw materials as residual material from printer's shops and as well as sorted paper grades. The price of the raw material varies and therefore an aim is to be able to run as low quality recovered paper grade as possible. The raw material differs a lot depending on several parameters among them:

- The regional composition of the paper and inks
- The age and history of the recovered paper

It is interesting to observe that the deinking process is actually dependent on the thermal history of the recovered paper. This is displayed by what is known as the summer effect. This ef-

---

[1] Ash constitutes the inorganic fraction of the paper, which mainly includes materials as fillers and coating pigments. The ash content is measured after incineration of the paper at high temperature. The temperature used varies between different standard but are usually in the range between 450 and 950 °C. According to new INGEDE (organisation for the European recycling mills) standard the ash should be measured at 450 °C due to that above this temperature the crystal water in clays evaporate.

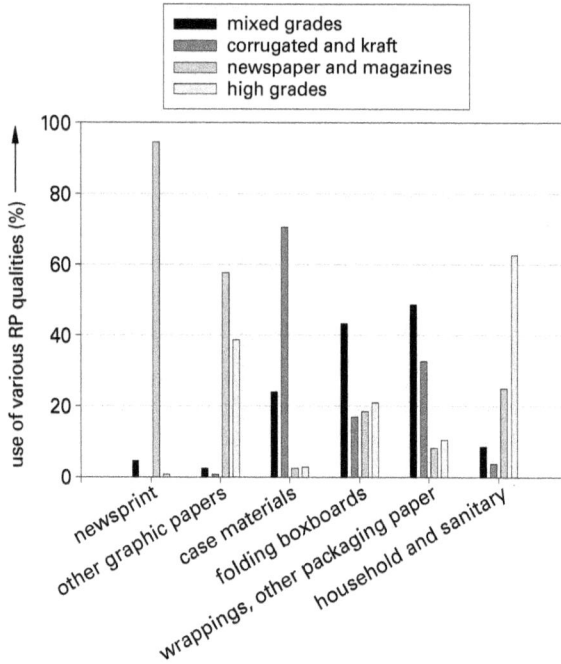

Figure 15.4. The use of different types of raw materials (recovered papers) for production of various types of paper . The data is based on Western European statistics from year 2000 compiled by CEPI.

fect is shown by an increased residual ink concentration in the deinked pulp during the summer months. The phenomenon is related to ageing of the inks.

In order to emphasise that recovered paper is not a waste fraction although it is often called waste paper some examples of pricing for recovered paper qualities are shown in *Figure 15.5*. The figure clearly shows how the prices are affected by the amounts of contaminants, quality of the fibre content and how well defined the raw material is. The most attractive recycled paper is well sorted, contains low amount of contaminants (ink and stickies) and high quality fibres. It is for example not possible from an economical sense to produce newsprint from computer print outs which cost approximately 250 €/ton. The price for the final newsprint product being approximately 550 €/ton (spring 2002) which cannot cover such a high raw material cost. The aim is to produce as high-grade product as possible with as low quality raw material as possible. This is why the choice of RP quality primarily is based on the paper quality produced. Mills producing several different qualities which for instance is common for the tissue mills often switch the recovered paper mix going into the pulper when they change the quality of paper produced.

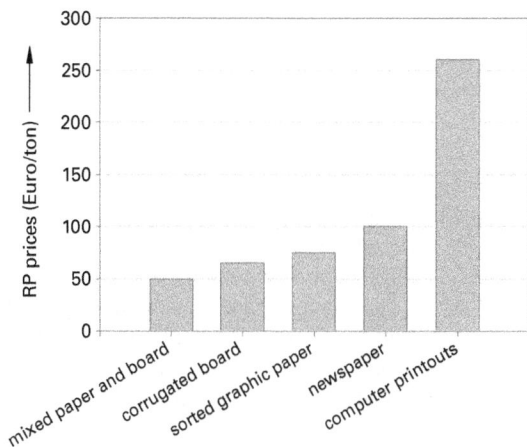

Figure 15.5. Approximate raw material prices during the spring 2002. Observe that the prices vary a lot in a similar way as for pulp and newsprint. These figures only act as a guideline to give an idea about the price differences for different types of recovered paper grades. As an example well-defined pre-consumer RP is expensive while post-consumer waste as our household collection is inexpensive.

15.3.1 Printing and Printing Techniques

The raw material not only consists of coated and un-coated papers, but also of different kinds of inks. The ink types mostly relevant for deinking can be divided into four major groups:

- Oil-based ink as offset lithography
- Toluene-based gravure inks
- Water-based inks as flexographic inks
- Toners and xerographic print

Oil- or toluene-based inks are characterised by the hydrophobic character of the pigment. The most common are offset lithography and gravure inks. The dominant ink type for newsprint printing is offset lithography (commonly called offset) this ink has three major components: oil, binder and pigment. The oil can either be mineral oil or vegetable oil where the vegetable oil has an increasingly market share mainly due to green labelling. The function of the oil vehicle is mainly to transfer the pigment onto the paper. After the print nip most of the oil penetrates into the paper leaving the pigment and binder on the paper surface. The binders can be of several types: resin acid metal salts, resin acids esters, modified resin acids, phenolic resins and alkyd resins. The reason to use binders is to improve the adhesion of the ink film to the paper and the cohesive strength of the ink film. This directly counteracts the ink removal process. The binders have a high surface affinity to the pigment surface and therefore the surface characteristics of the pigment are highly affected by the binders. The most common pigment used is carbon black, which is small carbon particles (approximately 10–400 nm) produced by collecting the smoke when burning hydrocarbon. However, this pigment only accounts for the black inks. The other colours consist of various types of pigment based on metal oxides, different aromatic or metal organic compounds with chromophore groups.

The toluene-based inks (rotogravure) behave rather similar to oil-based offset ink in the deinking process. This type of ink is used for high volume printing of various types of magazines. A scenario that may cause problems in the future is a conversion from toluene based to water-based gravure. However, this ink does not seem to fulfil the requirements of the printing-houses. A water-based gravure ink behaves rather similar to water-based flexographic ink which is described further in the next paragraph.

Water-based inks are easily dispersed in alkaline solution. The printing techniques used are, flexography, gravure and ink-jet. Easily dispersible pigments in this type of inks are very difficult to separate by flotation. Other techniques are required to remove this ink from the fibre stock. The flexographic ink on the other hand is already a problem during processing of recycled pulp. Today there are not any efficient ways to deink paper printed with flexographic inks. The problem is a too extensive fragmentation of the ink film during pulping leading to redeposition (lumen loading) and ink that has too small particle size to be flotated. The usage for flexography printing for ONP is limited to some geographic areas. It is a fairly common technique in US but it is also used in UK and in Italy. A more common usage for flexography in Europe is printing of packaging materials. This is however not a large problem since this material is often not deinked at all. Another type of water-based ink is ink-jet ink. Ink jet inks exist in a various different versions with a large variety of solvents. The water-based ink jet can either be pigment based of dye based. The pigment-based inks tend to behave in a similar fashion as the flexographic ink. The dye-based ink is on the other hand not a deinking issue. Dyes are handled by the addition of bleaching agents in particularly by reductive bleaching.

Toners and xerographic print can be considered as pigment in a cured polymer matrix. This type of ink is hydrophobic by nature and is therefore suitable for flotation. However, the main problem is to efficiently release this ink from the fibres since they tend to have a strong adhe-

Table 15.1. Deinking characteristics and the use of various types of printing techniques and inks.

Printing	Solvent	Main use	Corresponding recovered paper quality	Characteristics of ink type during deinking
Offset lithography (and other oil-based inks)	Mineral or vegetable oil	Newsprint and magazines	ONP, OMG, OCC	Easy to deink in conventional deinking processes
(Roto) Gravure	Toluene	Magazines and commercial adds	OMG	Easy to deink in conventional deinking processes
Flexography printing (often called flexo)	Water	Newsprint and packaging materials	ONP, OCC	Ink which is easily dispersed in water into very small particles is very hard to remove in conventional processes
Ink-jet	Water (among others)	Computer printouts	MOW	Pigment based water based ink-jet is hard to deink in a similar way as flexo. Dye-based ink jet inks can be bleached.
Toners and Xerographic ink	Polymer matrix	Computer printouts and photostatic copies	MOW	Easy to deink although a high energy input is needed in order to avoid ink specks.

sion to the paper substrate. Toners and xerographic ink are deinkable and are usually used for high brightness quality papers. The ink film is naturally hydrophobic and flotates easily. The problem in this case is poor ink detachment and particles that are too large to be efficiently removed (above 200 micron). This results in specks, which is large (visual) ink spots, in the final deinked pulp. In order to solve this problem higher amount of mechanical energy input is needed to obtain a higher degree of for ink detachment and fragmentation. *Table 15.1* summarises the different inks and printing techniques, their use and subsequent behaviour in deinking.

15.3.2 Fibre and Filler Content

The raw material contains not only fibres. Fillers and other additives are often a substantial part of that paper. The fibres are also of different types. MOW and OCC of course contain a lot of chemical fibres. ONP and OMG on the other hand primarily contain mechanical fibre.

The amount of ash varies a lot in different paper qualities. MOW usually contains a lot of fillers while newspaper usually much less. Coated paper of course has a mineral coating layer. In *Table 15.2* an approximate figure of the mineral content in these furnishes are described. It should be noted that the use of fillers and coating pigments are highly regional. The use of kaolin as filler is for instance more common in US than in Europe where the use of calcium carbonate is much higher.

An important point when discussing ash content of raw material is the different demands on ash content for different types of paper produced. Tissue paper production is sensitive to a high ash load in the deinked pulp. Tissue papers are usually made from pulp with ash content below 5 %. In newsprint production a high amount of ash in the deinked pulp means an increased yield which is positive for the economy. However, also the newsprint paper machine runs into problems with ash contents above 15 %. The newsprint mills in Europe use ONP/OMG mixtures ranging from a ratio of 80/20 to 30/70. The ash content in the incoming furnish load is ranging from 10 to 30 % ash.

Table 15.2. Ash / filler content of various qualities of recovered paper qualities.

Recovered paper quality	Ash / filler content (%)
ONP	3–15
OMG	25–35
MOW	15–35
OCC	5–15

The OMG fraction is known to have a large impact on the deinking process. OMG are favourable for the ink removal but on the other hand reduces the yield of the process. It is usually considered that a mixture of ONP and OMG is easier to deink than a sole ONP raw material. There are several reasons for this the most obvious being that the coated paper in the OMG fraction contains un-printed fibres.

15.3.3 Recycled Paper Collection Grades

Recycled paper is a big worldwide market just as the markets for pulp wood and market pulp. More paper is collected than used in countries with large population and paper consumption as USA and Germany. In many cases the distance from a collection region in one country is shorter to the paper mills in a neighbouring country than to the domestic paper industry. This has made the international recycled paper business huge and in order to define the different recycled paper qualities they are dived into 39 different well-defined qualities. The increase in the rate of recycling of paper and board is shown in *Figure 15.6*, below.

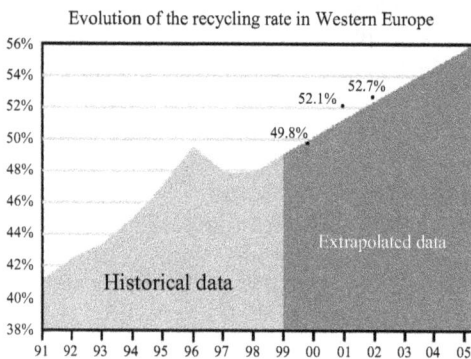

Evolution of the recycling rate in Western Europe

Figure 15.6. Diagram showing a long-term development of paper recycling in Western Europe (1991–2005 prognosis) according to data from European Declaration on Paper Recovery (ERPA).

15.4 Recycled Fibre Processes – General Targets

The principles for preparation of the recycled pulp for paper manufacture are fairly simple in theory but difficult in practise. First the recovered paper is pulped followed by various processes that have the objective to remove various types of contaminants as ink, plastic materials (stickies), metal objects and other types of garbage. Two methods for separating the ink from the paper fibres are used: washing and flotation. In both processes the first step is to detach the ink from the fibres using chemical and mechanical processing. In the washing process the difference in size between the dispersed ink and the fibres is used. The small ink particles are separated from the fibre suspension by the use of wires or screens (not to be mistaken for the separation technique that is called screening). The flotation process uses surface chemical differences between ink and fibre. The froth formed, in which the ink particles are entrained, is removed at the top of the cell. Contaminants with a density that differs a lot from that of the fibres are separated by various types of centrifugal separation techniques (cleaners) while large particles (larger than the fibres) are separated by screening. *Figure 15.7* shows the size of the contaminants removed in the different separation methods. Bleaching is not included in the separation processes but can be performed to various degrees depending on the demands on the final product.

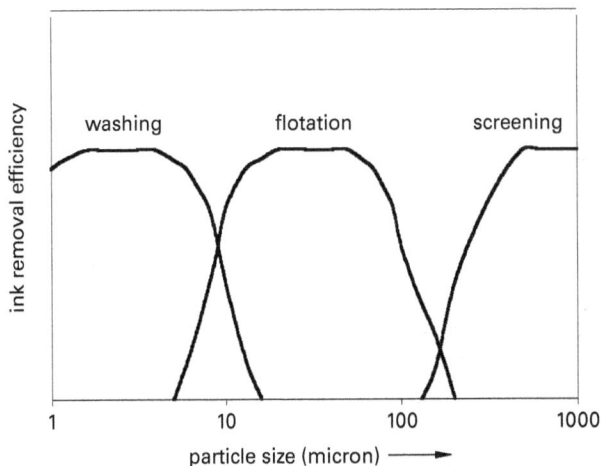

Figure 15.7. The removal of contaminants using various separation techniques. The importance of the particles size on the separation process and the efficient size intervals for the different separation techniques.

15.4.1 Product demands

The demands on the recycled pulp depend on the type of product produced and this in turn controls what raw material that can be used. *Table 15.3* gives an indication of the targets for various types of deinked pulps.

Optical properties are of course important for all paper grades except for pulp used for brown board. It is especially important for fine printing papers. Fine paper puts high demands on both brightness and a low dirt / ink specks count (amount of ink particles >50 µm in diameter, measured as mm^2/m^2 paper). Newsprint and tissue have also high demands on the optical properties although for tissue it varies a lot. In some markets a greyish tone to the hygiene paper has been considered as ecologically sound and therefore fairly low brightness tissue is produced. It is common that tissue mills produce several different qualities using the same process. The product qualities are controlled by the choice of raw material and the chemistry used, especially the use of bleaching agent.

The ash content is another important factor. In respect to the ash content of the DIP the tissue process is the most sensitive. The aim for most tissue mills is to reduce the ash content below 5 % during the pulp preparation which also to some extent explains the lower yield for this process. The reason for a low ash content requirement that a high load of ash results in problems during paper making but also that the end product itself needs to be low in ash. One very important part of tissue making is the sensitive paper making process. The adhesion to the Yankee cylinder is a key to tissue manufacturing. The need of low ash content is the main reason to why washing stages are used in tissue manufacturing. Newsprint on the other hand is not as sensitive to a high ash content in the pulp although also in this case a too high load of ash entering the paper machine with the recycled pulp may lead to runnability problems. Many newsprint mills actually want to retain a higher amount of ash in the recycled pulp. The reasons for this are: retention of the ash means an increased yield and a lower amount of sludge to deposit.

Table 15.3. The targets for different recycled fibre (deinked) pulp; optical properties, strength properties and filler levels for various paper qualities.

Paper produced	Brightness / ink content % ISO / ERIC* ppm	Ash / filler content (weight-%) in the deinked pulp
Newsprint	55–65	10–15
Magazine paper (SC and LWC)	65–75	5–15
Fine paper, market DIP	75–90	< 5
Tissue	45–80	< 5

Stickies is another material that is important to remove before the recycled pulp enters the paper machine. Stickies consists of various polymeric materials such as: glues, pressure sensitive adhesives (envelopes and post-it notes), hot-melt adhesives (used in book folders), latex (coating layers) etc. Stickies is a problem associated with all paper qualities produced from recycled pulp. The problem seems however to be especially large in the production of packaging material, since a lot of this material is not deinked. The techniques to remove ink also to a large extent remove the stickies.

Table 15.4. Yield during pulp production for various paper grades.

Paper produced	Yield over the pulp preparation process (%)
Newsprint	75–85
Magazine (SC, LWC)	70–80
Fine paper	65–75
Tissue	60–75
Packaging and board	90–95

The efforts to remove these contaminants to various degrees are also reflected in the yield of the various processes, see *Table 15.4*. Low prize end products as newsprint needs to be produced with a high yield and a low price raw material in order to have a cost efficient process. Product with a higher end price can afford to loose some yield on account of the pulp quality (for example fine paper). Brown grades of packaging and board not needed to be deinked can be produced at a very high yield which on the other hand also means that a large portion of the contaminants remain in the pulp.

Apart from the optical properties, ash content and stickies, there are also some important properties on the final paper product. All types of graphic papers need a good printability. The important parameters that affect the printability are the paper formation, the surface properties of the paper and strength of the paper. Production of tissue paper has other specific demands as adsorption characteristics, wet- and dry strength. In packaging production the focus is on paper strength. Therefore the properties of the recycled pulp affecting the paper strength is of importance. Contaminants of various types affect the strength of that paper. The strategy is either to fragmentize the contaminants into small particles well distributed in the paper or to keep them large and separate them in screens and cleaners.

15.5 Description of the Recycled Fibre Processes

The outline of a recycled fibre process depends on the goals for the end use of the recycled fibre pulp in combination with the available raw material (the recycled paper) as seen from section 15.4. As the goal for the recycled fibre process is to minimise the content of unwanted substances; everything from metal and glass particles to plastics, print inks and adhesives there is a need for a series of different separation steps along the process. In general the process outline and equipments are also connected to the cost gap between the value of the produced pulp and the sum of the costs for the recycled paper, production costs (chemicals, heat, electricity and labour) and finally capital costs for the process equipments together with buildings, controls, piping etc. The development of processes has always been driven by the possibility to produce a pulp as good as the virgin fibre pulp, but at a lower cost. As can be seen from the statistics in section 15.2 the percentage printing and packaging papers and board based on recycled fibres ha s grown at an enormous pace during the latest 10–15 years.

The collection systems and the sorting systems are of coarse similar for the different recycled fibre processes. Depending on country and region of different countries the recycling companies, or communities get the recycled news, magazines, fine papers, packaging papers, board and folding boxboard either relatively well separated, or contaminated and mixed. In most cases there is a need for sorting stations to further improve the separation into different grades as this makes it easier for the mills to use the recycled fibre. A good and even quality mix of recycled paper is very important for the mills regardless if they produce printing, tissue or packaging grades. This means that the value of the recycled paper is higher if it has been sorted to fractions of news, magazine, fine paper etc. as clean as possible. During the latest years automatic sorting plants also have been designed. One example is an automatic sorting plant close to the new (started 2002) greenfield newsprint mill at Hürth outside Cologne in Germany, designed and owned by Trienekens GmbH. Another example is the Swedish automatic sorting plant in Gothenburg built by IL Retur AB.

The Trienekens process use shredders, magnetic bands and large cyclones while the IL Retur process use an advanced image analysis system combined with air blowing techniques to move the different paper types to the correct direction. Both processes results in a better cleanliness than manual sorting on a mixed recycled paper raw material.

If the recycled paper is sorted by hand, or not sorted and comes in baled form, the recycled paper mill needs devices to take away the steel threads and to shred the bales into lose paper. The paper is then fed by means of band transporters to re-pulping. In modern recycled paper plants also magnetic bands are used in order to minimize the amount of iron containing objects reaching the pulper.

The dissolving and coarse screening systems for treating the recycled papers and boards are commonly used in all recycled fibre processes.

Pulpers are used both for recycled board, tissue and magazines while the drums are mainly used for news and magazine. The re-pulping conditions in pulpers can be much tougher than in a drum. The tougher conditions are often necessary for recycled board and packaging papers than for news, magazine and fine papers. At the same time mild conditions is a prerequisite not to cut synthetic plastic (polymeric) materials into too small pieces, as this will make it difficult to reject these particles in the screening unit operations along the process. As in many process steps this is a balance between achieving best possible pulping / defibration and minimum destruction of materials that should be rejected. Remaining synthetic polymers often create sticky particles that are very detrimental to the paper, tissue and board production, in principle the sensitivity increases with decreasing thickness and increasing paper machine speed.

The major part of the rejects consists of plastics, metals, glass and other not wanted materials are rejected by means of the screening included in the low consistency part of a pulping drum, the coarse hole screen system directly after drum, or pulper, the slot screen systems, light and heavy reject cleaning systems. It is common that the amount of material rejected here is close to 5 % of the incoming recycled paper, the total amount of unusable material is however somewhat less than this. The reject will unfortunately also contain some fibre material as the sorting process steps never can reach 100 % efficiency.

High consistency dispersion with preheating is also a common process step with the main goals to further disperse inks and sticky particles from the fibres, which also improves the mechanical properties of the pulp through its refining action.

Recycled pulp system processes for printing, writing, and tissue papers also include flotation and/or washing with the main goal to minimize the content of remaining print ink in the final pulp. Furthermore, oxidative bleaching and/or high consistency dispersion / mixing in a bleaching tower directly after the disperger is used for these products. Reductive bleaching may be used either with the high consistency dispersion bleaching (30 %), or as a medium consistency (10 %) bleaching step immediately before the paper machine. More details about the chemistry are explained in section 15.6 and information about the unit operations is found in section 15.7.

15.5.1 Deinking Processes for Mechanical Printing Papers

A common outline of the processes for deinking mechanical printing papers can be seen in *Figure 15.8.* The process system consists of a recycled paper feed and bale cutting system, re-pulping, coarse screening, pre-screening, flotation 1/preflotation, centricleaning, fine screening, dewatering on disc filter and presses, dispersing, high consistency bleaching, flotation 2/post-flotation, dewatering on disc filter 2 and presses 2 and medium consistency bleaching.

DIP system for newsprint and improved news

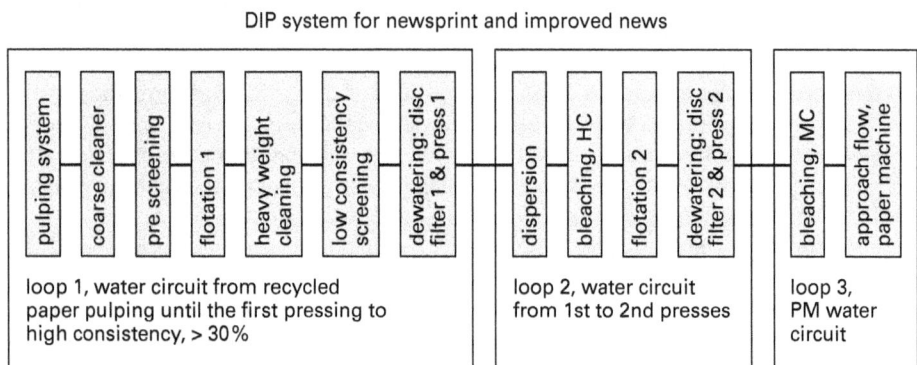

Figure 15.8. Typical process concept for newsprint and improved news. Typical data; 40–90 g/m² papers, yield approximately 83 %, effluent approximately 8 m³/t, filler; feed 22 % finished paper 13 %, brightness 63–66 %ISO. What is not shown in the process concept are the rejects from screening, cleaning and flotation, in total somewhere between 15 and 25 % of the incoming recycled paper.

Deinked pulps for newsprint need to have a brightness of 60 %ISO. This can normally be reached by means of using 5–10 kg/t alkaline peroxide in the pulping together with 10–20 kg/t sodium silicate and about 5 kg/t sodium soaps (oleic and steraic acids) or tensides. These are often charged both in pulping and flotation. The final medium consistency (10 %) bleaching with sodium dithionite is normally used for smaller adjustments in brightness. The amount of print ink is originally about 1,5–2 % of weight of the incoming newsprint and magazines. About 80–90 % of this is rejected by over the flotation steps, but some is also rejected due to the washing in the dewatering stages.

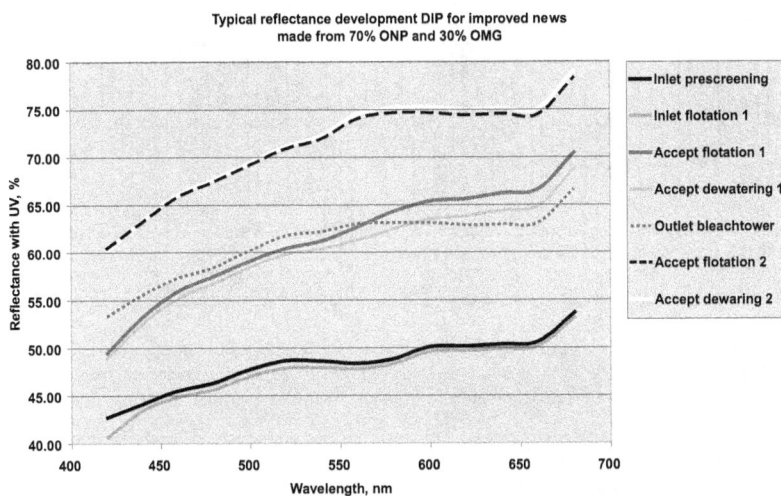

Typical reflectance development DIP for improved news made from 70% ONP and 30% OMG

Legend:
- Inlet prescreening
- Inlet flotation 1
- Accept flotation 1
- Accept dewatering 1
- Outlet bleachtower
- Accept flotation 2
- Accept dewaring 2

Figure 15.9. Normally only the brightness around 457 nm (ISO brightness), or 557 nm (Y-value) are measured. This figure however describes typical reflectance spectra along the whole visible spectrum measured on lab sheets (P. Engstrand).

When deinked pulps are used in magazine grades as SC and LWC papers the brightness demand is between 70 and 75 % ISO. In order to reach these values, normally higher amounts of recycled magazines (OMG = old magazines) and lower amounts of recycled news (ONP = old newsprint) in the raw material mix are used in combination with higher dosage of peroxide primarily in the high consistency bleaching step directly after dispersion. Sometimes it is also necessary to have a larger flotation capacity in flotation 2 and to use more dithionite in the final MC-bleaching tower. Some mills prefer to use reductive chemicals in the dispersion bleaching instead of peroxide.

Figure 15.9 describes the development of brightness / reflectance of a typical deinked pulp starting with samples from pre-screening, via samples from flotation 1, from outlet of bleach tower (here with peroxide bleaching 3 % charge) and finally samples from flotation 2 and dewatered pulp. It should be pointed out that the bleaching chemicals mainly reacts with the lignin and only to a small extent with coloured ink pigments. This is clearly shown in the figure above as it is only in the reflectance area below 600 nm that is bleached. It is also possible to see that when print ink is removed as in the two flotation steps the curves are moved upwards in parallel. More important is that the combined dispersion-bleaching actually decreases the reflectance level at wavelengths over 550 nm at the same time as the reflectance below 550 nm is increased.

This is due to the fact that the disperger reduces the size of the ink particles and this is interpreted by the measurement device as a reflectance reduction. The amount of ink directly after the bleach tower is still the same, but it is spread out over a larger surface in the pulp. The real brightness response is therefore in this case 4–5 units higher along the whole outlet bleach tower curve than can be seen. Flotation 2 will have a much more important role when producing higher brightness DIP as the relative effect of remaining print ink is larger when the pulp brightness is higher, i.e. the lower the light absorption coefficient of the pulp fibres.

DIP system for wood free pulp for fine paper

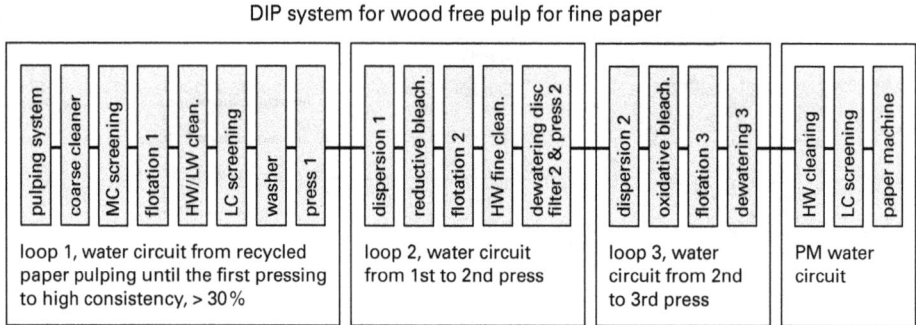

Figure 15.10. Advanced process concept for fine paper grades (high grade printing and writing papers) or market pulp based on Mixed Office Waste (MOW). What is not shown in the process concept is the rejects from screening, cleaning, flotation and washing steps, in total somewhere between 30 and 35 % of the incoming recycled paper.

15.5.2 Deinking Processes for Fine Paper and Market Pulp

The deinking process for fine paper pulps is pretty similar to the one for news and magazines, but there are some important differences. As the print ink to a large extent consists of xerographic inks and also inkjet inks there is a bigger need for tough conditions in order to detach enough amounts of inks from the paper / fibres. There is also the same need is due to reject plastic materials as in the process for newsprint and magazines described above. This results in a process concept often including pulper instead of the milder pulping drum and an extra focus on dispersion. In order to disperse the more difficult inks a dispersion unit of kneader type is often used. A typical process for deinking of fine paper pulps can be seen in *Figure 15.10* below.

Surprisingly, the brightness levels reached with the pulps based on recycled fine papers can reach the same, or even higher levels compared to virgin chemical pulps. This is due to the use of optical whiteners in the fine papers and because many of these additives survive the deinking process. If the UV part of the light entering the measurement is eliminated the DIP brightness will be lower also for MOW based pulps, see *Table 15.5* below, describing typical optical values from two European recycled fibre pulps (in France and UK).

Table 15.5. Typical efficiency af a MOW based DIP system. (Ref. Recycled Fiber and Deinking, p.227, Göttsching and Pakarinen).

Quality parameter	Pulp after coarse screening	Finished pulp
Brightness (without UV*)	60 %ISO	80–85 %ISO
Dirt Speck Area	2800 mm²/m²	< 10 mm²/m²
Ash content	21 %	5 %

* Without UV luminance

15.5.3 Deinking Processes for Tissue Papers

The deinking process for tissue papers has one main difference from the processes above and this is the goal to minimise the filler content. In order to minimise filler content the washing unit operation is used not only to reduce the amount of print ink but also to reduce the amount of filler, the drawback being that yield is lowered as not just fillers are removed but also fine material from the fibres. A typical tissue deinking line is outlined in *Figure 15.11.*

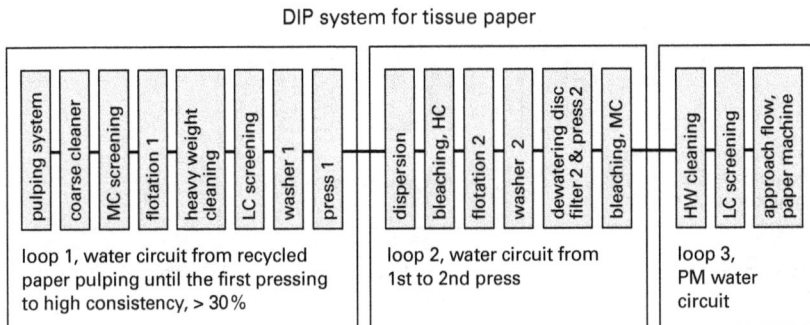

DIP system for tissue paper

pulping system | coarse cleaner | MC screening | flotation 1 | heavy weight cleaning | LC screening | washer 1 | press 1 | dispersion | bleaching, HC | flotation 2 | washer 2 | dewatering disc filter 2 & press 2 | bleaching, MC | HW cleaning | LC screening | approach flow, paper machine

loop 1, water circuit from recycled paper pulping until the first pressing to high consistency, > 30%

loop 2, water circuit from 1st to 2nd press

loop 3, PM water circuit

Figure 15.11. Typical process concept for tissue paper grades normally based on both Mixed Office Waste (MOW) and household collection of old news and old magazines (ONP / OMG). What is not shown in the process concept are the rejects from screening, cleaning, flotation and washing steps, in total somewhere between 30 and 37 % of the incoming recycled paper.

As the softness and water absorption of the tissue are the most important properties it is very important that the filler amount is reduced as much as possible. Depending on what tissue product the pulp is aimed for there are different demands on brightness and on absorption level.

15.5.4 Recycled Fibre Processes for Packaging Papers

For recycled pulps replacing virgin pulps in brown grades there is normally no demand to minimise the content of print inks, the most important issue is instead the strength properties and to some extent cleanliness. Liner produced mainly from recycled paper is called test liner. For virgin pulp based liner as well as for recycled paper based liner the most important properties are burst strength and ring crush strength. Even if it would be possible to get a recycled fibre from clean OCC (old corrugated containers) it would never be possible to get as high strength properties as from virgin fibre pulp. This is due to the fact that the OCC both contains liner (unbleached softwood and hardwood kraft) and the corrugated medium (NSSC, neutral sulphite semi chemical hardwood pulp) and that the chemical fibre suffers from less bonding ability after drying and curing. This is sometimes referred to as hornification and can be described as a state where fibres after drying are not as able to swell as in the original wet state.

In *Figure 15.12* below a typical outline of a recycled pulp system for packaging paper is outlined. This process is focused on re-pulping of the recycled corrugated containers and other packaging papers followed by cleaning and screening.

recycled pulp system for packaging paper as test liner

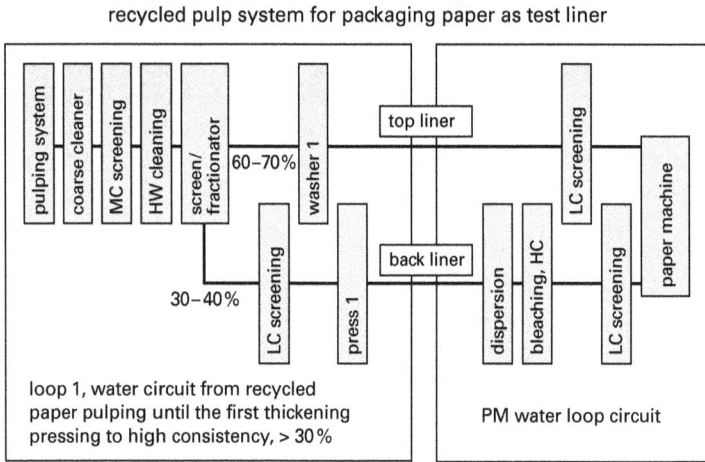

Figure 15.12. An example on a recycled pulp system for a two layer test liner, it should be pointed out that there are many versions of processes used in production of packing papers.

15.6 Process Chemistry

15.6.1 Removal by Flotation

The removal of ink in a flotation deinking mill can be divided into a number of different surface chemical sub-processes. The five main processes are described in *Figure 15.13*. First the ink has to be released from the fibres. Then the particles have to be agglomerated to a floatable size, because hydrodynamic forces prevent small ink particles from attaching to an air bubble, see *Figure 15.14*. The size range where flotation is an efficient separation technique is between 10 and 200 μm.

The agglomeration is accomplished with collectors. The most common are fatty acid calcium soaps. The collectors also improve the chemical interactions between ink particles and air bubbles, during the flotation.

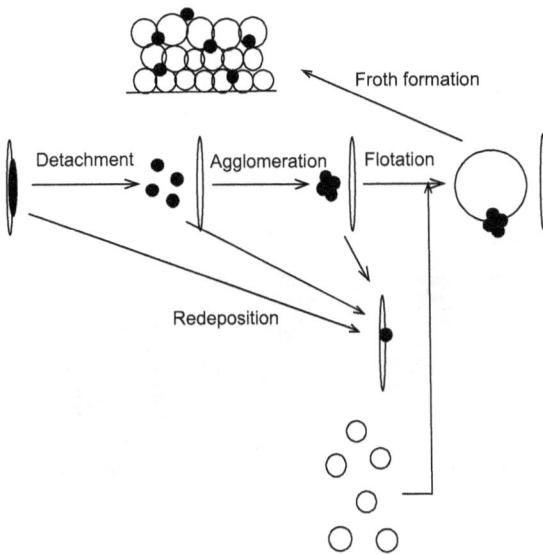

Figure 15.13. Simplified description of the sub-processes of flotation deinking chemistry.

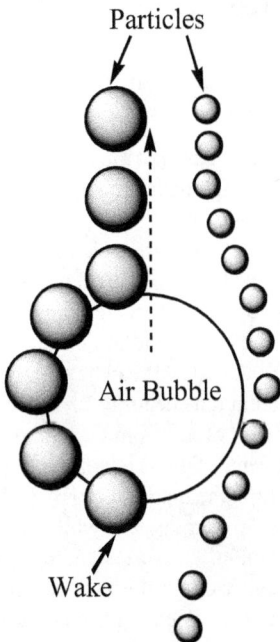

Figure 15.14. The reason to why small particles do not flotate. Small particles follow the stream lines around a rising bubble and therefore do not collide with the bubble. The momentum of the particle is to low. A larger particle on the other hand collides and can therefore attach if the interactions between the particle and the bubble are favourable.

The fourth process is redeposition, which means that detached ink particles re-attach to the fibres. This is something that should be avoided. Redeposition is an especially large problem for water-based inks. The small well-dispersed ink particles formed when pulping water-based inks tends to penetrate into the lumen of the fibres (sometimes referred to as lumen loading).

Instead of redepositing the particles are supposed to attach to the air bubbles that form a froth layer in which the ink particles are entrained. *Figure 15.15* shows how a mill size flotation cell with a froth layer is working. A picture on the froth formed in the first of 6 cells is shown as well as of the last cell where most of the ink has already been removed.

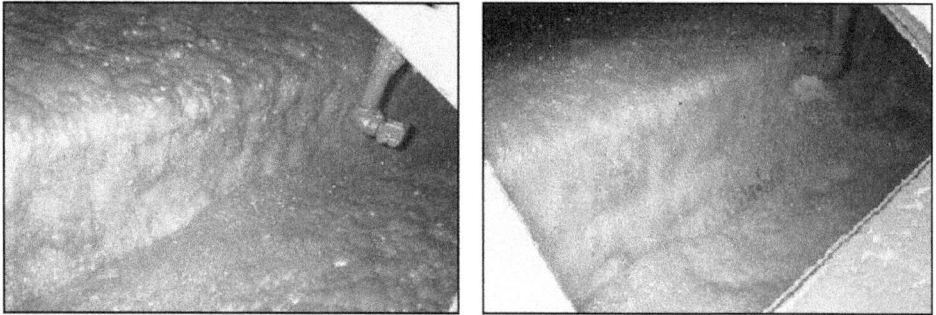

Figure 15.15. The picture to the left show the first and the one to the right show the last flotation cell of six in the pre-flotation. The cells on these picture are Voith cells which can be further studied in Figure 15.21.

15.6.2 Chemicals used for Flotation Deinking

* *Sodium hydroxide,* is used to increase the pH, which improves the swelling of the fibres and the emulsification/dispersion of ink. A high pH is important for a high detachment rate of oil-based-ink from fibres. The sodium hydroxide is added in the pulper.
* *Fatty acid soaps,* are used as collectors. The calcium ions precipitate the fatty acid anion. The actual collector is thus not the free fatty acid ions but the calcium soap. Calcium also contributes to increasing the ionic strength, which is necessary for an efficient agglomeration. The fatty acids are added as sodium salts or emulsions to the pulper and/or before the flotation. The effects of the fatty acids on the agglomeration of ink particles can be seen in *Figure 15.16.* In some mills with low water hardness extra calcium ions needs to be added.
* *Surfactants,* are sometimes used to further improve the detachment of ink from the fibres and to prevent re-deposition of ink onto the fibres after detachment. There is a large regional difference between Europe and North America. In Europe the market is completely dominated by fatty acids in various forms (fatty acids, soaps, emulsions) while in North America the use of surfactant systems are much more common. The surfactant in this case is either used on its own or as mixtures with fatty acids. The reason for this is of course related to the difference in process layouts in North America but may also be related to differences in the raw material. As discussed earlier the froth structure is very important during flotation and therefore froth control agents are often used to ensure that a sufficient froth is formed during flotation. The collector or dispersing agents used usually fulfil this function. The most common situation is too much frothing which is efficiently counter-

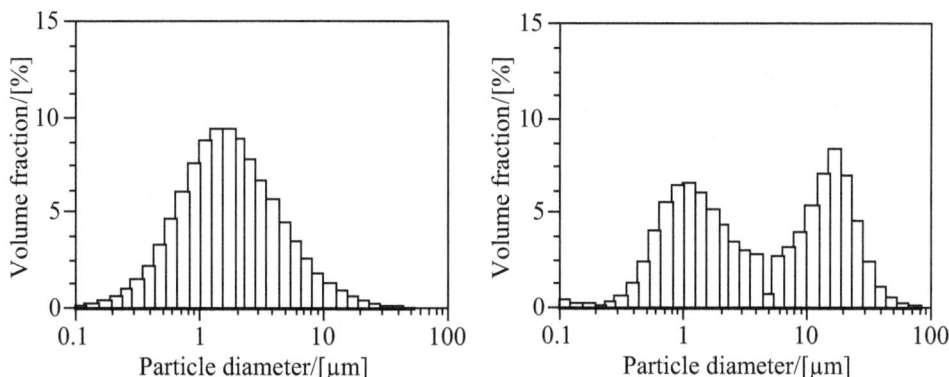

Figure 15.16. Examples of agglomeration of a suspension containing small ink particles (news cold-set offset ink) using fatty acid soap. <u>Left</u> before addition of fatty acid soap and <u>right,</u> after addition of soap. Observe that the optimal size range for flotation is 10–200 micron.

acted by an addition of fatty acid soap shortly before flotation. The calcium soaps formed in the process act as defoamers.

- *Hydrogen peroxide,* is used, as a bleaching agent to counteract the yellowing of the fibres at alkaline pH. There are three different positions where the peroxide can be added; in the pulper, before the disperger, or in a separate bleaching stage after deinking. The standard is to add some of the peroxide in the pulper in order to counter-act irreversible yellowing but addition in the disperger is also very efficient. Bleaching in a separate bleaching stage is used for papers with high demands on brightness.
- *Waterglass,* is used as a stabilising agent for the hydrogen peroxide and as a pH buffer. The waterglass is also supposed to have positive effects on ink detachment, ink agglomeration and prevent ink redeposition. Usually it is considered favourable to partly substitute sodium hydroxide with waterglass as a source for alkali.
- *Talc,* is used as an anti-stickies treatment chemical. The talc acts as a detackifier of the sticky material. The addition point is usually in the end of the deinking process.

There are also several speciality chemicals that sometimes are used in the deinking process. The complexing agent EDTA is sometimes used to control salt concentration in the pulper. It is, however, more commonly used to control some detrimental metal ions during bleaching. Enzymes may be used in some mill processing MOW to aid ink detachment. Fixation agents and other type of stickies treatment agents are also added to the pulp before papermaking.

15.6.3 Chemicals used for Deinking by Washing

In the washing process the difference in size between the dispersed ink and the fibres is used in the separation process. The small ink particles are separated from the fibre by passing the pulp over a wire or screens. The chemistry for this process is similar to the one used for flotation deinking. The main difference is a higher usage of surfactants instead of fatty acids.

Combinations of flotation and washing are very common and all mills have some sort of de-watering stage, which could be compared to a washing stage. In Europe the flotation technique dominates completely. In North America only a few large wash deinking mills remains although mills using a combination of flotation and washing is more common in North America than in Europe. The reason for this geographic difference needs a topic of its own but one main physical difference between the techniques is the washing technique removes much more filler than flotation. This is the reason why washing is used when processing pulps that need low ash content as tissue and pulps for paper machines running at acidic conditions.

15.6.4 The Differences between Printing Paper, Tissue and Packaging Grades

The normal process chemistry for printing paper is alkaline deinking with fatty acid or fatty acid emulsions as ink collectors. However, in North America the use of surfactants is common.

In tissue paper manufacturing normal surfactants or mixed fatty acid/surfactant systems are used. The use of washing in tissue mills clearly favours the use of surfactants to fatty acids as deinking agents. Another important difference is that neutral condition often is used,since the raw material for tissue manufacturing often is mixed office waste. This is especially the case in North America while the European tissue manufacturers use a larger variety of raw materials. This in turn is related to the differences in the quality demands in these two areas. The optical demands on the paper in North America are higher than in Europe, although European mills also produce high brightness qualities.

Both for tissue and fine paper MOW is a common raw material and this material needs special attention. Mixed office waste is a very special grade and therefore the chemistry has been adapted to suite this paper grade. The ink in this case consists of polymer films that adhere strongly to the fibres. This means that in order to increase the separation of ink from the fibres more energy input as well as surfactants is needed. Surfactant systems are standard when processing MOW. Enzymes have in some cases been used to further improve the ink detachment. It is also common that this type of mills run at a neutral pH, which is due to that as the swelling of the fibres is less important for MOW. This is partly due to that the fibres in this raw material fraction are chemical pulp fibres, which do not swell as much as mechanical pulp fibres.

The brown packaging grades are as already known usually not deinked and bleached. A minimum of chemicals is used. The recovered paper is pulped and some larger contaminants are removed after which new paper is made of this fairly dark pulp. Packaging grades that are deinked are treated in a similar way as pulp used for graphic papers.

15.6.5 Bleaching of Recycled Pulp

The bleaching of the pulp is another way of improving the optical quality on the pulp. The bleaching of the recycled pulp varies depending of the raw material used and the type of paper produced. Hydrogen peroxide is the standard bleaching agent since it is easy to use and has a low cost. In a newsprint deinking facility the raw material is usually ONP/OMG which means that a large amount of the paper contains mechanical pulp fibres. The hydrogen peroxide is very suitable in this application since it is non-delignifying. The paper should be bleached and not delignified. Lignin degrading bleaching would result in a lot of unwanted residual material

(COD etc) and a reduced yield. The oxidative bleaching is often combined with a reductive bleaching in the end of the recycled pulp preparation process. The sodium dithionite (hydrosulphite) and formamidine sulfinic acid (FAS) are two reductive bleaching agents, which are commonly used for bleaching and colour stripping. In Europe dithionite is the most commonly used although the two have very similar bleaching result. The reductive bleaching is especially important for pulp containing a lot of dyes either from dye based inks or from coloured paper. The newsprint mills usually use only hydrogen peroxide or a combination with hydrogen peroxide and a reductive bleaching agent. Tissue mills also use hydrogen peroxide while fine paper mills can have a more extensive bleaching sequence after the deinking process. These mills often use other raw material qualities as MOW and other high chemical fibre content raw material, which has a more positive response to delignifying bleaching due to a lower content of lignin. In Europe this would most probably mean bleaching with oxygen while in North America the choice may also be chlorine dioxide or sodium hypochlorite. The oxygen bleaching is usually combined with hydrogen peroxide bleaching. When discussing bleaching an important factor in hydrogen peroxide bleaching is an enzyme called catalase. Catalase is formed by bacteria in the process water as a defence towards the hydrogen peroxide added. The catalase acts as a catalyst for hydrogen peroxide degradation. This means higher hydrogen peroxide consumption than otherwise would be the case. There are several ways of dealing with this serious problem: cleaning of the system with regular intervals, use of biocides, chock dosages of hydrogen peroxide (kills the bacteria), dosage of the peroxide in the disperger instead of in the pulper (higher temperature in the disperger) and several other strategies.

15.7 Unit Operation

In order to understand the function of the many different unit operations used in the processes of recycle fibres, this chapter describes them based on the specific goals for the recycled pulp processes. Many of the unit operations discussed here are the same as used when processing chemical and mechanical pulps, but there are in many cases adoptions made to each unit operation in order to fit it in with the specific problems and goals associated with recycled fibres. In the cases when the machinery is specific or very different from those used in chemical or mechanical pulping there are more thorough descriptions and pictures added to the text.

15.7.1 Recycled Paper Collection and Handling

This is a very important and sometimes neglected part of the system and could well be compared with the wood handling systems for chemical and mechanical pulps, which if optimised in a correct way starts already in the forestry with planning and later after cutting, the wood is sorted depending on product demands from saw mills and from the pulp and paper producers. The important difference is of course that the quality of the recycled paper to a large extent depends on the paper recycling system and the habits of people in different regions and countries. The main recycled paper and board qualities are mechanical papers as newsprint (ONP = Old Newsprint) and magazine (OMG = Old Magazine), bleached chemical pulp based printing and writing papers (MOW = Mixed Office Waste) and finally the largest group, i.e. packaging papers dominated by the brown grades (OCC = Old Corrugated Containers). It is obvious that brown paper and board grades will disturb the quality when producing recycled pulps replacing bleached chemical pulp and de-inked pulps for printing papers or tissue. The recycled packaging paper grades are also disturbed by printed grades mainly due to their high content of fillers.

Figure 15.17. The picture shows typical Swedish household collection raw material, 70 % old news and 30 % old magazines.

15.7.2 Sorting and Feeding Systems

The world wide recycling degree has increased and that the recycling systems for paper, plastics, metal etc. have developed a lot though the years. There are in spite of this a lot of contaminants / not wanted substances still remaining in the different recycled paper assortments. Therefore there is a need for further sorting of the recycled paper coming especially from the household collection. This is normally done manually, but full or half automatic sorting procedures have lately been developed in Sweden (IL Retur) and in Germany (Trienekens). The recycled paper bales and loose paper arriving to the recycled paper plants still contains contaminants in the form of metal wire, staples, plastics, glass etc. Bigger contaminants are extracted by means of shredding of the paper and magnetic metal traps. The total amount of unwanted substances reaching the European DIP plants that are working with extensive control of the recycled paper is normally around 2 %, of 0,2–0,3 % are impurities and the rest consists of unusable papers, see *Figure 15.18*.

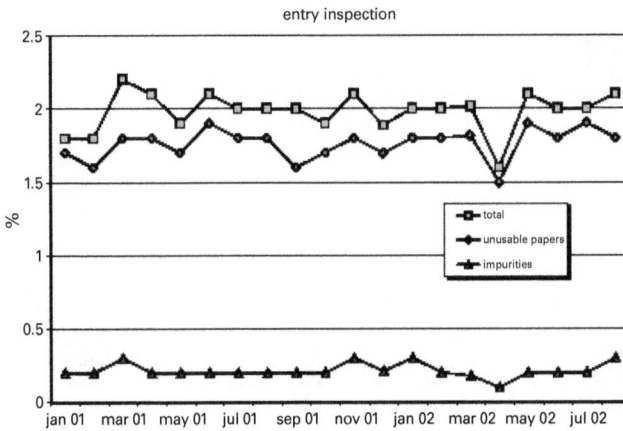

Figure 15.18. Typical follow up of recycled paper quality as mean values among 10 European DIP mills producing printing papers. Unusable papers are in this case typically brown carton board, envelopes with unsuitable glues, stickers and post it. Impurities can be anything from plastic bags and CD´s to cellular phones, toys and shoes.

15.7.3 Pulping

Recycled paper coming from the sorting system is slushed in conventional high or low consistency pulper, or in a pulping drum, see *Figure 15.19*.

The two main purposes with this unit operation are to dissolve the paper to separate fibres and to separate print inks and paper coatings from the paper surface. The mechanical strength of the paper is reduced by wetting, which makes the hydrogen bonds between fibres and fibre fines break. At the same time ink and coating on the paper surfaces is detached. The wetting time / strength reduction is only 1–3 min[1] and will depend on temperature and pH. The rate of wetting increases with the temperature and pH. If temperature and pH is too high there is a risk that hot melt glues in magazines and catalogues are softened and not possible to reject by the screen plates of the drum or the screens later in the system. Residence time in a drum pulper is about 20 minutes and the same, or somewhat shorter in conventional high consistency (15–20 %) and low consistency (5 %) pulpers. During the pulping the ink particles and sodium hydroxide, sodium silicate and in the case of printing and tissue products also hydrogen peroxide and in some cases a tenside is added. This normally improves the wetting and is important for the removal / dispersion of print inks paper coatings. When the pulp is diluted the fibres pass the holes in the screening zone of the drum or through the holes of the disk screen after the pulper. Pulping drums normally have a screen-hole size of 6–8 mm in the screening zone. Disk screens used after conventional pulper normally have a hole-size of 4–6 mm.

---

[1] H. Holik, „Ch. 5. Papermaking Science and Technology, Recycled Fiber and Deinking"

fiber flow drum pulper

consistency 2...5%

consistency 15 (...20) %

Figure 15.19. The FibreFlow® drum pulper (Andritz-Ahlstrom™) above is common as pulping equipment when repulping recycled paper for production of DIP for news and magazines.

15.7.4 Screening

The focus of the design of the screening operation in recycled fibre pulp production lines is to minimise the amount of detrimental substance / contaminants as; plastics synthetic, widely used polymers as catalogue backs, coatings, glues and other adhesives, metal clips and wet strength packaging papers can also be rejected. In order to succeed with rejecting heavy particles the coarse screens not just only contain rotors and hole screen baskets but also a cyclone function separating the heavy reject from the light reject. In advanced recycled fibre and deinking systems coarse screening, pre screening and fine screening is included to insure clean enough pulp for papermaking. One of the most important developments in the field of screening is the slotted screen baskets with slot widths down to 120 μm (in fine screens) that together with efficient design of the bars between the slots give good opportunity to reject very small detrimental particles. An example of this is shown in *Figure 15.20*. With smaller slots widths in fine screening there is a risk that also pulp fibres will be rejected. Screening is performed in 2–3 steps, i.e. the reject from the first step is screened again in a second step and the reject from this step is screened in a third step from which the reject goes out to the reject system.

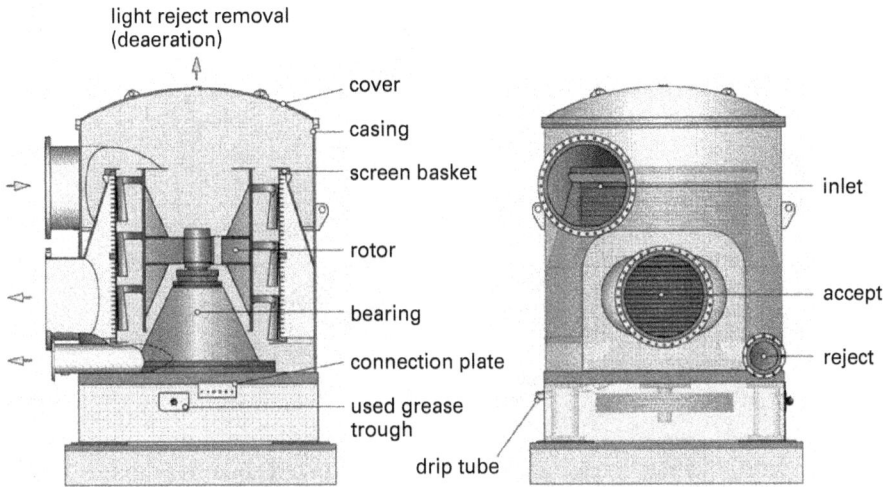

Figure 15.20. The screen above is a modern version of screens from Voith Paper a so-called MultiScreen™ MSS. This is in recycled fibre processes used for combined mostly for fine-screening at low consistency. The same type of screen can also be set up for fractionation.

15.7.5 Fractionation

Screens for fractionation are to a dominating extent based on the same technology as conventional screens. The goal is however different as the main objective here is to separate long fibres from fine fibres, fines, fillers and small contaminant particles as ink. Normally the best results are gained by means of hole-screen baskets with very smooth surface and mild conditions, i.e. smooth rotor and low enough fibre consistency not to press fibres through the holes of the basket. Fractionation is most commonly used in processing recycled pulp for paperboard and testliner to make it possible to produce a product with a bulky middle or base layer and fine top or outer layers. Fractionation combined with refining of only the coarse fibre fraction is evaluated as a method to improve recycled pulp strength and paper surface properties.

15.7.6 Flotation (Primary, Secondary – Pre and Post)

Flotation cells have many different designs even if the function is the same, one of the most commonly used is the Voith tank flotation system shown in *Figure 15.21*. Other constructions that should be mentioned are; Escher Wyss round flotation cells, Lamort round flotation cells and Beloit pressurized cells with froth suction systems. Recently Andritz and Metso Paper have developed new flotation systems. The chemistry of the flotation was shown in section 15.6. The flotation is performed in steps where accept from one cell enters as inject in the next cell until the pulp is free enough from print inks in order for the pulp brightness to reach the required level.

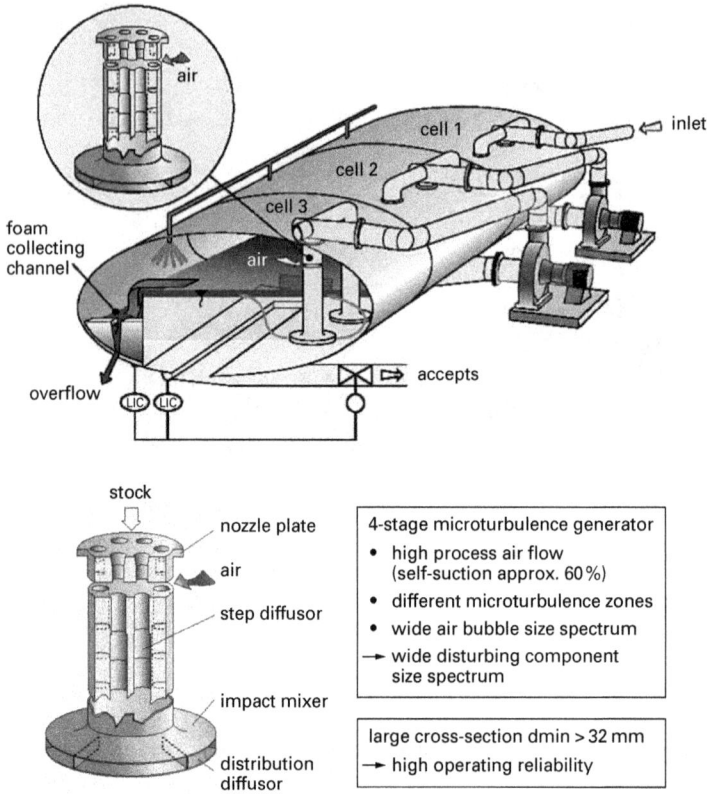

Figure 15.21. The flotation equipment above is a modern and very common set-up of Voiths tank flotation system with air injectors in the bottom part of the cells stepwise mixing air into the pulp at 1–1,5 % consistency. The foam created at the surface flows over an edge into a foam channel. In modern systems normally 5–6 steps are used in flotation 1 and 2–3 steps in flotation 2. The injectors consisting of nozzles and step diffusors is also shown.

The reject from the primary flotation cells contain a substantial amount of fibre material and are therefore treated once more in a secondary flotation that recovers fibres back to the primary flotation and increase the concentration of ink and fillers in the reject. Many modern de-inking plants contain two loops with one flotation operation in each loop. Normally between 10 and 20 % of the original print inks still remain in the pulp after flotation. Ink particles in the size range from 30 to 300 μm in diameter can be floated provided that they are hydrophobic enough to attach to the air bubbles created in the injectors in the bottom part of the flotation cells.

15.7.7 Dissolved Air Flotation

Dissolved air flotation (DAF) consists of simple process equipment using complicated surface chemistry and is very commonly used in DIP systems to make it possible to reuse the water transporting different rejects out of the system. Pressurized air is introduced in the contaminated stream that is pumped in to a simple flotation cell allowing the very fine bubbles to reach the surface. Normally a combination of a large cationic polymer as polyacrylamide (PAM) and fine dispersed bentonite clay is used to agglomerate all particles colloidal or larger. The goal is that all material of larger size than colloidal is rejected over a dissolved air flotation unit. This reject is mixed with the reject from the flotation and washing, dewatered in cyclones, presses and / or centrifuges.

15.7.8 Washing

Washing is in deinking systems used to reduce the amount of particles normally smaller than about 30 um, of coarse there is no sharp limit and the limit will depend on the type of machinery and the running mode. In North America washing deinking has been more widely used than in Europe also for production of DIP for newsprint. Due to differences in acceptance among customers / print houses regarding what paper density is acceptable there are different demands on how much filler the paper should contain. In the production of tissue and market DIP, washing is a necessity in order to achieve low enough filler content in the final product. In DIP for high grade printing papers as LWC there is also a need to reduce both filler content and to have a further reduction in ink. It should also be mentioned that the presses used in conventional deinking systems also to some extent work as washing equipments.

Specialized equipments that should be mentioned are; VarioSplit and GapWasher (VoithPaper and MetsoPaper), see *Figure 15.22*. Disk thickeners and Dynamic washer (screen type washer).

Figure 15.22. This washer is one of the latest developed and is used typically in recycled paper processes for tissue pulps, where it is important to minimise the filler and fibre fines coming from the recycled paper normally consisting of recycled office paper, magazines and newsprint.

15.7.9 Dispersing

Dispersing is also a key process in recycled pulp systems in order to further detach dirt specks, stickies and in some cased to refine the pulp. Most of the dirt specks and stickies still attached to the pulp after the first loop need more energy than what was introduced in the first pulping to be detached from fibres. These conditions can be achieved in a disperging unit as outlined in *Figure 15.23* below. The pulp is preheated up to 3 minutes at up to 120 °C at high consistency (> 30 %) and fed to the disperger where it is dispersed with 40–100 kWh/t energy input. Dirt specks and stickies in the size range above 50 μm (as diameter) are reduced normally by 50–80 %. At the same time the brightness value decreases somewhat (2–4 %ISO) as the same amount of ink is present as before but distributed into smaller ink particles. This makes the brightness measurement unit measure a lower value. The ERIC method (efficient residual ink concentration) is for the same reason showing an increase in the ink concentration. When post-flotation is used in loop 2 after the dispersion the brightness value will be at or above the level reached after flotation 1 in loop 1, but now the amount of ink is in reality further reduced.

Figure 15.23. The picture above describes a typical dispersion system, consisting of a conical plug-screw (in the back) feeding the high consistency DIP to the pressurized pre-heater (up to 120 °C, 3 min residence time) from where the pulp is fed to a feed screw and directly to the disperser, in this case a conical model.

15.7.10 Cleaning – Hydrocyclones

Hydrocyclons also called centrifugal cleaners are used both in order to reject particles with higher density than pulp fibres and particles with lower density than pulp fibres. Very important is the heavy weight cleaners rejecting sand, metal, glass and some shives as these particles are common contaminants in the raw material. Lightweight cleaners are used to reject plastic materials with low density. The specific surface of the particles is also important. It should be mentioned that the detrimental particles called stickies can have lower, same or higher density than water.

15.7.11 Bleaching

Bleaching can be performed both during pulping, high consistency dispersion and in a separate MC-bleaching tower finalizing the process. Due to that alkaline conditions normally are used in a pulping stage, peroxide is often used as high pH is required to get high enough consistency of peroxide anions in the pulp to achieve an efficient bleaching. If a high consistency (> 30 %) bleaching tower is installed after the disperger this position is the most efficient position for alkaline peroxide bleaching in a DIP process as the pulp has been washed in loop 1. When there is no high consistency bleach tower after disperger it is more common that reductive bleaching with dithionite or FAS is used. Normally a medium consistency (10 %) pump is placed before an upstream MC-bleach tower (stand pipe) where a reductive bleaching agent, normally sodium dithionite is used.

15.7.12 Refining

Low consistency (4 %) refining is also a common unit operation when processing recycled pulps. It is always used when pulps are produced for packaging grades and is becoming more common also for improved grades of deinked pulps intended for use in magazines (SC and LWC) and improved news. The goal is as mentioned in other chapters to improve strength properties, or optimise the fibre length distribution both for the recycled chemical and mechanical pulp fibres. If the objective is to improve strength the refining is optimised to fibrillate fibres without cutting while if the goal is to optimise the fibre length for some specific product generally that fibre cutting is allowed. In the case of deinked pulps low consistency refining can also be used as an extra disperser after or instead of the more expensive high consistency disperger to disperse print ink and other contaminants that can be flotated in the post-flotation.

15.7.13 Fibre Recovery

In the different cleaning steps of the recycled fibre processes there is always a risk that good pulp fibres are lost with the rejects. This is true for the screening stages, cleaning stages as well as for the flotation and washing stages. The techniques used are specially designed washers, filters and screens adopted for the purpose of recovering fibres.

15.8 Environmental Aspects

15.8.1 Water Use and Cleaning

In a large-scale process as the recycled fibre industry there is of course also a need for strict control of the use of water and the amount of wastewater. In general it can be said that the recycled fibre plant is almost totally closed. This means that the fresh water is added to the process in the paper machine and cascaded backwards into the process. This is also the way the water is handled in the deinked pulp plant. Water from the first dewatering stage is used to dilute the incoming recovered paper in the pulper etc. The water consumption varies for different papers produced from recovered fibre, *Table 15.6,* due to differences in process, lay-outs as well as the need for removal of contaminants. The water is usually treated in one or several water treatment facilities in the paper mill. The dominating water treatment technique for internal cleaning in the plant is dissolved air flotation (also called micro-flotation). A mill using recycled pulp for paper manufacturing may have from 1 up to 3 dissolved air flotation units. The chemistry used is flocculating systems with various types of polymers, betonite clays and precipitation salts. The contaminants need to be flocculated in order to be efficiently removed. Polymers are also used in order to improve the dewatering of the waste sludge coming from the dissolved air flotation unit. Observe that only a fraction of the water phase is cleaned. The total amount of water in the process is large and it would cost too much to treat the whole water phase. The sludge from the dissolved air flotation units is dewatered in a press and incinerated.

Table 15.6. The fresh water consumption used for production of various types of paper based on recovered fibre.

Type of paper produced	Fresh water consumption (m³/ton air dry pulp)
Newsprint	6–15
Tissue	10–15
SC paper	10–15
Packaging paper	1–15

15.8.2 Sludge Handling

The residual material after incineration of the sludge has to be taken care of. There are several ways of dealing with this material. In Sweden it is still common to deposit the material. The large problem with this procedure is of course the cost but it is nether desirable from an environmental point of view. The fee for deposition of this material varies a lot in different countries and there is a large need to find less expensive and more recycling correct ways of dealing with this material. For example, it can be used for land reclamation, filling out old mines and using the material as a component in concrete and brick production, it is also common to use the material in road building. Another very attractive solution is to reuse this material as paper filler although the trials to do so have not yet been successful.

15.9 Measurement Techniques

15.9.1 Specific Measurements for Recycled Fibre

There are some special techniques used for quality control of recycled fibre. The remaining ink is measured using two different techniques. The relatively large, visible ink particles (> 50 um in diameter) are measured using a scanner and image analysis software. The smaller ink particles (> 4 μm in diameter) are measured using camera technique with microscopy and image analysis software. The measurement of relatively large dirt / ink specks is usually called specks or dot count. This measurement reflects what we visually perceive as specks and dirt in a sheet of paper. The Effective Residual Ink Concentration (ERIC) is on the other hand a method to determine the remaining ink that we do not see but that reduces the brightness of the paper. The reason for this is that the smaller the size distribution is for a specific amount of ink the larger total surface this ink will cover and thus the reduction in the measured reflectance value. This ink fraction is an important part of the ordinary brightness value measured, but the problem is that the brightness also reflects the bleaching performance as this is measured around 457 nm and/or 557 nm. To avoid the interference from the bleaching, the reflectance of the paper pad is measured at a higher wavelength 950 nm where the impact of the colour eminating from the pulp fibres is infitecimal. The reflectance value at this wavelength correlates well with the amount of residual small ink particles as long as the ink particle size distribution is not changed. It is important to note that the ERIC value is based on a number of assumptions regarding the optical properties of the paper as well as the ink. In spite of this it is a very useful tool during recycled fibre processing. If the instrument for paper brightness / reflectance measurement is limited to a maximum wavelength of 700 nm this maximum wavelength can also be used with the drawback that the pulp fibres still have a small amount of colour in this range. It is also common to calculate a value based on this method that is called ink elimination (IE). This method compares the estimated amount of ink before and after the flotation stages based on measurements performed at 700 nm, or 950 nm. Measurement of the conventional brightness is of course also a standard analysis. The use of brightness and ERIC or IE are very strong tools to keep track on the amount of ink as well as the bleaching performance of the process.

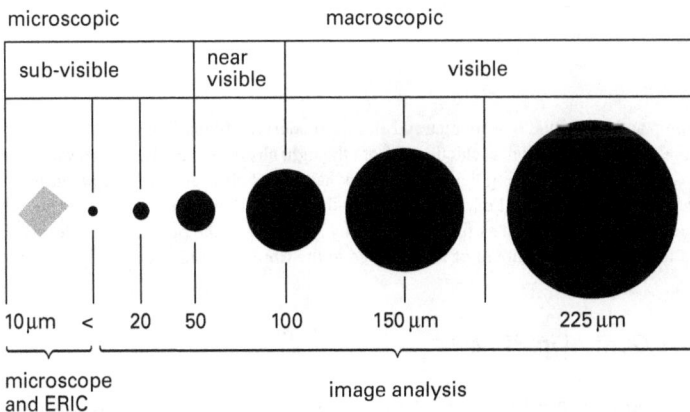

Figure 15.24. The measurement techniques used for ink particles of varying sizes.

Stickies on the other hand can be measured with a large number of methods all having their advantages and disadvantages. Large stickies are usually measured by image analysis. The stickies are deposited on a suitable surface by adsorption and coloured followed by size measurement by microscopy and image analysis. Smaller stickies are much harder to analyse but can to some extent be approximated by deposition tests. The stickies are deposited on a wire or hydrophobic film under stirring and are then determined by gravimetry or some imaging technique.

Concerning the mechanical aspects of deinked pulps an important parameter is the coefficient of friction (COF). The recycled pulp may contain increased levels of fatty acids and other components that reduce the COF. The COF is of course important for the reeling of the paper as well as the usage in high speed printing presses. The fatty acid content is therefore from time to time measured in the pulp. Fatty acid carried over from the recycling mill may also have the negative effect of contributing to deposits and paper breaks in the paper machine.

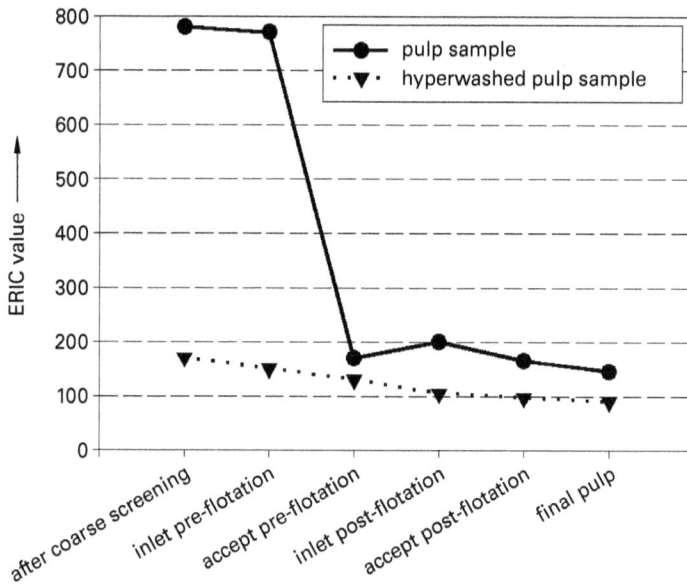

Figure 15.25. Ink content in pulp and ink bound to fibre material along a modern two-loop flotation-deinking process. ERIC = Effective Residual Ink Concentration, is calculated from the light absorption coefficient measured at 950 nm where paper don't absorb light and assuming that carbon black has a light absorption of 10000 kg/m2. The ERIC value mainly reflects the amount of small ink particles left in the pulp (<10 micron). Hyperwashing is a method to remove all detached ink particles as well as filler and fines by screening of the pulp during extensive washing. This method is used to determine the amount of ink attached to the fibre fraction.

15.9.2 On-line Process Control in RF Mills

Process control is a very important tool for recycled fibre production. The main reason for this is the raw material fluctuations that are a problem for recycled fibre mills. Good control of the optical properties, filler content and stickies levels are the main objectives for deinking mills and therefore on-line brightness, ERIC and filler measurement devices are very useful tools to control the process. Unfortunately the measurement tools for stickies have not reached a level well enough for process control. On-line brightness can be considered as standard tool in the modern deinking mills and on-line ERIC measurements are becoming more and more common. Also other on-line equipments are used in deinking plants as: fibre length distribution, CSF, pH, cationic demand, conductivity, turbidity and consistency that are less specific for recycled fibre.

15.10 How many Times can a Fibre be Reused?

During the last 30 years there has been a worry that the paper properties will successively get worse if too much recycled fibre is used. During the late nineties research was performed on this subject especially directed towards printing papers, as very large investments in new deinking capacity were discussed. The main existing knowledge in the early nineties, (practical experience published by Aurell and Munich experiments by Schweitzer) suggested that „not more than 5 % of the mechanical pulp fibres should be re-circulated more than twice if acceptable paper properties are to be retained" and that full cycles would cause losses in bonding properties to such an extent that newspaper made from this would cause very large problems in printing presses due to linting and even paper delamination.

In Europe evaluations were initiated by INGEDE, led by Prof. Lothar Götching, Darmstadt University (1992) and a very large combined pilot scale and full-scale evaluation (Multiple Recycling Project) by a constellation of European companies. These companies already used recycled paper and were hesitating on how much more of recycled paper that could be used in their newsprint production lines. The participants in this project were; Norske Skog, Enso-Gutzeit, Holmen Paper and Bridgewater Paper Company and the project was also supported the Nordic Industrial Fund. Earlier evaluations had shown that chemical pulp fibres loose bonding properties due to that the drying causes hornification, i.e. the fibres become much less prone to swell. This problem was also clearly shown in the packaging paper industry where packaging papers with recycled pulp never can replace virgin fibre based pulp at the same specific strength properties.

The Multiple Recycling Project was performed from 1994 to 1997 in order to cover pre-trials on different pilot plants and to have time enough to run the system five full recycling cycles. The finally run set of equipments used in the project were the pilot paper machine at Enso-Gutzeit R&D Centre in Imatra, Finland, pilot softnip calander at Valmet R&D Centre in Järvenpää, Finland, Aarendalspostens printhouse (commercial printing press), Norway and the pilot deinking plant with modern 2-loop-system at Escher Wyss R&D Centre in Ravensburg, Germany. The trials started with a carefully calculated mix of TMP, bleached TMP, SGW, bleached softwood kraft, filler clay, coating clay and coating carbonate according to models for Central Europe and UK year 2000. The future situations in Central Europe and the situation in UK could be tried according to the recycling models. Also paper run five full cycles was tried in the printing press in such a way that the development for each cycle could be studied. The main results were as follows for newsprint at constant filler content:

- Linting propensity decrease with increased recycling
- Specific strength properties are improved

- Bulk is reduced
- Light scattering coefficient is reduced
- Brightness can be kept at newsprint standard level (some ink always remain)

According to these results the worries regarding bonding related problems were exaggerated. On the other hand it was also found that the same bulk and light scattering properties would be difficult to reached with high degree of recycling. The reasons for the better results regarding bonding properties compared to earlier experience is believed to be a combination of the following aspects:

- Better control of fibre, fines and fillers than in earlier studies
- Improved deinking process compared to earlier studies
- Softnip calendaring preserves paper strength while the steel calendars reduce it

This evaluation was useful as a technical base for future investments for the participating companies and the investigation was published three years after it was finalised according to an agreement between Nordic Industrial Fund and the participating companies.

15.11 Future Developments

The global recycled pulp capacity was about 30 Mt year 2000 and as the recycled paper and board consumption as a mean value is 1,25 t/t, the recycled paper consumption was about 38 Mt. In year 1996 is estimated that 53.5 % of paper and board production was based on virgin fibres, 36,5 % on recycled fibres and 10 % on fillers. A recent prognosis for year 2010 tells us that the paper production will consist of 42,5 % virgin pulps, 42,5 % recycled pulps and 15 % fillers. It is impossible to recycle all paper and board products and it is probably possible to reach a maximum level of about 80 % in total recycling rate. A level above 80 % as a mean value is probably not economically and environmentally feasible even if it would have been possible to achieve.

The most important driving forces behind the increasing use of recycled pulps in many different paper products are the combination of lower production cost, technology development and increasing environmental legislations. As successively larger parts of the world.

The pulp and paper quality issue is of course a very central issue when deciding in what products it is possible to use recycled pulps. In the packaging and board area there are limits due to that the strength properties of pulp made from recycled paperboard is inferior compared to virgin chemical pulps. In the case of printing papers the strength issue is not an obstacle but instead brightness for the higher grades and also the bulk compared mainly to virgin mechanical pulps. In general this means that the higher more value added grades of both packaging, paperboard and printing papers will contain small amounts of recycled pulps, while the lower grades of packaging, paperboard and printing papers will to a dominating degree contain recycled pulps. Two typical examples of this are newsprint and fluting, products in which it will not be feasible to use any virgin pulp (mechanical or chemical). Other examples are high grade quality paperboard and high grade light weight coated papers where it will not be possible to use even the highest grades of recycled pulps as the bulk and brightness demands on the final products will be too high.

The yields in the different recycled pulp producing processes are in the range from 60–95 % depending on raw material, process and final paper or board product. Thus a simple estimate of maximum amount possible of recycled fibre world wide should be about 60 %, provided that the pulp quality will be high enough for the paper products.

Sweden is a big net importer of recycled paper and board as most other of old forestry countries. Knowledge in the area of recycled fibre processes is thus as important in Sweden and the other traditional forestry countries as the knowledge of chemical and mechanical pulping processes. The increase in paper and pulp production will in the future as right now be larger in areas with high population and large amounts of recycled papers than in the traditional forestry countries.

15.12 Further Reading

Göttsching, L., and Pakarinen, H. (eds.) (2000) *Recycled fiber and deinking. Book 7 in the Papermaking science and technology series.* Fapet Oy, Helsinki, Finland.

INGEDE, Internationale Forschungsgemeinschaft Deinking-Technik e. V. (International Association of the Deinking Industry), home page: www.ingede.com. This organisation includes most of the European deinking mills and has a lot of useful information on the homepage.

McKinney, R. W. J. (ed.) (1995) *Technology of paper recycling.* Blackie Academic & Professional, Glasgow.

Spangenberg, R. J. (ed.) (1993) *Secondary Fiber Recycling.* Tappi, Atlanta, Georgia, USA.

For recycled fibre statistics the CEPI home page are recommended: www.cepi.org.

16 Pulp Characterisation

Elisabet Brännvall
Department of Fiber and Polymer Technology, Royal Institute of Technology, KTH and
Göran Annergren
Mid-Sweden University

16.1 Introduction

By testing the properties of a pulp it is characterised and evaluated in comparison to other pulps. In the pulp mill, routine tests are performed for process control purposes as well as to characterise the product quality for the customer. The main purpose of this pulp characterisation is to ensure a stable pulp quality. Pulp characterisation is also a tool for improving existing and developing new products. However, it is not possible to take results from the pulp characterisation to predict the properties of a paper made on a specific paper machine.

To facilitate the communication between researchers, producers and customers, standard test procedures have been issued by the pulp and paper industry. Standardised tests make it possible to compare results in one laboratory to results from other laboratories. Industries within a geographical area have formed associations for the standardisation of tests.

- SCAN, Scandinavian Pulp, Paper, and Board Testing committee (Finland, Norway and Sweden)
- TAPPI, Technical Association of the Pulp and Paper Industry, USA
- CPPA-TS, Canadian Pulp and Paper Association, Technical Section
- APPITA, Australian Pulp and Paper Industry Technical Association

The standards set up by ISO, *the International Standardisation Organisation,* are used in connection with international trade and are officially recognised around the world. As ISO standards get available the corresponding SCAN methods will be withdrawn and replaced by the ISO standards.

16.2 Fibre Level Properties

16.2.1 Fibre Length

Fibre length is an important property of a pulp. Longer fibres, up to a certain point, generally enhance the strength properties of pulp. Very long fibres on the other hand have a negative effect on strength. Longer fibres more easily are entangled with each other and the fibre distribution of the sheet will be uneven, in other words, the formation of the sheet becomes poor. The importance of long fibres is predominant in sheets with low bonding strength, such as sheets made of only slightly beaten pulp or wet sheets. The length is in range 1–6 mm. Hardwood fibres have an average length around 1 mm whereas the average length of a softwood fibre is 3 mm.

The fibre length can be determined either directly or indirectly. The indirect methods divide the pulp fibres into different size fractions and give an indication of the average fibre length. One such fractionatingment is the Bauer-McNett classifier. The fractionation is accomplished by placing a pulp slurry of very low concentration in a vessel equipped with vertical slots. The slurry is circulated with high speed parallel to the slots, while a flow of much lower velocity passes through the slots. In this way, the fibres will be transported in their longitudinal direction towards the slots. The fibres short enough to pass through the slots will slip away to the next vessel where the slots to pass through are shorter. The method was originally developed for ground wood pulp as stiff fibres have more tendency to line up longitudinally toward the slots. Bauer Mc-Nett characterisation is not suitable for chemical pups.

Direct measurements of fibre length can be made by microscopic counting, image analysis or by using optical devices to make measurements fibres in a pulp slurry. Microscopic counting is very time-consuming. Image analysis, by projecting an image of fibres on a screen, is faster, especially if the fibre length is not manually measured but calculated automatically by a computer. However, for very fast measurements of a large number of fibres, automated optical devices using microscopy and image analysis are used. Polarised light is used to detect the fibre position and the image is recorded by a CCD camera. CCD stands for Charge Coupled Device and is an electronic memory that is charged by light. CCD can hold a variable charge and is therefore suitable to use in cameras to record variable shades of light. Through image analysis techniques, the signals from the camera are interpreted to give information of fibre dimensions. The number of fibres measured is about 10.000, this has been found as a suitable number for stable average measurements. The measurement takes about five minutes. The area and perimeter of the objects are measured. From these dimensions, length and width are computed. For objects to be defined as fibres the length-to-width ratio has to be 4:1 or higher. Objects below this value are defined as belonging to the fines fraction.

In order to get useful fibre length values the fibre length is weighted. The arithmetic average,

$$L = \frac{\sum_{i}^{n} L_i}{n}$$

is not practical. (L_i = *length of individual fibre*, n = *number of fibres*). The small objects add very little to the sum in the numerator although they count for as much as full-length fibres in the denominator. By weighting the fibre length by length, longer fibres have more influence.

$$\text{Length weighted length } L_l = \frac{\sum_i^n L_i^2}{\sum_i^n L_i}$$

16.2.2 Cell Wall Thickness

The most obvious difference in cell wall thickness is that between earlywood and latewood fibres, *Figure 16.1*. It is visible even to the human eye as the annual rings in a tree stump. The cell wall of earlywood fibres is thinner and the fibre diameter is larger than latewood fibres. The earlywood fibres collapse more easily, they are more flexible and conform more easily to other fibres.

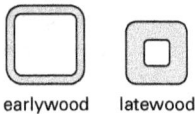

earlywood   latewood

Figure 16.1. The earlywood fibres have a thinner cell wall, bigger lumen and bigger cell width than latewood fibres.

The *fibre coarseness* is defined as the weight of the fibre divided by the length of the fibre. It is given in mg/m and is somewhere around 0.1–0.3 mg/m. The thicker the fibre wall the higher the coarseness value.

The cell wall thickness can be determined through microscopy. The optical analysis methods provide data of fibre width.

16.2.3 Fibre Deformations

In wood, fibres are tube-like and straight whereas fibres in pulps can be twisted, bent, compressed and so forth. The most familiar fibre deformations might be the so-called latency in mechanical pulps, the fibres are not straight but twisted and curled.

The fibre deformations can be of many kinds, *Figure 16.2*.

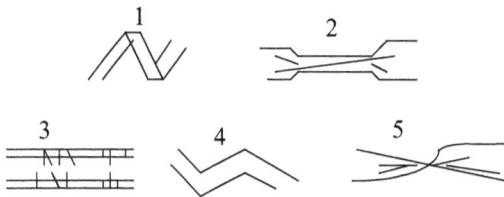

Figure 16.2. Examples of fibre deformations. 1. Folds. 2. Compression. 3. Microcompression (dislocation). 4. Knee. 5. Twist.

The effect of fibre deformations on paper properties may vary depending on the type of deformation. The sackpaper quality is actually improved by the introduction of microcompressions and folds, as this enhances the stretching ability of the fibres. Fibre deformations such as folds and knees decrease the ability of fibres to transmit load leading to a decrease in tensile strength. Cutting, i. e. decreasing the length of the fibre, can be classified as a deformation. Fibre cutting may reduces pulp strength.

To what extent the pulp fibres are deformed from the straight natural shape in wood can be quantified by the *curl index* or *form factor*. It is a measure of the fibres projected length in relation to the true length of the fibre, *Figure 16.3*.

Projected fibre length, l

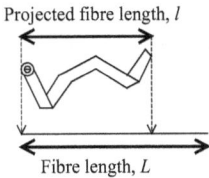

Fibre length, L

Figure 16.3. Curl index $= \dfrac{L}{l} - 1$.

The optical analysers usually provide data of the form factor. Fibre deformations can be detected with microscopy using polarised light and manual counting of the different types of deformations.

16.2.4 Fibre Strength

The strengths of individual fibres have an influence on the strength of the paper web they form. Additionally, the paper strength is also dependent on the bonds between fibres. Most strength properties, such as tensile and tear strength, are influenced of both the fibre and bonding. Knowledge of the fibre strength can provide information on for example fibre damages caused by a process step.

The strength of individual fibres can be estimated by the zero-span tensile strength test. The denomination zero-span derives from the distance between the clamps fastening the strip of paper to be tested. The two clamps grip the paper strip as close to each other as possible, in practice the distance between the clamps is zero. The idea behind the test is that fibres thus are fastened and pulled in both their ends and each fibre fastened by the two clamps will be broken. The force needed for rupture is recorded and defined as fibre strength, given in Nm/g.

However, it is not possible to accomplish absolute zero distance between the clamps. A residual span will always exist, however small it may be. Because of this, only straight, undeformed fibres will add to the zero-span fibre strength. Deformed fibres will generally not be under strain when the clamps pull the paper strip apart and it is therefore desirable that the fibres are as straight as possible. The influence of fibre deformations on the zero-span test can be minimised by beating the pulp lightly in a laboratory PFI mill, since this straightens the fibres.

Ideally, the zero-span test would only give the strength of the individual fibres and not the strength of the bonds between fibres. In dry sheets, however, deformed zones in the fibre may be over-bridged by bonds or a fibre, only clamped in one of its ends, may form a bridge to other fibres and thus be activated in the breakage. Zero-span testing may therefore also be performed on re-wetted sheets. The re-wetting reduces the influence of bonds between fibres and fewer fibres will participate in the process of fracture. Re-wetted zero-span strength is more sensitive to local fibre defects and fibre length than the dry zero-span strength.

16.2.5 Bonding Strength

The strength with which fibres adhere to each other contributes to the pulp strength, together with fibre strength. Bonding strength depends on both the bonded area as well as of the specific bonding strength between fibres. The bonded area is influenced by the flexibility of the fibres and the amount of fines. The specific bond strength is more dependent on surface properties of the fibres.

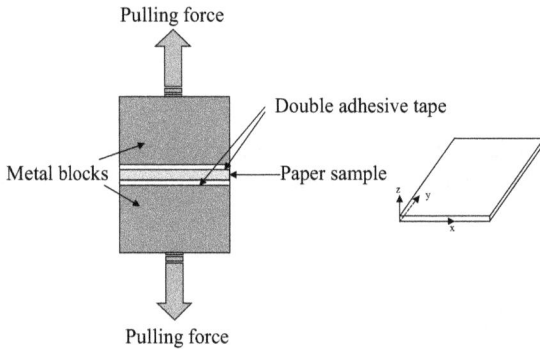

Figure 16.4. The principle for determining the z-strength by tensile force.

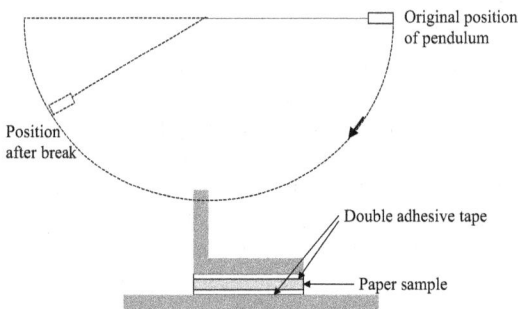

Figure 16.5. The principle of the Scott-Bond tester for bonding strength.

The bonded area can be estimated by the sheet density and the light scattering ability of the sheet. The more points of contact there is between fibres, the more densely packed the fibres in the sheet become. Increased density and decreased light scattering ability denotes more bonded area. However, notice that fibres having optical contact need not be so close so chemical bonding can occur.

The bonding strength can be defined as the strength in z-direction, i.e. at a 90-degree angle to the plane of the paper. Some different devices exist to measure the z-strength; one example is the z-directional tensile strength tester depicted in *Figure 16.4* Another is the Scott Bond tester, *Figure 16.5*. The principle behind the methods is to measure the strength needed to tear the metal blocks apart (z-directional tensile strength tester) and tear the metal gauge off the paper sample (Scott Bond) by delaminating the paper.

16.2.6 Fines Content

Apart from the fibres, which are the main constituents, pulp additionally contains a fraction of much smaller material, called the fines fraction. No strict definition of the size of fines exists. Commonly they are defined as particles smaller than 200 μm. The primary fines in pulp are made up of ray cells, smaller pieces of broken fibres and thin sheets from the fibre surface. Refining, also called beating, of the pulp creates secondary fines. The mechanical pulping method produces a high amount of fines. The positive feature of fines is their contribution to pulp strength, which is very pronounced for mechanical pulps. Their negative side is a decreasing dewatering capacity of the pulp with increased amount of fines.

The Bauer-McNett fibre classifier can be used to determine the amount of fines in pulp. Another method is to use the Britt Dynamic Drainage Jar (BDDJ). Slurry of the pulp to be tested is poured into a jar equipped with a sieve and a revolving propeller. The fines will be strained through the sieve and collected for gravimetric determination. The remaining coarse fraction in the jar is likewise collected, dried and weighed.

16.3 Properties on Molecular Level

16.3.1 Carbohydrate Chain Length

The cellulose chains in wood consist of thousands of glucose units, numbers from 7000 to 15000 have been reported. The hemicellulose chain length is only slightly above 100 sugar units. The length of polymer chains is important for the mechanical properties of all polymers. Polymers have to be long enough to form entanglements before they can carry any load. The strength of the pulp fibre is affected by the cellulose chain length, i. e. the molar mass of cellulose. It is therefore of greatest importance to monitor the decrease in molar mass during the course of delignification. Once a carbohydrate chain has been cleaved, the shortening of the chain can not be repaired. In order to ensure a pulp of good quality, the molar mass has to be above a certain level. The selectivity of a pulping or bleaching process can be examined by comparing the molar mass decline to the degree of delignification

The molar mass in a pulp can be estimated by dissolving the pulp and determining the viscosity of the solution. The viscosity value is related to the molecular weight of the carbohydrates or in other words, their degree of polymerisation. Cellulose, however, is not easily dissolved. It requires specific solvents, such as metal complexes of organic bases and the most common solvent used is CED, cupriethylenediamine. The solvent is alkaline so the presence of oxygen has to be avoided when dissolving the pulp sample or this would result in cellulose degradation. Different relationships between the viscosity of the pulp solution and the molecular weight of the pulp have been suggested, *Figure 16.6.*

For process control purposes, the CED viscosity together with lignin content are the most widely used analysis. Since the cellulose chains are so much longer than the hemicellulose chains, the viscosity value reflects the degree of polymerisation mainly of the cellulose.

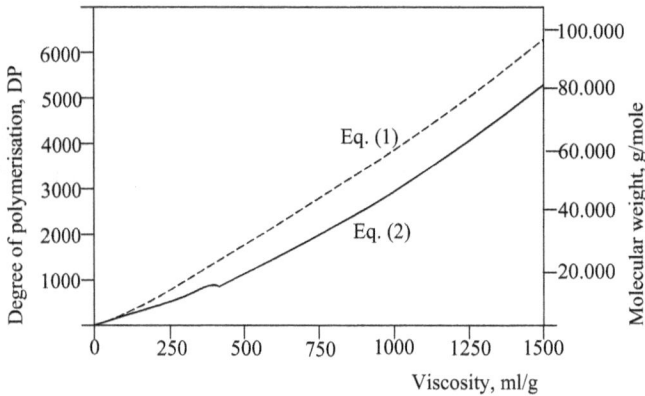

Figure 16.6. Degree of polymerisation and molecular weight vs. SCAN viscosity. Curvature according to Equation 1 is suggested by Evans and Wallis (1987) and Equation 2 by Marx Figini (1978) and Gruber and Gruber (1981).

16.3.2 Carbohydrate Content

Pulp strength can be related to the amount of cellulose as well as the ratio between the amounts of cellulose and hemicelluloses. In general, more cellulose gives a stronger pulp and lower cellulose/hemicellulose ratio reduces pulp strength. The hemicelluloses make the pulp easier to beat and increased content of xylan contributes to the bonding ability between fibres. The amount and distribution of cellulose and the various hemicelluloses is therefore of interest.

The most frequently used analysis of the carbohydrates is through acid hydrolysis. By treating the pulp with sulphuric acid of 72 %, the carbohydrate chains are degraded to the monomers they are made of. Cellulose ends up as glucose and the hemicelluloses as glucose, xylos, mannose, arabinose etc, depending on the hemicellulose in question. The monomers are recorded by gas chromatography and through the relative amounts of the monomers, the cellulose and hemicellulose contents are calculated.

A disadvantage of acid hydrolysis is that uronic acids can not be determined as they are dissolved and lost. Enzymatic hydrolysis may be used to degrade carbohydrates into monomers, without destroying the uronic acids. The amounts of the monomers is determined by HPLC, High Performance Liquid Chromatography.

When performing the analysis, the yield of the hydrolysis step has to be made with prudence. The yield of the hydrolysis, acid or enzymatic, is not to drop below 90 %. This leads to an uncertainty of the reproducibility and the deviation between determinations can be quite large.

16.3.3 Lignin Content

To measure the lignin content of the pulp is of interest for a number of reasons. For process control purposes it is vital to know the degree of delignification of the pulp produced and in the oxygen delignification and bleaching stages the lignin content determines the conditions to be used.

The lignin content can be determined either directly or indirectly. The most common direct method is to determine the Klason lignin. The wood or pulp sample is extracted with acetone to remove the extractives, whereupon concentrated sulphuric acid is added to the sample. The carbohydrates are thus hydrolysed and dissolved and the residue is lignin, which is determined gravimetrically. Part of the lignin is acid-soluble and thereby excluded from the Klason lignin. In contrast, some carbohydrates will be included in the Klason lignin analysis. For softwood and kraft pulps these effects are small and will rule out one another. For hardwood, sulphite pulps, CTMP and pulps being bleached, it is necessary to determine the acid-soluble lignin as well.

The Klason lignin determination is quite time-consuming and indirect methods offer much faster estimations of the lignin content. By far the most widespread method is the kappa number determination. In this procedure, a specified amount of potassium permanganate and sulphuric acid is added to a thoroughly disintegrated pulp sample. Lignin readily reacts with the permanganate as manganese is added to all three double bonds in the aromatic ring, *Reaction (16.1)*.

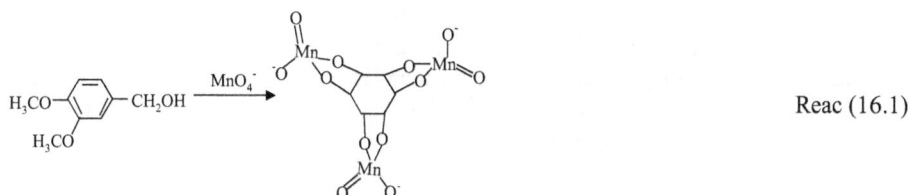

Reac (16.1)

According to the standard procedure, approximately only half the amount of permanganate should be consumed in Reaction (16.1) during the stipulated reaction time. To achieve this, an adequate amount of pulp has to be added which requires that an initial estimation of the kappa number of the pulp has to be made. The reaction is terminated after 15 minutes by addition of potassium iodide, *Reaction (16.2)*.

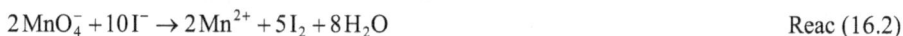

$$2\,MnO_4^- + 10\,I^- \rightarrow 2\,Mn^{2+} + 5\,I_2 + 8\,H_2O \qquad\qquad \text{Reac (16.2)}$$

The amount of reacted lignin is then determined by back-titration of the free iodine with sodium thiosulphate, *Reaction (16.3)*.

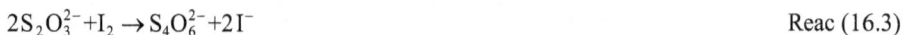

$$2\,S_2O_3^{2-} + I_2 \rightarrow S_4O_6^{2-} + 2\,I^- \qquad\qquad \text{Reac (16.3)}$$

The kappa number is applicable for a wide range of lignin contents, from semi-mechanical pulps with high lignin content, to semi-bleached pulps with less than 1 % lignin on pulp remaining. The kappa number can be related to the Klason lignin content in pulp. For softwood kraft pulps the amount of lignin in pulp can be calculated by multiplying the kappa number with 0.14. This relationship however, depends on wood specie and delignification method.

In *Table 16.1,* typical kappa number values for different pulp grades are given.

Table 16.1. Typical kappa numbers of different pulps. Lower kappa number denotes lower lignin content in pulp.

Pulp	Kappa number
Kraft pulp liner	85–100
Kraft pulp sackpaper	55–80
Kraft pulp grades for bleaching	20–35
Oxygen delignified pulp	10–20

In the kappa number standard, it is specified that about half the amount of the permanganate added should be left unreacted, in order to ensure that only the double bonds in lignin take part in the reaction. However, it is known that components originating from carbohydrates nevertheless contribute to the consumption of permanganate. Especially in the case of birch kraft pulps, the kappa number gives an overestimation of the lignin content. Degradation products from the xylan result in hexenuronic acids. The double bonds in these acids consume permanganate in the kappa number determination and thereby contribute to the kappa number.

Lignin also readily reacts with chlorine and this has been used to determine the lignin content expressed as the chlorine number or Roe-number of the pulp. The procedure is quite similar to the kappa number test, as the consumption of chlorine is ascertained by iodometric titration. However, since chlorine is hazardous to health and environment, the use of this method is limited today. As with the kappa number, a relationship exists between the pulp's chlorine number and its Klason lignin content. Multiplication with 0.9 of the chlorine number gives the percentage of lignin in pulp (Kyrklund, Strandell 1969).

16.3.4 Ionisable groups

The ionisable groups found in pulp are carboxylic acid and phenolic hydroxyl groups. In sulphite and CTMP pulp, sulphonic acid groups are found. The repelling forces between the negatively charged groups make the fibril network swell as illustrated in *Figure 16.7*. The various acidic groups have different pK_a values, not all are ionised in papermaking system, in particular the hydroxyl groups. A higher degree of swelling gives a stronger paper.

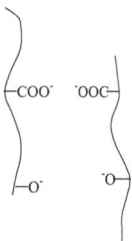

Figure 16.7. The negatively charged groups in lignin and on the carbohydrate chains create repelling forces between the fibrils. A higher amount of charged groups gives higher swelling of the fibre.

Table 16.2 lists general values for the total charge of different fibres. Unbleached pulp has a higher total charge than bleached pulp and hardwood pulps have higher charges than softwood pulps.

Table 16.2. The total charge for different types of fibres. (Wågberg, Annergren 2002).

Pulp	Sulphonic acid groups µmole/g	Carboxylic acid groups µmole/g
TMP unbleached/bleached	–	80–110/150–250
CTMP unbleached/bleached*	40–70/40–70	80–110/150–250
Unbleached softwood kraft**	–	60–150
Bleached softwood kraft	–	20–60
Bleached hardwood kraft	–	40–100
Unbleached sulphite***	100–120	10–25

* Values will depend on sulphite addition and degree of bleaching.
** Values will depend on the yield of the pulp, and values given in the table should be taken as typical range for unbleached fibres.
*** Kappa number < 32

The total charge is determined by conductometric titration with sodium hydroxide and sodium hydrocarbonate.

The surface charge of a fibre is important for its interaction with other fibres. The surface charge of the fibre can be determined by adsorption of a cationic polymer on the negatively charged fibre surface. The polymer has to have a molecular weight so large it is unable to penetrate into the fibre.

Usually, the surface charge is only a smaller portion (7–10 %) of the total charge.

16.4 Beating and Hand-sheet Forming

16.4.1 Beating

The terms beating and refining are used interchangeably. They refer to the mechanical treatment of chemical pulp to make it suitable for papermaking. However, refining also applies to the mechanical pulping process where chips are refined between disks to produce pulp. In this section, only beating of chemical pulp is discussed. Chemical pulp practically never is converted to paper without first being subjected to beating. Beating is an important process step since this type of pulp can be said to have no papermaking properties until modified by the mechanical action of a beater.

The appearance of cooked fibres differs from the native wood cells. Delignified fibres are more ribbon-like, not stiff tubes, since the cell walls have, to a vast degree, collapsed into the lumen. Much of the primary wall has been removed, thus revealing the fibrils of the S1-layer. The fibres are, however, unfit to form a strong and uniform paper since they lack the flexibility needed. Unbeaten pulp would result in a weakly bonded paper with an uneven distribution of fibres. It is thus necessary to treat the fibres mechanically to give them characteristics enabling them to form sheets with desired properties, *Fig 16.8.*

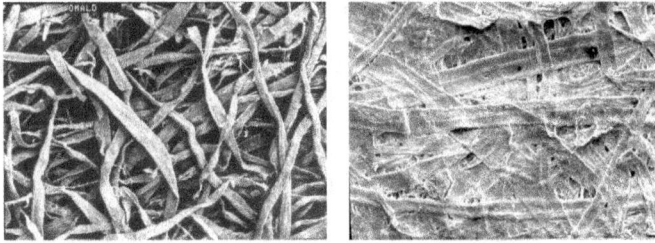

Figure 16.8. Unbeaten, left, and highly beaten pulp, right. The conformability of the fibres greatly increases with beating, resulting in a much better bonded sheet. Photos depicting dried pulp samples. (Photo STFI).

Beating have the following concequences:

- *Creation of fines.* In the beating process, the primary wall, or what is left of it after cooking, is torn off together with a great deal of the S1-layer. These fibre fragments constitute the secondary fines and are part of the smallest size fraction present in pulp, the fines fraction.
- *Internal fibrillation.* The beating moulds and breaks down the inner structure of the fibre. The fibrils in the primary wall have a criss-cross pattern and as long as the fibres are wrapped in this net-like structure, it is unable to expand radially. At the removal of the primary wall, the water penetrating into the cell wall gets between fibrils and breaks intrafibre hydrogen bonds and the fibre swells. Internal fibrillation softens the fibres and makes them more flexible, thus making them much more able to conform to other fibres. Inner fibrillation, together with the removal of the primary wall, is often considered as the most important primary effects of beating.
- *External fibrillation.* External fibrillation is achieved when the fibrils on the fibre surface are loosened. The fibrils will rise giving the fibre surface a somewhat hairy appearance and thereby increase the surface area of the fibres. Together with the increased flexibility of the fibres, this leads to better bonding between fibres.
- *Fibre cutting.* A usually negative effect, is the cutting of fibres. Tensile or shear forces above the fibre capacity to endure them are put through to fibres at positions where they can not avoid the load and the fibre is cut. The shortening of the fibre length is damaging for the paper strength.
- *Fibre deformation.* Depending on the beating conditions, fibres may be straightened or curled. Mild laboratory beating in general straightens the fibres, whereas harsher conditions in industrial refiners increases the fibre curl.

Many different laboratory standard refiners are in use. Probably the most common type is the PFI mill. It has a smooth housing into which the pulp is evenly distributed around the wall. The beating action is achieved by bars on the edge of a disk that is pressed against the housing with a stipulated load, *Fig 16.9*. During beating, both housing and disk rotate in the same direction but with somewhat different speed. The amount of refining is given as number of revolutions.

The Valley hollander beater is the original beater, still in use in some old pulp mills. A smaller sized hollander can be used for laboratory beating. It consists of a circulating beater roll and a stationary bedplate. The pulp is circulated continuously and receives the beating action when pressed between the roll and beater plate, *Figure 16.10*.

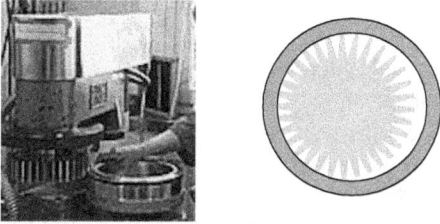

Figure 16.9. Putting a pulp sample into a PFI mill. When beating, the disk is sunk into the housing and pressed against the walls of the housing by a lever to apply the beating load. (Photo STFI).

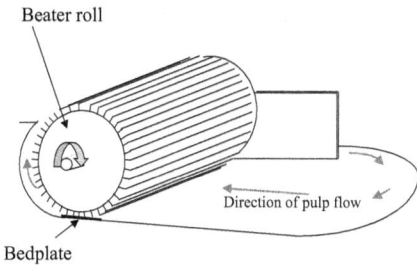

Figure 16.10. Principal representation of a Valley hollander.

In the Lampén mill, *Figure 16.11,* a rolling ball offers the beating action on the fibres.

Figure 16.11. The rolling ball Lampén laboratory mill.

The various beater types treat the fibres differently whereby the primary effects listed above are developed differently. Fines formation, by removal of the outer layers of fibres, is most efficiently achieved by PFI beating. A PFI mill removes material very equally over the entire fibre surface. The Valley beater modifies the fibre surface in a much more uneven manner. The Valley beater, however, is much better at inner fibrillation, which the PFI mill is unable to do effectively. Fibre cutting is most pronounced in the Valley beater. The Lampén mill also gives some fibre cutting although the most important primary effects obtained by this beater type are inner and external fibrillation.

16.4.2 Drainability and Water Retention

A noticeable effect of beating is the decrease of the drainability of the pulp suspension, as the development of all primary effects tends to make the pulp more difficult to de-water. The relative importance of the different primary effects on de-watering capacity depends to a great deal on the de-watering conditions. The creation of fines is the principal effect influencing the drainability when measuring it as the pulp's SR-number, Schopper-Riegel number, or its freeness, Canadian Standard Freeness, CSF. A low CSF signifies a pulp difficult to de-water. The SR-number is inversely graded – a high SR-number indicates a pulp difficult to de-water. Accordingly, the more a pulp is beaten the higher the SR-number and the lower the freeness. *Figure 16.12* shows the general features of a Schopper-Riegel or Canadian Standard Freeness tester. The CSF test was originally developed for the characterisation of mechanical pulps, whereas the Schopper-Riegel test is better suited for chemical pulps.

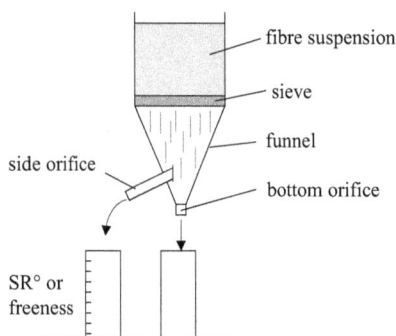

Figure 16.12. Drainability test. A fibre suspension is poured into a vessel with a sieve in the bottom. The faster the water drains through the sieve, the higher the water level in the funnel and accordingly more water flows through the side orifice. The water from the side orifice is collected and the amount gives the CSF or SR-number.

The degree of beating of a pulp is usually quantified as its SR-number or freeness. The SR-number/freeness, however, measures the development of only one primary effect, the creation of fines, whereas the beating impact is made up of several effects. There is no clear relationship between a certain SR-number/freeness of as pulp and the de-watering of the paper web on the paper machine. The de-watering conditions on a paper machine differ greatly to those of the drainage testers.

The Water Retention Value, WRV [g water/g pulp], primarily quantifies the inner fibrillation of the fibres and thus the swelling of the fibres. A pulp sample, soaked in water, is subjected to centrifugation for a specified time at a specified speed. The amount of water it is able to hold gives a measure of the degree of swelling. The WRV-value increases with increased beating. It gives a hint of the paper webs performance in the press section. The higher the WRV value, the more difficult it is to press out water from the paper web.

16.4.3 Sheet Forming

Many pulp properties are determined on laboratory made hand-sheets. The laboratory sheet formers consists of an upper section, (the container for the pulp suspension), a wire screen on which the sheet is formed and a lower section for the water to be drained into through the wire screen. The pulp is diluted to a low concentration, < 0.5 g/l, and a certain volume of the pulp slurry, sufficient to make one standard hand-sheet, is transferred to the upper section of the sheet former, containing water. The very low concentration at sheet forming results in an exceptionally even formation, the fibres will be uniformly distributed in the sheet. The wet sheets are pressed onto drying plates and dried at a conditioned atmosphere of 23 °C and 50 % humidity. The sheets stick to the drying plate whereby shrinking is prevented. Another option is to dry the sheets between blotter papers in order to have free, or relatively free shrinkage while drying.

Some different sheet forming equipment for laboratory use exists. They differ in concentration of the pulp slurry, sheet type (rectangular, square or circular) and sheet dimensions, type of wire screen etc. In general, sheets are made in an open white water system resulting in low retention of fines. The fibres are evenly and randomly spread out in all directions of the sheet resulting in a so-called isotropic sheet. If a hand-sheet is to bear more resemblance to commercial sheets from paper machines, some adjustments need to be made. A closed white water system accumulates the fines amount and improves the retention. A Rapid Köthen lab sheet former is useful in this respect and well suited for sheet forming of mechanical pulps. In order to have a sheet with fibres orientated in a certain direction, an anisotropic sheet, a somewhat more complicated sheet former needs to be employed. For example the French sheet former, formette dynamique, where the pulp slurry is sprayed from a nozzle toward a cylindrical rotating wire, creating a sheet where the fibres are more or less aligned parallel to each other.

16.5 Strength Properties

For many paper grades, pulp strength is an essential product requirement. Additionally, pulp strength is a prerequisite for the manufacture of paper. The paper web has to be strong enough to endure the forces it is exposed to, for example in the paper machines pressing and drying sections or in printing presses. Many different test methods exists, using forces that pull, tear, compress, burst, or otherwise deform or break the paper in order to determine different capacities of the paper. Strength is determined on hand-sheets with a given grammage (usually 60 g/m^2) and the tests are performed in conditioned environment.

16.5.1 Tensile Properties

The tensile strength is the greatest longitudinal stress a substance can take without breaking. Paper is often subjected to tensile forces, both in the manufacturing process and in product applications. Tensile strength is one of the basic strength properties tested on pulp and paper.

Tensile strength is tested by fastening a strip of paper (15 × 100 mm) between two clamps and pulling the paper apart. In the process, the paper strip is extended until it no longer can bear the load and it breaks. The force applied vs. stretch of the paper sample is recorded, *Figure 16.13*.

Force

Tensile
strength

Tensile
stiffness

TEA

Stretch

Stretch at break

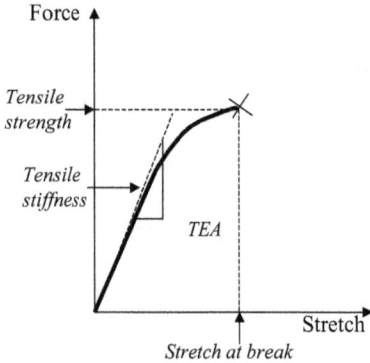

Figure 16.13. The stress-strain curve for tensile strength testing.

The tensile forces applied extend bonds between and within the fibre as well as the fibres themselves. In the rupture zone of the paper strip, some fibres are broken and others are pulled out from the network breaking bonds in the action. Tensile strength is dependent on both the bonding strength in the sheet and the strength of individual fibres.

- The *tensile strength* is defined as the force needed to accomplish the rupture divided by the width of the paper strip (N/m). Dividing the tensile strength with the basis weight of the sheet gives the *tensile index* (kNm/kg).
- The stretch at break is the increase in length of the paper strip at the moment of rupture and it is expressed as percentage of the original length.
- By integrating the area under the stress-strain curve, the total energy that can be absorbed by the sheet before failure is obtained. The *tensile energy absorption, TEA,* is defined as energy/area (J/m^2). Division by the grammage gives the TEA index (J/kg). It depends on the tensile strength and the stretching ability of the paper and is taken as an indication of the paper toughness. It is an important property for sackpaper.

Yet another property can be obtained from the stress-strain curve. The initial inclination of the curve gives the *tensile stiffness* (N/m), or *tensile stiffness index* (Nm/kg). This is a very important property for paperboard.

The tensile strength depends on fibre length, fibre strength, specific bonding strength and bonded area. Although beating may decrease both the average length and strength of the fibre, the dominating effect on tensile strength is the increased bonding. Tensile strength increases with increased beating energy until it reaches a plateau at high degrees of beating. The beatability of a pulp is evaluated as the amount of beating needed to reach a certain tensile strength. The less beating energy needed to reach the stipulated tensile strength, the easier the pulp is to beat.

16.5.2 Tearing Strength

The tear strength is the energy needed to extend a crack in the paper. By measuring the tear strength of paper, information is obtained of its ability to withstand cracks. Good tearing strength is important for the runnability of the paper machine as well as an important property for many products. Tearing can be made in different modes, *Fig.16.14.*

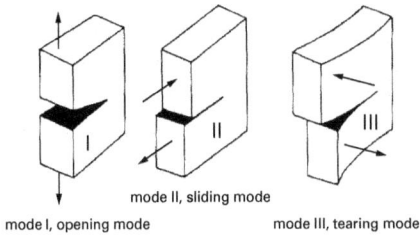

Figure 16.14. Tearing forces applied in different modes. (Illustration STFI).

Mode I is the most realistic mode in which tearing is propagated in a paper. However, the most common way to measure the papers tearing strength is by the Elmendorf tear resistance tester, which tears the test specimens according to mode III. An initial cut is made into the test strip and the force to continue the tearing through the strip is recorded. The tear index is the tear force divided by the basis weight of the paper sheet tested and expressed as Nm^2/kg.

The work of tearing is made up of two processes. One is to stretch fibres until they break, this process being governed by the strength of the individual fibres. The other process is to pull out fibres from the network and this part of the work depends on both breaking of bonds and frictional forces. The energy needed to pull out a fibre from the network is much larger than the energy needed to break a fibre. Whether a fibre will be broken or pulled out primarily on the degree of bonding. As bonding increases with increased beating, an increasing amount of fibres will be broken. The tear strength may increase at the beginning of the beating but continued refining decreases the tear strength. A tear vs. tensile strength diagram is commonly deployed to characterise a pulp, *Figure 16.15.*

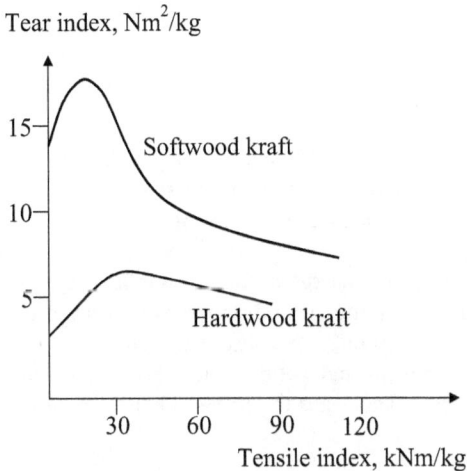

Figure 16.15. The refining energy input is varied for a softwood kraft pulp and a hardwood kraft pulp and the pulps tear and tensile strengths are recorded. Prolonged beating increases tensile strength whereas tear strength decreases.

The length of the fibre influences the tear strength. Longer fibres have higher tear strength, since longer fibres naturally provide more points of bonding and are pulled a longer average distance from the within the network.

16.6 Structural Properties

16.6.1 Density and Bulk

The sheet density (kg/m³), or its inverse bulk (m³/kg), is probably the most important structural property, since it has an influence on other properties. As a rule, higher density denotes better bonding in the sheet. The density of the sheet can thereby be used as a measure of the degree of bonding. High bulk is desirable for certain products since bulky paper in general is more adsorbent and more opaque.

Generally, density is determined by dividing the basis weight of the sheet with its thickness. The basis weight poses no difficulties as long as there is a ruler and a scale at hand. The thickness, however, is not quite as simple. It can be determined in a micrometer in which the sheet or sheets are squeezed between two metal plates, *Figure 16.16*. Since paper is elastic, the pressure will compress the sheet.

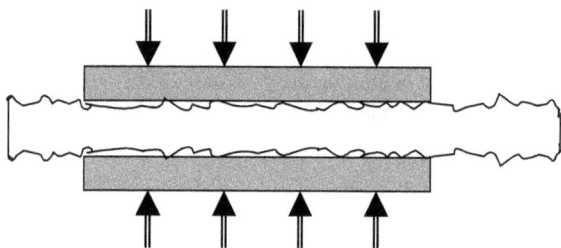

Figure 16.16. Thickness, or caliper, gauge according to SCAN standard.

Determining the thickness on a pack of sheets will give a different thickness value than the determination on a single sheet. The unevenness of the paper surface of one sheet will kind of fit in into the unevenness of the adjoining sheet, thus giving a lower thickness value compared to single sheet determination.

The STFI thickness tester gives a measure of the paper thickness without an influence of the paper surface roughness or the paper compressibility. In the method, spherical measuring tips register changes in thickness as the test specimen is taken through a measuring gauge.

The density of a sheet increases with increased beating and wet pressing. The more flexible beaten fibres adjust themselves easily to other fibres. This leads to increased bonded area and more densely packed fibres in the sheet.

16.6.2 Porosity and Air Permeability

So far, only the fibres in the paper sheet have been discussed. However, also the space in the sheet not occupied by fibres is of interest, namely the volume of air in the sheet, or the porosity of the sheet. The porosity is important for example for sackpaper, as for efficient filling of sacks, the air within the sack must be able to escape through the paper. Porosity is evaluated as the air permeability.

The test is performed by recording the time for a specified amount of air to pass through a paper sample at a certain pressure, temperature and humidity. The air permeability is given in [(m/s)/Pa]. The most common air permeability tester is the Gurley tester.

16.6.3 Roughness, Smoothness and Softness

The surface properties of paper are important, predominantly for printing paper but also for tissue grades. Roughness quantifies the paper surface irregularities, to what extent the paper surface deviates from a completely flat surface. Roughness is defined as the volume of air per time unit leaking between the paper and the gauge head due to a specified pressure difference and given in ml/s. Smoothness is the inverse of roughness.

Softness is a subjective property relating to how velvet-like the paper is and how easily it yields when crumbled.

Surface softness is also a subjective property and relates to the feeling of softness when stroking the paper surface gently with the fingertips.

16.7 Optical Properties

White paper is often considered as a prerequisite for many applications. The readability and printability usually improve with the whiteness, or brightness, of the paper. The bleaching operations are monitored by registering the brightness increase.

The important optical properties of a material are:

*   Fundamental:*Absorbency* –ability to absorb light
*   *Light scattering* –ability to scatter light in different directions
*   Applied: *Reflectance* –a materials ability to reflect light
*   *Transmittance* –ability to let light through the material

The best known and mostly used theory for describing the optical properties of paper is the Kubelka-Munk theory. This theory links together the above mentioned properties.

16.7.1 Light Scattering

The interfaces between air and fibre provide surfaces for the light to be reflected on. A sheet of stiff, tube-like fibres scatters light more than slender, collapsed fibres, well-bound fibres provide less surfaces for light scattering than fibres with fewer bondings. The ability of a sheet to scatter light is not affected by the degree of cooking or bleaching of the pulp, whereas beating, pressing or any procedure changing the physical appearance of the sheet highly influences its scattering capacity. Both beating the pulp and pressing a paper sheet result in a higher degree of bonding and a denser sheet with less light scattering interfaces. The light scattering is to some extent affected also by the difference in refractive index between the interfaces. The refractive index, and thus the light scattering, is bigger between air and fibres with a high lignin content than between air and fully bleached fibres. Another example of this is the higher light scattering for dried fibres compared to wet fibres.

In the Kubelka-Munk theory, the light scattering coefficient, s (m^2/kg), is introduced. For the Kubelk-Munk theory to be applicable, the light scattering coefficient is assumed to be additive, Equation (16.1).

$$s_{mixture} = x_1 s_1 + x_2 s_2 + x_3 s_3 + \ldots \qquad \text{Eq (16.1)}$$

x_1, x_2, x_3 etc are the weight percentage of the constituents 1, 2, 3 with their specific light absorption coefficients s_1, s_2, s_3. The constituents can be either different pulps or additives, such as clay.

16.7.2 Light Absorption

The light absorption coefficient, k, is dependent on the chemical composition of pulp, mainly the chromophoric groups in lignin. It is not affected by beating or pressing. Delignifying procedures, however, such as cooking and bleaching, affect light absorption, decreasing the k-value as lignin is removed. Apart from lignin, certain metal ion complexes in the pulp can contribute to the colour of pulp. As in the case of light scattering coefficient, the light absorption coefficient is assumed to be additive, Equation (16.2).

$$k_{mixture} = x_1 k_1 + x_2 k_2 + x_3 k_3 + \ldots \qquad \text{Eq (16.2)}$$

The light absorption can be measured either on paper sheets, using a diffuse reflectance spectrophotometer, or on pulp solutions, using UV spectroscopy. In either method, the wavelength used is 475 nm and the k-value is reported as m^2/kg. The light absorption is strongly dependent on wavelength.

The methods are valid as long as the k-value is between 1 and 10 m^2/kg. To be able to measure the light absorption of pulps with high lignin content, such as unbleached kraft pulps and mechanical pulps, certain modifications have to be made. One way is to lower the grammage of the paper sheet, thus lowering the k-value. It is also possible to use the additiveness of the k-value and dilute the pulp sample with fully bleached chemical pulp.

16.7.3 Brightness

Brightness is the intrinsic luminous reflectance factor, R_∞, measured at 457 nm. Determined by standard procedures using a reflectometer it is designated ISO-brightness and given in percentage (100 % ISO-brightness is the standard the sample is tested against). *Table 16.3* shows some typical values.

Table 16.3. Typical brightness values for different pulps.

Pulp	% ISO-brightness
TMP	~60–70 %
Unbleached softwood kraft	~40 %
Bleached softwood kraft	90 %

The purpose of the term brightness is to be able to measure the effect of bleaching and to quantify the amount of coloured substances in pulp. Nevertheless, brightness has the disadvantage not to be proportional to the quantity of coloured species in pulp. Moreover, it is affected by the structural appearance of the sheet. In the Kubelka-Munk equation for brightness, light scattering as well as light absorption appears Equation (16.3).

$$R_\infty = 1 + \frac{k}{s} - \left[\left(\frac{k}{s} \right)^2 + 2\frac{k}{s} \right]^{1/2}$$

Eq (16.3)

In simple terms, R_∞ is proportional to k/s. the brightness increases with decreasing k, signifying a decrease in coloured substances usually achieved by bleaching. However, brightness increases as well with increasing light scattering coefficient, e.g. drying of pulp increases brightness and pressing decreases brightness, although the amount of coloured substances is not altered in these operations. The standard procedure for brightness states that unbeaten pulp is used for the determination and unbeaten pulp has higher light scattering ability than beaten thus giving a higher brightness. Nor is the brightness value additive, which inhibits the calculation of the brightness from a mixture of different pulps or the resulting brightness on the addition of filler. A better measure would be light absorption coefficient, which directly quantifies the coloured substances.

16.7.4 Opacity

Opacity is defined as the paper sheets ability to conceal the printing on its backside. The opacity increases with increasing light scattering and is calculated according to Equation (16.4):

$$Opacity = 100\frac{R_0}{R_\infty}$$

Eq (16.4)

R_0 = the light reflected from a single sheet placed above a black cavity
R_∞ = light reflected from an pack of sheets thick enough to refrain light from passing through

Table 16.4. Summary of the main characteristics of chemical and mechanical pulp.

	Chemical pulp	Mechanical pulp
Pulp yield, %	40–55 %	90–95 %
Lignin content in pulp	0–3 %	20–28 %
Fibre character	Long, strong, collapsed, flexible	Short, weak, uncollapsed, stiff
Sheet density	High	Low
Porosity	Low	High
Fines content in pulp	5–15 %	20–30 %
Pulp strength	High	Low
Light scattering ability	Low	High
Yellowing and ageing	(May be) low	High

16.8 Properties of Pulps from Different Processes

Table 16.4 summarises the general differences between chemical and mechanical pulps. The two pulping methods give pulp of different character depending on chemical composition and physical appearance of the fibres. Chemical methods dissolve and remove lignin from the wood, as well as some carbohydrates, resulting in a pulp yield somewhere between 40 and 55 %. The lignin content of unbleached chemical pulp is approximately 3 % on pulp and bleaching removes practically all the remaining lignin. Mechanical pulping methods on the other hand preserve most of the wood material, the pulp yield being 90–95 %. Mechanical pulp has a lignin content similar to that of wood, 20–28 %, and bleaching removes the main part of the chromophoric groups in lignin but preserves the lignin itself. The high lignin contents of mechanical pulps makes them highly susceptible to yellowing, especially when exposed to light, and when the paper ages it becomes brittle.

The pulp fibres from the two pulping methods also differ in physical appearance. Mechanical pulp fibres are shorter, stiffer, and keep the tube-like form they had in wood, whereas chemical pulp fibres are collapsed and flexible. The sheets formed by these fibres with different character show distinct differences. The stiff mechanical pulp fibres have fewer bonds between each other and form sheets of low density, in other words of high bulk, offering many surfaces for the light to be reflected and scattered on. The fines content is much higher for mechanical pulps. The fines contribute to better bonding that to some extent improves the strength of the paper. Fines also contribute to the light scattering ability. The chemical pulp fibres are able to conform to other fibres and the bonded area is much bigger. They form sheets of high density with fines filling out the voids and contributing to better bonding. Chemical pulps give paper with good strength qualities.

As a rule, chemical pulps are used for their strength qualities whereas mechanical pulps offer good opacity and printability.

The two main chemical pulping processes, kraft and conventional acid sulphite pulping, result in pulp of somewhat different character. The advantage of kraft pulp is its excellent strength, unconquered by any other pulp. The sulphite process, however, produces a bright pulp that is easy to bleach. Sulphite pulp is also easier to beat.

There are a number of different mechanical pulping methods, but the main methods are Ground Wood Pulp (GWP), Thermo-Mechanical Pulp (TMP), and Chemi-Thermo-Mechanical Pulp (CTMP). Of the mechanical pulps, CTMP has the greatest proportion of long-fibre material and the lowest amount of fines. It gives the strength properties but has the lowest light scattering ability. The largest fraction of the GWP consists of fibre fragments and broken fibres. Together with the high amount of fines, this gives GWP excellent light scattering ability but poor strength.

16.9 Product Specific Pulp Characterisation

By choosing the raw material and the treatment of the fibres, a paper with desired properties can be obtained. In the following sections, different paper grades will be presented and the characteristics important for each grade.

16.9.1 Tissue

Products belonging to this category are for example toilet paper, paper towels, and napkins. Fluff grades are used in diapers and sanitary towels. The grammage for tissue paper is in the range 15–25 g/m^2, and usually converted to 2–4 ply products.

The raw material for the grade can be TMP, recycled paper or bleached kraft pulp. CTMP in the form of fluff pulp is used for diapers and sanitary napkins. Tissue grades require little or no refining. *Table 16.5* shows important properties.

Table 16.5. Properties of Tissue.

Important property	Why it is important
Softness	For the end-use the paper should be velvet-like and easily yield when crumbled
Smoothness	A rough surface, with fibres sticking up giving the surface an uneven appearance would feel rough against the human skin. Smoothness, the inverse of roughness, is therefore important.
Ability to absorb water	The most common use of tissue paper is for wiping up liquid media of some sort, usually water, and the ability to absorb relatively large volumes is probably the most important property. Oil, grease etc are also commonly wiped by tissue.
Wet strength	The tensile strength of wet tissue paper is important since in its end-use tissue will always be wetted. The tissue has to have sufficient wet strength not to fall apart at once when getting moist or soaked.
Dry strength	In the converting process, a certain dry strength is necessary.

16.9.2 Printing Papers

The papers in this category are newsprint and magazine paper for printed media. The grammage is around 40–45 g/m^2.

Newsprint is primarily produced from mechanical pulp and re-circulated paper. The necessity for reinforcement pulp (chemical pulp) has practically been eliminated due to the development of paper machines and the properties of mechanical pulp. The magazine papers consist of primarily mechanical pulp (50–90 %), possibly some chemical reinforcement pulp (0–30 %) and fillers (10–30 %). *Table 16.6* shows important properties.

Table 16.6. Properties of Printing paper.

Important property	Why it is important
Surface strength	Low surface strength will result in fibres torn off the web surface. This may lead to dusting, if the fibres are ripped off the web and end up in the air. Linting can occur if the fibres ripped off will stick to the printing press surface. Linting will reduce the runnability of the printing machine as it would have to be shut down and cleaned.
Fracture strength	Fracture strength is another property important for the runnability of the printing press. It gives the ability of the paper web to resist against fracture from a crack in the web. Low fracture strength can result in web breakage.
Opacity	It is not desirable to read the printing on the opposite side of the paper. Opacity is a measure of „the see-throughness" of a paper. High opacity means the print on the opposite side is well hidden.
Smoothness	A smooth surface is a prerequisite of good printing. A rough surface has fibres sticking up from the surface and may both lead to linting and dusting, as discussed earlier, and to uneven print quality.

16.9.3 Fine Paper

Fine paper is a group of papers in the grammage range from 50 to 100 g/m^2. A large item is A4 copy paper. Fine papers are made from a mixture of fully bleached softwood pulp, hardwood pulp and fillers. Good refining control is necessary to develop internal bonds and obtain smoothness and good formation. *Table 16.7* shows important properties.

Table 16.7. Properties of Fine paper.

Important property	Why it is important
Surface strength	Low surface strength will result in fibres torn off the web surface. The fibres ripped off will stick to surfaces within the copying machine, causing machine failure.
Dimensional stability	For example, when toner is applied onto the paper in a copying machine the paper should not curl.
Bending stiffness	The paper has to stay flat when held, not be like a piece of cloth, velvet-like and bendingn.
Smoothness	A smooth surface is a prerequisite of good printing. A rough surface has fibres sticking up from the surface and may lead to uneven print quality.

16.9.4 Board and Packaging

- *Sackpaper* has a grammage somewhere from 60 to 150 g/m². The raw material is unbleached kraft pulp in the kappa number range 35–55. The pulp is refined in high consistency refining in order to introduce microcompressions and crimps. These properties favour the stretching ability. *Table 16.8* shows important properties.

Table 16.8. Properties of Sackpaper.

Important property	Why it is important
Toughness	Sackpaper needs to endure applied forces of large magnitude by having a good stretching ability. Curled fibres with microcompressions are more stretchable
Porosity	Good porosity is important so air can escape from within the sack as it is filled.

- *Kraftliner* is a paper with a grammage around 100 to 400 g/m² and used as liner (outside layer) for corrugated fibreboard. It is mainly made from high-yield kraft pulp with a kappa number of 80–110. When the raw material is re-circulated corrugated fiberboard the paper grade is denominated testliner. The most important properties for liner are toughness and compression strength. Compression strength is also the most important property for fluting, which is the corrugated media in corrugated fibreboard. *Table 16.9* shows important properties.

Table 16.9. Properties of Kraftliner.

Important property	Why it is important
Toughness	Tough paper with good resistance against fractures is important for papers used for packages.
Compression strength	Boxes and cartons made of kraft liner are often piled on each other and thereby experience compressional forces.
Bending stiffness	The packages made from linerboard need to be rigid and not crumble when stresses are applied.

- *Paperboard* and cardboard grades are a big and heterogeneous group of packaging materials. They are usually made in 3 to 5 plies with a grammage of 250 g/m² or higher. Bleached or unbleached kraft pulp is used as well as mechanical pulps and re-circulated fibres. *Table 16.10* shows important properties.

Table 16.10. Properties of Paperboard.

Important property	Why it is important
Bending stiffness	Packages need to be rigid and not crumble when stresses are applied.
Compression strength	Boxes and cartons are often piled on each other and thereby experience compressional forces.
Smoothness	A smooth surface is a prerequisite of good printing. The top layers of paperboard boxes are often used to print messages of the box contents etc.

16.10 Laboratory Practice compared to Industrial Methods

Standardised procedures for sample preparation and testing are established in order to be able to compare results obtained at different times and different places. This makes it possible to check the properties of products to ensure the quality to be constant. They are also used to evaluate changes in the processes and comparing different pulps. However, the standard laboratory procedures differ from the paper making process in the mill. The differences can lead to misjudgements when applying results from the lab to the paper produced on a paper machine. Below, laboratory preparation of hand-sheets is compared to industrial practice.

16.10.1 Laboratory Sheet Forming vs. Paper Machine

Sheet forming in the laboratory and industrially are two quite different operations resulting in sheets with different properties. To begin with, the laboratory sheet former gives an isotropic sheet i. e. fibres randomly oriented in the paper. In paper made on a paper machine on the other hand, the fibres align themselves in the direction of the machine. The paper is anisotropic and has different properties in the machine direction (MD) compared to the cross direction (CD).

Secondly, there is a difference in formation or in other words the uniformity of the fibre distribution. For laboratory sheets, the formation is practically ideal as the fibres distribute themselves evenly all over the sheet. On the paper machine, the fibres have a tendency to catch on to each other and form groups or flocs of fibres. Formation is important for the strength of the paper. When exposed to pulling or tearing forces the paper web will break at its weakest point, which would be between flocs, where the fibres to take up the force are few. The tendency to flocculate depends on the type of fibre. In general, longer softwood fibres have a greater tendency to flocculate than hardwood fibres. It can be misleading to give judgement on the paper strength based on laboratory sheets if a pulp with good formation on the paper machine is compared to a pulp with inferior formation. Laboratory sheets of softwood fibres give much higher strength compared to sheets of hardwood fibres. However, the tendency of softwood fibres to flocculate reduces the strength of the paper formed on the paper machine and when comparing paper machine made papers, the hardwood and softwood will have more or less equal strength.

In order to get a laboratory sheet resembling the anisotropy of a paper from a paper machine, a dynamic laboratory former may be used, such as the French *Formette Dynamique*.

The closed system of a paper machine gives good retention of fines. Generally on the laboratory, the sheets are made in an open system whereby the fines are lost to a certain degree. How much of the fines the sheet is able to retain depends of the type of pulp. Mechanical pulps have stiff fibres that form into a bulky sheet with an open structure that is ineffective in catching up and retaining fines. Since the fines of mechanical pulps are very important for the strength of the paper, it is necessary to recycle the water used for sheet forming in order to accumulate the concentration of fines in the water and thereby increase the retention of fines.

The grammage of laboratory made hand-sheets is stipulated at 60 g/m^2. At this level, the dependence of for example tensile index on grammage is negligible. At lower grammage, the strength properties decrease with decreasing grammage. However, the stipulated grammage may not be relevant considering the actual grammage of the product neither the strength properties determined at this grammage.

16.10.2 Laboratory Wet Pressing vs. Industrial Press Section

The removal of water by pressing is accomplished for laboratory sheets by applying a static pressure in a flat plate pressing machine in one or two steps for a relatively long time. The press section of the paper machine consists of 2–3 sets of pressing rolls between which the paper web experiences the pressure for a short time. In the laboratory, an increased pressure load increases the tensile strength of the paper due to an increase in sheet density. This is not the case on a paper machine where the load is so much higher.

16.10.3 Laboratory Sheet Drying vs. Drying Section of Paper Machine

After the press section of the paper machine, the paper web enters the drying section with a dry solids content of 40–55 %. Most of the free liquid around the fibres has been removed and fibre-fibre bondings start to form. The paper starts to shrink, the shrinking continues until a dry solids content of 80 % is reached. The fibre length decreases 1–2 % and the fibre width is reduced by 20–30 %. Fibres crossing other fibres are compressed and reduced in length giving rise to microcompressions, *Figure 16.17*.

Figure 16.17. When the underlying fibre shrinks in width, the fibre bonding to it will shrink in length and get microcompressions.

On the laboratory the sheets are dried on drying plates, so shrinking is constricted. Conventional laboratory sheets can thereby not be used to evaluate a papers stretch properties.

16.11 Pulp Properties affected by Industrial Operations

Inherent features, for example fibre length and cell wall thickness, determine to a vast degree the properties of fibres. As much as 80 % of the fibre properties depend in fact on the raw material. This does not, however, imply that the operations at the pulp and paper mill are insignificant for the final paper quality. All steps involved in the transformation of wood into pulp and subsequently paper will in some way make an impact on the fibres.

Laboratory chips, made by hand cutting of wood slabs, generally have no cracks, whereas technical chipping easily gives cracks. Cooking chemicals can more easily enter the chip through cracks, whereby impregnation is improved. The effective chip thickness is decreased by the cracks, as shown by *Figure 16.18*. The chip thickness of technically prepared chips to laboratory prepared chips is compared. As an example, if laboratory chips with a thickness of 4 mm give 1 % in shives content, the thickness of the corresponding technical chips can be more than 6 mm with the same amount of shives.

Technical chip
thickness, mm

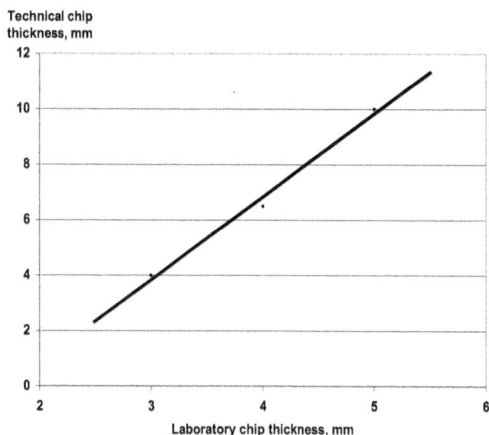

Figure 16.18. Kraft cooks with conditions resulting in a shives content of 1 % have been performed on laboratory and technically prepared chips. Shives are intact, undelignified fibre bundles, usually originating from the core of the chip due to insufficient impregnation.

The chipping operation may introduce damages to the chips. The chip end in contact with the cutting knife-edge is subjected to compressing forces. This leads to a plastic deformation in the shape of cracks and compression damages on the macroscopical scale, *Figure 16.19*.

Microscopically, in the fibre wall, misaligned zones and folds are introduced. In the damaged zone, the hydrogen-hydrogen bonds are largely broken and the whole zone is dislocated. Usually, this does not weaken the individual fibres, but the damaged zones are more accessible to chemicals. In the pulping operation, the damaged area run the risk of being overcooked and thereby weakened. In the beating operation, damaged fibres tend more easily to be cut and produce more fines compared to undamaged chips.

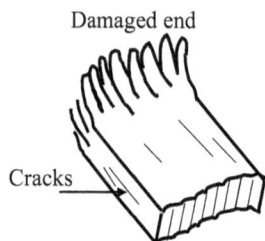

Damaged end

Cracks

Figure 16.19. Technically prepared chip with a compression damage at the end that has been in contact with the chipper knife.

Laboratory produced pulps are generally stronger than corresponding industrially produced pulps, also when technically prepared chips are used in the laboratory cooking. One reason could be the homogeneity of the cook. Ideally, the concentrations of chemicals and the temperature should be without gradients inside the chips as well as between chips. Gradients lead to chips or part of chips that are either too much delignified or not sufficiently delignified. In the laboratory it is easier to remove over-size chips and chips with knots that would end up as shives. The chemicals are also easier to distribution more evenly with better control of the impregnation in the laboratory.

Since no cooking chemicals are entirely selective towards lignin, all delignification leads to degradation of the carbohydrates. The cellulose fibrils are the load bearing entities of the fibre and the purpose of hemicelluloses is to distribute the load evenly. The fibre strength has a clear correlation to the amount of cellulose. The length of the cellulose chains is also of importance, *Figure 16.20*. As long as the D.P. (degree of polymerisation) is high enough, the pulp strength is unaffected. However, when the cellulose degradation has reached down to a level below the „long enough" point the pulp strength steeply decreases with decreasing chain length. In pulping and bleaching operations, a limit of the lowest acceptable viscosity is set up. A bleached pulp should not have a viscosity below 800 ml/g. A decrease in the hemicellulose content makes the distribution of stress among the fibrils less effective.

Dry zero span, kNm/kg

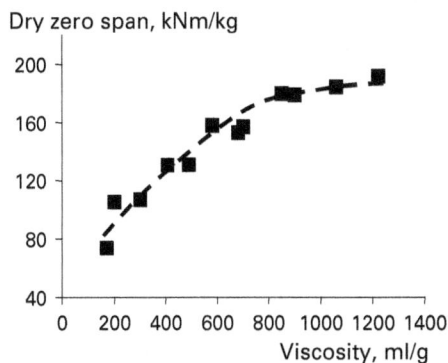

Figure 16.20. The viscosity value correlates to the degree of polymerisation of the cellulose chains. Above a certain chain length, the pulp strength is practically unaffected by a decrease in chain length. However, below the critical point, the pulp strength is rapidly reduced with further reduction in viscosity.

Higher amount of lignin gives stiffer fibres. The Neutral Sulphite Semi–Chemical process results in fibres with the highest stiffness.

The bonding ability of fibres depends on both the lignin and hemicellulose content. Lignin on the fibre surface prevents good bonding between fibres. Hemicelluloses, on the other hand, contribute to form strong bonds. Removal of lignin thus promotes the bonding ability of fibres. However, at a certain point the amount of lignin on the fibre surface is no longer a limiting factor to the fibre bonding ability. Prolonged delignification leads to dissolution of hemicelluloses and the bonding ability decreases, *Figure 16.21*.

The discharge of cooked chips from the digester is a much rougher operation industrially compared to the laboratory. The action is called to blow the contents of the digester, which suggests that it is not a mild procedure and it has been shown to cause a decrease in pulp strength. The pressure drop from the pressure within the digester to atmospheric pressure can be quite considerable and at high alkalinity and high temperature, the fibres are damaged. Addition of cool filtrate to the digester bottom decreases the temperature of the stock before blowing. The practice is called cold blowing and gives improved pulp strength. What the damaging action of the hot blow is has not been completely established. Some of the strength loss has been attributed to a reduction of fibre length.

Aslo oxygen delignificationa nd bleaching influence pulp properties. Fibre deformations have a great influence on pulp strength.

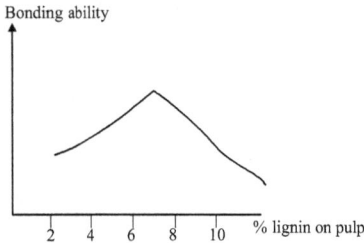

Figure 16.21. Lignin on the fibre surface prevents good bonding between fibres and hemicelluloses promote it. At a certain point, the lignin on the surface has been removed and further delignification dissolves hemicelluloses and the bonding ability starts decreasing.

Fibre deformations such as twists and compressions, compiled under the term curl, increase after oxygen delignification.

The bleaching processes increase significantly the fibre deformations. The biggest increase is in the number of microcompressions.

16.12 Laboratory Test Methods and their Relevance

ISO brightness is used as a measure of the pulp whiteness. An increase in ISO brightness should reflect a decrease in the amount of coloured substances in the pulp. However, the ISO brightness test does not give satisfactory information for a number of reasons. It is not solely dependent on the amount of coloured substances, also the sheet structure affects the ISO brightness value. In optical terms, the ISO brightness depends of both light absorption, given by the light absorption coefficient k, and the sheets light scattering ability, measured as light scattering coefficients. A simplified description of the dependence gives that the ISO brightness equals k/s.

Processes eliminating substances that absorb light decrease k. Delignification in pulping and bleaching stages increase the ISO brightness of the pulp. However, also processes increasing the light scattering ability, increase the ISO brightness. Increased refining and pressing decreases brightness, as the light scattering surfaces becomes less, although these processes does not affect the coloured substances. Increasing the grammage of the sheet would increase the ISO brightness.

In the ISO brightness test method, unrefined pulp is used and the brightness measured at 475 nm. Unrefined pulp has a higher light scattering ability than refined pulp. The human eye is most sensitive at 575 nm.

The light absorption coefficient k is a better measure of the pulp whiteness as it is a direct measure of the coloured substances. A lower k represents a brighter pulp. *Table 16.11* lists the k value ranges of some pulps.

The Kubelka- Munk equation is not able to completely describe a material such as paper and the light absorption coefficient determination has some limitations. k values above 5 m²/kg are not to be fully trusted.

Tear index is commonly used to evaluate the strength properties of pulp. The fracture mode of common Elmendorf tear strength test is perpendicular to the plane of the sheet. This mode of tearing is not the kind a paper is normally subjected to. More important is the ability of the paper to resists a propagation of a crack from either the edge of the paper web or somewhere in the

Table 16.11. Approximate k values of different pulps.

Pulp	k_{457}, m²/kg
Unbleached kraft pulp	20–70
Unbleached sulphite pulp	2–15
Mechanical pulp	5–12
Bleached mechanical pulp	2–10
Bleached kraft pulp	0.1–2

web when the tearing mode is in the plane of the web. The fracture toughness test is thereby more realistic in its approach to determining relevant paper properties.

Refining pulp usually increases tear strength slightly at lower degrees of beating whereas prolonged beating decreases the tear strength. Tensile strength on the other hand increases with beating until it reaches a plateau value. In order to optimise the refining, it has to be stopped at a sufficiently high tensile strength without arriving at too low tear strength. The more realistic tearing mode of the fracture toughness test however, responds somewhat differently to beating. It continues to increase up to very high tensile index values, *Figure 16.22*.

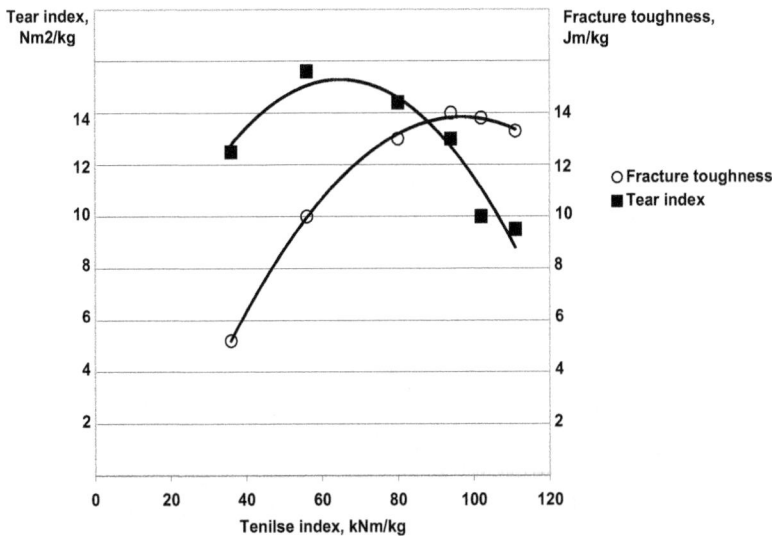

Figure 16.22. By using the fracture toughness measure, it is seen that beating to higher tensile index values is possible without risking low fracture toughness.

Judging by the tear index curve, a much lower tensile index is possible than what the fracture toughness strength permits.

Properties measured on *hand sheets* have been shown to be misleading. The information received from hand-sheet tests is not always relevant to industrial practice and product demands. It would therefore be better to determine fibre properties that are known to affect the final pulp and paper properties.

Index

DE GRUYTER

Volume 63 · 2009 · Issue 1
ISSN 0018-3830
CODEN HOLZAZ

HOLZFORSCHUNG
International Journal of the Biology, Chemistry, Physics and Technology of Wood

This issue contains selected papers from the
2nd International Cellulose Conference (ICC 2007)
Tokyo, Japan, October 25-29, 2007

Walter de Gruyter
Berlin · New York

HOLZFORSCHUNG

International Journal of the Biology, Chemistry, Physics, and Technology of Wood

Editor-in-Chief: Oskar Faix, Germany

Publication frequency: bi-monthly (6 issues per year).
Approx. 700 pages per volume. 21 x 29.7 cm
ISSN (Print) 0018-3830
ISSN (Online) 1437-434X
CODEN HOLZAZ
Language: Englisch

Holzforschung is an international scholarly journal that publishes cutting-edge research on the biology, chemistry, physics and technology of wood and wood components. High quality papers about biotechnology and tree genetics are also welcome. Rated year after year as the number one scientific journal in the category of Pulp and Paper (ISI Journal Citation Index), *Holzforschung* represents innovative, high quality basic and applied research. The German title reflects the journal's origins in a long scientific tradition, but all articles are published in English to stimulate and promote cooperation between experts all over the world. Ahead-of-print publishing ensures fastest possible knowledge transfer.

Indexed in: Academic OneFile (Gale/Cengage Learning) – Aerospace & High Technology Database – Aluminium Industry Abstracts – CAB Abstracts – Ceramic Abstracts/World Ceramic Abstracts – Chemical Abstracts and the CAS databases – Computer & Information Systems Abstracts – Copper Data Center Database – Corrosion Abstracts – CSA Illustrata – Natural Sciences – CSA / ASCE Civil Engineering Abstracts – Current Contents/Agriculture, Biology, and Environmental Sciences – Earthquake Engineering Abstracts – Electronics & Communications Abstracts – EMBiology – Engineered Materials Abstracts – Engineering Information: Compendex – Engineering Information: PaperChem – Journal Citation Reports/Science Edition – Materials Business File – Materials Science Citation Index – Mechanical & Transportation Engineering Abstracts – METADEX – Paperbase – Science Citation Index – Science Citation Index Expanded (SciSearch) – Scopus – Solid State & Superconductivity Abstracts.

All de Gruyter journals are now hosted on **Reference Global**, de Gruyter's new and integrated platform.
Please visit www.reference-global.com for more information and free TOC alerts.
Electronic sample copy at www.degruyter.com/journals/holz

W
DE
G
de Gruyter
Berlin · New York

www.degruyter.com

www.ingramcontent.com/pod-product-compliance
Lightning Source LLC
Chambersburg PA
CBHW051114200326
41518CB00016B/2504